CALIFORNIA

The Geography of Diversity

Second Edition

CRANE S. MILLER

University of Redlands

RICHARD S. HYSLOP

California State Polytechnic University, Pomona

MAYFIELD PUBLISHING COMPANY
Mountain View, California
London • Toronto

Library of Congress Cataloging-in-Publication Data

Miller, Crane S.
California, the geography of diversity / Crane S. Miller, Richard S. Hyslop.—2nd ed.
p. cm.
Includes bibliographical references and index.
ISBN 0-7674-1345-8
1. California—Geography. 2. California—Historical geography. 3. Human geography—California. I. Hyslop, Richard S. II. Title.
F861.M55 1999
911′.794—dc21 99-19807
CIP

Manufactured in the United States of America

10 9 8 7 6 5 4 3

Mayfield Publishing Company
1280 Villa Street
Mountain View, California 94041

Sponsoring editor, Richard Greenberg; *production editor*; Carla White Kirschenbaum; *manuscript editor*, Tom Briggs; *text and cover design*, Terri Wright; *design manager*, Jean Mailander; *art editor*, Robin Mouat; *manufacturing manager*, Randy Hurst. The text was set in 9.50/12 Palatino by UG / GGS Information Services, Inc. and printed on 60# Millenium Matte by Banta Book Group.

Cover Image: Derek/O'Hara/Tony Stone Images

Contents

Preface

As many of our loyal readers have pointed out, the second edition of this text has been a long time in the making. We are very happy to present this updated and expanded second edition of *California: The Geography of Diversity*, a text that we hope reflects the many suggestions and comments we have received over the years from both professors and students who have used the original edition. As with our first edition, we have continued to espouse the unique and dynamic nature of our state, in both its physical and cultural manifestations.

We have maintained the topical format in this edition, believing that encyclopedic coverage of this state is an impossibility at best and a confusing mistake at worst. In the process, the most crucial and compelling aspects of the state's physical and cultural landscapes have been emphasized. The text covers the expected areas of climate, landforms, regions, population patterns, and physical processes. Also presented are such less common topics as personalities of places, geo-historical developments, modern political and demographic phenomena, literary perspectives, and legal-geographic interfaces. As much of the final polishing of this edition occurred during the sesquicentennial year of California's statehood, we have been sharply aware of the rapid changes and developments that have emerged throughout the state's history (both recent and not so recent). The sweep and drama of the "Golden State" in time and space provide a compelling geographic chronicle that continues unabated as we forge ahead into the opening years of a new millenium.

We have both seen many, many students pass through our classes and understand the constantly changing demographics of these students who see and experience their surroundings from different perspectives. Therefore, our desire has been to provide a text that is relevant and vital to these readers, conveying the excitement of various aspects of the geography of California while providing a solid base of information for future use.

Our sincere appreciation must be expressed to the many individuals who have encouraged and prodded us along during this grand adventure in rewriting. The folks at Mayfield have been patient and supportive. In particular, our production editor, Carla Kirschenbaum, has shepherded the project very closely. Robin Mouat provided guidance and support as we prepared the art for this text. Jean Mailander oversaw the design and gave the book a new look. The task of avoiding jargon and unnecessary pomposity in our writing—was made easier by the stellar efforts of our exceptional copy editor, Tom Briggs, who did his best to keep us honest.

We have been fortunate to have a collection of reviewers who spent careful time and effort helping us to write a better edition. Our thanks and acknowledgement are extended once again to the following reviewers of the first edition: James D. Blick, San Diego State University; Susan W. Book, Cosumnes River College; Robert W. Christopherson, American River College; Howard F. Gregor, University of California, Davis; Donald G. Holtgrieve, California State University, Hayward; Ronald F. Lockman, University of Southern California; Richard F. Logan, University of California, Los Angeles; Gertrude M. Reith, California State University, Fullerton; and Christopher L. Salter, University of California, Los Angeles. We are also grateful to the reviewers of this second edition: Albin Kwolek, Sacramento City College; Carolle J. Carter, San Jose State University; Stephen F. Cunha, California State University, Chico; Marsha Dillon, California State University, Sacramento; the late Donald Floyd, California Polytechnic State University; David Larson, California State University, Hayward; Paul Lehman University of California, Santa Barbara; Stanley Norsworthy, California State University, Fresno; and Nancy Lee Wilkinson, San Francisco State University.

Many other individuals have provided valuable assistance, including our colleagues at California State Polytechnic University, Pomona and other campuses. Our students have been an amazing source of information and advice—sometimes without realizing it. We would be remiss if we did not mention the patience and constant help of two other significant contributors. Professor Sally Boyes-Hyslop, of California State Polytechnic University, Pomona, has frequently provided a marketing perspective of California's geographic phenomena, and has also served as a patient sounding board for different ideas and concepts. Likewise, Barbara Miller has seen another edition born and has assisted in the delivery at crucial moments with advice and support. We cannot attempt to name all the other people who have helped bring this project to fruition. Suffice it to say that we recognize that such a work is only as good as the people who inspire it.

CRANE S. MILLER
RICHARD S. HYSLOP

CHAPTER

1

Lotus Land Revisited

Contemporary California appears to be at a crossroads in terms of both its development and its reputation. Entering the twenty-first century, the Golden State still earns more geographic superlatives than any other states, but the negative phenomena associated with California's geography have accelerated noticeably in recent years. Noted for its mildness, California's climate has no duplicate in the entire nation, but smog and drought increasingly temper its appeal. And when the rains finally return in full force to scrub the atmosphere and break a long drought, as they did in 1993 and 1998, they do so with a fatal fury. California's spectacular coastal headlands (Figure 1-1), lofty mountains (Figure 1-2), and deep valleys are surpassed in dimension and grandeur only by those of Alaska and Colorado. Yet this diverse terrain is made less inviting by the very geological events that created it, natural cataclysms still echoed in such destructive modern manifestations as the Loma Prieta and Northridge earthquakes.

With 33 million residents, California is the most populous state, but ancillary to California's growth are mounting environmental, economic, and social problems. California continues to rank first among the states in agriculture and among the leaders in mining, forest products, manufacturing, retail trade, technological innovation, tourism, and finance. Yet economic recession accompanied by shrinking tax revenues and mounting government deficits, layoffs in the defense and construction industries, a changing energy base, keen competition from other states for a shrinking number and variety of federal contracts, and increasingly stringent environmental constraints tend to erode these standings. Trendsetting in architecture, art, education, entertainment, fashions, and recreation endures as a tradition in California—but one now shared by a growing number of states. Still, if there is any one aspect of California's wealth of people, cultures, and skills that, more than anything else, tarnishes its reputation as the golden land of goodwill and opportunity, it is that it is a melting pot that all too often boils over. California boasts an unparalleled ethnic mosaic, but tension is evident every time graffiti is sprayed on a wall (Figure 1-3) or there is a drive-by shooting or violence breaks out in the streets. Evidently the lessons of the Watts riots of 1965 remained largely unlearned, for the 1992 riots in South Central Los Angeles remain the deadliest and most destructive to occur in an American city in this century.

California's demographic dynamics and economic slumps and surges during the last decade of the twentieth century displayed a decided regional bias, one that many Southern Californians would just as soon forget but that residents of one part of Northern California would like to see go on forever. In the early 1990s, the end of the Cold War signaled major cuts in defense spending and led to the loss of hundreds of thousands of jobs. As a focal point of the national aerospace industry, Southern California bore the brunt of the cutbacks. By mid-decade, unemployment was pushing double digits, real estates sales and prices were plummeting, foreclosures and short sales (in which real property sells for less than is owed on it) were rising, more people and

Figure 1-1 *Entering the Golden Gate, California's Pacific gateway. The view is northward with Marin Peninsula connected to San Francisco Peninsula in the foreground by Golden Gate Bridge. This photo was taken from near the Pacific side of the Presidio, a 1,500-acre U.S. Army base, which was closed during the 1990s. As the Spanish military had intended when it established a garrison here in 1776, the site commands a sweeping view of the ocean, coastline, and San Francisco Bay.* (Crane Miller)

Figure 1-2 *Returning to California from Oregon via Interstate 5, California's main transportation link with the Pacific Northwest. The valleylike lowland separates 14,162-foot Mount Shasta (extreme left) and the Southern Cascades Range to the east from the Klamath Mountains to the west (right).* (Crane Miller)

Figure 1-3 *Graffiti at the eastern entrance to the 6th Street bridge over the Los Angeles River, with the downtown Los Angeles skyline in the background. The on-the-wall graphics reflect frustration with society, as well as being a "signature" for individuals and gangs. In this scene, the graffiti-marred tower contrasts sharply with the gleaming high-rise office buildings, monuments to society's progress.* (Crane Miller)

businesses were vacating California than moving in, and Nevada and Arizona were drawing more retirees nationwide than coastal California or Florida.

Meanwhile, having already established itself as the computer reasearch and development capital of North America, a valley fronting on the southern edge of San Francisco Bay was defying the California recession. Officially but less well known as the Santa Clara Valley, Silicon Valley had an unemployment rate half that of the rest of the state and enjoyed rising real estate sales and prices as more people and businesses relocated to the prosperous region. By the late 1990s, the whole state was on the economic mend, yet nowhere more so than in Silicon Valley and environs, where negative unemployment (more job openings than jobless workers) was now the rule and median home prices averaged $100,000 higher than in Southern California ($300,000 in Santa Clara and San Mateo counties versus $200,000 on the average in Los Angeles and Orange counties). Population growth, too, was back on track by the end of the century, so that the 1993 projection by the Demographic Research Unit of the state's Department of Finance of a population of nearly 49 million by 2020 once again seemed conservative.[1]

Of course, some see population growth as the principal problem in and of itself, especially given the ten-

[1] California Department of Finance, Demographic Research Unit, *Projected Total Population of California Counties: 1990 to 2040*, Report 93-P-3 (Sacramento, May 1993).

dency of Californians to crowd onto a relatively small proportion of the state's land area located as close to the ocean as possible. The migration of people from around the world to California, and the resulting overcrowding and attendant social and environmental problems, probably will not diminish, at least in the foreseeable future. But, as these seaside population centers grow, they increasingly rely on imported resources, mainly water and energy, to sustain them. But those resources are dwindling, and their providers are ever more reluctant to export them.

If there is, in fact, reliable cause for optimism about California's growth into the twenty-first century, at the heart of this hope is the state's geographical diversity. Unparalleled diversity and wealth of natural and human resources have made California what it is today and hold out the promise for tomorrow. At a glance, Figure 1-4 offers a sample of that diversity.

TRAVELOGUE TALES AND ROMANTIC MUSINGS

There has always been a certain mystique about California. The name itself evokes thoughts of golden sunshine, golden opportunities, and the good life. This is hardly a purely contemporary phenomenon; much of the written history of the area abounds with these images. Whether in references to the fabulous Seven Cities of Gold, the glowing reports of the Forty-Niners, the frequently glamorized TV commercials, or the Hollywood images that have flashed on the screens of the nation, California is seldom portrayed as anything less than magical. Somehow, California seems a natural home for "lifestyles of the rich and famous."

How did such a glowing identity emerge? It is astonishing how consistent this enthusiastic boosterism has been throughout the written history of the region. Before Europeans ever set foot in California, a tourist tale equal to any travelogue promotion had been concocted on the Iberian Peninsula. During the sixteenth century, romance writers in Spain indulged in many fanciful and exotic literary creations. (Cervantes' tale of Don Quixote was a classic parody of these imaginative stories.) One such popular tale, by Ordóñez de Montalvo, spoke of an idyllic kingdom on an island near the Indies called "California." This mythical (or not-so-mythical) paradise came complete with many of the characteristics we now readily associate with the Golden State: balmy climate, beautiful women, easy wealth, and eternal happiness. The tale was a travel agent's dream, a classic early advertisement for a mythical land that really existed—and one that would try to live up to these early fantasies three or four centuries later.

Speculation also exists as to whether this mythical land was known to the Chinese and described as "Fusang" (literally, "rich mountain") well before the Spanish tales emerged. Although this hypothesis has not been definitively established, at least one respected historian, Charles Edward Chapman, has concluded that the adventurous Chinese did indeed reach California as early as the fifth century A.D. Whether actually true or false, the tale merely lends further gloss to the romanticization of the region's history.

California Through the Eyes of the Spaniards

Among the first eyewitness accounts of California were those provided by the Spaniards. Indeed, some of the earliest written historical remembrances of the land are found in the diaries and reports of these early Spanish explorers and missionaries, and, for the most part, they reinforce rather than contradict the Spanish literary myth. Working their way northward from Mexico beginning in the mid-sixteenth century, the explorer-missionaries made some initial superficial observations about California, particularly its coastal regions, where they discovered both "beautiful valleys and groves" and flat and rough country. As for the native inhabitants, many of the explorers described them in glowing terms as mostly peaceful, gentle, and attractive people. However, "gentle" did not mean "weak," and several Spanish soldiers described them as brave and determined when threatened or pushed. Other comments on the climate, water, and fertility of the soil were frequent and effusive. In fact, these explorers and missionaries described a California familiar to most of present-day America: a beautiful, sun-drenched land with friendly natives lolling about, usually grateful recipients of culture from the rest of the world. The expected and familiar result of such enthusiastic reports was the establishment of missions and permanent settlements by Hispanic (Spanish and Mexican) colonists in this remarkable land—initiating what was to become a common theme throughout the history of the state. This is not to say that there were no detractors of the area. Indeed, one crusty Spaniard, Gaspar de Portolá, speculated on whether it would have been better to let the Russians have the country. In all fairness, Portolá's disposition may have been somewhat soured by several weeks of dieting on mules that he had been forced to kill along his route to feed his expedition. His journey of exploration from Baja (now Mexico) into Alta California (now the U.S. state) in 1769 was fraught with perils. Moreover, Portolá's dim view of Baja California probably carried over into Alta California; his expedition occurred at the tail end of a long period of exploration, after the Spaniards had finally decided to settle Alta California.

Over the course of approximately 230 years, beginning in 1542 with Juan Rodríguez Cabrillo, a respected

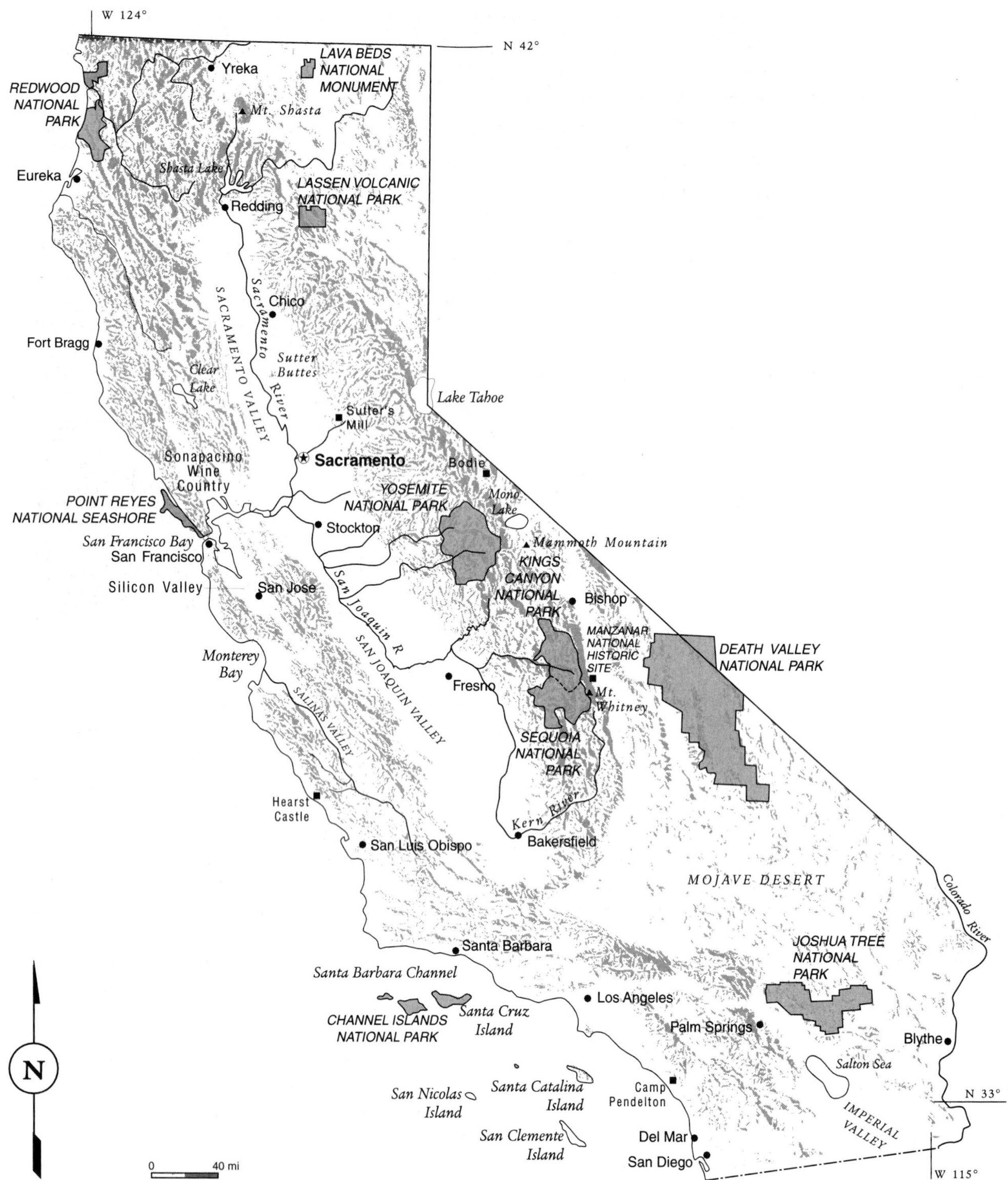

Figure 1-4 *California's geographical diversity. Some of the Golden State's environmental contrasts are highlighted in this reference map. California spans nine degrees of latitude and longitude and features everything from ancient redwood forests in the northwest, to the fertile Central Valley, to desert in the southeast.*

Figure 1-5 *San Luis Rey Mission, in the Peninsular Ranges north of San Diego. Established in 1798 by the Franciscan Order of the Roman Catholic Church, this was the seventeenth in their 21-mission system in coastal Alta California. The missions proved to be the effective vanguard of Spanish settlement in the far-flung frontier province, with the* presidios *(military garrisons) and* pueblos *(civilian towns) playing less significant roles than originally anticipated.* (Union Pacific Railroad)

sailor and navigator, sea and land expeditions were sent out from Mexico on a sporadic basis to explore and report back. But it was not until the 1770s that the Spaniards made any significant efforts to colonize Alta California. Perhaps this delay was partly due to the fact that 2,000 miles separated this area from the Spanish population centers in Mexico—certainly not a casual stroll around the plaza! Once Spain perceived a threat to its territory by other powers, however, it began its three-part (pueblo–presidio–mission) colonization of Alta California, and by the close of the eighteenth century it had established the area as Spanish Alta California. The attitudes and commentaries of the various mission fathers, settlers, and officials during the subsequent period reflected the desires of the Spanish government to encourage and expand the scope of settlement (Figure 1-5). The people who most benefited from the largesse of the government—the Californian rancheros, soldiers, and priests—were most often those who wrote extolling the virtues of the land. Through their eyes, pastoral Spanish California was transformed into a romantic "Spanish Arcadia," an image that has had amazing staying power.

The Spanish settlers of California took little active note of the revolt against Spain taking place in Mexico. When they learned that an independent government had been formed in Mexico, however, they soon adopted the new political loyalty. The predecessor Spanish period lasted until the early 1820s, when it was succeeded by this new Mexican period. Mexican rule in California was quite abbreviated—a brief quarter of a century notable for three significant changes. The first was the completion of the secularization of the missions, with resulting economic and social dislocations that fell especially heavily upon the indigenous inhabitants of California, whom the Spaniards dubbed "Indians." The second was the rise of the cattle rancho, replacing the mission as the focus of settlement. The third was the increase of foreign influence and importance in California, particularly of American interests. The Americans who trickled into California during this brief period sang the praises of the locale and certainly eased the way for acceptance of American rule in 1848. Although the Treaty of Guadalupe Hidalgo made California a part of United States territory in 1848, it was the discovery of gold the same year that made statehood a reality in 1850.

Early American Images

Although the Spaniards were the first serious white colonizers and "tourists" of California, they were not alone in their efforts. Others, including the Russians and the British as well as the Americans, were active in their tentative incursions into the area.

Following the American Revolution, the British navy indulged in some exploration, mapping, and surveying of the California coast. In a somewhat stuffy vein, the British explorers frequently described the manner of life they observed as being too lazy and nonproductive. For example, British naval officer George Vancouver remarked on the Spanish Californians' "habitual indolence and want of industry" and concluded that true civilization would probably be a long time in coming.

The British observations were often reinforced by various Yankees who frequented the ports of California for purposes of trade. Primarily focusing upon the Spanish and Californian officials with whom they dealt, the American seamen often found little to admire. The Californians were frequently described as pompous or cowardly, greedy or lazy, disorganized or frivolous, or all of these things. Indeed, many other Americans seemed to share the opinion of U.S. Naval Lieutenant Charles Wilkes: "Although the Californians are comparatively few in number, yet they have a distinctive character. Descended from the old Spaniards, they are unfortunately found to have all their vices, without a proper share of their virtues."

Other visitors, however, found a people eager to socialize, ready to play, and quick to display their talents in dancing, riding, and other physical activities. Rather than uniformly greedy, pompous, or irritating, the Californians described by commentators such as the respected trader Alfred Robinson were gracious, generous, and kind. Although these comments tended to be *by* men *about* men, complimentary remarks about the California women were also common. The presence of many other Americans in California during the Spanish and Mexican periods attests to the fact that the California way of life held strong attractions for at least some Americans.

However mixed their comments about the inhabitants may have been, the descriptions of the climate, land, and physical features of the area were hardly discouraging. One of the first American writers, Captain William Shaler, in distinctly noneffusive terms, described the land and climate as "dry and temperate, and remarkably healthy," while another, Alfred Robinson, perhaps more typically spoke of the "fairy spots . . . met with so often in California."

The initial American coastal settlements contacts with California were soon followed by a series of reports, comments, letters, and enthusiastic tall tales from explorers who crossed the continent seeking the golden clime and opportunities that beckoned from the land of the dons. Beginning with mountain man Jedediah Smith, Americans gradually began to reach California by the overland route. After their arduous trek across mountain, plain, and desert, California did indeed appear a paradise to many of these travelers. As men and women began to settle and establish themselves in the area, they composed enthusiastic letters and comments on the wonders of California. For example, John Bidwell described his "Journey to California" to the folks back home and detailed the fabulous character of soil, flora, and fauna in the early 1840s. Seeking to encourage more Yankees to settle in the area and thus increase the American influence, these promoters described a fertile, verdant area ripe for settlement—that is, until 1848, when such promotion was no longer necessary. The cry of "gold" became the overnight stimulus to the growth of California, beginning an odyssey of fortune seeking that has continued in various forms up to the present.

Literary Visions

Although truth is often stranger than fiction, the "factual" reports of California frequently came dangerously close to being outright fantasy. Likewise, the fiction writers who used the region as a backdrop in their works may have been no more imaginative or creative than the authors of "true accounts." In any event, the images of California that emerged from the pens of some of America's most popular literary figures did not vary greatly from their previous (and contemporary) "truthful" colleagues. Continuing in the now-established positive and colorful vein, these literary visions usually perpetuated the myth of the Golden State—and certainly did nothing to discourage new immigration to and interest in the region. Although the number of writers who described California is too great to permit exhaustive treatment here, a stellar few bear mentioning.

Richard Henry Dana is noted not for his promotion of California, but for his crusade for the rights of sailors. When his classic work *Two Years Before the Mast* was published in 1840, however, it did act as a major influence in attracting American settlers to California. His vivid descriptions of handsome people, attractive land, pleasing climate, and boundless opportunities helped lure ambitious Americans to this sleepy economy.

Another well-known figure, Washington Irving, wrote of "the plains of New California, a fertile region extending along the coast, with magnificent forests, verdant savannas and prairies that look like stately parks." The great regional authors of the mid-1800s also found eager audiences for their tales of the Far West. Bret Harte, Artemus Ward, and Mark Twain spun romantic tales of the gold rush towns, colorful characters, high Sierra country, and unusual occurrences of the California scene. Twain's story "The Celebrated Jumping Frog of Calaveras County" was only one of many amusing fictions whose pages provided enticing whiffs of the golden land. Even Walt Whitman created a tribute to one aspect of California in "Song of the Redwood Tree"; and although Rebecca of Sunnybrook Farm never visited California, her creator, Kate Douglas Wiggins, sang the praises of the state, and particularly of Santa Barbara, which she called her "Paradise on Earth." Even more widely read were the romantic crusades of Helen Hunt Jackson, who in *Ramona* and other works about the old Spanish period both shed further light upon the development and historical roots of the state and created more romanticized myths about California.

As means of transportation and communication improved, writings by and about Californians became even more popular with the American reading public. Of major importance to the naturalistic school of American fiction were several California authors now considered important writers in their own right. Frank Norris effectively utilized the diversity of the California landscape as a dramatic backdrop for his epic novels such as *McTeague* (with scenes ranging from glittering San Francisco to a dramatic climax in Death Valley) and *The Octopus* (which dealt with the wheat production of the San Joaquin Valley and the farmers' battle with railroad interests). Likewise, Jack London derived inspiration and mood from his Bay Area birthplace. His descriptions of fog in the Golden Gate and of the slums and dives of San Francisco, Sausalito, and Oakland, and the characters he had known there, were particularly vivid. Indeed, he began one of his best-known works, *The Sea Wolf*, with a tremendously evocative depiction of the clammy, blanketing fog of San Francisco Bay—a description instantly recognizable to anyone who has experienced that phenomenon. Other well-known writers, such as Upton Sinclair (*Oil!*), portrayed the scattered and sometimes tawdry development of the state, and Sinclair Lewis had his run at exposing San Francisco to the public eye. Many other popular and respected authors, poets, and journalists also paid tribute to the Golden State in their works, either directly or through the settings used in their works. For example, mystery writer Raymond Chandler provided one of the most evocative passages yet penned to explain the effect of the famous Santa Ana winds. In his short story "*Red Wind*," he described the "desert wind blowing that night. It was one of those hot dry Santa Anas that come down through the mountain passes and curl your hair and make your nerves jump and your skin itch. On nights like that every booze party ends in a fight. Meek little wives feel the edge of the carving knife and study their husbands' necks. Anything can happen."

In all, California provided a rich source of material for the literary endeavors of the nation. This literary activity effectively drew attention to the state, as did *sunset* magazine and publications by the Southern Pacific Railroad, further enhancing its mystique and nurturing the desire in multitudes of people to see firsthand the areas that had been so vividly described. Thus, both directly and indirectly, the literary visions of three more centuries further expanded the almost hypnotic appeal first created in Ordóñez de Montalvo's sixteenth-century tale of a mythical California.

A Taste of Today

The fascination of Americans with the state of California has not substantially subsided even to this day. There is still an aura of otherworldliness to the region. While temperatures plummet in the East and Midwest, millions of Americans enviously watch the sun shine on scantily clad cheerleaders and spectators at the Rose Parade and Rose Bowl game televised every January 1. Shivering through another winter, many people cannot help but view the balmy surroundings pictured on their television sets as a slice of paradise.

Modern writers have continued to express this interest in their writings, wherein California's unique position (climatically, geographically, and historically) plays a key role. Certainly the works of John Steinbeck must top the list in terms of influence, style, and importance; his great *Grapes of Wrath* described a California able to produce winter crops that were impossible to grow in many other parts of the country (at a time when no one could afford to buy them). In this and other works, Steinbeck exposed the very soul of much of the state, particularly the rich farm areas such as the Salinas Valley, and in the process became a modern California classic himself.

Nor is Steinbeck alone. William Saroyan's Fresno comes alive in his works; Joseph Wambaugh's Los Angeles gains in notoriety through his widely read police stories; and Eugene Burdick has exposed politics, California style, to millions of readers. Furthermore, national magazines and newspapers continue to examine the phenomenon that is California, with "special California issues" and with famous writers regularly and vigorously praising (or panning) the state. From *Popular Science* to *Newsweek* and from *Westways* to *American Photo*, the fascination continues to be expressed. Thus, through these and other current sources, the myth-making process continues. Television shows, movies, magazines, and newspapers constantly extoll the virtues and mystique of the golden land, in the process keeping intact the august tradition of travelogue tales and romantic musings about California.

Recently, however, a slightly different tone has begun to creep into some of the iconology of California. It is not uncommon now to find negative commentary interspersed in media depictions of the region. Graphic reports of urban gang violence and freeway shootings sound an ominous note in parts of the state. Official recognition that the South Coast Air Basin, which spans the greater Los Angeles area, contains the worst air quality in the nation certainly removes some romance from the image of glamorous Southern California. Widely broadcast "fictionalized" accounts of the "great Los Angeles earthquake" give pause to possible immigrants. The now-famous 1989 World Series broadcast from Candlestick Park in San Francisco, which was interrupted by the Loma Prieta earthquake, demonstrated that fact could be as dramatic as fiction—especially as the number of casualties soared into the hundreds.

Yet, even in the face of these negative depictions, the population of California has continued to grow. According to 1990 census figures, in the decade from 1980 to 1990, the population increased by over 26 percent, and by early 1997 the population numbered over 32.6 million people. As the undisputed largest state in the country (by population), California can continue to claim its place as the destination of choice for millions of newcomers. Although the image may be somewhat tarnished, the state retains its golden aura.

DIVERSITY IN PERSPECTIVE

Few regions on earth can lay legitimate claim to the degree of geographical diversity that California boasts. Where else in a single area of some 160,000 square miles are found glaciated mountains, verdant river valleys, countless seascapes, snowless winters, redwood rainforests, seemingly barren deserts, deep lakes, plentiful petroleum, bountiful agriculture, and 33 million people with national and ethnic origins representing all of humanity? We could go on and on with such superlatives, some of them seemingly paradoxical, and thus render the question even less answerable, but the inescapable conclusion is that California is in fact *uniquely* diverse.

Environmental Diversity

One need not look outside California for either a quick change of scenery or a better physical environment. There is sufficient diversity in California's physical geography to satisfy just about anyone's environmental needs (except for the person seeking the humid tropics). For the Angeleno who suddenly feels the urge to commune with nature rather than commute on the Santa Monica Freeway, the majestic isolation of 14,000-foot mountains in the Sierra Nevada (Figure 1-6) is only a few hours away by car (and foot). For the Mojave Desert dweller who needs more relief from summer's dry heat than an air conditioner can provide, Lake Tahoe's cool alpine setting (Figure 1-7) provides the needed environmental diversion and is only a half-day's drive away. For the San Joaquin Valley farmer overdue for a vacation by the seashore, more than 1,000 miles of Pacific coastline are available just over the Coast Ranges to the west. For the San Franciscan yearning for fogless skies, a trip south to balmy Palm Springs is likely to provide them. Perhaps this environmental diversity explains why Californians are sometimes said to be defensively parochial. "After all," they say, "why travel out of the state when it's all right here?"

Figure 1-6 *Ansel Adams* Rock and Lone Pine Peak *photographed in the southeastern Sierra. Like John Steinbeck, Ansel Adams was born in 1902 in California (San Francisco), but Adams captured the California landscape with camera rather than pen.* (Crane Miller)

Besides furnishing Californians with seemingly boundless recreational opportunities, the state's environmental diversity is also the basis for an abundance of natural resources. The geography of California's natural resources is uneven but predictable, when the state's 158,693 square miles (third largest in the United States) and the variety of geologic, climatic, and biotic processes that formed it are taken into consideration. For example, gold is created in the Sierra and is subsequently weathered, eroded, and carried down by countless streams to *placer* (laid down by water) deposits in the foothills. Petroleum and natural gas are found in the sedimentary formations of the Coast Ranges and offshore on the continental shelf. Borates and potash accumulate in the basins of interior drainage that dominate the arid eastern deserts. High-grade iron ore is formed in the mountains of those deserts. Water, the most vital mineral of all, is stored in the snows and lakes of the Cascade, Klamath, and Sierra ranges, whose towering peaks originally forced the moisture from winter storms. Soil formed from deep alluvial deposits in the San Joaquin, Sacramento, Salinas, Imperial, and numerous lesser valleys is the very foundation of California's agricultural supremacy. Timber comes from the coastal redwood rainforest, the Douglas fir stands of the northwest, and the yellow pine belt of the western Sierra Nevada. Augmenting this enviable resource base, California has an ocean and freshwater fishery rivaled in few other states and a developed geothermal energy resource, at the Geysers and in the Imperial Valley, that no other state except Hawaii possesses.

Exploitation of these varied natural resources fluctuates with supply and demand, official perception of environmental impact, technological capability, govern-

Figure 1-7 *Lake Tahoe from Emerald Bay in the northeastern Sierra. The lake derives its name from the Washo tribe's word for water or lake. Presently shared by California and Nevada, Lake Tahoe measures 22 miles long by 12 miles wide, lies 6,229 feet above sea level, and is about 1,600 feet deep, which helps the lake maintain its outstanding clarity in spite of the impacts of intense urban development along and upslope from its shoreline.* (Crane Miller)

ment constraints, natural phenomena, and other concerns both evident and not so evident. For instance, though it was once the national leader in gold production, today California is unranked even though its reserves are still plentiful. One of several deterrents to a revival of gold mining is that most of the existing ores are of insufficiently high grade to be profitably mined. Petroleum production, in which California ranks with the Gulf of Mexico states and Alaska, provides another case in point: Local demand far outstrips local supply. Furthermore, imports of petroleum will increase as local reserves of natural gas near depletion. But no matter where California ranks in individual resource categories, the state remains unchallenged in resource versatility. And if any other constraint besides actual depletion is placed on this resource potential, it is likely to be the limits of human ingenuity.

Cultural Diversity

Debatable though the point may be, the thesis that environmental diversity influences cultural diversity bears strong witness in California. Moreover, environmental variety has fostered a distinctive heritage for California through its successive cultures. Not only were native Californians the most populous aboriginal group for an area the size of California in pre-Columbian North America, but their ecological and linguistic diversity compared with that of the entire continent. The aboriginals' habitat adaptation and survival strategies ranged from fishing for salmon and steelhead trout in the prolific streams of the northwest, to gathering oak acorns as a staple food source in the Great Central Valley, to practicing simple flood plain agriculture along the lower Colorado River. And their linguistic diversity—six language families composed of dozens of languages and dialects—could be likened to that of Europe and Asia together.

The Hispanic cultures that superseded aboriginal culture contributed relatively few immigrants to California but noticeably modified the cultural landscape with the development of a 600-mile, 21-station coastal mission system and numerous ranchos whose land areas were often larger than those of small states. This condition of sparse population abruptly ended with the Gold Rush, when hordes of Yankees invaded the Mother Lode. Some of these prospectors left the new state (admitted to the Union September 9, 1850) once the placers played out, but many stayed on and pursued more reliable livelihoods than picking and panning offered.

Perhaps more important than generating California's first genuine population boom, the Gold Rush marked the initial unlocking of a vast treasure of natural resources. Wells were drilled for water and oil, and irrigation agriculture and petroleum drilling boomed. Logging of seemingly inexhaustible stands of virgin redwood and Douglas fir began in earnest. Whalers, sealers, and fishers tapped the bounty of the Pacific. The railroad came, linking California to eastern markets eager for its products. The news of the state's perpetual spring also disseminated eastward, starting a "Health Rush" that continues to this day. The infant motion picture and aircraft industries, perceiving an essential natural resource in California's mild climate (for landing planes and shooting movies could both best be done on sunny days), gravitated to the state and subsequently expanded. Industry in general continued to grow and diversify in California, lending credence to a notion that has persisted since the Gold Rush: "California is the land of economic opportunity."

Environmental amenities thus attracted millions of migrants to the Golden State from all over the globe. The new Californians came from Africa, Asia, Europe, and Oceania, as well as the Americas. Thus the ethnic diversity of today's 33 million Californians is obvious, if not always clearly definable. The 57 percent majority, variously referred to as white or Anglo or something else, is itself a veritable ethnic jumble of western and eastern European origins to which such labels hardly have universal application. Groups among the 43 percent ethnic "minority," however, are distinctly defined. For instance, of all Californians, nearly 26 percent are Hispanics; less than 7 percent are blacks; more than 9 percent are Asian Americans, including Chinese, Filipinos, Japanese, Koreans, and Vietnamese; and less than 1 percent are Native Americans. These minorities tend to be concentrated in the larger cities and may well become the majority in the twenty-first century, especially in light of the increasing influx of Asians and Hispanics. Whether or not much of urban California is perceived as a melting pot, the existing degree of ethnic variety reflects the already unparalleled cultural diversity.

TABLE 1-1 California population trends, 1850–1990.

CENSUS YEAR	POPULATION	PERCENTAGE CHANGE
1850*	92,597	
1860	379,994	+310.4
1870	560,247	+47.4
1880	864,694	+54.3
1890	1,214,498	+40.3
1900	1,485,053	+22.4
1910	2,377,549	+60.1
1920	3,426,861	+44.1
1930	5,677,251	+65.7
1940	6,907,387	+21.7
1950	10,586,223	+53.3
1960	15,717,204	+48.5
1970	19,971,069	+27.1
1980	23,667,764	+18.5
1990	29,760,021	+25.7

* The 1850 total of 92,597 is no doubt understated for a number of reasons, including the following: (1) Contra Costa and Santa Clara counties' census returns were lost; (2) San Francisco County's census returns were destroyed by fire; and (3) although Native North American populations in California had been in steep decline since the advent of Spanish settlement barely 80 years earlier, members of surviving tribes were in many cases not counted in the 1850 census.

Source: U.S. Department of Commerce, Bureau of the Census, *1990 Census of Population and Housing* (Final Report) (Washington, DC: U.S. Government Printing Office, January 1991, and previous U.S. censuses.

Population Trends

If mild climate is singled out as the chief physical attraction of California, the most apparent cultural response to it—spectacular population growth—has been the major *human* event in the state. The influx of people since California achieved statehood in 1850 has variously flowed and trickled, proving that climate, a none-too-reliable constant in itself, is only one of several factors governing population trends.

Consider for a moment the decennial (by decade) changes in population shown in Table 1-1. From 1920 to 1930, for instance, California's proportional growth rate was nearly 66 percent, the second highest of any single decade since 1850. Yet, in the following decade, the rate fell to less than 22 percent, the lowest decennial rate until the 1970s. Socioeconomic conditions affecting the entire nation were at play here. The Great Depression of the 1930s came close on the heels of the economic prosperity of the Roaring Twenties. Americans standing in breadlines and selling apples were simply less mobile and less likely to have children than when they had "a chicken in every pot and two cars in every garage."

Two other decades worthy of examination are the 1950s and 1960s, during which combined time a near doubling of population took place, from more than 10.5 million to almost 20 million. This greatest of numerical increases over any other two decades can be attributed to a number of factors, not the least of which was the rise of the aerospace and electronics industries to national prominence in California. Growth in these industries was rapid during World War II and was sustained well into and beyond the postwar period by such varied events as the Korean War, the Cold War, Sputnik, the Jet Age, and the Vietnam War. But what accounts for the growth slowdown of the 1970s? One explanation concerns the aforementioned industries: Although in the 1950s and 1960s they provided many jobs for newcomers to California, in the 1970s they began laying off workers as firms in Texas and other Sun Belt states outbid California companies for contracts.

The final decade of the Cold War, the 1980s, saw a resurgence of population growth, with 6 million new residents as compared with less than 4 million for the 1970s. But in the 1990s, as in the 1970s, defense plants closed, workers lost their jobs, and the state's population growth slowed; this time, however, world events had a more lasting economic influence as the Soviet Union

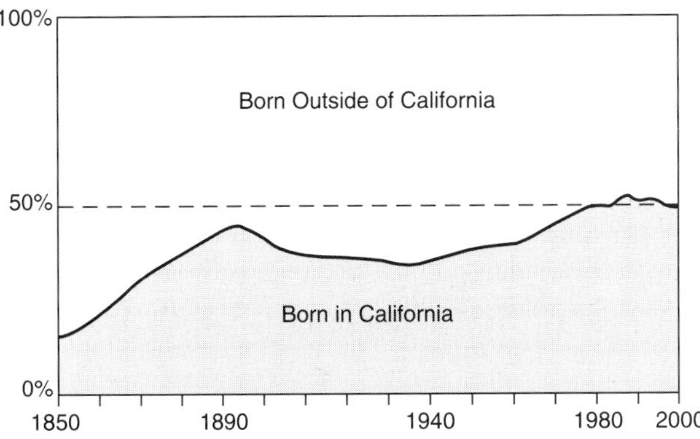

Figure 1-8 *Origin of California's population, 1850–1980. From 1980 to 1990, the proportions of those born in and outside of California remained at 50 percent each. From 1993 through 1996, however, the "born in California" category may have increased beyond 50 percent in light of more people moving out of California than in, as this data shows: 1993, net migration +5,000; 1994, −205,000; 1995, −10,000; 1996, +29,000.* (Source: California Department of Finance, Demographic Research Unit, *California Statistical Abstract 1997.* [Sacramento, November 1997], Table B2. To access the California Department of Finance Web site, type www.dof.ca.gov.)

and the Communist nations in eastern Europe faded into history.

Two regional population variables, *migration* (includes immigration from foreign countries, in-migration from other U.S. states, emigration to foreign countries, and out-migration to other U.S. states) and *natural change* (the difference between births and deaths, or natural increase or decrease) underline these population trends in California. Apparently migration since statehood has contributed more to population growth than natural change (Figure 1-8). Since the late 1960s, however, the pattern has reversed, so that by 1980, the number of native-born Californians equaled those from out of state. Numerically speaking, annual migration to California fell dramatically between 1950 and 1980, from more than 300,000 during the 1950s and 1960s to less than 100,000 during the 1970s. Yet, due in large part to an increasing influx of Latin Americans and Asians, foreign immigration alone averaged 150,000 people per year between 1976 and 1990. A single-year record was set in 1988 when 421,000 people moved to California.

Age and sex distributions also influence household formation and overall population trends. California is clearly past its frontier days when men greatly outnumbered women. A population pyramid would now show the number of males and females as being roughly equal, but with a slightly higher proportion of males than females in the lower age brackets and a lower proportion of males than females in the senior citizen years. In the 1980s, the relatively high proportion of people in the most productive child-rearing years of 20 to 35 temporarily put the damper on talk of ZPG (zero population growth) in California. By the 1990s, however, the baby boomers of the 1980s were a decade older, and their baby-booming days clearly were behind them. In fact, as a percentage of total population, the 20- to 35-year-old group declined from 28 percent in 1980 to 24 percent in 1990, while the proportion of people in over-60 brackets increased, as did life expectancy. These trends indicate that *vital rates* (birth and death rates) have stabilized at a relatively low level, placing California in a nearly steady-state stage in its *demographic transition* away from rapid population growth.

A county-level (Figure 1-9) examination of the spatial dynamics of California's human population, or its *geodemographics,* reveals trends over the past three decades (1960–1990) that may well carry into the next 30-year period (1990–2020). During the 1960s, as shown in Figure 1-10 and Tables 1-2 and 1-3, Orange and Ventura led all 58 counties in rate of growth. Then, as today, there was a preference among new-home buyers for seaside location—if they could afford it! And afford it they could. By the end of the decade, for instance, one could still purchase a new tract home within 2 miles of the Pacific in Orange County for the same price ($30,000–$40,000 range) as a like home deep in the eastern interior of Los Angeles County, and in both cases commuting time by automobile to and from downtown Los Angeles was about the same. Add to the attraction of living by the sea, competitive land prices, and acceptable commuting distances the spillover of development from the recently built-up southeastern (e.g., Downey and Norwalk) and northwestern (e.g., the San Fernando Valley) portions of Los Angeles County into neighboring Orange and Ventura counties, and it's no wonder the latter two grew by 102 percent and 90 percent, respectively, between 1960 and 1970. Similar forces would propel Santa Clara County at the southern end of San Francisco Bay into the position of fourth fastest-growing county by 1970 and also influence events that would cause the Santa Clara Valley to become better known as Silicon Valley. Notable in Orange and Santa Clara counties' growth is the fact that both their populations topped the 1 million mark.

By the end of the 1970s, though, bargain-price land near the ocean in metropolitan California was only a memory, with new-home prices skyrocketing tenfold or more in the coastal portions of Orange and Ventura counties. Population growth continued at a brisk pace, with Orange County pushing the 2 million mark by 1980 (see Table 1-3); however, urban development would shift to the inland areas of the counties where land was plentiful and cheaper. Notably, not a single coastal county was among the 10 fastest-growing counties in 1970–1980; rather, all 10 were landlocked (see Table 1-2). For example, coastal Santa Cruz County grew by more than 50 percent in 1970–80, but it and Placer and

Figure 1-9 *California counties and topography. Of 58 counties, 15 front the Pacific Ocean, from Del Norte in the north to San Diego in the south, and another 13 are in the Sierra Nevada, from Butte and Plumas in the north to Kern in the south. The coastal and Sierra regions comprise California's two principal outdoor recreation areas. The Great Central Valley, which includes parts of 20 counties stretching from Shasta in the north to Kern in the south, is the state's agricultural heartland.*

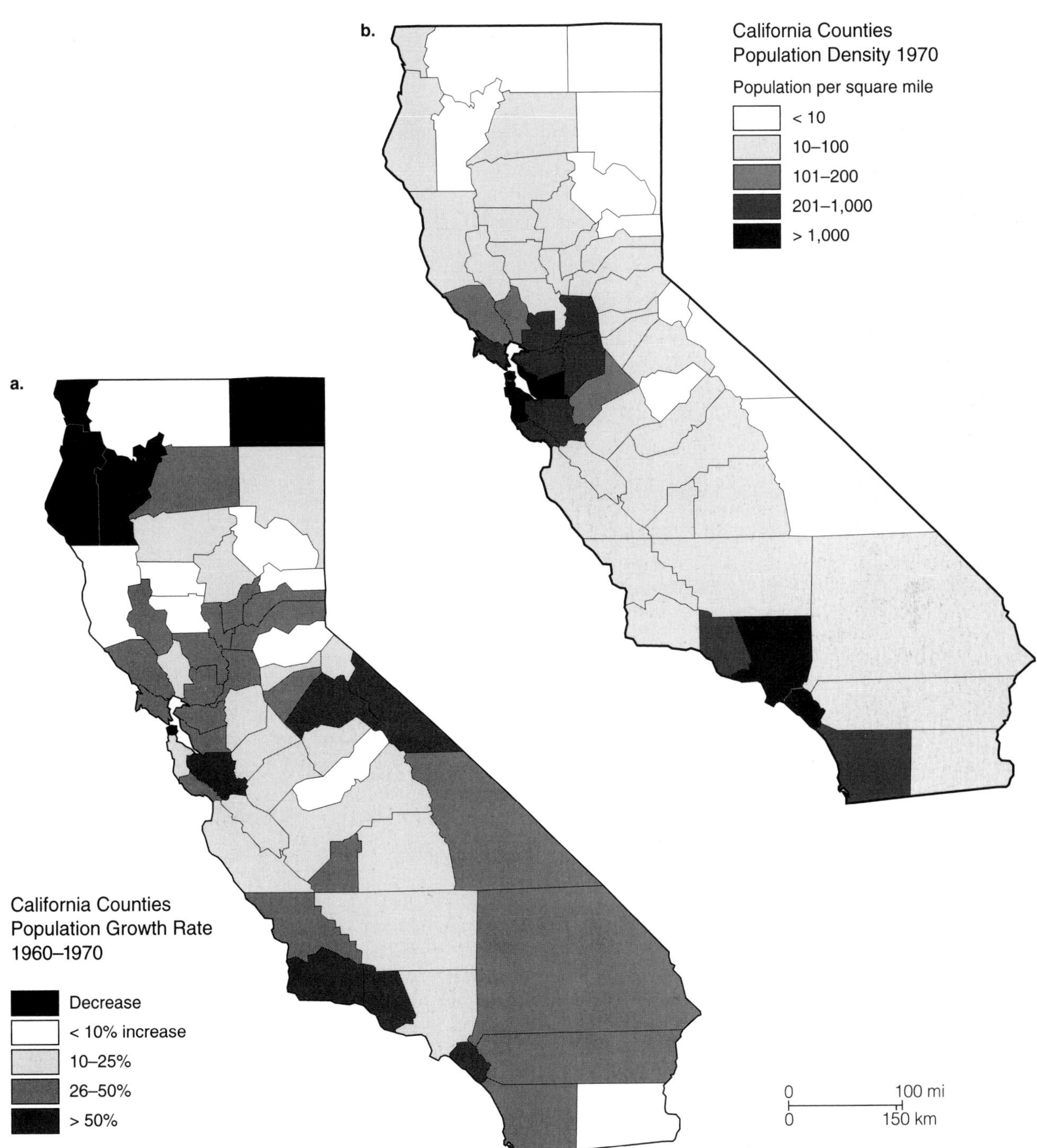

Figure 1-10 *California population growth rate, 1960–1970, and density, 1970, by county.* (Source: U.S. Department of Commerce, Bureau of the Census, *1990 Census of Population and Housing* [Final Report] [Washington, DC: U.S. Government Printing Office, January 1991]. Crane Miller and Michelle Gutiérrez.)

TABLE 1-2 California's ten fastest- and slowest-growing counties by rate of growth by decade, 1960–2020.

RANK	1960–1970	1970–1980	1980–1990	1990–2000	2000–2010	2010–2020
1	Orange*	Alpine*	Riverside	Calaveras	Calaveras	Sutter
2	Ventura	Mono*	San Bernardino	Del Norte	Sutter	Calaveras
3	Mono	Nevada	Amador	Riverside	Riverside	Riverside
4	Santa Clara	El Dorado	Calaveras	Madera	Amador	San Bernardino
5	Santa Barbara	Lake	Nevada	Kern	Fresno	Fresno
6	Tuolumne	Mariposa	Placer	Placer	Merced	Merced
7	Riverside	Amador	El Dorado	Fresno	San Benito	Amador
8	El Dorado	Trinity	San Benito	San Bernardino	San Bernardino	Kern
9	Santa Cruz	Tuolumne	Solano	Nevada	Stanislaus	Yuba
10	Marin	Madera	Tuolumne	Amador	Kern	Stanislaus
49	Madera	Santa Barbara	Napa	Plumas	Santa Cruz	Humboldt
50	Glenn	Yuba	San Mateo	Alameda	Plumas	Plumas
51	Siskiyou	Kings	Trinity	San Mateo	Humboldt	Alameda
52	Plumas	Humboldt	Humboldt	Santa Clara	Orange	Orange
53	Mendocino	Los Angeles	Siskiyou	Napa	Santa Clara	Napa
54	San Francisco†	San Mateo	Sierra	Siskiyou	Alameda	Santa Clara
55	Humboldt†	Marin	San Francisco	Inyo	Sierra	San Mateo
56	Modoc†	Alameda	Marin	Marin	San Mateo	Sierra
57	Del Norte†	Colusa	Inyo	San Francisco	San Francisco	San Francisco†
58	Trinity†	San Francisco†	Apline	Sierra	Marin†	Marin†

* Signifies more than 100 percent growth during the decade.
† Signifies population loss during the decade.

Sources: U.S. Department of Commerce, Bureau of the Census, *1990 Census of Population and Housing* (Final Report) (Washington, DC: U.S. Government Printing Office, January 1991); California Department of Finance, Demographic Research Unit, Report 93-P-3, *Population Projections for California Counties 1990–2040 with Age/Sex and Race/Ethnicity* (Sacramento, 1993).

TABLE 1-3 California area and population change, rate of change, and density by county, 1970, 1980, and 1990.

		1970			1980			1990		
COUNTY	AREA	POPULATION	% CHANGE	DENSITY	POPULATION	% CHANGE	DENSITY	POPULATION	% CHANGE	DENSITY
Alameda	737	1,071,446	+27	1,454	1,105,379	+3	1,507	1,279,182	+16	1,735
Alpine	739	484	+22	<1	1,097	+127	<2	1,113	+1	<2
Amador	593	11,821	+18	20	19,314	+63	33	30,039	+56	51
Butte	1,640	101,969	+24	62	143,851	+40	88	182,120	+27	111
Calaveras	1,020	13,585	+32	13	20,710	+52	20	31,998	+55	31
Colusa	1,151	12,430	+3	11	12,791	+3	11	16,275	+27	14
Contra Costa	720	556,116	+36	772	656,380	+18	912	807,372	+23	1,116
Del Norte	1,008	14,580	−18	14	18,217	+25	18	23,460	+29	23
El Dorado	1,712	43,833	+48	26	85,812	+96	50	125,995	+47	74
Fresno	5,963	413,329	+13	69	514,621	+25	86	667,490	+30	112
Glenn	1,315	17,521	+2	13	21,350	+22	16	24,798	+16	19
Humboldt	3,573	99,692	−5	28	108,514	+9	30	119,118	+10	33
Imperial	4,175	74,492	+3	18	92,110	+24	22	109,303	+19	26
Inyo	10,192	15,571	+33	<2	17,895	+15	<2	18,281	+2	<2
Kern	8,141	330,234	+13	41	403,089	+22	50	543,547	+35	67
Kings	1,390	66,717	+34	48	73,738	+11	53	101,469	+38	73
Lake	1,259	19,548	+42	16	36,366	+87	29	50,631	+39	40
Lassen	4,558	16,796	+24	4	21,661	+29	5	27,598	+27	6
Los Angeles	4,060	7,041,980	+17	1,734	7,477,503	+6	1,842	8,863,164	+19	2,183
Madera	2,138	41,519	+3	19	63,116	+52	30	88,090	+40	41
Marin	520	208,652	+42	401	222,568	+7	428	230,096	+3	443
Mariposa	1,451	6,015	+19	4	11,108	+84	8	14,302	+29	10
Mendocino	3,509	51,101	0	15	66,738	+31	19	80,345	+20	23
Merced	1,929	104,629	+16	54	134,560	+28	70	178,403	+33	93
Modoc	3,944	7,469	−10	<2	8,610	+15	2	9,678	+12	<3
Mono	3,045	4,016	+81	1	8,577	+114	<3	9,956	+16	>3
Monterey	3,322	247,650	+25	74	290,444	+17	87	355,660	+23	107
Napa	763	79,140	+20	104	99,199	+25	130	110,765	+12	147
Nevada	958	26,346	+26	28	51,645	+96	54	78,510	+52	82
Orange	790	1,421,233	+102	1,799	1,932,709	+36	2,446	2,410,556	+25	3,053
Placer	1,404	77,632	+36	55	117,247	+51	84	172,796	+47	123
Plumas	2,554	11,707	0	<5	17,340	+48	<7	19,739	+14	<8
Riverside	7,208	456,916	+49	63	663,166	+45	92	1,170,413	+77	162
Sacramento	966	634,373	+26	657	783,381	+24	811	1,041,219	+33	1,078
San Benito	1,389	18,226	+18	13	25,005	+37	18	36,697	+47	26
San Bernardino	20,062	682,233	+35	34	895,016	+31	45	1,418,380	+59	71
San Diego	4,205	1,357,854	+31	323	1,861,846	+37	443	2,498,016	+34	594

(continued)

TABLE 1-3 *(continued)*

		1970			1980			1990		
COUNTY	AREA	POPULATION	% CHANGE	DENSITY	POPULATION	% CHANGE	DENSITY	POPULATION	% CHANGE	DENSITY
San Francisco	48	715,674	−3	14,910	678,974	−5	14,145	723,959	+7	15,502
San Joaquin	1,399	291,073	+16	208	347,342	+19	248	480,268	+38	344
San Luis Obispo	3,305	105,690	+30	32	155,435	+47	47	217,162	+40	66
San Mateo	449	557,361	+25	1,241	587,329	+5	1,308	649,623	+11	1,447
Santa Barbara	3,739	264,324	+56	97	298,694	+13	109	369,608	+24	135
Santa Clara	1,291	106,531	+66	825	1,295,071	+22	1,003	1,497,577	+16	1,160
Santa Cruz	446	123,790	+47	278	188,141	+52	422	229,734	+22	515
Shasta	3,786	77,640	+31	21	115,715	+48	31	147,036	+27	39
Sierra	953	2,365	+5	>2	3,073	+30	>3	3,318	+8	<4
Siskiyou	6,287	33,225	+1	5	39,732	+20	6	43,531	+10	7
Solano	828	171,989	+28	208	235,203	+37	284	340,421	+45	411
Sonoma	1,576	204,885	+39	130	299,681	+46	190	388,222	+30	246
Stanislaus	1,495	194,506	+24	130	265,900	+37	176	370,522	+39	248
Sutter	603	41,935	+26	70	52,246	+25	87	64,415	+23	107
Tehama	2,951	29,517	+17	10	38,888	+32	13	49,625	+28	17
Trinity	3,179	7,615	−22	>2	11,858	+57	>3	13,063	+10	4
Tulare	4,824	188,322	+12	39	245,738	+26	51	311,921	+27	65
Tuolumne	2,236	22,169	+54	10	33,928	+53	15	48,456	+43	22
Ventura	1,846	378,497	+90	205	529,174	+40	287	669,016	+26	362
Yolo	1,012	91,788	+40	91	113,374	+24	112	141,092	+24	139
Yuba	631	44,736	+32	71	49,733	+11	79	58,228	+17	92
Total	155,793	19,971,069	+27	128	23,667,764	+19	152	29,760,021	+26	191

Note: Area is dry land in square miles. Percentage change is based on the previous decade's census (1960), as well as on the decade shown. Density is in persons per square mile. Zero change represents less than 1 percent increase or decrease from the previous decade. Totals may vary from those derived from mathematical addition of the columns; however, in all cases the totals are those published in the U.S. censuses from 1970, 1980, and 1990.

Source: U.S. Department of Commerce, Bureau of the Census, *1990 Census of Population and Housing* (Final Report) (Washington, DC: U.S. Government Printing Office, January 1991).

Calaveras counties wound up in a virtual tie for tenth place with the listed tenth-place finisher, Madera County. As a comparison of Figures 1-10 and 1-11 reveals, by 1980 Santa Clara became the sixth county to have a density of more than 1,000 residents per square mile; Orange County had achieved that density by 1970. San Diego County posted the highest numerical increase during the 1970s, adding a half-million people to its population, yet the bulk of suburban development increasingly occurred in Rancho Bernardo, La Mesa, and other locations several miles inland from the expensive properties along the Pacific shoreline. Two eastern Sierran counties, Alpine (+127 percent) and Mono (+114 percent), claimed the highest percentage gains for 1970–80, though the former would drop from first to fifty-eighth place during 1980–90 in rate of growth and claim only 1,113 residents by 1990 (see Table 1-3). The numbers were more impressive just to the south of Alpine with Mono County more than doubling to over 8,500 by 1980 and pushing 10,000 by 1990. Development of a world-class ski resort at Mammoth Mountain added permanence to Mono County's growth. Eight counties in the western Sierra (Nevada, El Dorado, Mariposa, Amador, Tuolumne, Madera, Calaveras, and Placer)

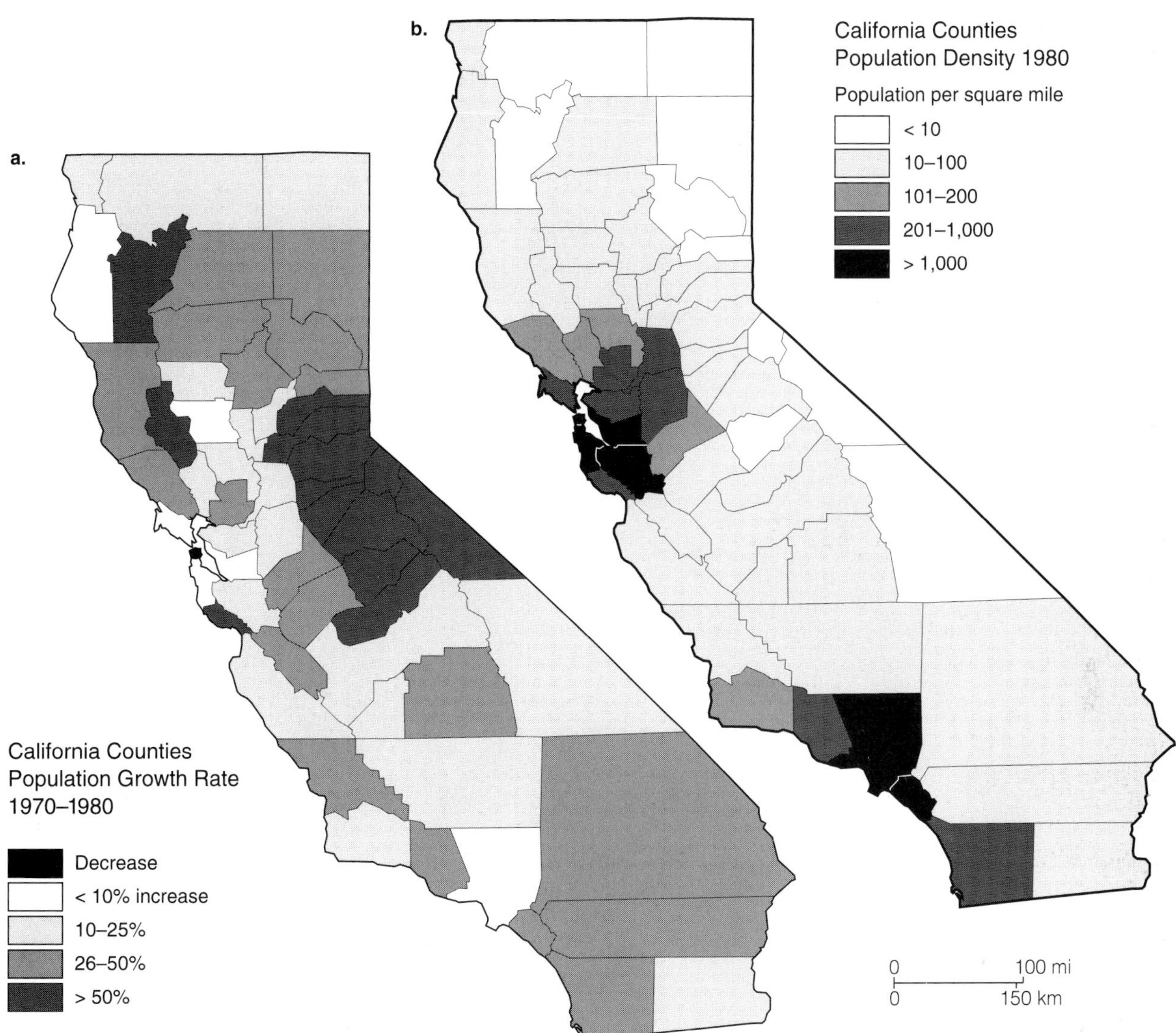

Figure 1-11 *California population growth rate, 1970–1980, and density, 1980, by county.* (Source: U.S. Department of Commerce, Bureau at of the Census, *1990 Census of Population and Housing* [Final Report] [Washington, DC: U.S. Government Printing Office, January 1991]. Crane Miller and Michelle Gutiérrez.)

and two landlocked counties in northwestern California (Lake and Trinity) rounded out the list of counties growing by more than 50 percent during the 1970s. All could fairly claim the benefits of low land prices, rural living, an alpine environment, and distance from the congestion of the cities and suburbs; Lake County could boast an additional attraction in Clear Lake, the largest natural, freshwater lake entirely within the boundaries of California. El Dorado and Placer counties border Lake Tahoe, but this deepest lake in California is shared with the state of Nevada. Finally, for the second decade in a row, San Francisco County lost population. Besides high real estate values and rents, location just east of the San Andreas Fault zone and a density of more than 14,000 people per square mile proved oppressive to some former residents of the county.

In the last full decade of record, 1980–90, five counties had more than 50 percent growth and eight counties with densities of more than 1,000 population per square mile (Figure 1-12). What all these maps bring into sharp focus are geodemographic trends that will likely be sustained through 2020: eastward migration deep into the Inland Empire of southern California and continued growth of several Central Valley–western Sierran counties. With Riverside (+77 percent) and San Bernardino (+59 percent) counties leading in growth and each claiming populations well in excess of 1 million by 1990, the Inland Empire seems poised for significant population expansion well into the next century. Consistent with the abundance of relatively low-priced, developable land located along major road and rail arteries in the eastern Inland Empire—namely, Interstates 10 and

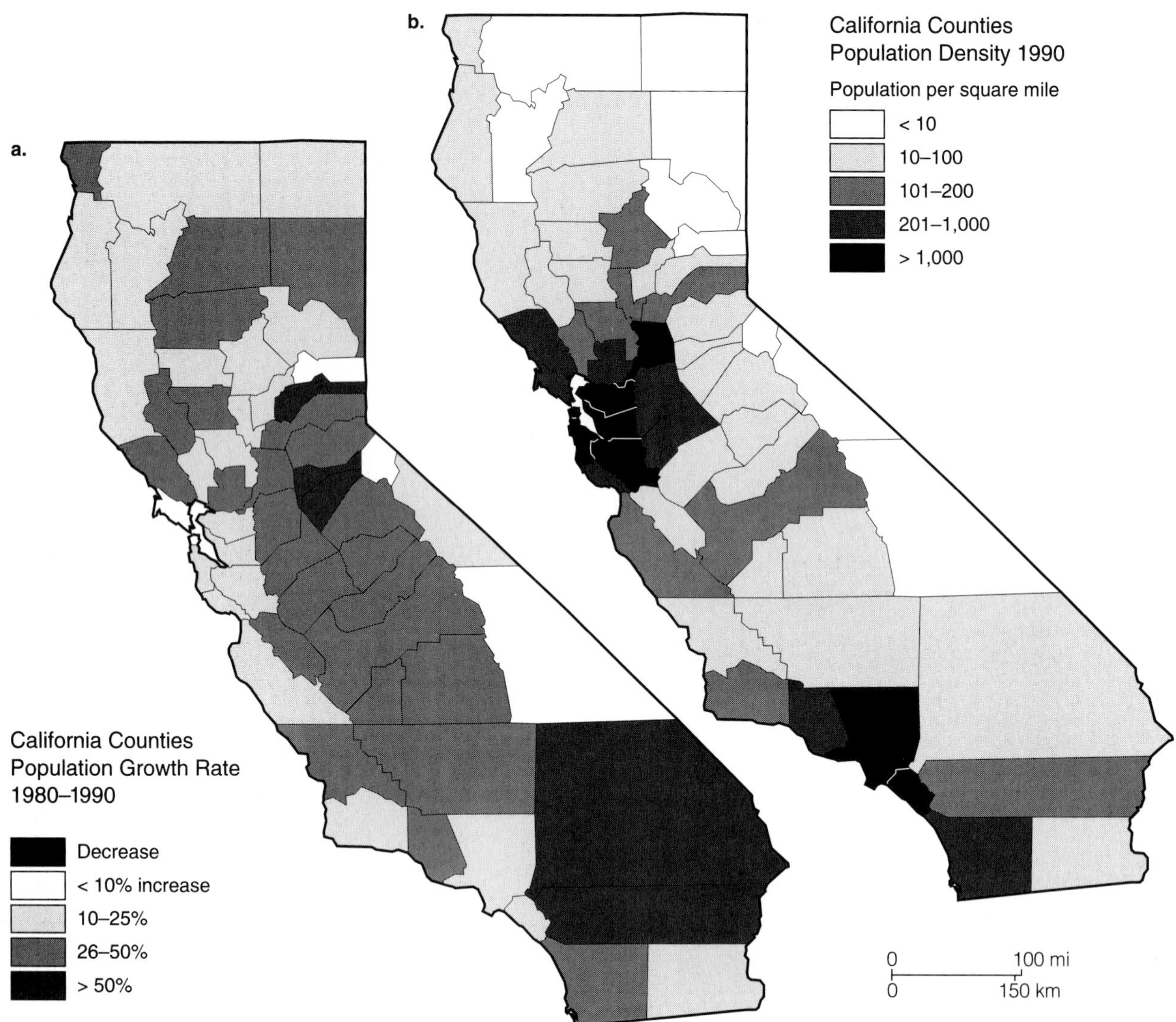

Figure 1-12 *California population growth rate, 1980–1990, and density, 1990, by county.* (Source: U.S. Department of Commerce, Bureau of the Census, *1990 Census of Population and Housing* [Final Report] [Washington, DC: U.S. Government Printing Office, January 1991]. Crane Miller and Michelle Gutiérrez.)

15 and Union Pacific, Southern Pacific, and Santa Fe rail lines—growth will continue to be two-pronged geographically. One prong will reach ever northeastward from Victorville and Barstow toward the Nevada border and Las Vegas, while the other stretches ever closer to the Colorado River boundary with Arizona. As Table 1-4 and Figures 1-13 through 1-15 show, populations are projected to reach 3.1 million in Riverside County and 3.4 million in San Bernardino County, for a combined total of 6.5 million, by 2020.

Similarly, in the Central Valley–Sierran counties, the geographical expression of burgeoning population growth by 2020 will be one of movement eastward from established metropolitan centers—but in this case from Sacramento, Stockton, Modesto, and Fresno rather than from Los Angeles and Orange or the Bay Area counties. And, as with the Inland Empire, it's as though the new suburbanites will follow the admonition, "Go east, young people, for ye shall find detached dwellings at lower prices than you'll pay for crowded condos closer to the city." Yet, although financial motivations may be similar for people gravitating to lower-density living deeper in the interior of the Inland Empire or the eastern Central Valley and western Sierran foothills, the physical environments of the two regions are quite different and thereby offer contrasting attractions and choices. For instance, comparing the desert landscape of the interior Inland Empire to the verdant farmscape of the Central Valley or the oak–pine woodlands of the Sierran foothills (see Chapter 7) is akin to comparing apples and oranges. Even though both regions become quite hot every summer, the latter offers by far the friendlier over-

TABLE 1-4 California area and population projections, rate of change, and density by county, 2000, 2010, and 2020.

		2000			2010			2020		
COUNTY	AREA	POPULATION	% CHANGE	DENSITY	POPULATION	% CHANGE	DENSITY	POPULATION	% CHANGE	DENSITY
Alameda	737	1,457,409	+14	1,977	1,561,851	+7	2,119	1,664,155	+7	2,258
Alpine	739	1,411	+27	2	1,685	+19	>2	2,013	+19	3
Amador	593	42,455	+41	79	56,089	+32	95	70,059	+26	119
Butte	1,640	228,708	+26	139	269,579	+18	165	310,921	+15	196
Calaveras	1,020	54,382	+70	53	78,151	+44	77	102,243	+31	100
Colusa	1,151	21,228	+30	18	26,421	+24	23	32,864	+24	29
Contra Costa	720	971,282	+20	1,349	1,096,253	+13	1,523	1,212,788	+11	1,684
Del Norte	1,008	38,255	+63	38	48,796	+28	48	59,532	+22	59
El Dorado	1,712	155,881	+24	91	188,108	+21	110	220,097	+17	129
Fresno	5,963	945,908	+42	159	1,237,432	+31	208	1,589,665	+28	267
Glenn	1,315	31,071	+25	24	38,369	+23	29	47,233	+23	36
Humboldt	3,573	139,744	+17	39	152,147	+9	43	164,940	+8	46
Imperial	4,175	152,313	+39	36	182,994	+20	44	219,278	+20	53
Inyo	10,192	20,863	+10	2	24,089	+15	>2	28,613	+19	<3
Kern	8,141	801,991	+48	99	1,037,673	+29	127	1,310,050	+26	161
Kings	1,390	135,222	+33	97	168,926	+25	122	207,506	+23	149
Lake	1,259	69,950	+38	56	88,830	+27	71	108,548	+22	86
Lassen	4,558	35,652	+29	8	39,886	+12	9	43,960	+10	10
Los Angeles	4,060	10,180,868	+15	2,508	11,441,900	+12	2,818	12,916,552	+13	3,181
Madera	2,138	133,976	+52	63	171,802	+28	80	214,097	+25	100
Marin	520	248,571	+8	478	245,454	0	472	240,010	−2	464
Mariposa	1,451	20,251	+40	14	24,855	+24	17	29,577	+19	20
Mendocino	3,509	98,224	+22	28	116,719	+19	33	136,041	+17	39
Merced	1,929	238,985	+34	124	313,616	+31	163	401,947	+28	208
Modoc	3,944	11,459	+18	3	13,002	+13	3	14,499	+12	<4
Mono	3,045	12,179	+22	4	15,313	+26	5	18,698	+22	6
Monterey	3,322	414,014	+16	125	485,297	+17	146	574,082	+18	173
Napa	763	125,337	+13	164	139,911	+12	183	147,836	+6	193
Nevada	958	110,386	+41	115	139,488	+26	146	168,372	+21	176
Orange	790	2,866,832	+19	3,629	3,104,100	+8	3,929	3,306,383	+7	4,185
Placer	1,404	247,119	+43	176	312,267	+26	222	369,050	+18	263
Plumas	2,554	22,751	+15	9	24,869	+9	10	26,800	+8	>10
Riverside	7,208	1,775,042	+52	246	2,406,655	+36	334	3,146,936	+31	437
Sacramento	966	1,329,062	+28	1,376	1,579,339	+19	1,635	1,839,529	+16	1,904
San Benito	1,389	50,658	+38	36	66,454	+31	48	83,212	+25	60
San Bernardino	20,062	1,993,762	+41	99	2,621,482	+31	131	3,356,444	+28	167
San Diego	4,205	3,018,363	+21	718	3,476,093	+15	827	3,980,473	+15	946

(continued)

TABLE 1-4 *(continued)*

		2000			2010			2020		
COUNTY	AREA	POPULATION	% CHANGE	DENSITY	POPULATION	% CHANGE	DENSITY	POPULATION	% CHANGE	DENSITY
San Francisco	48	774,011	+7	16,125	781,735	+1	16,286	777,391	0	16,196
San Joaquin	1,399	620,322	+29	443	778,404	+25	556	956,456	+23	684
San Luis Obispo	3,305	263,209	+21	80	306,781	+17	93	351,400	+15	106
San Mateo	449	740,370	+14	1,649	787,291	+6	1,747	825,627	+5	1,839
Santa Barbara	2,739	435,798	+18	159	484,765	+11	177	536,509	+11	196
Santa Clara	1,291	1,703,936	+14	1,320	1,839,696	+8	1,425	1,958,603	+6	1,517
Santa Cruz	446	263,974	+15	592	291,762	+11	654	322,329	+10	723
Shasta	3,786	196,754	+34	52	231,640	+18	61	267,226	+15	71
Sierra	953	3,446	+4	<4	3,669	+6	4	3,851	+5	4
Siskiyou	6,287	49,354	+13	8	55,719	+13	9	61,409	+10	<10
Solano	828	477,727	+40	577	557,403	+17	673	625,347	+12	755
Sonoma	1,576	468,601	+21	297	534,335	+14	339	580,903	+9	369
Stanislaus	1,495	517,618	+40	346	670,009	+29	448	840,191	+25	256
Sutter	603	89,885	+40	149	124,095	+38	206	168,600	+36	280
Tehama	2,951	63,782	+29	22	72,983	+14	25	82,950	+14	28
Trinity	3,179	14,980	+15	5	16,999	+13	>5	18,941	+11	6
Tulare	4,824	417,314	+34	86	521,231	+25	108	644,357	+24	134
Tuolumne	2,236	65,753	+36	29	81,177	+23	36	97,095	+20	43
Ventura	1,846	782,688	+17	424	905,622	+16	491	1,040,456	+15	564
Yolo	1,012	192,608	+37	190	237,828	+23	235	285,883	+20	282
Yuba	631	76,827	+32	122	96,481	+26	153	121,759	+26	193
Total	155,793	36,443,857	+22	234	42,408,137	+16	272	48,976,518	+15	314

Note: Area is dry land in square miles. Percentage change is based on the previous decade's census (1990), as well as on the decade shown. Density is in persons per square mile. Zero change represents less than 1 percent increase or decrease from the previous decade. Totals may vary from those derived from mathematical addition of the columns; however, in all cases the totals are those projected in Department of Finance, Demographic Research Unit Report 93-P-3, as of July 1, 2000, 2010, and 2020.

Source: California Department of Finance, Demographic Research Unit, Report 93-P-3, *Population Projections for California Counties 1990–2040 with Age/Sex and Race/Ethnicity* (Sacramento, 1993).

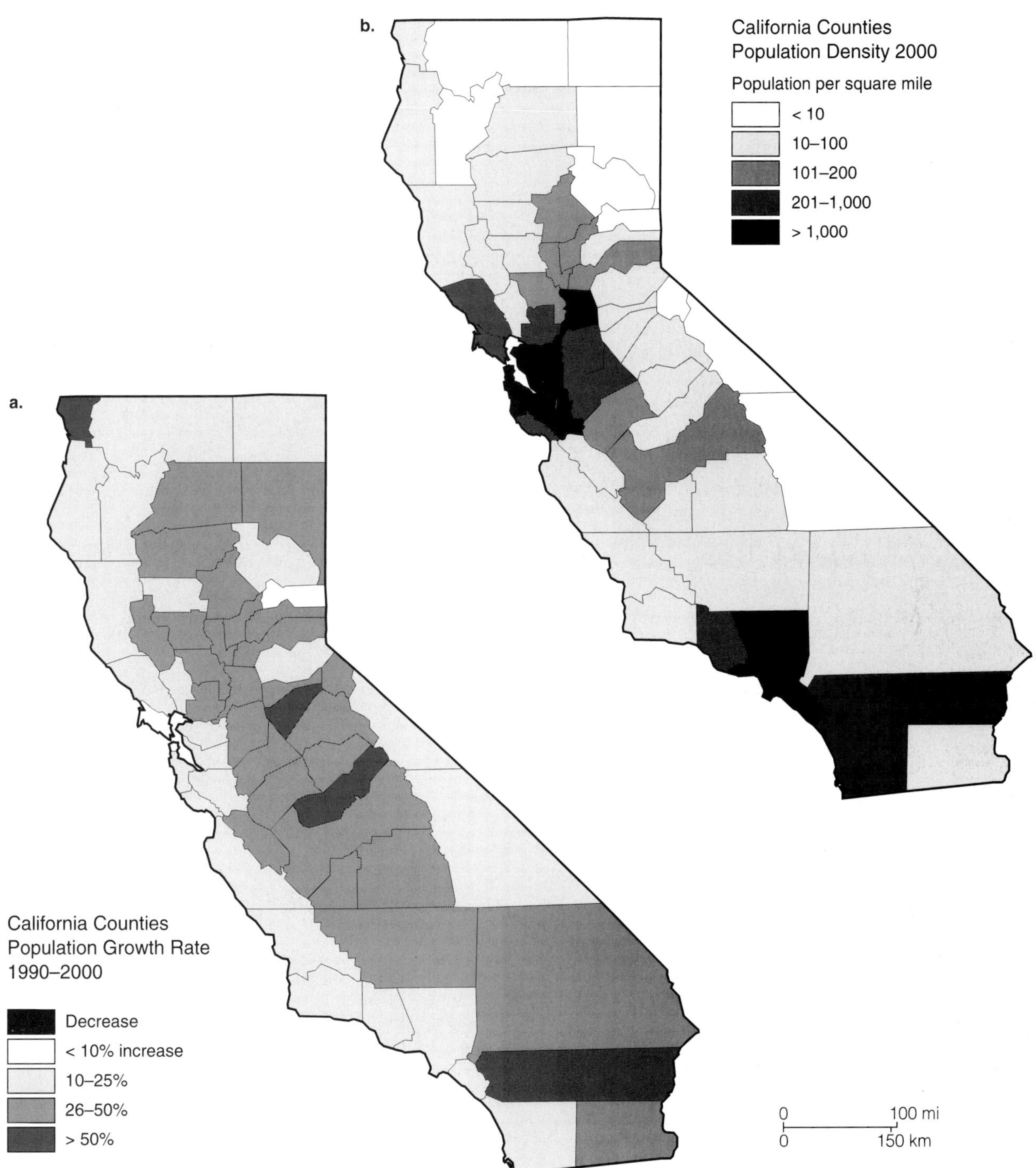

Figure 1-13 *California population growth rate, 1990–2000, and density, 2000, by county.* (Source: California Department of Finance, Demographic Research Unit, Report 93-P-3, *Population Projections for California Counties 1990–2040 with Age/Sex and Race/Ethnicity* [Sacramento, 1993]. Crane Miller and Michelle Gutiérrez.)

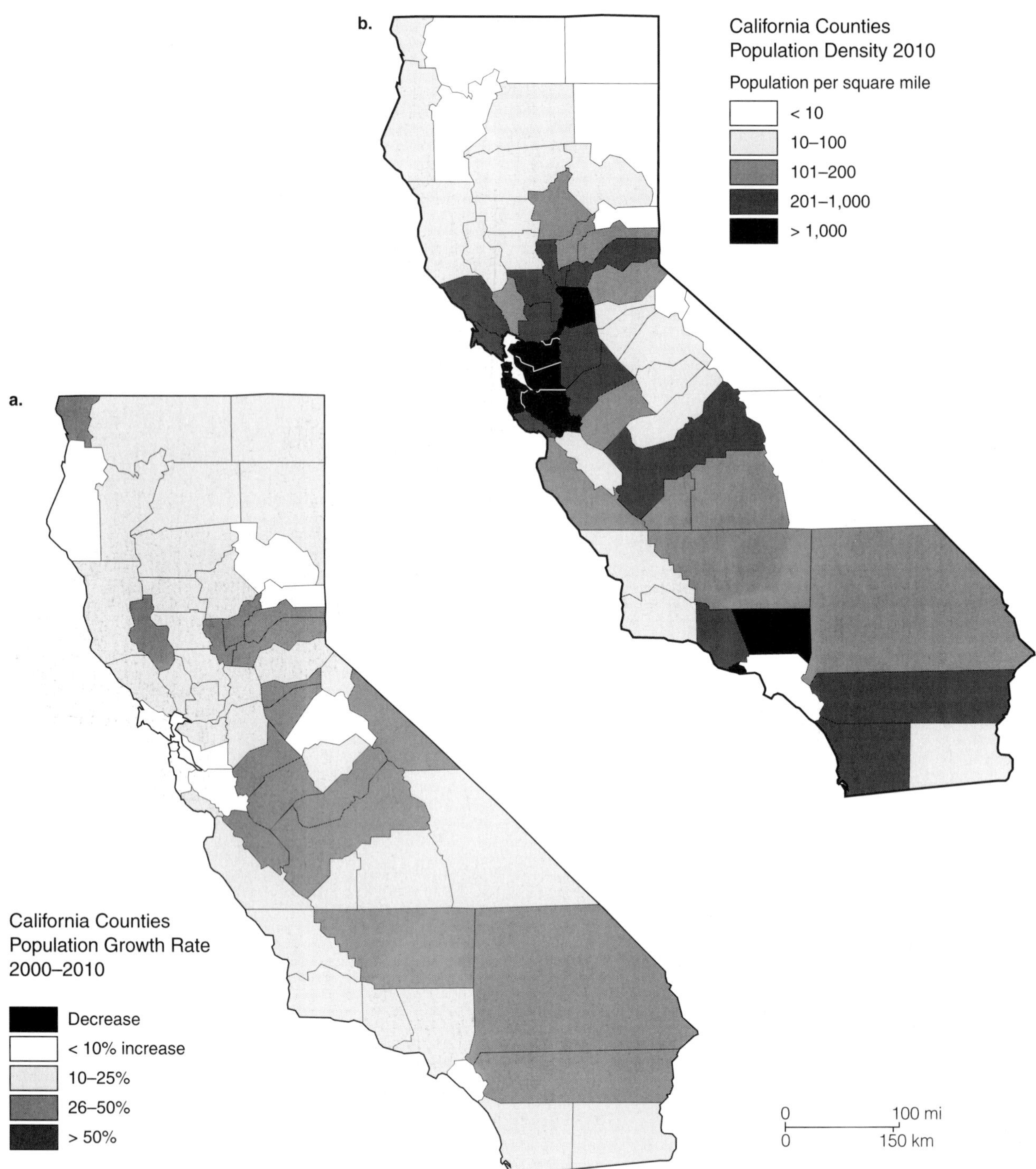

Figure 1-14 *California population growth rate, 2000–2010, and density, 2010, by county.* (Source: California Department of Finance, Demographic Research Unit, Report 93-P-3, *Population Projections for California Counties 1990–2040 with Age/Sex and Race/Ethnicity* [Sacramento, 1993]. Crane Miller and Michelle Gutiérrez.)

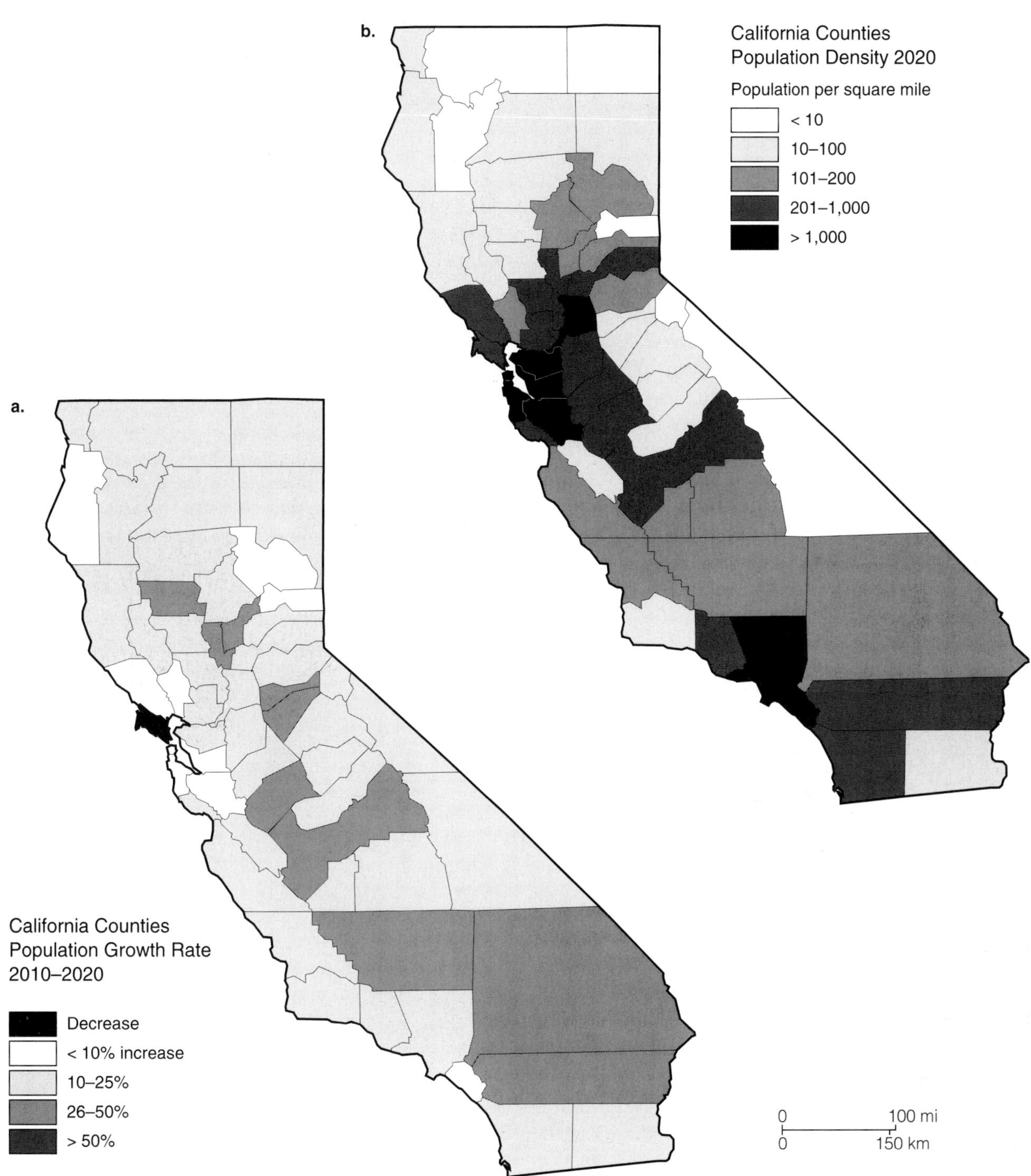

Figure 1-15 *California population growth rate, 2010–2020, and density, 2020, by county.*
(Source: California Department of Finance, Demographic Research Unit, Report 93-P-3, *Population Projections for California Counties 1990–2040 with Age/Sex and Race/Ethnicity* [Sacramento, 1993]. Crane Miller and Michelle Gutiérrez.)

all environment (see Chapter 3). Add to this the alpine wonder of the High Sierra and it's not at all surprising that people by the hundreds of thousands are "headin' for the hills"—but this time for homes rather than gold as they did 150 years ago (see Chapter 8).

Legislation to preserve prime agricultural land, such as the California Land Conservation Act of 1965 (see Chapter 11), may increasingly serve to direct suburban development away from the valley floor and into the western foothills of the Sierra Nevada over the next 25 years. By the same token, increasingly stricter environmental regulations will serve to limit growth in the higher reaches of the mountains. For instance, the region's two counties (El Dorado and Placer) bordering Lake Tahoe, where development has long since stagnated due to sewage treatment facilities that long ago reached their maximum capacities, both drop out of the 10 fastest-growing counties lists for the first two decades of the twenty-first century (see Table 1-2). Given these forces, development will continue to be concentrated in the foothills, where plenty of open space still remains and the constraints on development are fewer. Also, this is where most people would prefer to live, as opposed to down on the valley flats where intensive agriculture fights to maintain its last extensive stronghold in California (see Chapter 10) or high up in the mountains where the winter snows drift deep. As we have already seen in various tables and figures, several of the region's counties appear in the lists of the 10 fastest-growing counties in every decade from 1960 through 2020. Most consistent is Calaveras County, which places first for two decades in a row, 1990–2000 and 2000–2010, and finishes second in the last decade covered, 2010–2020. Incidentally, Calaveras County is forecast to pass the 100,000-mark in total population by 2020.

Figure 1-16 *The Owens Valley. The valley and the rest of the Great Basin region of eastern California are well isolated from urban California both in distance and by significant landform barriers. But sparseness of population and abundance of open space can also be attributed to the aridity and seasonal temperature extremes characteristic of regions in the lee of high mountain ranges. In the case of the Owens Valley, the export of its surface and ground waters by the city of Los Angeles has all but precluded growth.* (Crane Miller)

ISOLATION OR UNIQUENESS

Isolation is a frequently mentioned theme in studies of California. At least as far back as the time of Ordóñez de Montalvo's mythical island, the region has attracted continued references to its relative isolation from the rest of the country (and at times the world).

In 1579, the famous English adventurer Sir Francis Drake's explorations and subsequent efforts at cartography, as well as those of other European geographers, resulted in the widespread conviction that California was indeed an island—and so it was depicted in the maps of the early period. Not until the explorations of Father Eusebio Francisco Kino over a hundred years later was California recognized to be a part of the North American mainland.

Although the Spaniards discovered that they could reach the area by a land route, this did not substantially affect the concept of California's isolation. Removed from the settled areas of Mexico, the area was difficult to get to by either sea or land owing respectively to adverse coastal headwinds and currents and forbidding, barren desert wastes. Thus, once settlers arrived in California, they were not much inclined to engage in active travel back and forth to Mexico. This very real *insulation* contributed to their sense of *isolation*.

The Mexican period saw little change in the patterns of settlement in the area first established by the Spaniards. Not until the discovery of gold did outsiders make a concerted effort to reach the region. Yet, even with the lure of gold to draw them on, travelers from the States had to contend with the arid expanse of the Great Basin (Figure 1-16) between the Rocky Mountains and the Sierra Nevada, the imposing barrier of the Sierra itself, the barren desert wastes of the Mojave, or the semiarid regions of the Central Valley. The alternative was a long, costly, and time-consuming ocean voyage around Cape Horn or an ocean–land–ocean route ultimately ending in San Francisco. In either case, the trip to California constituted a significant separation in terms of time and distance. Without the impetus of gold, the journey took on even greater overtones of sacrifice. Thus, until the completion of the transcontinental railroad, the isolation of the California settlers was very real and very marked.

The initiation of overland mail, the development of the telegraph and the railroad, and eventually the invention of radio, television, airplanes, and autos did reduce the physical isolation of the West Coast, but often not the psychological isolation. The fact remains that

California still is somewhat distant from the centers of eastern and midwestern population. Even today, time and distance may be factors of importance to many visitors to the state. Likewise, most Californians must give serious thought to a vacation trip back East, especially when everything they could desire, from beach to mountain and desert to river recreation, is readily available within the state itself. If air travel continues to expand and become more accessible to all socioeconomic levels, the time and distance factor may subside. However, as oil prices push the costs of air travel upward, fewer travelers will opt for this rapid connection, and surface journeys eastward will continue to involve extended commitments of time and energy.

Thus, as some authorities have noted, California in many ways continues to be "an island upon the land," especially that portion of it west of the San Andreas rift or fault zone. Actually, this elongate sliver of coastal California apparently is moving about 2 inches per year northwestward with respect to the rest of the continent, a migration that could conceivably place it offshore from the British Columbia mainland and the Alaska panhandle about 50 million years from now. Californians, now and for generations to come, won't actually see this inevitable breaking-up of their land, but they will feel it every time an earthquake rumbles through their unstable landscape. We turn our attention now to that unstable landscape.

CHAPTER

2

The Unstable Landscape

California forms part of the *Pacific rim* (or *ring) of fire,* an ominous collection of active volcanoes and earthquake faults that more or less encircles the entire Pacific Basin. Like it or not, membership in this unsettling outer circle is unavoidable if one resides almost anywhere on the Pacific side of Asia, Japan, the Philippines, Melanesia, New Zealand, South America, Central America, or North America. This places "Pacific World" residents in the unenviable position of living on the most unstable terrain on earth. For Californians, there are such twentieth-century reminders of this uneasy instability as the Bay Area earthquakes of 1906 (San Francisco) and 1989 (Loma Prieta); the L.A.-area quakes of 1933 (Long Beach), 1971 (San Fernando), 1987 (Whittier Narrows), and 1994 (Northridge); those that have shaken the eastern Sierra region variously through the 1980s and 1990s; the 1983 quake that devastated downtown Coalinga in the San Joaquin Valley; and 1992 temblors that occurred at nearly opposite ends of the state, one in the northwest that damaged buildings in Ferndale and may have uplifted a 15-mile stretch of the nearby Lost Coast beach as much as 4 feet, and the others in the southeast at Landers in the Mojave Desert and Big Bear in the San Bernardino Mountains (Figure 2-1). The latter two events occurred on the same day, June 28, with the Landers quake registering higher on the Richter scale than any other in the entire state since the Tehachapi quake of 1952, and the Big Bear shaker (4 hours later and 35 miles west), a belated aftershock of the Landers quake.

Apparently, though, the hazards of living where the earth regularly shakes are far outweighed by other environmental amenities the state has to offer. To conclude otherwise is to ask why California is the most populous state and still growing. Earthquakes and earthquake disaster films may remind Californians of their perpetual peril, but these lessons are soon forgotten and detract little from the state's lotus land reputation.

A QUARTET OF DEADLY TEMBLORS: The San Fernando, Whittier Narrows, Loma Prieta, and Northridge Quakes

It was barely 6 o'clock on the morning of February 9, 1971, when an ominous rumbling abruptly awakened many Southern Californians. Within seconds, walls and floors in their homes began vibrating and glass could be heard shattering as mirrors shook loose from walls and dishes rattled from shelves and crashed to the floor. Instinctively, parents rushed to children's rooms, pulled them from their beds, and then scurried to the nearest doorway or crawled under the piano or dining room table to wait out the first shock of the San Fernando earthquake.

Although the mainshock was of short duratrion, it was sufficiently severe for residents almost anywhere in Los Angeles, Orange, and Ventura counties to assume that the quake was centered near their home. Yet Southern Californians soon learned that the *epicenter* (surface point of origin) of the earthquake was located just north of Sylmar in the extreme northern portion of

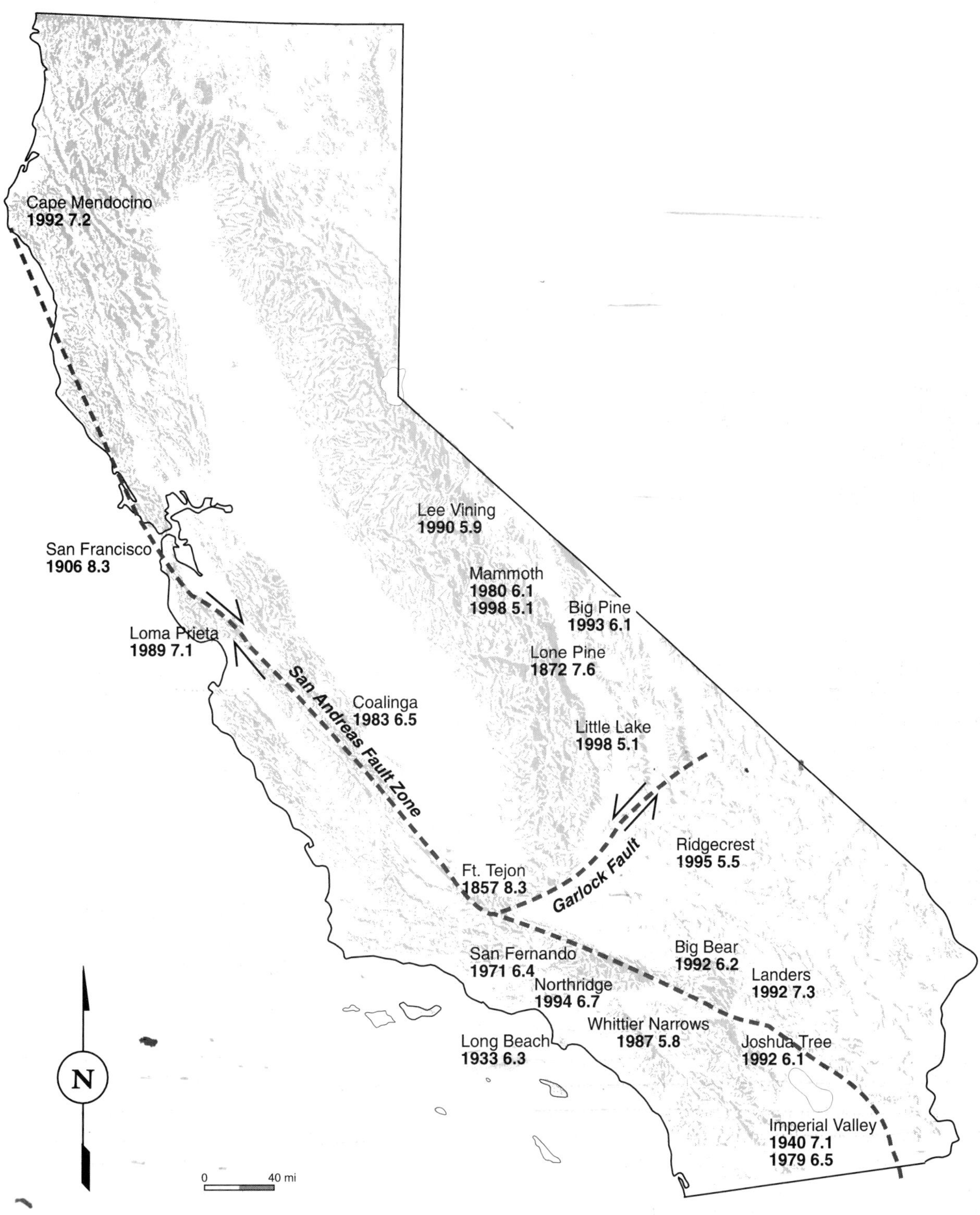

Figure 2-1 *Locations, dates, and Richter magnitudes of major earthquakes in California, and lineaments of the San Andreas and Garlock fault zones. A lineament is a large-scale linear feature on the land surface, such as a trough marking the trace of a fault line.*

Figure 2-2 *Collapse of the overpass connecting Foothill Boulevard and Golden State Freeway, caused by the February 9, 1971, San Fernando earthquake.* (U.S. Geological Survey)

the San Fernando Valley, or about 25 miles north of downtown Los Angeles.

Compared to other major California quakes, as shown in Figure 2-1, the *magnitude* of the San Fernando earthquake at 6.4 on the logarithmic *Richter scale*[1] is considered only moderately high. The 8.3 ratings assigned the 1857 Fort Tejon and 1906 San Francisco quakes are much higher, representing seismograph readings many times that of the San Fernando quake. The amount of actual energy released by an earthquake when measured by Richter magnitude is even more impressive, for with every whole-number increase on the scale, there is a 32-fold increase in energy.

Nonetheless, 6.4 was sufficiently strong to topple hospital buildings at two sites in Sylmar, severely damage the Pacific Intertie converter station at Sylmar, collapse parts of the Golden State Freeway (Figure 2-2), crack the earthfill Van Norman Dam (causing the evacuation of flood-threatened valley communities until the reservoir was drained), and damage scores of homes, many severely enough to cause their condemnation. Had the mainshock lasted longer and/or come later during the morning rush hour, the damage to property and loss of life would have been considerably greater. As it was, the human toll was 65 lives and the cost exceeded half a billion dollars.

Most of the loss occurred in the northern San Fernando Valley near the epicenter of the quake, which was located in a zone of minor faults a little farther north in the San Gabriel Mountains but still some 15 miles south of the San Andreas fault.[2] Obviously, the damage potential of a quake diminishes as distance from the epicenter or point of surface rupture (if different) increases and as density of urban development decreases. Sylmar, a residential community composed predominantly of single-family, detached homes of no more than two stories, lies in the northern suburban fringe of the valley away from major apartment, commercial, and industrial developments. If the density of urban development

[1] Where each whole number represents a 10-fold or exponential change; for example, 8.0 would be 10^4 or 10,000 times as great as 4.0.

[2] Faults are fractures in rock where movement of one side with respect to the other side occurs. Such movement can be up or down (*vertical* or *normal faulting*), horizontal (*lateral faulting*), or diagonal (*overthrusting*) in direction and is often caused by the release of elastic strain that has built up along the fault line.

A

B

Figure 2-3 *(A) Demolition and renovation of the uptown Whittier central business district (CBD) in the aftermath of the October 1, 1987, Whittier Narrows earthquake and aftershocks. (B) Collapse of the reinforced concrete parking structure adjacent to the former May Company department store in the Whittier Quad mall resulting from the Whittier Narrows earthquake.* (Crane Miller)

intensifies in Sylmar, as it inevitably seems to do in most California suburbs, then there will simply be more people and more structures of greater height subject to potential devastation in the event of another 6.4 earthquake.

Note that Richter magnitudes are not used to estimate damage, for a 6.4 quake in a barren, unpopulated region may do nothing more than frighten some wildlife. How much this higher risk is counterbalanced by more stringent structural safety codes, seismic safety elements in general plans (for example, prohibiting development on or near active faults), and other precautionary regulations is unknown. After all, even supposedly earthquake-resistant buildings can collapse in a 6.4 earthquake, as did the Olive View Hospital in Sylmar in 1971,[3] and inactive faults can suddenly release long built-up strain. Worse yet, the southern San Andreas fault, which was not the direct source of the 1971 earthquake but which is still too close for comfort, is apparently capable of producing an 8.0+ earthquake![4]

In the case of the Whittier Narrows earthquake of October 1, 1987, a hitherto undiscovered fault was to blame. At first, it was thought that the 5.8-magnitude temblor was epicentered on the Whittier fault, an active and well-documented fault zone extending southeastward from the Los Angeles Basin through the Puente Hills–Brea Canyon area. In close proximity to the Whittier fault, the newfound fault also produced two 5.0+ aftershocks, the first on October 4 and the second in February 1988. The mainshock and aftershocks were felt over the entire Los Angeles Basin. But most of the 10 deaths, the many injuries, and the heaviest property damage occurred on the campus of California State University–Los Angeles, a few miles east of downtown Los Angeles; southeastward through the Whittier Narrows *watergap* (where the San Gabriel River has eroded a southward path at the western foot of the Puente Hills); and in the communities of Whittier, La Habra Heights, La Habra, and Brea. Uptown Whittier was hardest hit, with some 4,000 commercial buildings and houses sustaining damage (Figure 2-3). As though to add injury to insult, the aftershock four days later irreparably damaged some structures. In the aftermath, many longtime Whittier residents described the central business district (CBD) of their city as a "war zone." Damage in Whittier itself amounted to $90 million, with another $280 million worth occurring outside the city. Miraculously, none of the eight deaths on October 1 nor the two on October 4 occurred in Whittier. By 1998, with the city rebuilt and the quake a distant memory, uptown Whittier was enjoying a retail business boom the likes of which it had never seen before the quake. All the more remarkable was that the city's CBD was bucking the trend of nearby, newer shopping malls outcompeting long-established

---

[3] According to the California Division of Mines and Geology, "San Fernando, California, Earthquake of 9 February 1971," *Bulletin* 196 (1975), p. 341, not all of the many buildings in the County of Los Angeles' Olive View Hospital complex were seriously damaged. However, excessive damage did occur to the newly completed (1970) "earthquake-resistive Medical Treatment and Care Building" and three lives were lost therein. Although the magnitude (M) of the San Fernando earthquake was tentatively given at 6.6, the Foreword in the *Bulletin* notes an M of 6.5 and later (p. 260) a reassignment of M 6.4.

[4] The San Andreas is best (or worst?) thought of as a fault zone or system—that is, the San Andreas fault itself plus auxiliary faults such as the one from which the San Fernando earthquake originated.

CBDs for retail trade. In 1999, nearly a dozen years after the Whittier Narrows quake, the location of the guilty fault was finally pinpointed. The fault system extends for 25 miles southeastward from downtown Los Angeles into the Coyote Hills just south of La Habra and poses the threat of up to magnitude 7.0 earthquakes for the southeastern Los Angeles Basin region.

Homeowners whose property is damaged by an earthquake are usually in for a long wait before they receive financial compensation and get their homes repaired. In the case of the moderately destructive Whittier Narrows quake, the affected region was declared a federal disaster area, and the Federal Emergency Management Agency (FEMA) set up shop to administer relief. The FEMA acted quickly to provide cash grants to remedy potential hazards, such as removal of brick chimneys teetering on the brink of collapse. But loans to cover costlier, permanent structural repairs were not handled with such dispatch. Victims applied for low-interest loans through the FEMA from the federal Small Business Administration (SBA), but many applicants found the process to be an exercise in futility when they learned that they did not qualify for lack of sufficient income. Fortunately, many unsuccessful applicants automatically qualified for loans from the State Earthquake Rehabilitation Assistance (SERA) program, which imposed no means test as had the SBA. But most SERA borrowers had to wait months, if not years, before getting their low-interest loans. Victims with earthquake insurance generally fared better than those without it, but again, many insured had to wait months for their money, and some belatedly discovered that their losses were not covered. Large deductibles were still another problem for the insured, with many policies carrying a minimum 10 percent deductible based on the total replacement value of structures.

That California is the country's undisputed earthquake leader was further substantiated by the 7.1 Loma Prieta quake of October 17, 1989, and its thousands of aftershocks (Table 2-1). As the table shows, Loma Prieta aftershocks numbered 4,760 barely 3 weeks after the main event. Although most of the aftershocks were "not felt," more than 80 ranged from "perceptible" to "strong," and as in the Whittier Narrows aftermath, two aftershocks measured at least 5.0 in magnitude and caused further damage. Comparing the Loma Prieta and San Fernando quakes, both caused almost the same number of deaths (63 and 64, respectively) and in both instances, the failure of a single structure was the principal cause of death. In the Loma Prieta quake, the collapse of the elevated Cypress Street Viaduct of the Nimitz Freeway (I-880) in Oakland was to blame; here alone, 42 people perished. Otherwise, the extent and amount of damage of the Loma Prieta quake was far greater than either the San Fernando or Whittier Narrows events, with 18,306 homes and 2,575 businesses damaged all the way from Oakland and San Francisco south through Santa Cruz and Watsonville, a distance of about 80 miles. In addition to the freeway disaster, the Loma Prieta caused a 30-foot span of the San Francisco Bay Bridge roadbed to break away; several multistory buildings in the Marina District of San Francisco to collapse due to the *liquefaction* (soil and earth materials invaded by groundwater and therefore assuming a slurrylike consistency) of sand and debris used to fill a former lagoon for the Pan-Pacific Exposition of 1915; and damage throughout the Pacific Garden Mall in Santa Cruz, which lies on old river deposits that suffered liquefaction during the great San Francisco earthquake of 1906. Estimates of the cost of damage range upwards of $7 billion, or 14 times that of the San Fernando quake. Perhaps most disconcerting for Bay Area residents, though, is the consensus among earthquake experts that another "big one" is still to come. The 7.1 quake was epicentered along the San Andreas fault and did cause the Pacific plate to move 6.2 feet to the northwest (Figure 2-4) and 4.3 feet upward and over the North American plate, but 8.0+ earthquakes are expected for both Northern and Southern California.

Shortly before dawn on January 17, 1994, the most powerful earthquake in the recorded history of the Los Angeles Basin struck a largely sleeping populace. The Northridge earthquake was named after the northwestern San Fernando Valley portion of the city of Los Angeles where it was epicentered. The powerful quake registered a magnitude of 6.7 on the Richter scale; took 57 lives; caused several thousand injuries; left tens of thousands of people homeless; severed several major freeway arteries and damaged hundreds of surface streets; ruptured natural gas, power, water, and communication lines; and shut down schools, including Cal-

TABLE 2-1 Loma Prieta earthquake aftershocks, October 17–November 7, 1989.*

RICHTER MAGNITUDE	QUANTITY†	EFFECT
5.0	2	Damaging
4.0	20	Strong
3.0	65	Perceptible
2.0	384	Not felt
1.0	1,855	Not felt
<1.0	2,434	Not felt

* By the first anniversary of the Loma Prieta earthquake, on October 17, 1990 at 5:04 P.M., more than 7,000 aftershocks had been recorded.

† A total of 4,760.

Source: U.S. Geological Survey, *The Loma Prieta Earthquake of October 17, 1989* (pamphlet) (November 1989, rev. January 1990).

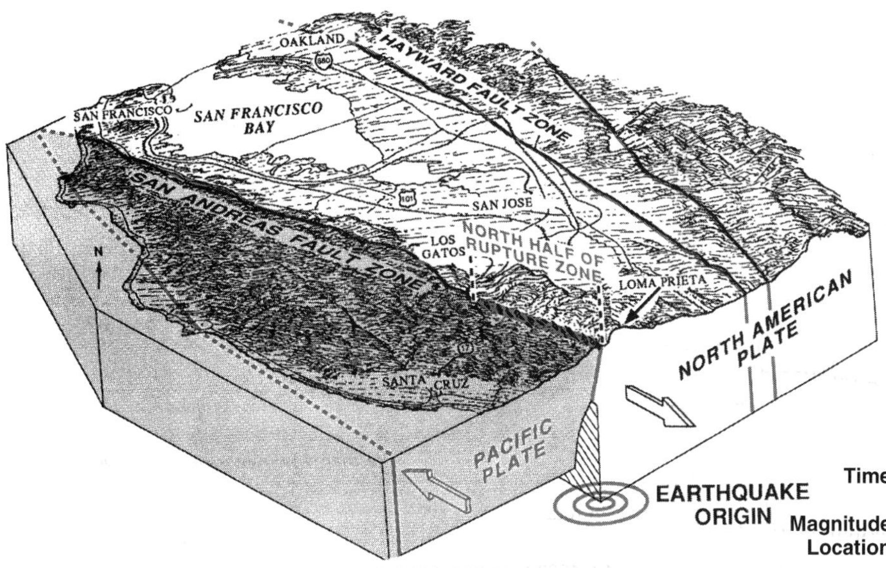

Figure 2-4 *Three-dimensional or isometric diagram of the surface rupture zone, directional movements, epicenter, and hypocenter (earthquake origin) of the Loma Prieta earthquake. Note the location of Oakland and I-880 (the Cypress Viaduct section), which is more than 70 miles from the epicenter.* (Source: U.S. Geological Survey, *The Loma Prieta Earthquake of October 17, 1989* [pamphlet] [November 1989, rev. January, 1990].)

A

B

Figure 2-5 *(A) Apartment complex where the second and third floors caved in on sleeping residents on the first floor, causing the largest single loss of life (16 died) in the Northridge quake. (B) Collapsed parking garage at California State University–Northridge and in a nearby shopping mall.* (Sarkis Maissian)

ifornia State University–Northridge. Estimates of the eventual costs range from $15 to $30 billion. From apartment buildings and parking structures collapsing to office buildings splitting apart (Figure 2-5), the drama that unfolded before a worldwide television audience on January 17 was one of epic proportions. If there was a positive aspect to that bleak day, it was that the 1994 Northridge quake, like its 1971 predecessor in the San Fernando Valley, occurred well before the morning rush hour traffic jammed the freeways, many of which were destined to crumble. Moreover, many Angelinos had already planned to take the Monday off from work in observance of Martin Luther King's birthday, a national holiday. As with the San Fernando, Loma Prieta, and other major earthquakes, thousands of aftershocks spread over many years followed the initial Northridge event. For instance, in January 1998, four years later to the month, a 3.2-magnitude aftershock hit Canoga Park in the southern San Fernando Valley.

Another similarity between the 1971 and 1994 seismic events, but one that portends future troubles for the San Fernando Valley, is the *thrust faulting* as a root cause of both earthquakes. Contributing to the severity of moderate-level Richter quakes in the San Fernando Valley is that the valley is a deep, cauldron like basin filled with sediments. Not only are seismic shock waves trapped and reflected back upon themselves within the "bowl," they are further amplified where liquefaction is occurring.

FAULTS, FOLKLORE, AND FACT

A good deal of folklore surrounds the subject of earthquakes, especially their predictability or lack thereof and what to do if one happens. So-called earthquake weather is said by some self-styled prognosticators to be a warning that a quake is imminent. But there is no real consensus on what constitutes earthquake weather. Some say it's when the air is stagnant and muggy; others contend that earthquake probability increases following a rainstorm. It so happens that on February 9, 1971, a mildly windy Santa Ana condition prevailed in the San Fernando Valley. Admittedly, that merely helped to complete the long list of different types of weather that precede or accompany earthquakes. In short, earthquake weather can be *any* kind of weather.

Perhaps more believable are certain forms of aberrant behavior among animals. For example, domesticated animals have been observed to become unusually fidgety immediately before an earthquake. Among wild animals, there is an alleged ominous silence of neighborhood birds, sometimes preceding a tremor by several minutes. Because separating fact from folklore is difficult in determining links between animal behavior and earthquake prediction, scientific investigation continues.

In looking beyond the earth itself for ways of predicting earthquakes, some scientists hypothesize that the alignment of the planets in our solar system in 1982 might have triggered major quakes in California. The last such alignment occurred in 1803, when historical record keeping of seismic events was limited to a few widely separated Spanish settlements along the coast, and none of these reported major quakes. Noteworthy, though, was the recording of ground shaking and damage to some missions and presidios three years earlier in 1800, evidently due to major tremors. Perhaps it is the other way around, and the *approaching* alignment of the planets signals major earthquakes, the day of final alignment being well past the peak period of danger.

Seismic and related events from 1979 to 1982 that may have been symptoms of the planets coming into alignment include the 6.5 Imperial Valley earthquake; a spate of 6.0 jolts around Mammoth Lakes; the Mount St. Helens eruptions in Washington State; movement in the Hayward fault zone east of Oakland; a 5.3 quake epicentered offshore from Los Angeles on that city's two hundredth birthday, September 4, 1981; and a series of quakes near Eureka. The Eureka area episodes seem extraordinarily significant in that they started on November 8, 1980, with a 7.2 jolt, the most powerful earthquake to be recorded in northwestern California in half a century. Fortunately, property damage (to highways and buildings) and injuries (to five people) were minimal, and no lives were lost.

A dozen years later, another series of earthquakes rocked the Eureka region. Although an offshore temblor registered 7.1 the previous year, the main event occurred on April 25, 1992, about 35 miles southwest of Eureka at Cape Mendocino at 7.2. Closest to the epicenter and hardest hit by the quake was Ferndale (see Figure 2-1), a 140-year-old town of 1,500 famous for its well-preserved Victorian homes, bed-and-breakfast inns, and dairy industry. No deaths were attributed to the quake and its aftershocks; however, some three dozen people were injured, and damage to homes, commercial buildings, and roads approached $3 million.

Understanding the mechanics of earthquakes themselves, rather than other supposedly related environmental phenomena, seems most likely to provide the key to predicting when they are going to happen. Seismologists have known for some time that the sudden rupture and displacement of rock along a fault line due to the release of accumulated strain has triggered many earthquakes. In the case of the San Fernando quake, the subsurface focus of initial energy release, or *hypocenter*, was located 5 miles underground and more than 7 miles north of the point where ground shaking was most severe in Sylmar. This indicated that the fault motion was brought to the surface along a fault inclined at an angle that broke the surface in Sylmar. The Northridge quake was hypocentered nearly a dozen miles beneath the surface of the San Fernando Valley. Since no surface rupture was evident, the Northridge quake is considered a *blind thrust* form of thrust faulting.

By comparison, the hypocenters of the Whittier Narrows and Loma Prieta quakes were at depths of 7.5 and 11.5 miles, respectively. These and other California quake hypocenter depths are all relatively shallow, which may account for the surface severity of the quakes. When built-up elastic strain in shallow rock formations is released and an earthquake ensues, the formations break apart and slip, and surface ruptures may result. When quakes are hypocentered dozens of miles deeper, as they commonly are in other parts of the world, such surface and near-surface fracturing is impossible because of the great thickness of overlying rock formations and the weight or pressure they exert. For example, studies of earthquakes hypocentered at close to 200 miles beneath the surface show rock formations compressing rather than fracturing because of the intense pressure from above. Rarely, if ever, is a California quake's hypocenter more than 20 miles beneath the surface.

Quite often, several days or even weeks before an earthquake, the ground surface on one side of a fault line will increasingly tilt and even change the direction of its tilt, then revert to its original position following an earthquake. This change is measurable with a *tiltmeter* and thus may be a reliable indicator of an impending earthquake. If the rate of ground movement accelerates rapidly, as it sometimes does before an earthquake, an *accelerometer* can monitor the change.

Other techniques that may bring us to an earlier realization of accurate earthquake prediction include monitoring fluctuations in the earth's magnetic field, studying the periodicity of the moon's gravitational influence on seismic events originating on earth, measuring fault movements with lasers, and utilizing the visual-historical record provided through aerial photography and other forms of *remote sensing*. Especially promising are chronological series of stereopairs of vertical airphotos from which changing surface fault patterns can be monitored three-dimensionally. (*Stereopairs* are overlapping pairs of photos in which the overlapped portion shows the same scene but photographed from two different perspectives or positions.)

Gradual movement along a fault, or *fault creep*, appears to be more predictable than an earthquake but does not necessarily presage an earthquake. Rather, a creep episode involving a mere fraction of an inch of movement may last several days and be nonviolent, although creep episodes spanning several years are capable of cumulatively causing costly damage in built-up areas. It is significant that "the first successful prediction of ground motion on the San Andreas fault"[5] was of fault creep in the late 1960s. Predicting creep episodes is one matter and earthquakes quite another, however. We still appear to be years away from predicting the latter with precision as to time, location, and severity. Furthermore, earthquake aftershocks and creep, once thought to be related, now appear to be independent processes for relieving stresses along an existing fault surface.

So far, much has been said about earthquake mainshocks and aftershocks, but what about foreshocks as precursors of mainshocks? For instance, 2 and 15 months prior to the Loma Prieta mainshock, earthquakes of 5.1 and 5.2 magnitude, respectively, were recorded in the Loma Prieta region. The California Office of Emergency Services quickly issued a public advisory after each quake, stating that they could be foreshocks of mainshocks to come within 5 days. In both cases, no mainshocks followed within the specified time frames, the advisories were withdrawn, and neither quake was identified as a foreshock. Yet, as shown in Table 2-2, the Loma Prieta mainshock followed shortly thereafter on one of the fault segments considered capable of producing a 6.5+ quake within a 30-year span between 1988 and 2018 by the Working Group on California Earthquake Probabilities of the U.S. Geological Survey. Given this chain of seismic events and forecasts, could the 5.1 and 5.2 quakes in any way be thought of as long-term precursors of the Loma Prieta mainshock?

Whatever the answer, the U.S. Geological Survey has focused its earthquake forecasting work on tiny Parkfield (pop. 34), 115 miles south of the Loma Prieta epicenter on the San Andreas fault, where 6.0-magnitude quakes occur about every 21 years (see Table 2-2), or with greater regularity at that magnitude than anywhere else in California. Still farther south in densely populated southern Orange County, two minor quakes (4.6 on April 7, 1989, and 4.0 on October 18, 1990) in the Newport–Inglewood fault zone raised short-lived fears of a major temblor striking the region in the near term. However, by 1998, long after the 1989 and 1990 quakes could reasonably be considered foreshocks, no such quake had materialized. A 4.3 quake did shake a large portion of northeastern Orange County on January 5, 1998, but its epicenter was quite remote from faults in the southern part of the county. The last severe quake to originate in the Newport–Inglewood fault zone was the 6.3 Long Beach event in 1933, which killed 120 people.

Another body of earthquake superstition has grown up around what to do if one happens. Suggestions for self-preservation if a severe earthquake strikes one's own community range from the practical to the outrageous. Where individuals are when the ground first starts shaking is of paramount concern. If they are in an open area, such as a city park, they would be well advised to stay there; the ground is not likely to open up and swallow anybody. Besides, when people start scurrying to the nearest building, there is a good chance they will be hit by bricks, glass, or other falling debris. But people already inside a building shouldn't rush outside and expose themselves to all those airborne hazards. Rather, they should stand in a doorway or climb under a desk, piano, or some other strong piece of furniture, unless an orderly evacuation can be accomplished or is recommended.

Increasing proportions of buildings are meeting stringent earthquake-resistant standards in California. The chances that a person will be living or working in

[5] California Division of Mines and Geology, *Mineral Information Service* 22, 4 (April 1969), p. 68.

TABLE 2-2 Major earthquake probabilities on segments of the San Andreas fault.

FAULT SEGMENT ON THE SAN ANDREAS	YEAR OF MOST RECENT EVENT	EXPECTED MAGNITUDE	ESTIMATED RECURRENCE (YEARS)	PROBABILITY OF RECURRENCE, 1988–2018 (%)	RELIABILITY LEVEL*
North Coast	1906	8.0	303	<10	B
San Francisco Peninsula	1906	7.0	169	20	C
South Santa Cruz Mountains	1989†	6.5	136	30	E
Central creep zone	—	—	—	<10	A
Parkfield	1966	6.0	21	>90	A
Cholame	1857	7.0	159	30	E
Carrizo	1857	8.0	296	10	B
Mojave	1957	7.5	162	30	B
San Bernardino Mountains	1812‡	7.5	198	20	E
Coachella Valley	1680‡	7.5	256	40	C

* A is most reliable, E least.
† Updated (original date was 1906) to reflect the 1989 Loma Prieta earthquake.
‡ Dates are approximate.
Source: U.S. Geological Survey, Open-file Report 88-398 (1988).

such a structure if and when a "big one" comes are improving. Steel-frame and reinforced concrete structures, even high-rise buildings, are likely to fare well in a major earthquake, as will older and newer wood-frame low-rise houses that are securely attached to their foundations. Brick buildings and others with masonry facades can be hazardous, especially to people immediately outside the walls. Deadly proof of the latter was evident in December 1988 in Armenia, where a 7.1 quake claimed 25,000 lives. Most of the victims were trapped in and around crumbling masonry structures. By comparison, the 7.1 Loma Prieta quake took 63 lives.

Of course, there is no guarantee that any of these steps is necessarily proper in the given circumstances or will save people from injury or death. This disclaimer is not meant to imply that one should not take shelter in a safe place. But in a city or a suburb, where is the safest place when an 8.0+ earthquake hits? Probably the best thing to do is maintain a sense of humor—plan on moving to Oklahoma and learning how to cope with tornadoes. Above all, it is important to remain calm and think through the consequences of any contemplated action. Better yet, plan what to do well before the next shaker, because during a quake, one has little or no time to contemplate a course of action. Human panic merely creates another set of hazards.

Preparations made before an earthquake, even inadvertently, may be of indispensible survival value after the quake has occurred. A classic example is found among those Sylmar residents who before the 1971 earthquake had stocked up on camping gear. Those equipped with camp stoves and fuel, first-aid kits, flashlights, water-purifying tablets, tools, canned food, portable radios, and other artifacts of outdoor living were prepared to ride out any extended disruption of utility services to their homes.

Fortunately, the San Fernando and Northridge earthquakes struck early enough in the morning that most residents were still at home where such survival paraphernalia was close at hand. It would seem advisable to provide such equipment at one's place of business.

In closing our discussion of earthquake faults, folklore, and fact, it seems fitting, albeit sobering, to ponder some highlights of the California Division of Mines and Geology 1982 report entitled "Earthquake Planning Scenario for a Magnitude 8.3 Earthquake on the San Andreas Fault in Southern California."[6] Estimates of the

[6] California Division of Mines and Geology, "Earthquake Planning Scenario for a Magnitude 8.3 Earthquake on the San Andreas Fault in Southern California," *Special Publication* 60 (1982). Studies indicate at least eight major earthquakes have taken place along the southern San Andreas (from Cajon Pass northwestward approximately 200 miles) in the past 1,200 years, the interval between quakes averaging about 140 years and the most recent being the Fort Tejon earthquake of 1857 (see Figure 2-1).

death and destruction in Los Angeles, Orange, Riverside, San Bernardino, and Ventura counties in the event of such an earthquake are staggering. Highways, railroads, pipelines, power lines, and aqueducts crossing the San Andreas into the five-county region would be damaged if not destroyed, resulting in severe curtailment of the flow of imported goods, food, oil, gas, electrical energy, and water upon which urban Southern California is so dependent. It might take months for some of these lifelines to be reopened. Within the region itself, 3,000–14,000 people would be killed, 12,000–55,000 would require hospitalization, and property damage would surpass $20 billion. Hardest hit would be heavily populated areas in alluvium-filled valleys (particularly San Bernardino, San Gabriel, San Fernando, and Simi valleys), river basins and floodplains (notably of the Santa Ana, Santa Clara, and Santa Clarita rivers), and sandy and clayey coastal lowlands (especially the Oxnard Plain, Marina Del Rey, and from Long Beach southeastward to Newport Beach and inland to Anaheim and Tustin). The dismemberment of structures as the ground beneath them failed would likely be most rampant in the aforementioned areas where water tables were relatively high and liquefaction was occurring, the ground itself taking on the properties of a liquid as a result of earthquake shaking. By contrast, structures anchored to solid bedrock and thus subject to lower levels of vibration might survive intact even if located closer to the earthquake epicenter. Unfortunately, most urban Southern Californians do not live, work, commute, or play over bedrock and thus appear to be in harm's way if this potentially greatest of all natural disasters ever strikes.

CALIFORNIA ADRIFT

The narrow part of California west of the San Andreas fault zone shown in Figure 2-1 is drifting laterally northwestward with respect to the rest of North America. This movement started at least 30 million years ago, when an oceanic rise in the Pacific Basin literally "hit" the continent, and it has continued ever since at an average annual rate of about 2 inches. At that rate and going back in time to 30 million years B.P. (before the present), the central Coast Ranges of California were located more than 700 miles to the south along the west coast of what is now Mexico. In another 30 million years, the Coast Ranges will have moved from their present position northwestward some 700 miles, putting Santa Cruz abreast of Seattle, Washington.

These movements are dramatic in the context of geologic time but represent a comparative snail's pace in human time. In other words, San Diegans, Angelenos, San Franciscans, and other coastal Californians reading these pages are not likely to find themselves suddenly being flung into the Pacific Ocean by some cataclysmic earthquake emanating from the San Andreas system. Yet each 3.0-magnitude earthquake, the smallest normally felt by humans, will at least remind them that they are inching northwestward and that California is merely a small part of a grander, worldwide design popularly referred to as *continental drift*. Continental drift, sea floor spreading, and seismic and volcanic activity all fall within the unifying concept of *plate tectonics*, which we will explore in the next section. We will also examine what may be an aberration of plate tectonics, the shifting of the San Andreas.

Plate Tectonics

The results of earth movement, including earthquakes and the motions of continents and ocean basins, ultimately emanate from the relative movements of huge sections of the earth's solid rock crust, or *lithosphere*, referred to as *lithospheric plates*. Stresses acting on the rock strata in these 15- to 100-kilometer-thick plates cause them to compress into the major features of *structural relief* or *landforms* of the earth's surface, such as mountain and valley systems. The building of these structural features and their internal causes are collectively referred to as *tectonics*. Figure 2-6 depicts this internal energy *convecting* upwards through a partially molten, 75- to 175-kilometer-deep *asthenosphere* to the overlying lithosphere, bringing about plate tectonics at the earth's surface.

The ultimate source of this geothermal heat energy is believed to be elements undergoing radioactive decay deep within the core of the earth. The heat energy (estimated by some at 4,200° C) is transmitted from the core, which has a radius of about 3,500 km, through approximately 2,700 km of mantle before reaching the athenosphere (at about 1,200° C or near the melting point of rock). As shown in Figure 2-7, mantle-derived *magma* (molten rock) then rises up through a rift in the relatively thin oceanic plate, causing the plate to spread apart in opposite directions. In the process, the magma cools and *lithifies* (becomes solid rock), forming an oceanic rise or ridge of *igneous rock* (rock formed by fire or literally by being "ignited").

In the case of California, the eastward-moving Pacific Plate probably collided with the North American Plate about 250 million B.P. and then proceeded to plunge under it, creating a trench that filled with sediments carried down by streams eroding the continental surface. The Pacific Plate and trench sediments then descended deeper under the continental plate, eventually reaching the athenosphere and melting. From the melting zone shown in Figure 2-7, magma intruded upward into the continental plate to form a giant *batholith* (deep rock mass) that became the granitic backbone of the Sierra Nevada and other mountain ranges composed of

Explanation of Tectonic Forces Responsible for the Creation of Califonia:

- *Magma* (1a) rises from lower mantle (1b) through less rigid *Asthenosphere* or upper mantle (1c).
- Magma rises to surface to form *oceanic ridge* (2) composed of volcanoes and falut block mountains.
- *Sea-floor spreading* (3). Newer rock created by cooling lavas forces older rock to "slide" across less rigid Asthenosphere and away from the oceanic ridge.
- *Subduction* (4). Lighter plate made of lavas is forced under adjacent plate.
- Marine sediments are deposited by undersea currents in *subduction trench* (5).
- Rock forced under continental plate melts (6) due to crustal pressure from above and higher temperatures of Asthenosphere.
- Melted rock either: (7a) cools to form granitic rock, (7b) returns to mantle as magma, or (7c) rises to surface as lavas.

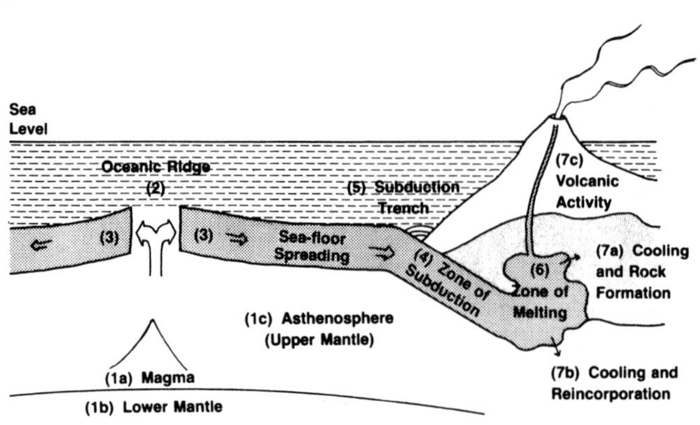

Figure 2-6 *Idealized plate tectonics diagram.* (Richard Crooker)

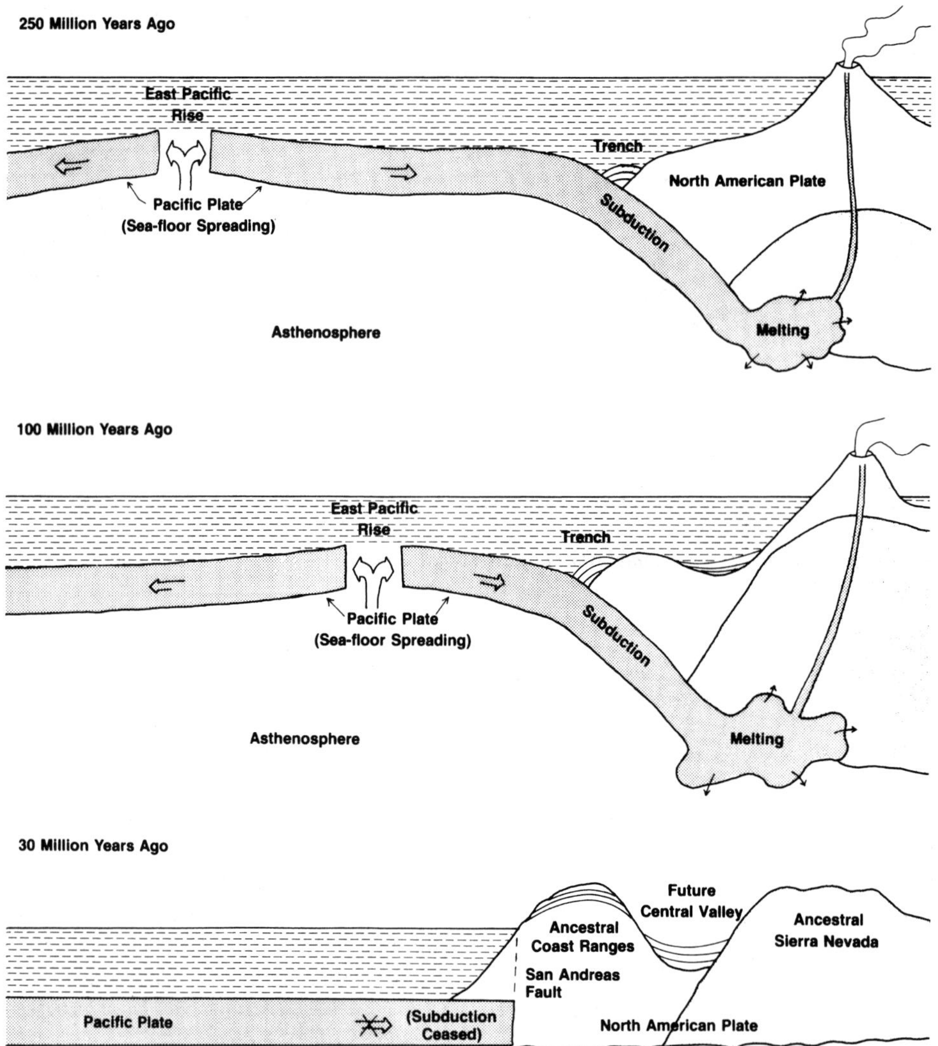

Figure 2-7 *Early geologic evolution of California.* (Richard Crooker)

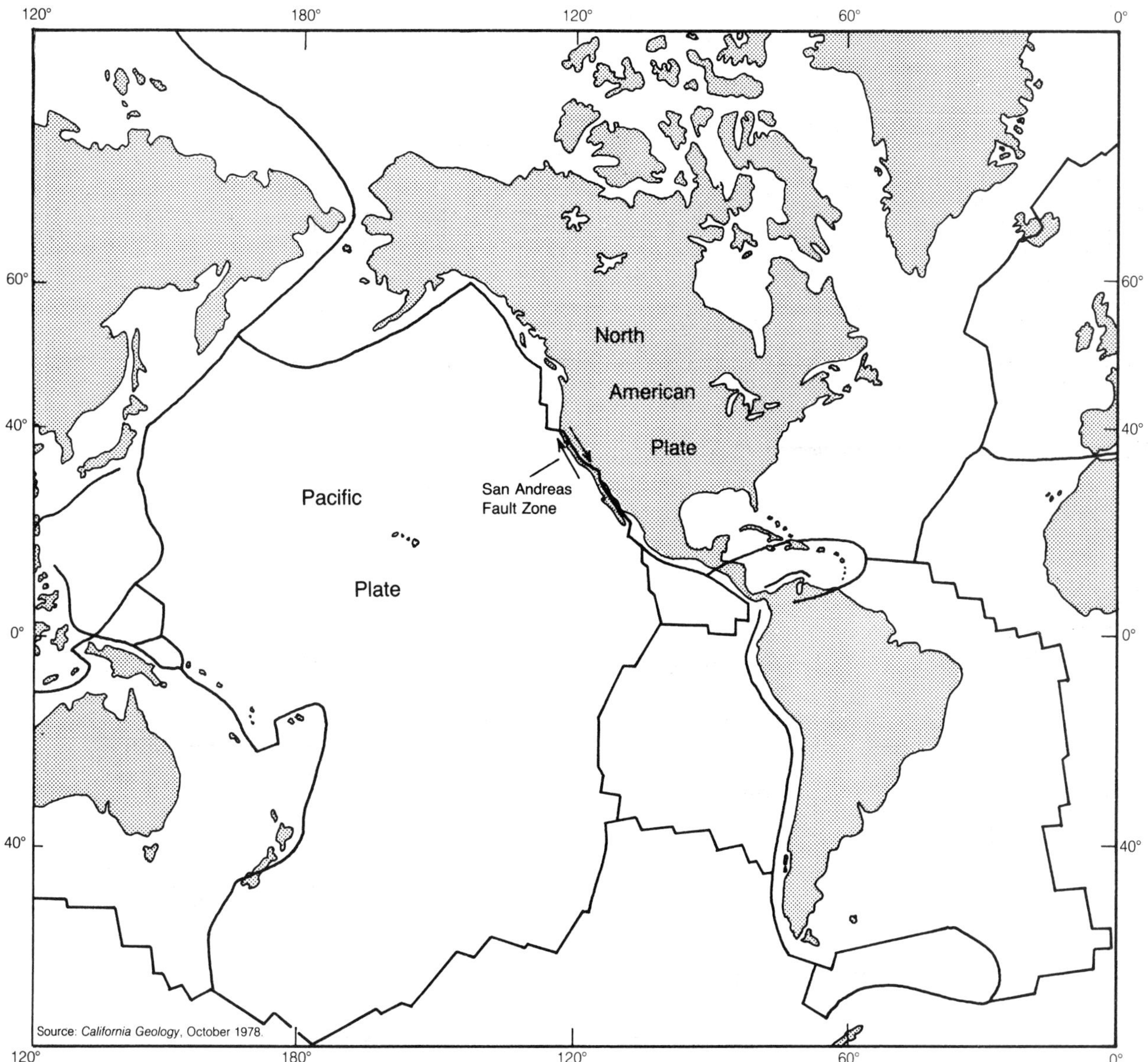

Figure 2-8 *Pacific and North American plates.* (Richard Crooker)

igneous and capping *metamorphic* rocks (rocks changed by heat and pressure from the original sedimentary rocks).

By about 30 million B.P., the east Pacific rise (see Figure 2-7) intersected with the continent, and the collision course of the two plates gave way to a sliding of one past the other that continues to this day. The San Andreas fault zone marks the interface of this shearing. Thus it is the Pacific Plate, which includes that small part of California west of the San Andreas, that appears to be "rotating" counterclockwise or northwestward, as seen in Figure 2-8. This horizontal motion along the San Andreas is referred to as *right-lateral faulting*, for as one looks out across the fault from either side, the opposite side appears to be moving to the right (Figure 2-9). This effect can be seen in the form of *stream channel offset* to the right along the San Andreas fault.

The northwestward movement of the Pacific Plate portion of California may not be uniform throughout. Rather, it appears to be fragmented. Evidently, where the northwestward-moving Peninsular Ranges meet the batholithic roots of the Sierra Nevada, the San Andreas fault zone and the Transverse Ranges are forced to bend to the west or left. As evidenced by several feet of vertical displacement occurring during the San Fernando earthquake and other recent seismic events in the region, the east-west-trending Transverse Ranges appear to be rising. The "growing up" of this young mountain range just north of the Los Angeles Basin is probably attributable to its location in a zone of compression be-

Figure 2-9 *San Andreas fault 48 miles due west of Bakersfield. The marked linearity (or lineament) along the trace of the fault is quite apparent in this photograph. Where streams cross the San Andreas, their courses are offset or displaced to the right on the downstream side of the fault line.* (U.S. Geological Survey, *Earthquake Information Bulletin)*

tween the Peninsular Ranges to the south and the Coast Ranges to the north.

A Bermuda Triangle for California and Rerouting of the San Andreas?

The fault lines associated with the Landers earthquake (7.3) and the Big Bear aftershock (6.2) of June 28, 1992, reveal two sides of a triangle,[7] whose apex points northward. To the south and forming the base of this ominous geometry is the San Andreas fault zone (see Figure 2-1). Although the Landers event took only one life, while 3,500 died in earthquakes worldwide in 1992, it was the highest-magnitude quake recorded anywhere that year. By the 1993 anniversary of the Landers/Big Bear quakes, more than 50,000 aftershocks had occurred in the vicinity of the triangle. In June 1994, just short of the second anniversary of the Landers quake, a 5.0-magnitude aftershock jangled nerves in the Yucca Valley portion of the triangle. Even more worrisome is the growing prospect of an 8.0 or greater quake occurring along the base of the triangle. If this segment of the San Andreas were to move with that much force, it could devastate heavily populated cities like Palm Springs (40,000+) and San Bernardino (165,000+).

Contrary to prevailing theories on the future of the San Andreas fault zone, there is a growing school of thought that the interface between the Pacific and North American plates is shifting to a more northerly than northwesterly route. The point of departure from the existing path of the San Andreas could possibly be from the northeast or Landers side of the triangle northward through the eastern California shear zone, which roughly straddles California's border with Nevada, as shown in Figure 2-1. Some fear that the directional shift of the San Andreas has already begun with events like the Landers quake and will intensify with the next "big one" along the southern segment of the San Andreas fault zone. They note that minutes after the June 28, 1992, temblor, dozens of small quakes were felt all the way up the shear zone from Mammoth Lakes in the eastern Sierra to Mount Shasta and beyond in the Cascade Ranges. It's purely speculation, but could more recent

[7] In Kerry Sieh et al., "Near-Field Investigations of the Landers Earthquake Sequence, April to July 1992," *Science* 260, 9 (April 1993), pp. 171–176, it is noted that although the National Earthquake Information Center gave surface wave magnitudes (M_s) of 7.6 and 6.6 for the Landers and Big Bear quakes, respectively, Richter "magnitudes calculated from seismic moment (M_w) are considered more representative of the size of an earthquake than M_s values." Landers is therefore listed in the literature at M_w 7.3 and Big Bear at M_w 6.2. The "Landers earthquake sequence" began with a M_w 6.1 preshock on April 23, 1992, known as the Joshua Tree preshock or Desert Hot Springs quake. Although neither is a standardized term, *preshock* differs from *foreshock* in that the former occurs over a greater span of time and on a broader spatial or geographical scale than the latter.

quakes in the shear zone, like the 1993 magnitude-6.1 east of Big Pine or the 5.8 event 11 miles northeast of Ridgecrest in 1995 be part of the scenario? Moreover, what do we make of moderate aftershocks of the Landers and Ridgecrest quakes continuing well into 1998? We may not find definitive answers for some time to come, but the notion that the triangle is a focus of California's ever-changing geography gains credence every time another earthquake strikes the area.

Now that we are more aware of tectonic activity on both sides of the San Andreas rift, be it the eastern edge of the Pacific Plate, in the San Fernando Valley, or well within the North American Plate in the Mammoth Lakes area, let us turn our attention next to the land forms these tectonic forces have built. With a volcanic tableland in one part of California, block-faulted mountains in another, and a flat-surfaced, 25,000-square-mile valley in still another region, it is obvious that tectonics have created unique landscapes, albeit ones that are none too steady. These landform or geomorphic provinces and their human settlements are the subject of the next chapter.

CHAPTER

3

Landform Provinces and Their People

Due in large part to the tectonic forces described in Chapter 2, California's landform diversity is unmatched by any other state in the "lower 48." The spectacular contrasts its physical landscape provides are reflected in the nine distinctively different landform or geomorphic provinces that make up the Golden State. The scenic display begins along the 1,000 miles of coastline, where, from south to north, the granitic Peninsular Ranges, the east-west-faulted Transverse Ranges, the sedimentary Coast Ranges, and the complex Klamath Mountains all front on the Pacific (Figure 3-1). Although they are not steeply truncated at an ocean's edge, the mountainous provinces of the interior also attain impressive heights, reaching upwards of 14,000 feet in the volcanic peaks of the Southern Cascades, the glaciated crests of the Sierra Nevada, and the up-faulted ridges of the Great Basin mountains. Down-faulted, below-sea-level basins such as Salton Trough give even greater vertical dimension to the Great Basin and southeastern deserts. Completing this array of landform contrasts are the volcanic tablelands of the Modoc Plateau in northeastern California and the down-folded, alluvium-filled Great Central Valley at the geographic heart of the state.

Given their dramatic diversity, California's landforms are a varied resource. Huge reserves of petroleum lie trapped in the sedimentary formations of the Coast Ranges and offshore on the continental shelf. There is still gold in the hills of the Sierra Nevada and the Klamath Mountains. More significantly, the Sierra, Klamaths, and other high ranges capture and hold moisture from air masses blowing in from the Pacific while at the same time casting a vast rain shadow over Arizona and Nevada. These mountain watersheds assure the state of near self-sufficiency in water supply, perhaps its most vital natural resource. The Great Central Valley provides thousands upon thousands of square miles of the most productive farmland on the face of the earth. The mountains of the Great Basin yield high-grade metallic ores, and the intervening basins are rich in alkaline and saline minerals. As an aesthetic resource, California's landform provinces contain, in an area of 158,693 square miles, just about as many scenic contrasts as one could ever hope to see, paint, or photograph.

Above all, though, the nine landform provinces of California provide a seemingly infinite array of human settlement sites. For reasons varying from the presence of spectacular seascapes and cool, breathable marine air to a relative abundance of certain biotic and mineral resources, the coastal provinces have proved the most attractive.

Archaeological evidence suggests that California's first cultural landscape took form along the Pacific shores of the Peninsular Ranges some 50,000 years ago. Beginning barely 200 years ago, European settlement displayed a decided preference for the seaward sides of the Peninsular, Transverse, and Coast ranges. Today, about four-fifths of the state's 33 million inhabitants live in these three landform provinces. Most of these Californians reside in the San Diego, Los Angeles–Long Beach, and San Francisco–Oakland metropolitan areas, within an hour's drive of the beach, and most would probably choose seaside residence were it not for pro-

Figure 3-1 *California's landform or geomorphic provinces.*

hibitive real estate costs, stringent environmental and zoning restrictions, remoteness from the workplace, and other locational constraints. By the turn of the century, one would have done well to find a modest three-bedroom house on a small ocean-view lot priced under $700,000 in or near one of these metropolitan areas.

Even California's contemporary Native Americans, though their ancestors were largely dispossessed of their coastal lands by the mid-nineteenth century, claim their largest holding—the Hoopa Valley Reservation and Extension of some 94,000 acres, mostly along the lower Klamath River—on the Pacific side of the Klamath

Mountains. Yet the rightful aboriginal heirs to California have generally fared none too well in the allotment of lands anywhere in the state.

By tradition, most California Native Americans were nomadic hunters and gatherers and so essentially did not participate in the sweep of agricultural settlement over the Great Central Valley, the mineral exploitation of the Great Basin, or the recent recreation-retirement rush to the Sierra Nevada. Thus most of their reservations are found in the landlocked provinces of the interior, off the beaten path on marginally productive land. Nonetheless, these Native Americans are descendants of the first Californians and as such merit special mention in our examination of the geomorphic provinces and contemporary settlement of the land.

THE PENINSULAR RANGES

The northwest-southeast-trending Peninsular Ranges region, the southernmost of California's four coastal landform provinces, is a northward extension of the peninsula of Baja California. As such, the province was at one time part of the continental plate and lay to the southeast of its present location. Then, several million years ago, the peninsular mass became welded to the northwest-tracking Pacific Plate and began rifting away from the mainland of Mexico. The waters of the Pacific filled the rift in the form of the Gulf of California, which has steadily widened ever since. Today, the peninsula, whose batholithic backbone stretches discontinuously from the southern tip of Baja California 1,000 miles north to near Los Angeles, continues to inch northwestward and farther away from the continent.

The Interior Highlands

The general structure of the Peninsular Ranges province is that of a massive, western-dipping, granitic fault block[1] dramatically uplifted on its east side. The abruptness of the eastern escarpment culminates in Mount San Jacinto at nearly 11,000 feet above sea level in the northeast corner of the province. As Figure 3-2 shows, internal faulting of the block along the San Jacinto, Elsinore, and other faults has helped shape a varying topography of northwest-southeast-oriented mountain ranges of moderate elevation and intervening upland plains, plateaus, and small valleys. *Weathering* of rock, occasional *mass wasting* incidents (usually landslides triggered by earthquakes or excessive rainfall), and *differential erosion* (mostly by streams over rocks of differing resistance to erosion) have further sculpted these highland landforms.

The rugged terrain of the highlands has discouraged both agricultural and urban settlement and has done little for recreational development. Although they attain impressive elevations in a few places, the higher mountain ridges are nowhere extensive enough to provide a significant recreational resource. For example, major ski resort complexes are precluded because of a dearth of slopes sufficiently high in elevation to assure long-lasting snowpacks. Idyllwild at the southwest edge of Mount San Jacinto State Park is the only genuinely "alpine" settlement in the Peninsular Ranges. Farther south in the high back country of San Diego County, the small villages of Julian and Warner Springs provide a semialpine setting with the help of an occasional winter snowfall.

Rural Native American settlement in the Peninsular Ranges, as well as in all the coastal landform provinces south of San Francisco Bay, is today almost exclusively found in the interior highlands of Riverside and San Diego counties. Here some 30 reservations comprising a quarter million acres accommodate a few thousand Native Americans. Actually, a majority of the southwestern California Native Americans live and work outside the reservations. The Morongo Reservation near Banning, with some 32,000 acres, is the third largest reservation area in California.

Agricultural settlement in the San Jacinto Basin and in the mountains to the west is much more extensive than in the eastern highlands because of more suitable terrain, better soil conditions, and longer growing seasons. For example, in the coastward mountains of San Diego County from Fallbrook south to Escondido is found California's optimal environment for growing avocados; indeed, the region is first in avocado production in the state.

To the northeast, in the flatter San Jacinto Basin of Riverside County, all types of agriculture exist, including citrus orcharding, viticulture, truck farming, and field cropping. Many basin towns, such as Hemet, Perris, and Temecula, originated as agricultural service centers. By contrast, in recent decades, retirement, military, and other nonagricultural development has attracted people to the communities of Rancho California and Sun City, as well as the March Air Reserve Base area. Murrieta Hot Springs and Lake Elsinore are among the Basin's principal recreational attractions. Nowhere in the San Jacinto Basin, though, has recent urban growth been more impressive than in Moreno Valley, which passed 100,000 in population in 1988. In that year, Moreno Valley was not only California's second fastest growing city but also a major factor in Riverside County having the

[1] A *fault block mountain* is a block or part of the earth's crust that has been uplifted by vertical or normal faulting over an extended period of time, usually several million years. Although faulting can prevail on any side of a mountain range, fault block mountains in most of California's provinces are sharply faulted on their east sides (the scarp side or escarpment) and slope gently downhill to the west (the dip side).

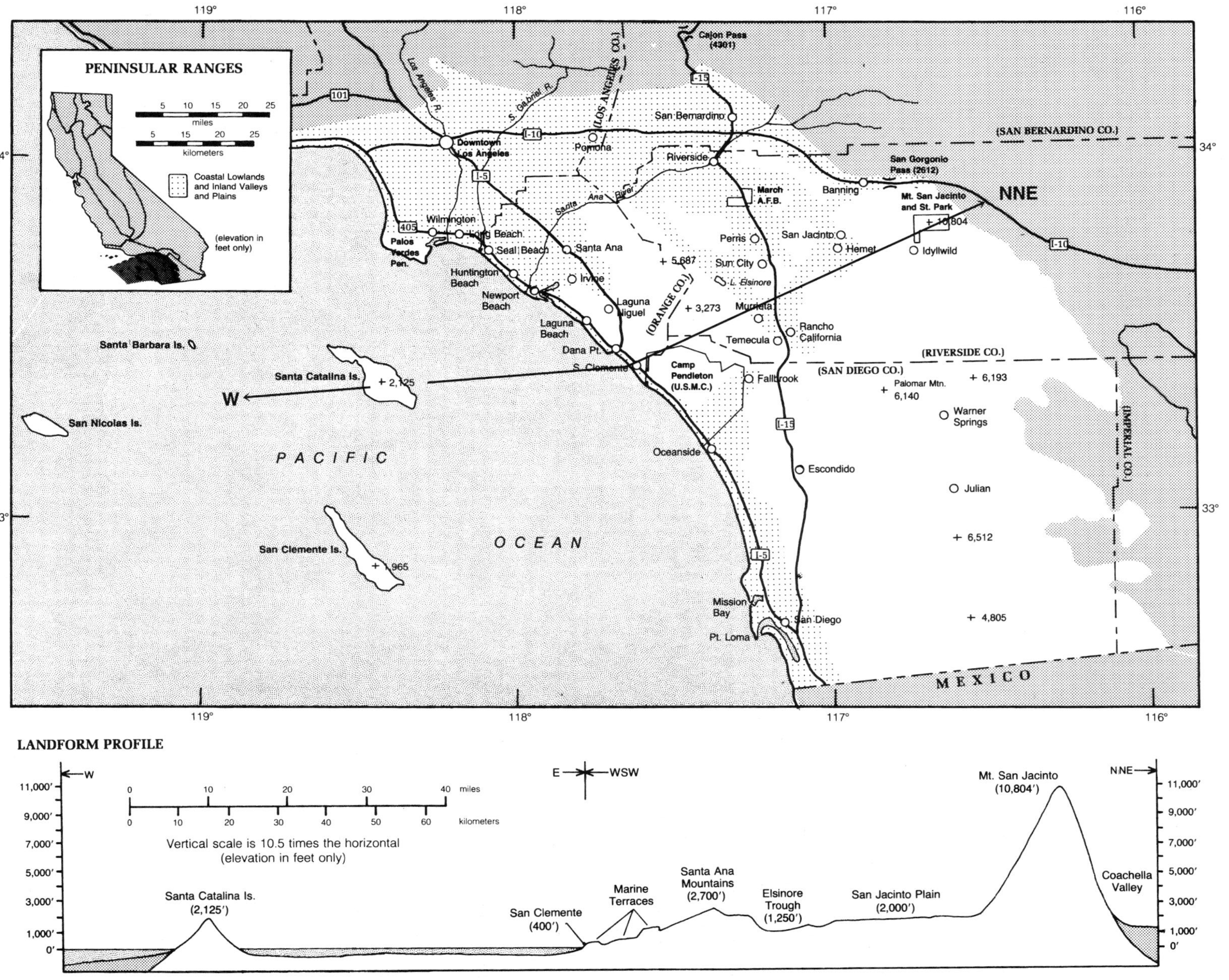

Figure 3-2 *Peninsular Ranges and landform profile.* (Richard Crooker)

highest growth rate among all 58 counties by the end of the decade. Just to the north, San Bernardino County was second in rate of growth in the state. During the 1980s, exorbitantly high housing costs in Los Angeles and Orange counties drove thousands of families eastward to seek affordable housing in the two Inland Empire counties. Once there, many of the new residents then had to turn around each weekday and commute 50 miles west to work. By the mid-1990s, however, an economic recession and layoffs had tempered the rush to reside in far-off suburbia.

Coastal Features

Westward toward the ocean, flat-topped but stream-dissected *marine terraces* (also referred to locally as *mesas*) pervade the landscape. Before the ice ages of the Pleistocene epoch, when the sea level was much higher, subsurface ocean wave and current action planed off shallow, seaward-sloping marine platforms. As the Pleistocene began and the Pacific retreated, a steplike pattern of marine terraces was revealed, with some rising above others because of vertical faulting. When ocean water became locked up in ice caps and glaciers, thus lowering sea level, streams cut *terrace canyons* or *barrancas* much deeper than they now appear.

Mission Valley, where today some of San Diego's newest regional shopping centers and hotel complexes, a stadium, and other urban attractions are located, is one of several such canyons that refilled with stream sands and gravels as the sea level rose during interglacial periods. Mission Valley's broad alluvial surface enticed not only modern urban development but also California's first Spanish mission settlement, San Diego de Alcalá. It is the gently sloping terraces, however, that have enabled San Diego, now the state's second largest city, to spread out many miles in nearly every landward direction from the central business district along San Diego Bay (Figure 3-3). San Diego Bay itself, protected from winds and tides by Point Loma Peninsula, North Island–Coronado, and Silver Strand isthmus and used primarily by the United States Navy, is one of the Pacific coast's premier deep-water harbors.

A

B

C

Figure 3-3 *San Diego. (A) Aerial view showing, from left to right, Mission Bay, Point Loma, North Island–Coronado Island, San Diego Bay and port facilities, city and suburbs, and a faint line (extreme right, or south) demarcating the border with Mexico: (B) Historic U.S. Grant Hotel from Horton Plaza in downtown San Diego. (C) San Diego Marina with hotels and the bridge to Coronado Island in background.* (A: NASA; B and C: Crane Miller)

The 118-mile coastline of Orange and San Diego counties is varied in both its geomorphological characteristics and its human use. In northern Orange County, the coastal landforms range from tidal wetland at Seal Beach, southward to wide, sandy beaches backdropped by low-lying *palisades* (cliffs) at Huntington Beach and extensive yacht harbors in Newport and Balboa bays. Sedimentary formations beneath Huntington Beach and offshore continue to yield petroleum; thus oil derricks, grasshoppers (oil pumpers), and drilling platforms pervade much of the landscape and seascape.

From the Newport–Irvine coast south through Laguna and Laguna Niguel, the beaches narrow appreciably and in some places are replaced by rocky tide pools and barely submerged rocks and reefs. This wave-battered shoreline is abruptly backed by steep coastal bluffs and marine terraces. Perched on these terraces is much of California's most expensive residential development; the Laguna coastline is often likened to the French and

Figure 3-4 *La Jolla beach, Scripps pier, and sea cliffs or palisades. A dozen miles northwest of downtown, yet well within San Diego's corporate limits, La Jolla (from* la joya, *"the jewel") is one of California's wealthiest residential communities and also boasts the world-famous Scripps Institution of Oceanography.* (Crane Miller)

Italian Rivieras, and justifiably so. South of the Dana Point Marina and Doheny Beach, the San Clemente coastline presents a somewhat less dramatic and less densely developed version of the Laguna coast. This terrace landscape, dissected by narrow beaches and bluffs, continues intermittently down the San Diego County coast to La Jolla (Figure 3-4), with major interruptions in the form of lagoons, estuaries, and tidal flats, such as those seen at Del Mar.

Several cities have grown up rapidly along this coast, with the northernmost, Oceanside, the largest after San Diego. The large United States Marine Corps base at Camp Pendleton has spurred the growth of Oceanside while at the same time acting as effective green belt separation between the Los Angeles–Orange County and San Diego metropolitan areas. Mission Bay, a reclaimed tidal flat in the northwestern part of San Diego, is an outstanding recreational area with attractions such as Sea World and Vacation Island.

A diminishing supply of beach sand is the most pressing environmental problem facing shoreline inhabitants. Not only is there less and less area for sunbathers, surfers, beachcombers, and others, but there is a narrowing buffer of protection for beachfront homes and sea cliffs against the periodic pounding of storm waves. The normal removal of beach sands by *longshore* (parallel to shore) or *littoral* (coastal) currents to submarine canyons would not be of concern were it not for a decline in sand replenishment every time a new dam is built upstream from an estuary or an artificial harbor is developed. The California Coastal Plan, as mandated by state voters in approving Proposition 20 in 1972, addresses itself to saving this fragile shoreline recreational resource. Still, one wonders what will be left of the beaches, lagoons, and sea cliffs once coastal settlement has merged Los Angeles and San Diego into one megalopolis.

The Los Angeles Basin

The Peninsular Ranges province extends northwestward to include the most densely settled portion of the Los Angeles Basin, highly developed Long Beach, the Palos Verdes peninsula, and the Santa Monica Bay portions of the "South Coast." Several streams, including the Los Angeles, San Gabriel, and Santa Ana rivers and Ballona Creek, built the broad alluvial coastal-plain surface that now accommodates California's largest urban population. Many of the channels have been lined with concrete to provide better flood control for the 11.3 million residents of the Los Angeles–Orange County portion of the coastal basin.

Underlying the basin's sedimentary fill to a depth of 31,000 feet below sea level is a giant *syncline* (downfold) of basement rock. The basin's generally smooth surface is broken here and there by faults such as the Palos Verdes and Newport–Inglewood, which have uplifted Palos Verdes Point, Dominguez Hills, and the Baldwin Hills. The folded and faulted sedimentary formations have yielded generous petroleum deposits, with the Wilmington oil field ranking first in cumulative production in California. This petroleum field underlies much of Long Beach and the adjacent artificial harbor.

The Islands

The Peninsular Ranges province does not terminate at the Pacific coast but rather submerges westward, forming a broad continental shelf with occasional mountain outliers in the form of islands. San Clemente, currently restricted to naval use and otherwise uninhabited except by a unique flora and fauna, and Catalina, whose nearly 2,000 residents live in or around the tourist node of Avalon harbor, are the largest islands. Both islands project at their highest about 2,000 feet above sea level and have hilly surfaces underlain by igneous rocks with abrupt, palisaded coastlines and few sandy beaches. San Clemente appears as a fault block island dipping downward to the west with a steep eastern escarpment, whereas Catalina displays no particular surface symmetry. Both islands measure a few miles wide and are elongated more than 20 miles northwest-south. While Catalina is usually visible on a clear day from anywhere along the south coast, San Clemente Island, at twice the distance (60 miles) offshore, is barely visible under the clearest of atmospheric conditions. Tiny Santa Barbara Island (part of the Channel Islands National Park) and somewhat larger San Nicolas Island are also not visible from the mainland and are uninhabited.

THE TRANSVERSE RANGES

The Transverse Ranges are Southern California's highest and most imposing mountains and, like the Peninsular Ranges, reach westward into the Pacific in the form of several islands. Unlike the Peninsular Ranges, however, the axis of the Transverse Ranges is east-west rather than north-south. This position in the northern part of Southern California has led many to the opinion that the Transverse Ranges demarcate the southland from the rest of the state. For statistical and other purposes, the Kern and other east-west county lines are probably preferred, but the notion that the Transverse Ranges are the real hurdle between north and south persists.

Onshore

The only major landform province oriented east-west, the aptly named Transverse Ranges are the most rugged and, for the most part, the least densely settled coastal province. Seen in Figure 3-5 stretching from the coast at Point Arguello eastward for 250 miles through the Santa Ynez, Topatopa, San Gabriel, and San Bernardino mountains, the Transverse Ranges steadily rise in elevation, attaining 11,485 feet in Mount San Gorgonio near their eastern end. Folding and faulting of sedimentary formations in the west and block faulting of metamorphic and igneous rocks in the east have created deep canyons lined with long, steep slopes throughout the ranges. Drainage follows these tectonic features in some places, such as the east and west forks of the San Gabriel River along the San Gabriel fault.

Such topography is obviously not conducive to settlement. Thus, only a small proportion of the province is intensively developed either for agricultural or urban uses—principally from the Santa Monica Mountains and San Fernando Valley (mostly incorporated in the city of Los Angeles) northwestward to the Oxnard Plain and lower Santa Clara River Valley and along the coast to Santa Barbara. Mount Baldy village and Wrightwood in the San Gabriels and Green Valley, Arrowhead Lake, Big Bear Lake, and communities between the lakes in the San Bernardinos all have small populations that are temporarily swelled during the skiing season. Altogether, however, they represent only a few thousand permanent residents.

Although rugged and impressive in elevation, the Transverse Ranges lack the great coniferous forests, glaciated valleys, and ridges of the Sierra and the Klamaths. The Transverse Ranges are scenically less attractive than other California mountain systems, and they occasionally act as an impenetrable landform barrier. During the heavy rains and snows of the winters of 1978, 1993, and 1997, for instance, there were occasions when all surface routes northward from the Los Angeles metropolitan area were closed, with the Pacific Coast Highway (PCH) blocked by mudslides and the Ridge Route (I-5) and Cajon Pass (I-15) closed by rock slides and/or snow.

Plate tectonics continues to involve the Transverse Ranges in a squeeze play resulting in crustal compression and ongoing uplift, especially in the western ranges. Here, thrust faulting responsible for the 1971 San Fernando and 1994 Northridge earthquakes may be part of this scenario. In the eastern portion, the San Andreas fault slashes the ranges in two, forming the northern boundary of the San Gabriels but cutting southeastward through Cajon Pass to become the southern edge of the San Bernardino and Little San Bernardino mountains. Were it not for displacement along the San Andreas and San Jacinto faults, the San Gabriels and San Bernardinos would compose one mountain range. Obviously, though, they will continue to pull farther apart as sections of two different crustal plates slide laterally in opposite directions. Furthermore, the San Bernardinos are known to be a deeply rooted granitic mass that, along with the Sierra Nevada batholith, deflected the Pacific Plate and created the big bend in the San Andreas fault zone.

Offshore

Like the Peninsular Ranges, the Transverse Ranges extend offshore in the form of a chain of islands known as the Channel Islands, which jut due westward into the Pacific as a prolongation of the Santa Monica Mountains. Geologically, they belong to the Santa Monica Mountains and were uplifted by folding and faulting along the east-west-trending Santa Monica fault just a few million years ago. At that time, they were connected to the land as part of the Ancient Cabrillo Peninsula. Later, the peninsula began to subside and the chain separated from the mainland. Santa Cruz and Santa Rosa, both roughly the same size, height, and distance offshore as Catalina, became the largest of the islands.

Unlike Santa Catalina, however, neither of these islands has been developed to accommodate tourism, although there is existing military and pastoral use and primitive camping is permitted on Santa Cruz and Santa Rosa, as well as Anacapa, San Miguel, and Santa Barbara islands. Santa Cruz Island has probably escaped development by virtue of the sale by Dr. Garvey Stanton of most of the island to the Nature Conservancy and the inclusion of the island in Channel Islands National Park, which was upgraded from national monument status. Somewhat smaller San Miguel Island and the tiny Anacapa chain (the other part of Channel Islands National Park) round out the Channel Islands group. In all, there

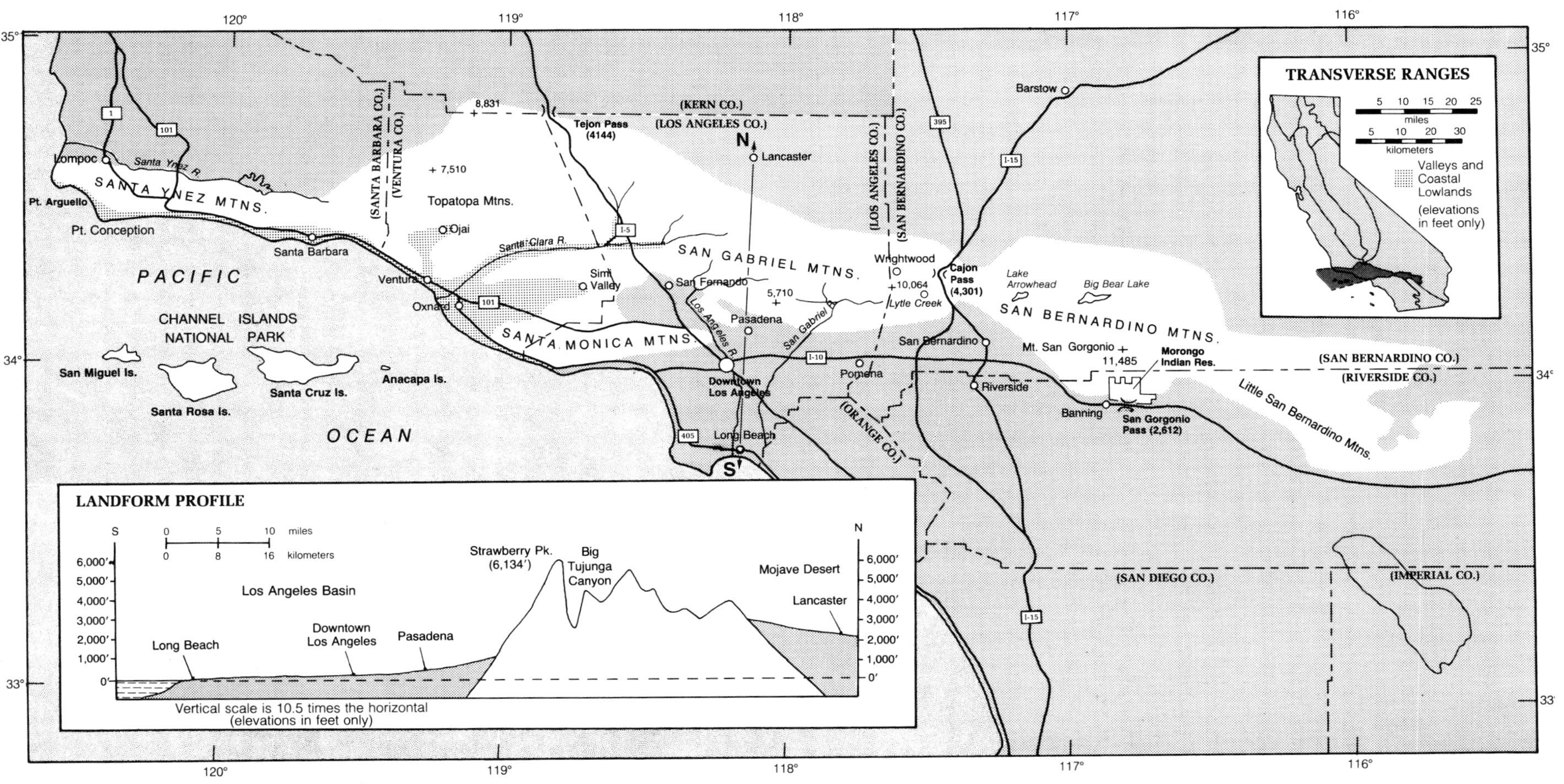

Figure 3-5 *Transverse Ranges and landform profile.* (Richard Crooker)

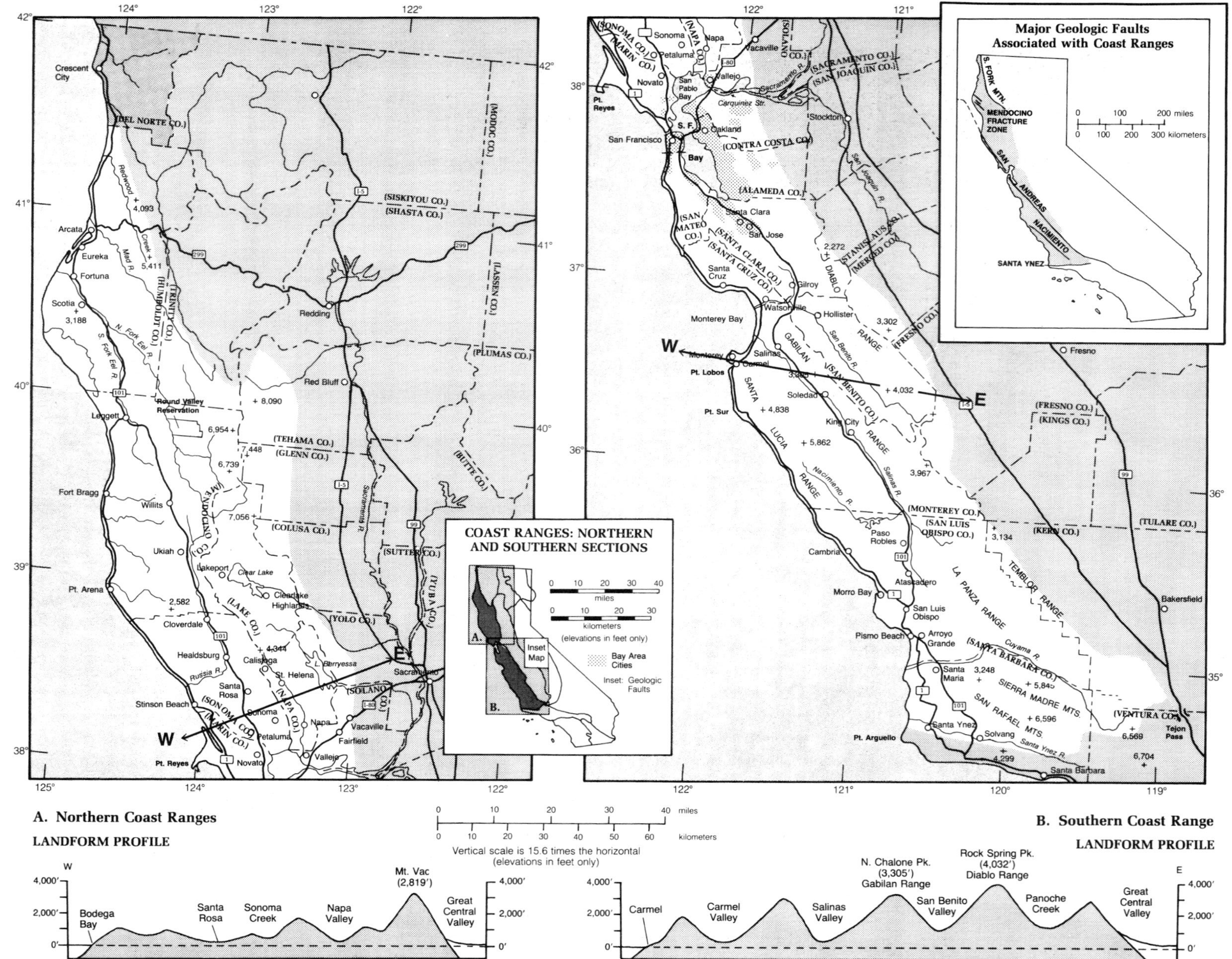

Figure 3-6 *Northern and southern Coast Ranges and landform profile. Some sources show the northern Coast Ranges extending narrowly into Oregon as in the two insets here. Others depict them being replaced by the Klamath Mountains about 20 miles south of Oregon as in map A above and Figure 3-1.* (Richard Crooker)

are eight islands in the Channel Islands group, five of which (the five mentioned in regard to primitive camping) make up the national park. Travel to the islands is restricted and must be arranged through park headquarters in Ventura.

THE COAST RANGES

The Coast Ranges, easily California's largest and longest coastal landform province, measure more than 400 miles in length from the Transverse Ranges northwestward to near the Oregon border and average 50 miles in width from the Pacific to the western side of the Great Central Valley. In fact, the San Andreas fault lies just west of the boundary between the southern Coast Ranges and the Central Valley,[2] the former sliding northwestward a couple of inches a year as part of the Pacific Plate. From San Francisco, the San Andreas runs northwestward along the Pacific side of the Coast Ranges until Point Arena, where it returns to sea. As the "Geologic Faults" inset in Figure 3-6 shows, the San Andreas continues northwestward another 110 miles to a point offshore from Cape Mendocino, where it turns westward out into the Pacific.

Northern Faults

North of the Mendocino fracture zone (see Figure 3-6), another great fault zone, the South Fork Mountain, forms the eastern boundary of the Coast Ranges as they narrow down to only a few miles in width south of the Oregon border. The South Fork Mountain fault interfaces the granitic basement of the Klamath Mountains and the sedimentary and intruded Franciscan formation of the Coast Ranges, the latter formation having originally accumulated as ocean trench sediments washed down from the continental slope. Later, starting about 150 million B.P. and continuing until about half a million years ago, intensive folding and faulting uplifted the Franciscan formation to prominence as a rock series throughout the Coast Ranges.

In the last few hundred thousand years, stream erosion and other gradational processes have outpaced tectonic processes, resulting in the wearing down of the Coast Ranges' highest ridges to maximum heights rarely exceeding 8,000 feet. The satellite view of the central Coast Ranges in Figure 3-7 shows the axes of the mountain ridges and intervening valleys as generally parallel to the major northwest-trending fault zones. This structural pattern dictates drainage of most of the rivers; for example, the Eel and the Mad in the north and the Salinas in the south flow northwestward into the Pacific.

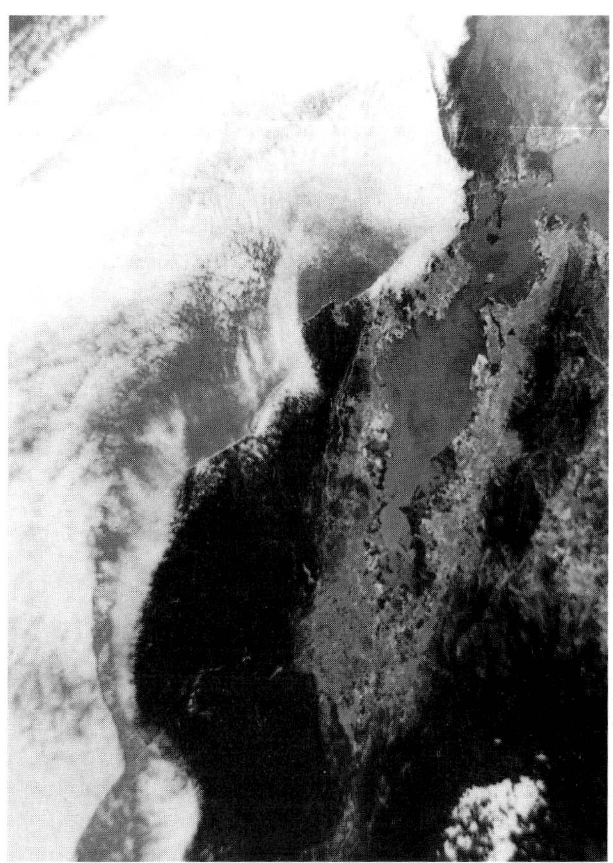

Figure 3-7 *Aerial view of central Coast Ranges focusing on, from north to south, Marin Peninsula and San Pablo Bay, San Francisco Bay and Peninsula, and the Santa Clara Valley. Golden Gate and Carquinez Strait (upper right or northeast), the latter carrying the Sacramento River into San Pablo Bay, are the only two water gaps found in the entire 400-mile length of the Coast Ranges. These gaps permit the only penetration of ocean-going vessels deep into the interior of California.* (NASA)

Valley Settlement

Contemporary settlement of the Coast Ranges has favored the structural depressions, principally valleys fronting on coastal plains and embayments. Of course, the mildest coastal climate to be found anywhere in North America has also drawn people to the west side of California's Coast Ranges. But even though the mild climate dominates the entire coast, these landform amenities are few and far between; thus, settlement has concentrated in the choicest of the valley and plain locations.

[2] The eastern edge of the southern Coast Ranges, including the Temblor and some other low-lying ridges, is separated by the San Andreas from the main part of the range. Remember that the San Andreas fault zone rises from the Gulf of California and is oriented northwestward through several landform provinces, including the southeastern Coast Ranges.

Figure 3-8 *Inside the Golden Gate, looking north across San Francisco and the East Bay (upper right) and Angel Island (upper left) from Twin Peaks. Where the ancestral Sacramento River originally cut its way to the Pacific through the Golden Gate, the bay is 300 feet deep; much of its 400 square miles, however, especially toward the southern end, is less than 10 feet deep at low tide. Because of bay filling and other human modifications, in the last 100 years the bay has shrunk by some 100 square miles to its present size.* (Crane Miller)

Along this lengthy and rugged coast, singularly lacking in natural deepwater harbors, San Francisco–San Pablo Bay (Figure 3-8) stands out as a dramatic exception. Nowhere else in California has nature provided so large, so deep, and so well protected an embayment. Furthermore, the bay is flushed not only by ocean tides entering through the Golden Gate but also by freshwater from the Sacramento River flowing in from the northeast through Carquinez Strait. However, because of the activities of several ports and dozens of small boat harbors, bay silting by the Sacramento River, toxic waste infusion, and the sundry pressures of some 6 million people residing in nine surrounding counties, the bay today is a threatened ecosystem, but one still supporting a variety of wildlife.

To both the north and the south of the San Francisco–Oakland–Vallejo metropolitan complex, settlement thins out in two long strings of widely separated towns and small cities—one stretching along the slow-going but eminently scenic State Highway 1 coastal route, the other through the interior along speedier U.S. 101. If one were to choose which direction to head from the Bay Area to escape congestion, however, it would be along the north coast. Nowhere to the north except for Fort Bragg is there a city of more than 5,000 inhabitants until far past where Highway 1 joins 101 and the Humboldt Bay cities of Eureka and Arcata are reached. Even these cities, with only about 42,000 residents between them, are small enough to fit the rural New England atmosphere of the north coast.

Indeed, the economic landscape of this rustic northern third of the California coastline is dominated by commercial fishing, dairying, lumbering, and summer tourism. Also adding to the settlement diversity of the northern ranges are several Native American *rancherías* (small village reservations) and the 18,543-acre Round Valley Reservation along the east banks of the upper Eel River.

A southward journey from San Francisco on Highway 1 will also leave urban congestion behind, but to a significantly lesser degree. Montara, Half Moon Bay, and their state beaches are among the first attractions encountered about an hour's drive from San Francisco. Another hour south, Santa Cruz (pop. 50,000), the first of the several moderate-sized cities that ring Monterey Bay, is reached. South of Carmel and Point Lobos (Figure 3-9), the picturesquely rugged and nearly empty Santa Lucía Range abruptly fronts the Pacific for about 100 miles until the Cambria–Morro Bay–San Luis Obispo area is reached. Southward from San Luis Obispo, Highway 101 is the principal route to Pismo Beach and Santa María, the southernmost city of any appreciable size (pop. 55,000) in the Coast Ranges.

Fronting on the Pacific Ocean a few miles south of the Monterey Peninsula, the Carmel Valley (see the map and landform profile in Figure 3-6B) came into the national limelight in 1999 when the Carmel River appeared as the only California river listed in *America's Most Endangered Rivers of 1999*. Published by the American Rivers organization based in Washington DC, the list included ten rivers located throughout the lower 48 of the United States. From its headwaters high in the northern Santa Lucia Mountains of the Southern Coast Ranges, the 36-mile-long river weaves its way westward through a basically rural landscape, but one that is displaying signs of suburban sprawl as the valley's population approaches 15,000. The Carmel River and the groundwater beneath it (see Chapter 5) serve as the major source of water for the Monterey Peninsula cities of Carmel, Monterey, Pacific Grove, Pebble Beach, and Seaside. As pointed out in Chapter 1, Monterey County, and the peninsula region in particular, is one of the fastest growing areas in California. The rapid growth and development in and near the Carmel Valley watershed puts the valley environment at risk in terms of spreading sprawl, increasing water withdrawals, and the impacts of existing and proposed dams. For example, the river is one of the very few streams in the Southern Coast Ranges that is still home to the steelhead trout, a sea-going rainbow trout (see Chapter 7). But, the Carmel River steelhead's days may be numbered, for it was recently listed as threatened, along with other aquatic species, under the federal Endangered Species Act. Implementation of effective land use and water conservation planning at the county, city, and district levels may assure the survival of the Carmel Valley's unique wildlife. A big step toward achieving this end was taken with the formation of the Carmel River Watershed Council, which focuses on local residents' participation in the planning process.

Figure 3-9 *Point Lobos near Carmel. A state reserve since 1933, its 1,250 acres encompass dramatic coastal headlands and a forest of extremely rare Monterey cypress* (Cupressus macrocarpa). *Offshore rocks and sea are frequented by pelicans, sea lions, migrating grey whales, and the sea otter, the last having made a remarkable comeback.* (Union Pacific Railroad)

In all, the counties of Santa Cruz, Monterey, and San Luis Obispo, comprising the Central Coast region, have a combined population approaching 1 million (excluding Santa María, which is in Santa Barbara County). The north coast counties of Del Norte, Humboldt, and Mendocino, in contrast, have a total population of less than 225,000.

Landlocked Counties

Landform and settlement patterns in the sparsely populated eastern interior of the Coast Ranges may best be exemplified by contrasting two entirely landlocked counties within the province: Lake County in the north and San Benito County in the south. Each county has about the same area (125–140 square miles) and population (37,000–51,000). Both are equally mountainous, with elevations exceeding 5,000 feet in places, and both are about equally distant from San Francisco.

But here the similarity ends. Whereas Lake County lives up to its name with Clear Lake (Figure 3-10), the largest natural freshwater lake (85 square miles) entirely within the borders of California, San Benito County is entirely lacking in natural lakes and must depend on reservoir and underground storage of water. Lake County lies well east of the San Andreas fault zone and is rarely affected by it; San Benito County, however, is adjacent to the fault and its towns, like Hollister, sometimes seem to move a little bit each day. Conditions of milder and moister climate, lusher vegetation, and lesser seismic hazard appear to favor Lake County for tourism and future growth, although agriculturally San Benito County, with expanding orchard and vineyard acreage, may be catching up to Lake County and its famous Bartlett pears. Another drought could cramp San Benito County's agricultural development, but the prospect of nearby California Aqueduct water becoming available diminishes this concern. San Benito is second only to Im-

Figure 3-10 *Clear Lake looking south toward volcanic Mount Konocti (4,100 feet). Although the lake is impressive from almost any viewpoint, its shallowness and the indifferent commercial and seasonal residential development ringing it detract from Clear Lake's overall appeal. Moreover, El Niño–sparked storms in the winter of 1998, which set precipitation records for February in many parts of northern California, caused the lake's level to rise and its waters to inundate many lakeside properties.* (Crane Miller)

perial as the California county in the proportion of Hispanic population, with 45.8 percent for the former versus 65.8 percent for the latter according to the 1990 U.S. census. This human resource is vital to the dominance of an agricultural economy in both counties. Given that the commute from the southern Bay Area now extends southward from Silicon Valley to Hollister, San Benito County is developing a suburban landscape as well. The county is expected to have 50,000 residents by the year 2000, or double the 1980 population, with most of the newcomers employed outside San Benito County.

The southeastern Coast Ranges do enjoy one major economic advantage over the northeastern ranges, namely, nearly all of the oil reserves found in the province. The upper Salinas and Cuyama valleys may be poor in local water resources, but they are rich in petroleum-bearing sedimentary formations.

THE KLAMATH MOUNTAINS

Although by no means the highest or largest of California's landform provinces, the Klamaths do claim the most complex physiography, vying with the Sierra for the most singularly spectacular terrain. As shown in Figure 3-11, the Klamaths are many mountain ranges whose ridges are oriented in seemingly all different directions and whose upper limits approach 9,000 feet. Their outer bounds nearly reach the Pacific in the west, extend into Oregon to the north, border the Southern Cascades in the east, and front on the Sacramento Valley to the southeast.

Compared to these surrounding provinces, the Klamaths are not only strikingly more rugged but much older and exceedingly more complex in their geology. In the broad view, the Klamaths appear as an *upwarp* or arc of ancient metamorphic and *intrusive* (lithified beneath the surface) igneous rocks that plunge downward under the much younger formations of the adjacent provinces. For instance, the Klamath Mountains are a northwestern extension of the Sierra Nevada, but the surface connection between the two provinces was buried by Quaternary lava flows born in the Southern Cascades.

Erosion and Deposition

In the last 60 million years, the Klamath Mountains have undergone extensive erosion. Many great rivers emanating both from inside and outside the province have incised steep V-shaped valleys and deep, sheer-walled gorges. Together, the Klamath River from the Columbia Plateau of southern Oregon, the Sacramento River from a spring at the base of Mount Shasta, and the Salmon, Scott, and Trinity rivers from within the Klamath ranges have created the most dissected mountain landscape for its size in all of California.

High above the narrow valley bottoms, Pleistocene glaciers sculpted the granitic and other erosion-resistant ridges of Castle Crags (Figure 3-12), the Marble and Salmon mountains, and the Trinity Alps into jagged peaks. Where the advancing alpine glaciers encountered serpentine and other rocks less resistant to the erosional work of ice, U-shaped valleys were formed. At the upper end or head of many of these troughs, concave *cirques* (deep, steep-walled basins) were scoured out by glacial erosion. Cirques and lakes at their bases are very much in evidence on north-facing slopes above 5,500 feet, where *ablation* (melting or wasting of ice) is minimal. Down-valley and along the sides of the troughs, glaciers deposited piles of sand, gravel, rocks, and other eroded materials known as *moraines*. Since the last of the Pleistocene glaciers melted more than 10,000 years ago, many of the glacial landforms of the Klamaths have been modified beyond recognition by subsequent stream erosion, chemical and mechanical weathering of rock, mass wasting, and forest cover. Nevertheless, the various ranges of the Klamath Mountains still present spectacular glacial landscapes surpassed in California only by those of the Sierra Nevada.

Getting Away from It All

The high, rugged mountains and deep whitewater canyons that lure thousands of hikers, campers, and fishermen every summer at the same time deter all but the sparsest of permanent settlement in the Klamath Mountains. Lack of any interstate, other major highway, or rail

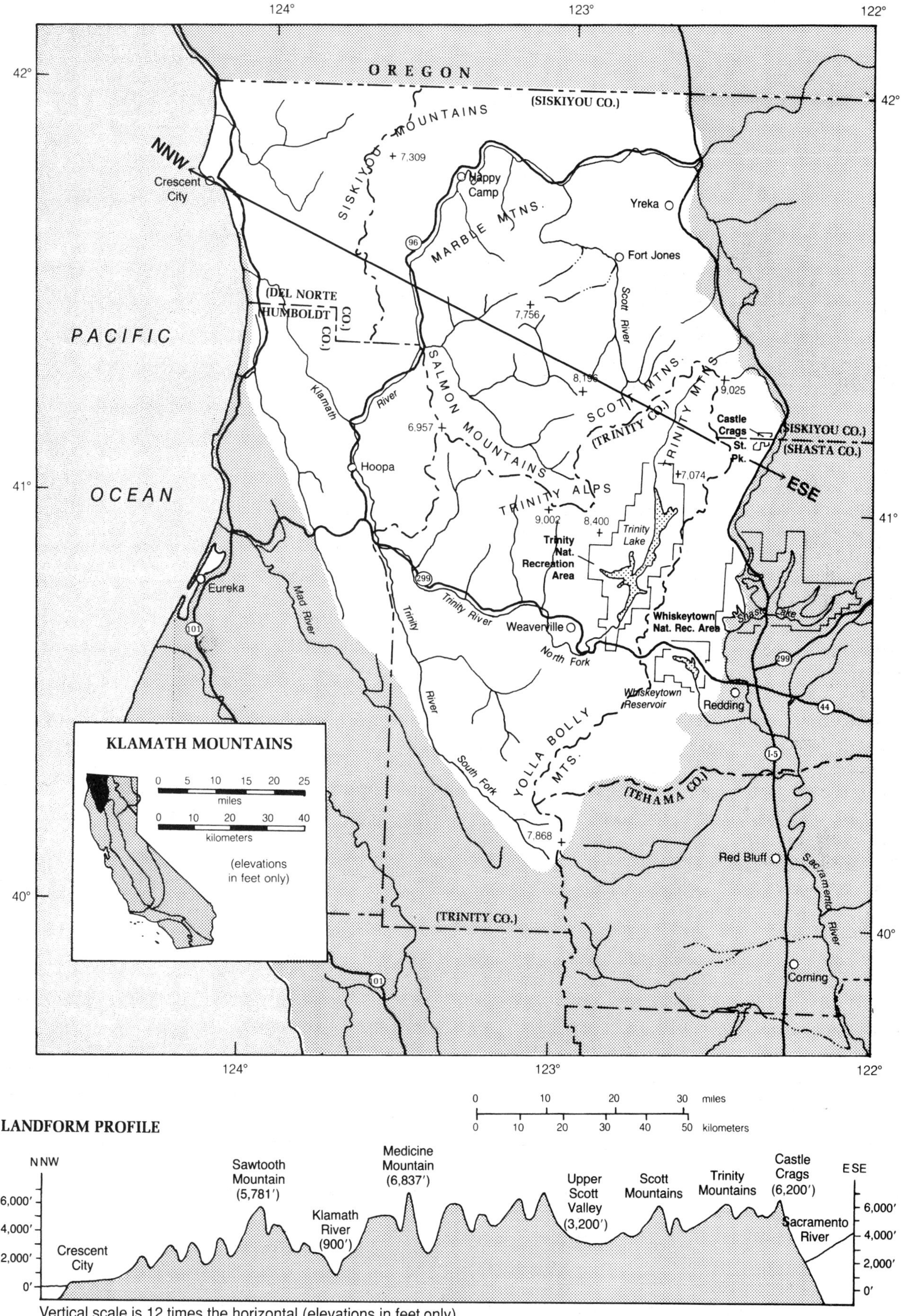

Figure 3-11 *Klamath Mountains and landform profile. As here, some sources depict the Klamaths as bordering the Pacific, while others show a narrow band of the Coast Ranges intervening, as in the Figure 3-6 insets.* (Richard Crooker)

route through the region and its isolation from urban California all keep would-be residents away. Although Interstate 5 skirts the eastern edge of the Klamaths, most of its travelers are concerned only with getting to and from the Pacific Northwest as quickly as possible. The two state highways that cross the region from I-5 to the Pacific are scenic, forest-lined canyon routes along the Klamath and Trinity rivers; but both are narrow and winding and must be closed occasionally in winter because of heavy snowfalls.

For vacationers who want both easy driving access to campgrounds and a variety of sites to choose from, the Shasta–Trinity–Whiskeytown national recreation areas are favorite destinations. These U.S. recreation areas feature three huge reservoirs of the same names, which are located in the southeast corner of the Klamath region. For the more ardent recreationist bent on hiking and fishing away from crowds, the Marble Mountain and Salmon–Trinity wilderness areas provide a chance to escape from it all in the high back country.

Figure 3-12 *Castle Crags, a more than 6,000-foot-high granitic formation dated from about 170 million* B.P. *The present rugged appearance of the formation owes somewhat to the erosive work of ice during the relatively recent Wisconsin Age (60,000–15,000* B.P.*). Today, Castle Crags State Park is easily reached via I-5 about 40 miles north of Redding.* (Crane Miller)

There are no cities as such in the Klamath Mountains, and even towns (none with populations over 3,000) are few and far between. The lumber industry is the economic mainstay of many communities, but recreation and retirement are now the principal investment attractions in the region. Lack of suitable terrain has limited agricultural settlement except in the alluviated, relatively flat Scott Valley, where ranching and hay farming are significant land uses. A few thousand Native Americans, who lay time-honored claim to the salmon and other fauna and flora of a bountiful natural environment, reside throughout the region both on and off reservations. The Hoopa Valley Reservation, comprising 85,445 acres along the lower Klamath and Trinity rivers, is the largest (in area) and one of the most prosperous Native American settlements in all of California. More than 2,100 reside on the reservation.

THE SOUTHERN CASCADES AND MODOC PLATEAU

From almost any vantage point in northeastern California, Mount Shasta, at an elevation of 14,162 feet, dominates the landscape (Figure 3-13). Except for Mount Rainier (14,406 feet) in Washington, Shasta is the highest of a long string of volcanoes, known as the Cascade Ranges, which extend from Mount Baker (10,750 feet) near the Canadian border southward through Washington, Oregon, and northeastern California to Mount Las-

Figure 3-13 *Mounts Shasta (14,162 feet) and Shastina (to the north or left, 12,336 feet) after an early autumn snowfall. Late every summer, long after winter snows have melted away, one can see several small glaciers on the higher flanks of Shasta, but none on Shastina. As in the Sierra Nevada to the south, the glaciers here developed during the Little Ice Age of the early 1700s but were eliminated from the northern sides of Shasta when Shastina erupted later in the century.* (Crane Miller)

sen (10,453 feet). To the east, Shasta overlooks the lava tableland of the Modoc Plateau, which is merely the southern tip of the vast Columbia Plateau. Shasta and Lassen, by virtue of their great heights and composite massiveness, stand out visually in northeastern California; but the dividing line between the Southern Cascade Mountains and the Modoc Plateau is otherwise obscure, and both landform provinces (Figure 3-14) share recent volcanic origins.

Lava Almost Everywhere

The Southern Cascades and Modoc Plateau are the only provinces in California almost completely covered with young basaltic lavas and other *extrusive* igneous rocks, such as seen at Lava Beds National Monument in the northern portion of the region. Almost all of these extrusives erupted through vents and fissures in the region's surface over the past few million years, and some, as in the case of Mount Lassen, as recently as the twentieth century.

It is noteworthy that intricate knowledge of the lava caves and other volcanic formations at Lava Beds enabled Captain Jack (Chief Kientepous) and his 50 Modoc warriors to hold out for more than 3 months in 1873 against a U.S. Army force with 16 times as many soldiers and superior armaments before being starved into surrender. (Precedents for the Modoc War are examined in Chapter 8.)

Evidence that volcanism continues to this day in the province can be seen in the steaming *fumaroles* (steam and gas vents), hot springs, and boiling mud pots of Lassen Volcanic National Park and felt in the occasional minor earthquake swarms that occur around Mount Shasta. Should these thermal and tectonic activities intensify, they may be the precursors of another violent eruption of Lassen or Shasta. Both volcanoes, although considered dormant, are nonetheless active.

Another indication of the recency of volcanism is the general lack of visible features of glacial erosion and deposition. Although five small glaciers exist on the upper slopes of Mount Shasta and signs of glacial scour can be seen in the Shasta Ski Bowl and U-shaped Avalanche Gulch, much evidence of the great glaciers that once emanated from Pleistocene Mount Shasta has been obliterated by recent volcanic activity. For instance, Shastina, Shasta's 12,336-foot satellite volcano, erupted and formed less than 250 years ago and in the process doubtless did away with preexisting glacial features.

Mount Lassen also mothered several late Pleistocene glaciers that converted V-shaped stream valleys into U-shaped glacial troughs. From the time when Lassen Peak initially formed from eruptions 11,00 years ago until the twentieth century, intermittent lava flows, cinder and pumice showers, and mudflows buried much of the evidence of earlier glacial erosion and deposition. Nevertheless, the glacially scoured sides of Mill Creek Valley, Blue Lake Canyon, Warner Valley, and other valleys in Lassen Volcanic National Park indicate that the ice exceeded 1,000 feet in thickness in some places.

Except along the higher western flanks of the Southern Cascades, northeastern California is not a well-watered land, and consequently fluvial erosion is minimal. Compared to the Klamath Mountains, the region is noticeably devoid of rivers, lakes, and forests. Originating from the Warner Mountains, the Pit River, which cuts a diagonal course from northeast to southwest right through the lava plateau and mountains and into the Sacramento River at Shasta Lake, is the only relatively long stream in the province. Some of the so-called short streams, however, flow from among the largest natural springs found in North America. The springs above Burney Falls alone gush forth hundreds of millions of gallons of water daily from the porous plateau basalts. Yet Burney Falls is not the most prolific of several springs in the region. Like the Pit River, the larger streams flow into manmade lakes. Eagle Lake, the only large natural lake in northeastern California, is the exception.

People Almost Nowhere

Any way one drives through northeastern California, even along its western edge on busy I-5, one can't help but notice that this is a land full of scenic wonders but virtually empty of people. Yreka, with less than 7,000 residents, is the largest town. It and the other towns strung out along the interstate service travelers and truckers mostly just passing through, as well as the local ranching and lumber industries.

Eastward on California 89, 139, or 299, the towns become smaller, fewer, and farther between until the junction with U.S. 395 at Alturas. With about half the population of Yreka, Alturas is the second largest town in the region. In terms of distance from California's population centers, Alturas and its environs make up the most isolated corner of the state.

Scarcity of people living in the northeast is not hard to fathom when the previously discussed geomorphology of the northeast and its climate are taken into account. Other than a limited amount of arable land in the Tule Lake and Pit River basins and other small valleys, the volcanic landscape provides practically no opportunity for agricultural development and mineral extraction. Cold, dry winters and short growing seasons make farming and ranching all the less attractive. Relatively scant annual precipitation, coupled with past overexploitation of virgin pine and fir stands, allows for only a meager existing timber reserve. In all, the only natural resource that presently lures appreciable numbers of people to the northeast is 13,000 square miles of magnificently secluded scenery. Of course, these transitory tourist populations are attracted principally to the geo-

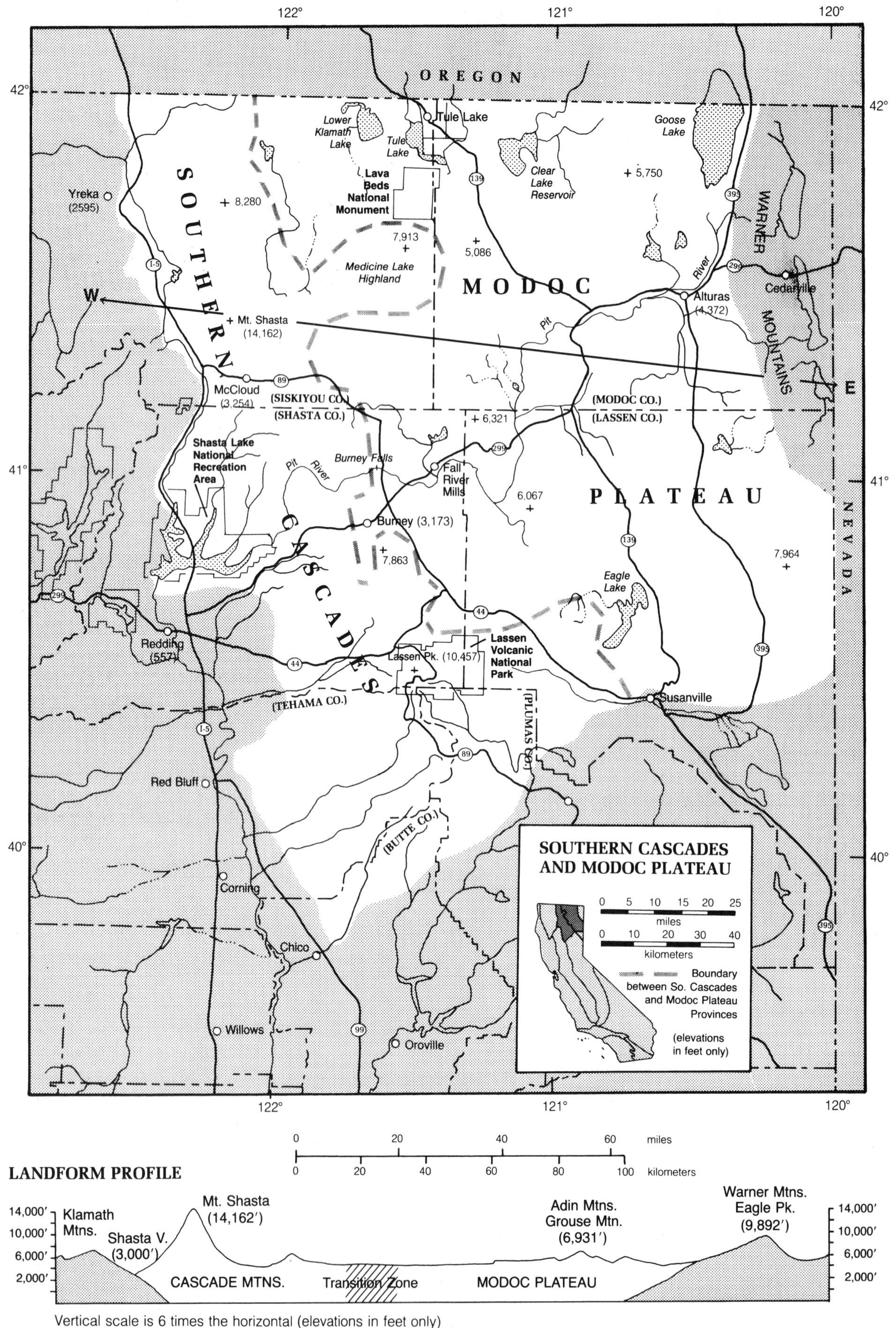

Figure 3-14 *Southern Cascades and Modoc Plateau and landform profile.* (Richard Crooker)

thermal wonders of Lassen Volcanic National Park in summer and Shasta's long, snow-packed slopes in winter.

Although white settlement never reached land-rush proportions in the northeast, it did precipitate a decline in the native population that has carried on from statehood to the present. Several thousand Native Americans, representing four distinct linguistic stocks, once lived in the region. Today, 1,000 or so of their descendants eke out a meager existence in the towns and on reservations. Of the tribes that have long since disappeared, the Yahi are probably the best known, thanks to an engrossing book and film about their last survivor, Ishi (see Chapter 8). It is noteworthy that in prehistoric times, as many as 4,000 members of the Yahi and three other tribes spent every late spring and summer hunting and fishing in Ishi's "first world," in and around what in the year of his death in 1916 became Lassen Volcanic National Park. By the middle of the century, less than 400 Native Americans resided in the same general area. Similar depredations befell the Pit River tribes; members of their tribe, however, survive to this day to claim 3.4 million acres of Modoc Plateau land and a casino built in the 1990s in Pit River (Burney). (The spread in the 1990s of gaming facilities throughout California owned and managed by Native Americans is discussed in the next section.)

THE GREAT BASIN AND SOUTHEASTERN DESERTS

The basaltic Warner Range, shown in all its lonely splendor in Figure 3-15, is lithologically part of the Modoc Plateau, but structurally it typifies the mountains of the Great Basin of the American West. The Warner Range is but one of seemingly countless north-south-trending fault block mountain ranges that dominate the landscapes of several western states, including the eastern and southeastern sides of California (Figure 3-16).

Between the uplifted ranges are down-dropped basins of interior drainage from which the Great Basin province derives its name. Mountain-born streams, usually dry except when filled with runoff from the rare winter storm or summer thundershower, drain down into the *sinks* or *playas* (intermittent or terminal lake basins), never finding their way to the sea. The fine sediments that have accumulated in these hydrologically enclosed basins in some cases yield high concentrations of borax, potash, soda ash, and other minerals of value to the playa mining industry. The heavier stream-eroded materials, such as rocks, sand, and gravel are deposited as *alluvial fans* (Figure 3-17) sloping gently from the steep flanks of the surrounding mountains to the flat surfaces of the playas.

Figure 3-15 *Warner Range in the very northeast corner of California. With two peaks (Eagle and Warren) at nearly 10,000 feet in elevation, the range captures enough precipitation from eastward-moving winter storms to feed the headwater forks of the Pit River and sustain coniferous forests and summer pastures.* (U.S. Geological Survey, photo by W. A. Duffield)

The geologic history of the intermountain desert region of California is long and complex. The record as preserved in the surface rocks of the province indicates lengthy periods of invasion by the sea and consequent marine erosion and deposition, stretching or extension of the earth's crust causing briefer periods of faulting and folding carrying on to the present, ancient granitic intrusion, more recent volcanic extrusion, and Pleistocene periglacial or *pluvial lake* formation (Figure 3-18.)

A Focus on Death Valley

Nowhere in the Great Basin are geomorphic contrasts more forebodingly dramatic than in Death Valley. Imagine what the thoughts of 100 or so gold seekers were in December 1849 when they first caught sight of the unfriendliest-looking piece of real estate they were ever likely to come upon. For one thing, they probably wished they'd never strayed from the traces of the Mormon Trail in an attempt to find a shortcut to the Mother Lode. For another, these Forty-Niners must have wondered where this below-sea-level, 140-mile valley ended or whether a passable gap in the 11,000-foot wall of mountains on its west side even existed. Apparently making matters worse, most of the argonauts realized that these mountains we know as the Panamints were not the Sierra Nevada—that an even greater barrier still lay beyond. Even before venturing out from their Furnace Creek Wash viewpoint to try to solve their quandry, at least some of the Forty-Niners must have already decided upon the morbid name they would give this barren valley.

To be strictly accurate, Death Valley is a basin rather than a valley. True, the Armagosa River does eventually terminate in Death Valley, but this is a down-faulted basin, not a stream-cut valley. In fact, these are adversarial processes, the tectonic deepening of the trough

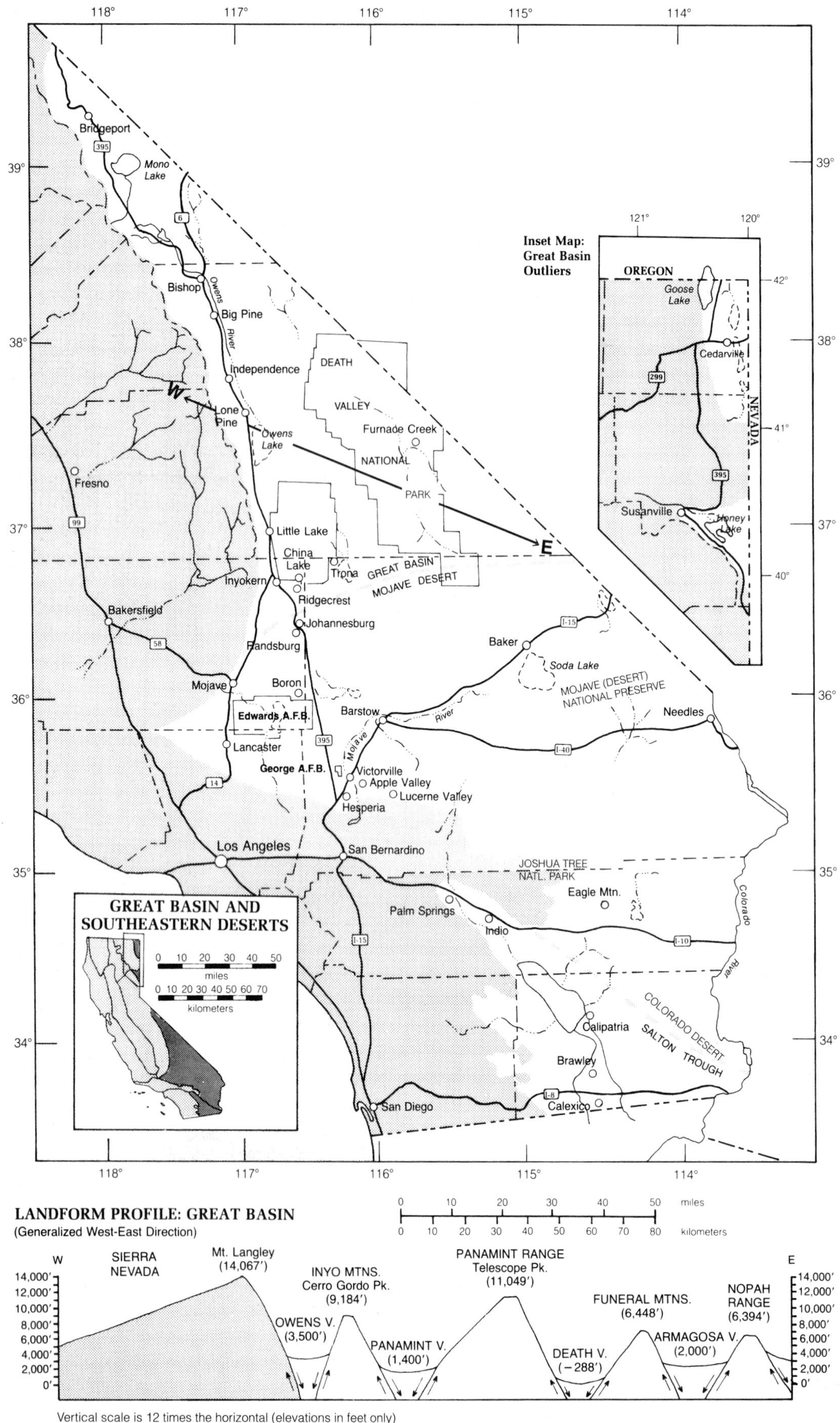

Figure 3-16 *Great Basin and southeastern deserts and landform profile.* (Richard Crooker)

Figure 3-17 *Coalesced alluvial fans, or* bajada, *at the base of the eastern front of the Sierra Nevada in the foreground. At the base of the Inyo Mountains, in the background to the east, alluvial deposits are smaller and still preserve their fan shape.* (Crane Miller)

Figure 3-18 *Soda Lake, near Baker in the eastern Mojave Desert. Only for a short while after a cloudburst or abnormal winter precipitation is there likely to be any water in this and most other Great Basin playas. During the Pleistocene, however, annual precipitation was greater in the region, and eastern Sierran and other mountain glaciers sent their meltwater far out into the Great Basin creating pluvial lakes such as Ancient Lake Mojave, of which Soda Lake was a part. Emanating from the San Bernardino Mountains, the Mojave River flowed into the ancient lake, but today its surface waters stop far short of Soda Lake, and only its groundwater reaches all the way to Soda Springs on the western side of the playa. Reached by Zzyzx Road off of I-15, the site briefly flourished as Zzyzx Mineral Springs resort several decades ago.* (Crane Miller)

surpassing its fluvial infilling. At 282 feet below sea level, Badwater (the little bit of standing water in eastern Death Valley is undrinkable) will undoubtedly sink even lower as the basin continues its structural collapse.

The birth of the basin and range landscape we see today in Death Valley National Park—indeed, throughout the Great Basin of Nevada, southeastern Oregon, southern Idaho, western Utah, and southern Arizona and New Mexico, as well as eastern California—dates to 17 million B.P. At that time, long after the forces of tectonic subduction illustrated in Chapter 2 had supposedly subsided and the *continental compression* (squeezing) of the western North American Plate had ceased, there existed broad lava plains pockmarked by the occasional high-rimmed volcanic crater. Next began a pronounced stretching of the earth's crust beneath the relatively subdued volcanic landscape. This stretching, which led to an east-west expansion of the original Great Basin surface ranging from 10 to 100 percent, caused north-south-oriented cracks in the crust. As this *continental extension* or *expansion* proceeded and the crust was pulled farther apart, huge crustal blocks, whose axes ran north-south, were variously down-dropped and uplifted along the entire east-west extent of the newly forming basin and range topography. Although faults are not normally thought of as cracks, but rather as vertical, horizontal, or diagonal displacement interfaces in the earth's crust, cracking and faulting of the Great Basin follow a consistent north-south alignment. Every time an earthquake rumbles through a part of the Great Basin and the edge of a basin is down-faulted or the flank of a mountain is up-faulted, we are reminded that crustal extension continues in the basin and range landform.

Block faulting of the Panamints to the west of Death Valley and the Black Mountains to the east, right-lateral shifting along the San Andreas, left-lateral shifting along the Garlock fault, and right-lateral faulting elsewhere in the southern Great Basin have all contributed to the pulling apart of neighboring Pananmint and Saline valleys, as well as of Death Valley. Normal vertical faulting has been most intense in the last 3 million years, with fault scarps or traces thereof often appearing as the truncated middle or lower part of an alluvial fan.

It is hard to believe that in a basin where temperatures sometimes exceed 135° F in the shade and precipitation often evaporates before ever hitting the ground, there was once a 600-foot-deep lake. Evidence of Ancient Lake Manly's heyday 11,000 years ago has largely been lost in subsequent deposits of younger fan gravels. Some irrefutable signs of that last Pleistocene lake's presence, however, are still to be seen in Death Valley at (1) Manly Bar, where longshore currents formed a gravel bar; (2) Mormon Point, where beach lines in alluvium are visible; (3) Shoreline Butte, a basaltic lava flow whose "steps" are actually wave-cut benches; (4) the mountains behind Badwater, where greyish bands of *tufa* (porous limestone or calcium carbonate) precipitated as a result of contact with an alkaline body of water; and (5) a shore-line terrace on a ridge of faulted gravels just north of the National Park Service residential area at Furnace Creek.

Considering that even during the Pleistocene, Death Valley was the hottest and driest place in North America, with the surrounding mountains providing the meagerest of watersheds, how did all that water get there? Its ultimate sources, in fact, were melting glaciers in the eastern High Sierra. Whenever climatic warming set in, such as during the waning millennia of the Pleistocene (15,000–10,000 B.P.) and at the end of previous interglacial periods (e.g., 75,000–70,000 B.P.), Great Basin valleys would begin to fill and then overflow with glacial meltwater. First, Owens Valley would spill its excess into China and Searles lakes at lower elevations to the south; in turn, they would empty into Panamint Valley to the east. Once Ancient Lake Panamint neared a depth of 1,000 feet, it would send water gushing through Wingate Pass into Lake Manly.

Another pluvial lake system, whose waters originated in the San Bernardino Mountains and were carried via the Mojave River and Ancient Lake Mojave (now Soda Lake; see Figure 3-18), also replenished Lake Manly during the late Pleistocene. However, much of Soda Lake's overflow may have drained eastward through Bristol and Danby basins into the Colorado River, with little left for Lake Manly.

Although essentially devoid of human settlement today, Death Valley historically has witnessed a long and varied parade of cultures occupying its landscape. While Lake Manly was still in existence, its shores were graced by the first culture in the region, that of the hunting, fishing, and gathering Nevares Spring peoples some 10,000 years ago. About 1,000 years ago, long after the lake had disappeared, the ancestors of the most recent aboriginal occupants of Death Valley arrived. Belonging to the Uto-Aztecan linguistic family and related to the present-day Paiutes of Owens Valley, these nomadic peoples were gatherers and pottery makers. A comparison of the linguistic diversity of California Native Americans with that of the entire continent is mapped in Figure 3-19. Note that except for the inclusion of coastal Southern California, the Uto-Aztecan region in the western United States corresponds with the Great Basin landform province; hence, the native peoples are sometimes referred to as Great Basin Indians.

Prior to the arrival of the first non-native settlers in 1769, there were some 300,000 Native Americans speaking more than 100 different tongues in six different language superfamilies in what we know today as California. California's aboriginal linguistic diversity was challenged by no other area its size in North America. By 1994, however, fewer than 1,000 Californians could speak in the handful of aboriginal languages and dialects that had escaped extinction over two-plus centuries. Among the largest remaining language groups or tribes at the turn of the century are the Great Basin Mono tribe, who number just under 200 and whose language belongs to the Uto-Aztecan family. The 1990 federal Native American Languages Act serves to preserve these languages, but it will have little impact until funding legislation is passed by Congress and signed by the president.

It was borate mining that ushered in the first semblance of non–Native American settlement in Death Valley. White mine and refinery operators and Chinese so-called cottonball carriers (bearers of calcium and sodium borate nodules) launched production of borax at Eagle Borax Spring, Cottonball Basin, and the Harmony Borax Works in the early 1880s. The refined borax was shipped by wagons pulled by teams of 20 mules apiece 165 miles southwest to the railroad town of Mojave. Harmony folded in 1889, but borate mining continued off and on in the region for another four decades. For a time during the 1920s, the Death Valley area was producing 80 percent of all the borax consumed in the nation.

Eventually, though, the discovery of borate deposits in the Mojave Desert near major rail lines and the establishment of Death Valley National Monument in 1933 dealt death blows to the borax industry in the region. Since then, tourism has become the lone industry. In 1927, borax crew quarters at Furnace Creek Ranch were converted to resort use and construction began on the luxurious Furnace Creek Inn. Although not originally

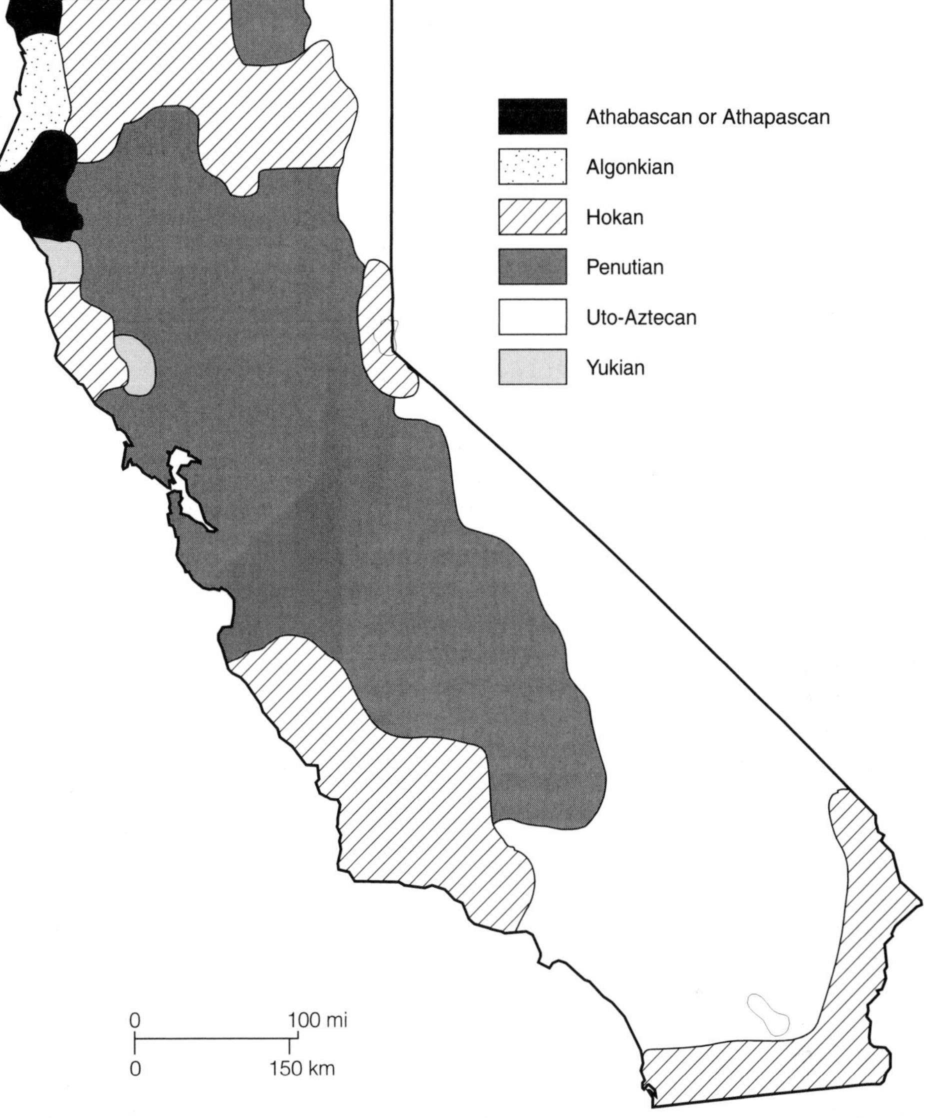

Figure 3-19 *Native language families of North America. Six of North America's seven aboriginal language families, or superfamilies, are found within California. Such linguistic diversity is found nowhere else on the continent in an area that matches California's size. Further diversity is evident in the dozens of linguistic subfamilies, languages, and dialects found among native California tribes (see Chapter 8 and Figure 8-1).*

intended as a tourist attraction, Death Valley Ranch, popularly known as Scotty's Castle, also was built during the Roaring Twenties. Since federal acquisition in 1970, Scotty's Castle and grounds, along with nearby volcanic Ubehebe Crater, have been the major tourist attractions at the northern end of the national park. Even so, Furnace Creek, with year-round accommodations, a park visitors' center, and Park Service headquarters, remains the focus of tourism and settlement in Death Valley.

Reservation and Rancheria Roulette

A century ago, when all seemed lost to Native Americans, some foretold of the buffalo one day returning to their lands and thereby regaining for them the basis for economic self-sufficiency. The bison may not be back, but for native peoples in several states and provinces in the United States and Canada, the prophecy may be fulfilled in an unforeseen way: owning and managing their own gambling casinos. Although by no means a new enterprise in California, reservation gaming for outsiders, for profit, is picking up steam. In 1997, there were 40 reservation/rancheria casinos in California (Figure 3-20). From the Agua Caliente band of the Cahuilla tribe in Palm Springs, who developed a casino–hotel complex, to the Rumsey rancheria in Yolo County, whose casino management is looking at a pricey expansion, there soon will be Native American gaming facilities in 19 of the state's 58 counties.

A Manmade Sea

Unlike Death Valley's Lake Manly, the Salton Sea is an unusual case, for it is neither a vestige of the Pleistocene nor a cutoff extension of the Gulf of California, although the gulf's waters did invade the Salton Trough several million years ago. Instead, the Salton "Sea" is a man-

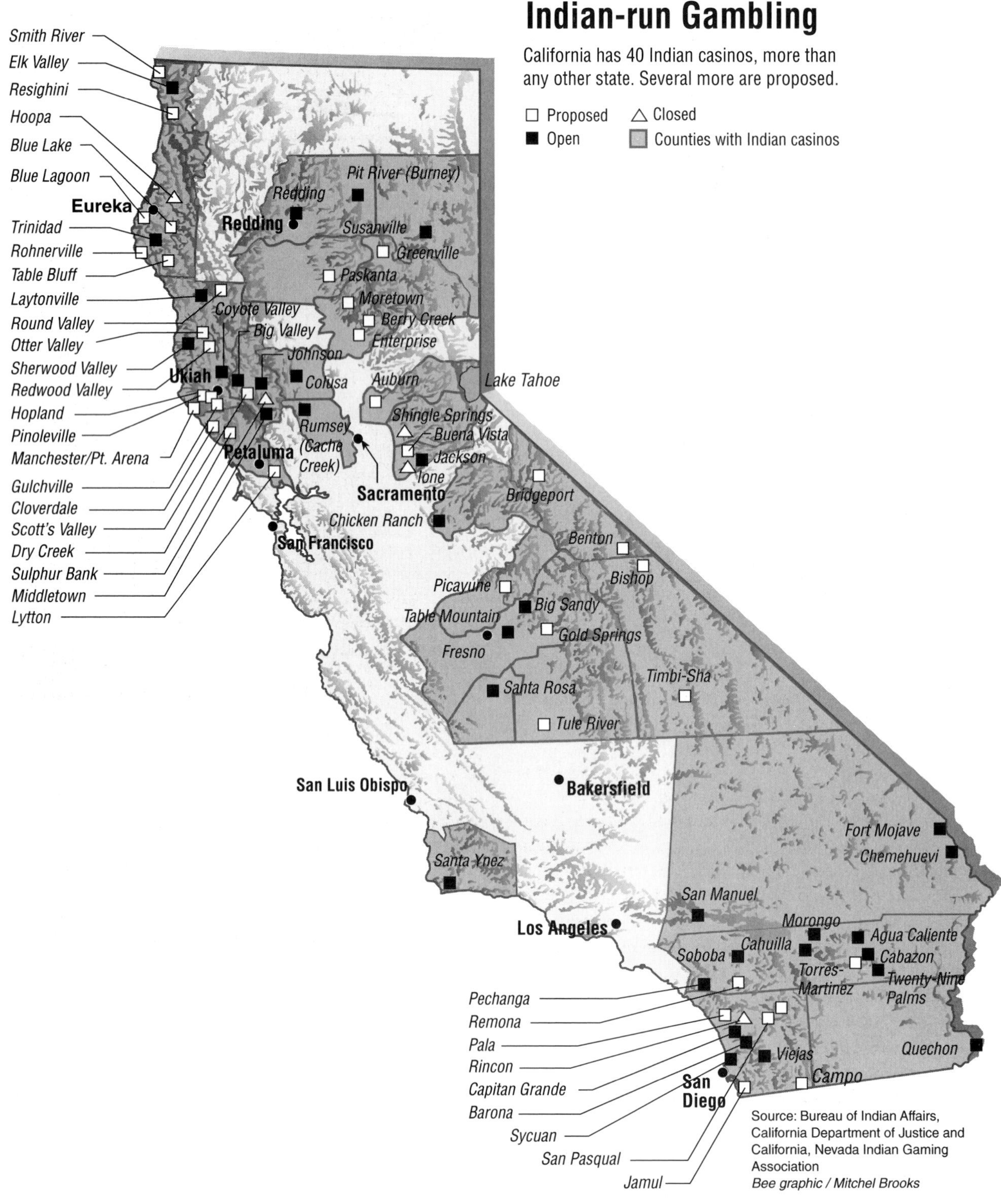

Figure 3-20 *Reservation gaming in California. As of 1997, there were 40 reservation/rancheria casinos in the state, more than are operated by Native Americans in any other state. Their combined annual gaming revenue was estimated at $1 billion—again, a record among the 50 states. Yet only 5 percent of California's 320,000 Native Americans belong to gaming tribes.* (*Source:* © The Sacramento Bee, July 2, 1997, p. A12 from series on "California's Lost Tribes" June 29–July 2, 1997.)

made accident of the early twentieth century resulting from an ill-fated irrigation canal temporarily diverting the flow of the Colorado River into the Salton Trough instead of into the Gulf of California.

Currently, the Salton Sea is kept from evaporating away by inputs of used irrigation water from the Imperial Valley. The sea acts as a sump for water that originates in the Colorado River and comes to the Imperial Valley via the All-American Canal. Before reaching the sea and as part of the irrigation process, the water flows through land that is *ditched* and *tiled* (Figure 3-21), so that crop-damaging dissolved salts are piped away and thus not allowed to accumulate in irrigated soils. Ditching and tiling in the Imperial Valley not only has maintained the Salton Sea but also has allowed agriculture to prosper where it would otherwise could not have.

The Salton Sea has become one of the state's largest wildlife refuges, albeit one that is on the brink of ecological calamity because of the very water that sustains it. Salton Trough farms annually pour about 3 million tons of salt into the sea. With the sea already 30 percent more saline than the Pacific Ocean, this volume of salt intrusion alone steadily reduces its viability as a habitat for an estimated 380 different species of birds, as well as several species of fish and other wildlife. News of salt buildup also keeps people away, as fewer come to fish, hunt, swim, and otherwise recreate. Of greater concern than the salts themselves, though, are the relatively minor volumes of artificial and natural toxicants that accompany the salts. A case in point is selenium, a naturally occurring, nonmetallic element of the sulfur group that is a beneficial micronutrient for humans and animals when ingested in minute amounts and a debilitating toxin when consumed in larger doses. Selenium is used to make everything from computer chips and photovoltaic (solar) cells to fungicides and paper. Although it can be emitted by power plants and occurs naturally in some foods, selenium is most commonly found in sedimentary rocks of marine origin. It originated in ancient seas that once covered much of the interior west, a region that the Colorado River and its tributaries now incise on their way southwestward. Via first the Colorado River, then the All-American Canal, and finally the ditched and tiled drainage systems of the Imperial and Coachella valleys, only some 50 pounds of selenium ultimately enter the Salton Sea each year. Even so, that is sufficient to have deformed fish in the Salton Sea. Though selenium-deformed birds have yet to be found in the Salton Trough, it is only a matter of time before the toxin accumulates sufficiently at one level of the food chain to jump to the next level. Birds feeding on tainted fish and in selenium-laced muds elsewhere in California, such as at Kesterson National Wildlife Refuge in the San Joaquin Valley since the early 1980s, have given birth to stillborn or deformed offspring by the thousands. Short of severely curtailing or altogether shutting down irrigated agriculture in the Imperial and Coachella valleys, which in turn would diminish California's ability to help feed an overpopulating planet, selenium *bioaccumulation* will no doubt spread to ever higher levels of the Salton Sea food chain.

Figure 3-21 *Ditching and tiling in the Imperial Valley. The checkerboard pattern (Fig. 3-32 of agricultural land use seen in the satellite view would likely not exist were it not for the removal of salt-laden irrigation water. The grid pattern of settlement on the U.S. side of the border owes to a rectangular land survey, in which property lines, irrigation systems, hedgerows, roads, and so on are laid out in conformity with the latitudinal baselines and longitudinal meridians of the survey. South of the border, where land uses are similar but a metes-and-bounds survey is employed, no such checkerboard landscape appears.* (NASA)

Thinly Populated Deserts

Stretching discontinuously nearly 1,000 miles from Oregon to Baja California and lying in the water-starved rain shadow of the Cascades, Sierra, Transverse Ranges, and Peninsular Ranges, the Great Basin and southeastern deserts of California are understandably a thinly populated land. The few cities of any appreciable size exist mostly in the western portions of the Mojave and Colorado deserts, where they are closest to the Los Angeles–San Diego metropolitan complex and where the terrain is leveler and the mountain passes are less imposing.

Wedged in the apex where the Garlock and San Andreas faults join, the western Mojave's Antelope Valley contains the desert's largest city, Lancaster, which, with neighboring Palmdale, claims nearly 160,000 residents, many of whom are connected with the local Lockheed–Martin plant or Edwards Air Force Base. As noted in Table 3-1, between 1980 and 1990, Palmdale was the state's second fastest-growing city with a 432 percent increase, and Hesperia, some 40 miles due east of Palmdale, was third at 268 percent.

The Antelope Freeway (California 14) puts these desert communities within an easy 45-minute drive of the San Fernando Valley. Another 120,000 people are associated by location with the Mojave River, Interstate 15, and the main Union Pacific–Santa Fe transcontinental rail lines in the communities of Adelanto, Apple Valley, Barstow, Hesperia, and Victorville. George Air Force Base (now closed), the Boron open pit borate mine, two Portland cement plants, light manufacturing plants, Solar One (the nation's first solar electric plant; see Chapter 6), and retail services broaden the employment base of the area.

Much larger in area and correspondingly smaller in population, the central Mojave sustains some 20,000 inhabitants. Most of them are retirees in the Yucca Valley–Twenty-Nine Palms area, with incorporated Twenty-Nine Palms alone boasting a 1990 population of nearly 13,000. Kaiser Steel's iron mines at Eagle Mountain closed down in the early 1980s, and the company town was abandoned. Because an existing rail line makes the site attractive as a waste-by-rail landfill, Eagle Mountain may soon suffer the ignominious fate of becoming a desert dump. Proponents claim that the open-pit mine can handle up to 20,000 tons of waste per day that would otherwise go to harder-to-find sites closer to crowded metropolitan areas. Opponents seeking to derail the trash train project argue that its development would increase air pollution and endanger wildlife in neighboring Joshua Tree National Park. Immense federal holdings in the region, which include the monument administered by the National Park Service, the Marine Corps Training Center, and Bureau of Land Management (BLM) land, dictate a major role for the federal government in the landfill controversy.

Settlement of the eastern Mojave's 15,000 residents is almost entirely along the banks of the Colorado River in the agricultural service centers of Blythe and Needles and the Chemehuevi Valley, Colorado River, and Fort Mohave Indian reservations. Although most of the Colorado River reservation lies on the Arizona side of "The River," its 42,696 acres in California render it the third largest reservation in area in the state. Recreational development of "The River," notably upstream from where Parker Dam impounds Lake Havasu and farther north into Nevada where Laughlin has become a competitor of Las Vegas as a hotel and casino center, currently provide the principal stimulus for new settlement.

TABLE 3-1 Ten fastest-growing California cities by population, 1980–1990.

CITY AND COUNTY	1980	1990	PERCENT CHANGE
Cathedral City, Riverside	4,130	28,959	+601
Palmdale, Los Angeles	12,277	65,357	+432
Hesperia, San Bernardino	13,540	49,818	+268
La Quinta, Riverside	3,328	10,723	+222
Apple Valley, San Bernardino	14,305	45,651	+219
Perris, Riverside	6,827	20,835	+205
Adelanto, San Bernardino	2,164	6,499	+200
Lake Elsinore, Riverside	5,982	17,920	+199
Victorville, San Bernardino	14,220	39,462	+177
Fontana, San Bernardino	37,111	85,281	+129

Source: U.S. Department of Commerce, Bureau of the Census, *1980 Census of Population and Housing* and *1990 Census of Population and Housing* (Preliminary Reports) (Washington, DC: U.S. Government Printing Office, 1981, 1991).

Though much smaller in area than the Mojave Desert, the Colorado Desert to the south boasts a rapidly growing population that now almost equals that of the entire Mojave region. The Coachella and Imperial valleys harbor most of the residents. Palm Springs, now a world-renowned winter resort and mecca for retired presidents and entertainers, is among the three largest cities in the Coachella Valley. Cathedral City, which topped the list of California's 10 fastest-growing cities between 1980 and 1990 (see Table 3-1), and Indio are the other two cities. By 1994, the combined population of the three cities stood at 110,000. Add La Quinta, Indian Wells, and the smaller, agriculturally oriented settlements of the Coachella Valley and the population of the valley region swells to 175,000. As Table 3-1 shows, 7 of the 10 cities (all but Perris, Lake Elsinore, and Fontana) are in the desert, and 9 of 10 are in Riverside and San

Bernardino counties. The dry, sunny climate and low land prices are obvious drawing cards.

South of the settlement void of the Salton Sea and Anza-Borrego Desert State Park, intensive irrigation agriculture of a variety of crops provides the major source of income for most of the Imperial Valley's 110,000 residents. Imperial ranks among the nation's leading counties in beef, feed crop, and winter vegetable production and thus relies heavily on its resident population, two-thirds of which is Hispanic, and migrant workers to sustain its economy. Imperial has the highest proportion of Hispanic population among the state's 58 counties. The agribusiness centers of Brawley and El Centro and the border town of Calexico are the population nuclei. At the northern end of the valley, Calipatria, at −183 feet, can rightfully claim to be lower in elevation than any other community of 3,000 or more residents in the nation. Underlying the Imperial Valley is one of the world's potentially most productive geothermal fields, and facilities are producing power from this natural source of steam energy.

Although a small tribe owns and occupies some valuable real estate within Palm Springs, Native American settlement in this area is mostly to the south on two Cahuilla reservations in the Coachella Valley and the Fort Yuma Indian Reservation in the irrigated Bard Valley of the lower Colorado River. The Yumas were among the few tribes of native California Indians who practiced flood plain agriculture in prehistoric times. They have carried this tradition through to the present, but today the Yumas lease much of their land to outsiders and increasingly pursue nonagricultural endeavors.

Basin Settlement and Transient Tourism

North of the low deserts of the Colorado and the high deserts of the Mojave, the mountain ranges become more numerous, the basins smaller, the distance from Los Angeles greater, and the people fewer. Except for about 30,000 residents of China Lake, Inyokern, Ridgecrest, and Trona, many of whom are economically sustained by the China Lake Naval Ordnance Test Station and by playa mining and processing operations at Searles Lake, there are no urban nodes in California's share of the Great Basin. The naval weaponry range itself is devoid of settlement and extends northwestward through most of the Coso Range almost to the Owens Valley. Searles Lake yields a wealth of potash (for fertilizer), soda ash (for glass making), and various other dry lake brines (for everything from detergents to pharmaceuticals). The Trona plant, purchased in December 1990 by North American Chemical Company from Kerr-McGee Corporation, employs more than 1,000 people. Owens Lake is also an important source of soda ash and has seen playa mining firms come and go throughout much of the twentieth century, the most recent venture starting in 1993. By providing a local source of electricity, a recently developed geothermal power plant in the China Lake Basin (see Chapter 6) rounds out the economic geography of the region.

Figure 3-22 *Scene from the Lone Pine Film Festival, with Lone Pine Peak (center) and Mount Whitney (right) in the background. Although most of the scenes actually are shot in the Alabama Hills, the majestic background formed by the highest portion of the Sierra Nevada, with several 14,000-foot peaks, is often used to depict a towering mountain range in another part of the world, such as the Himalayas in* Gunga Din. *Obviously, California's diversity of landforms helps account for the state's status as the film capital of the world.* (Crane Miller)

The Owens Valley is the eastern gateway to the Sierra and as such maintains a population of several thousand in four communities strung out along U.S. 395. Bishop and Lone Pine, the largest towns, provide a variety of overnight accommodations for skiers heading for Mammoth or Tahoe, fishermen seeking trout from Lake Crowley or the lakes and streams of the eastern Sierra, and other vacationers. The Los Angeles Department of Water and Power (which owns much of the valley land and most of the water reserves), the California Department of Fish and Game, the U.S. Forest Service, the U.S. Bureau of Land Management, and other government agencies employ many local residents. Native Americans, mostly Paiutes, live on reservation land both in and outside the towns and are employed by the local service industries and governments. Mining and ranching are relatively insignificant employers of the local working force.

The Alabama Hills, which afford a ruggedly scenic backdrop immediately to the west of Lone Pine (see Figure 1-6), have long been used as locations for films, ranging from the 1939 classic *Gunga Din*, whose setting was India under British colonial rule, to the 1994 full-length version of the old TV series *Maverick*. To celebrate its movie heritage, Lone Pine instituted the Lone Pine Film Festival in the 1990s (Figure 3-22). Held on the second

Figure 3-23 *Randsburg mines. Located a mile off of U.S. 395 in the Rand Mountains of eastern Kern County, these mines produced more than $10 million worth of gold between 1895 and the end of World War I. Gold-bearing quartz veins in granitic rocks were the source of the precious metal. The towns of the Rand district, including also Atolia, Johannesburg, and Red Mountain, have also serviced silver and tungsten mining operations.* (Crane Miller)

weekend of October, the festival temporarily swells the population of the southern Owens Valley by thousands with the arrival of movie buffs from around the world, as well as actors and crews who worked on location there.

Settlement north of the Owens Valley to the Nevada border, although a distance of more than 100 miles on U.S. 395, is limited to a few tiny communities. The towns of Lee Vining and Bridgeport service travelers on the lonely but magnificently scenic stretch of highway.

Mountain settlement in the Great Basin has come and mostly gone. A century ago, gold and silver drew thousands of miners to boomtowns like Bodie in the mountains east of Bridgeport and Cerro Gordo (Fat Hill) high in the southern Inyo Mountains. Darwin in the Argus Range southeast of the Inyos and Randsburg (Figure 3-23) and Johannesburg in the Rand Mountains of the northern Mojave are of more recent vintage, but they too appear to be on the brink of becoming ghost towns.

In the two tongues of the Great Basin that project from northwestern Nevada into northeastern California, Susanville is the only community of significant size. The city's several thousand residents earn a living from lumber mills, retail stores, a state penal facility and other government functions, the railroad, agricultural and livestock service industries, motels, and auto service stations. In short, Susanville is a service center for a remote part of the state and is economically more closely tied to Reno, Nevada, than any large city in California. The California Correctional Center at Susanville was seventh in the state in prison population in 1991, holding just under 5,800 inmates. Like the Susanville facility, many of the state's 21 prisons are located in out-of-the-way, rural areas; penitentiaries at Vacaville (California Medical Facility), Soledad (Correctional Training Center), Jamestown (Sierra Conservation Center), and Tehachapi (California Correctional Facility)—the first, fourth, fifth, and eighth largest, respectively—are examples. As exemplified in the early 1990s by the vehement opposition of Los Angeles residents to new prison construction within city limits, most urban and urban fringe dwellers would just as soon see prisons remain as distant as possible. But in smaller communities, like Susanville, sentiments are quite the opposite. Residents not only depend on the penal institution as a significant driver of the local economy but no doubt would fight any move to close it down. The state's prison population as a whole increased from about 25,000 in 1982 to more than 100,000 in 1992, which has put pressure on both urban and rural California to expand the state prison system. Assuming city dwellers remain opposed, will country folk continue to accept the bulk of such expansion?

THE SIERRA NEVADA

By any measure, the Sierra Nevada is more impressive than any other landform province in California. Late in the nineteenth century, naturalist John Muir calculated the Sierra to be "about 500 miles long, 70 miles wide, and from 7,000 to nearly 15,000 feet high." Partly because of Muir's vivid descriptions of the Sierra, tens of millions of visitors have since ventured into this massive expanse of mountains where "glaciers are still at work in the shadows of the peaks, and thousands of lakes and meadows shine and bloom beneath them, and the whole range is furrowed with cañons to a depth of from 2,000 to 5,000 feet, in which once flowed majestic glaciers, and in which now flow and sing a band of beautiful rivers."[3] Yosemite Valley is the most dramatic of the Sierran canyons carved out by "majestic glaciers" (Figure 3-24). During glacial periods over the past 2.5 million years, the most recent of which until about 10,000–15,000 years ago, large valley glaciers covered much of the High Sierra. The Yosemite Valley glacier was among the largest, measuring dozens of miles in length and about a mile in depth. Long before and long since the advent of the glaciers, the Merced River has incised a path through the valley.

John Muir's portrayal of the Sierra Nevada (which translates as "Snowy Mountains") is just as reliable as it is graphic, save for an exaggeration in length of about 100 miles by his inclusion of the Southern Cascades. His description is all the more remarkable in that he wrote

[3] John Muir, *The Mountains of California* (Berkeley, CA: Ten Speed Press, 1977), pp. 2–3. The quotations are from this facsimile edition of John Muir's original 1894 classic.

Figure 3-24 *Aerial view eastward of Yosemite Valley. The sheer granite face of El Capitan appears in the foreground on the north (left) side of the valley, with Half Dome and Clouds' Rest on the south (right) side, where the main valley grades into upper or Little Yosemite Valley.* (Crane Miller)

it nearly a century ago without benefit of either aerial or satellite information. Obviously, though, he had lengthy firsthand experience in the Sierra and knew the range perhaps better than any other person in recorded history. Muir's favorite spot was Yosemite Valley, and it was his decades of intimacy with the landscape that first prompted him to suggest glacial origins for the deep, sheer-walled valley. Muir was not a geologist by training, and his view ran counter to the then-prevailing theory that the valley was down-dropped by normal faulting. Yet Muir eventually was proved correct, and as a consequence we know the Sierra Nevada as one of the most glaciated mountain ranges in the world.

Ancestral Formations

Pleistocene glaciation was merely the latest of a series of major geologic events that gave birth to the Sierra Nevada that we know today. Figure 3-25 traces that history beginning 135 million years ago when the ancestral Sierra was emerging after 300 million years beneath the sea. Indeed, the oldest rocks now found in the Sierra date from about 435 million B.P., when they first formed at the bottom of the sea. These ancient rocks can be seen overlying granitic rocks forming *roof pendants*. One would expect these roof pendants to contain fossils since they originated in a sea basin teaming with life. The heat and pressure later exerted by granitic intrusion from below, however, transformed marine fossils, other sedimentary rock materials, and even submarine volcanics into metamorphic rocks. For example, shale and basalt metamorphosed into schist, shale into slate, sandstone into quartzite, and limestone into marble. The metamorphosis also caused gold, silver, scheelite (tungsten ore), and molybdenite (ore of molybdenum, a steel alloy) to form at contact points between the older rocks and the granitic intrusion.

About $3 billion worth of these valuable minerals has been extracted from the Sierra. Gold, mined from placer stream deposits and *lodes* (gold-bearing quartz veins, for example), accounts for the bulk of this value; but today it is only of historical interest (see Chapter 8). The steel alloy ores, on the other hand, are presently mined in the eastern Sierra, notably at the nation's largest molybdenum–tungsten mine, near Pine Creek and west of Bishop.

Granitic Intrusion

The next great episode in the building of the Sierra was the intrusion of a massive granitic batholith that would eventually become the rock core of the entire range. Presently, 10,000 square miles of the Sierra region is covered by this speckled, off-white crystalline mass. Before 135 million years ago, the mass was largely fluid magma convecting upward and intruding into the older surface rocks that by then were being exposed to atmospheric weathering and erosion as the sea retreated.

In the ensuing 70 million years, most of the magma cooled beneath the surface to become granite, although some extruded onto the surface as volcanoes and lava flows. Yosemite Valley granites lithified about 80 million B.P. As erosion ravaged the overlying rocks, more and more granite became exposed. Mild uplift from normal faulting raised the ancestral range to about 3,000 feet above sea level, but the constant erosion of a tropical climate kept the highest peaks from surpassing these heights. Gold veins were also exposed as streams and rivers carried the precious metal downstream to the placer deposits of the western foothills.

Following the relative quiescence of ongoing stream erosion from about 65 to 30 million B.P., a period of violent volcanism set in that persisted until almost the present. At first, volcanic explosions rocked the northern Sierra, scattering steam, ash, and *bombs* (rocks and other volcanic debris falling from the sky) over the landscape. The erupting volcanoes spent themselves and eventually disappeared, but outcrops of their light-colored rhyolitic ash can still be seen near the western and north-

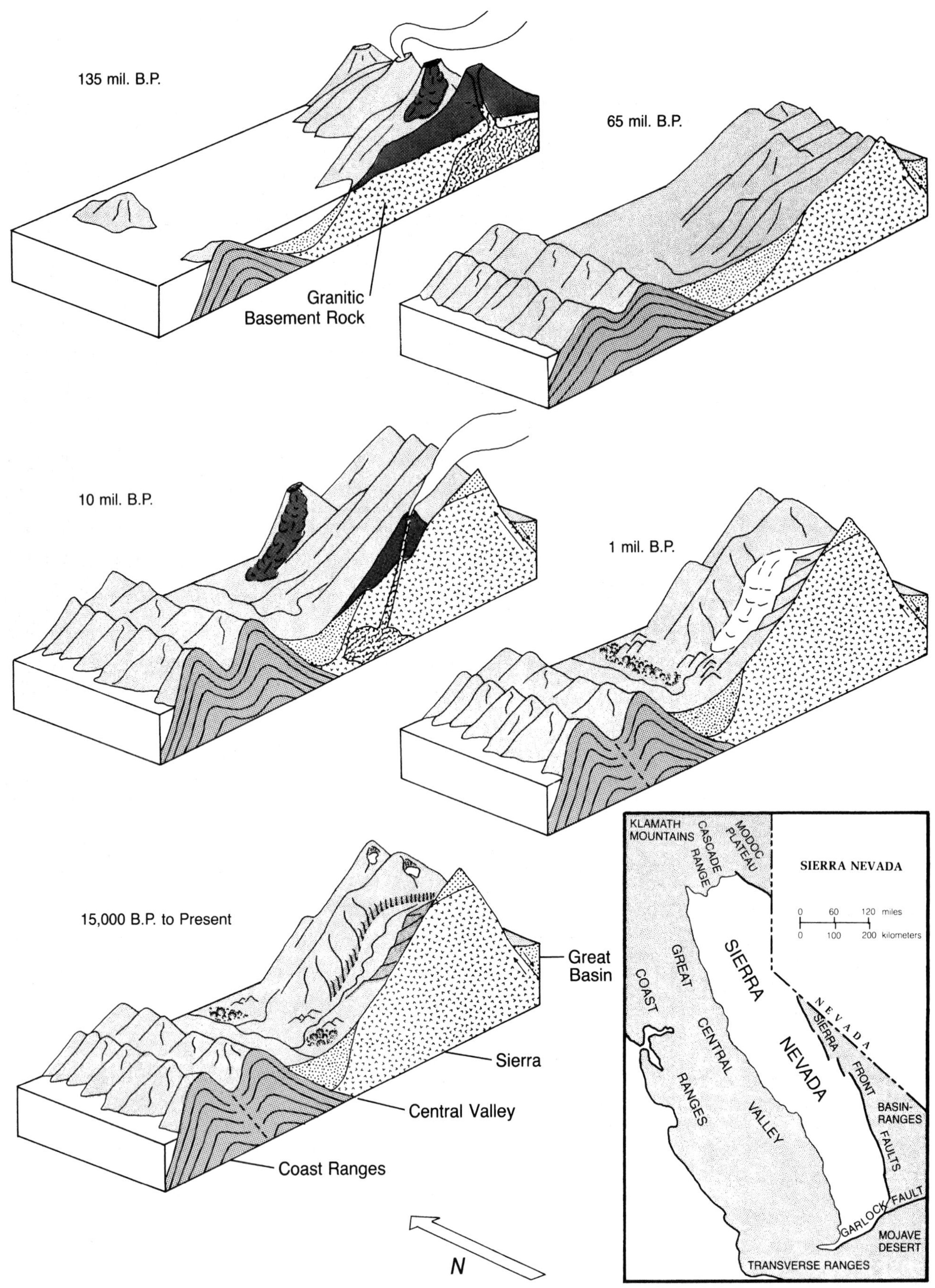

Figure 3-25 *Geological history of the Sierra Nevada in cross-section.* (Richard Crooker)

Figure 3-26 *Devil's Postpile, part of a 600,000-year-old lava flow that originated a few miles to the southeast near Mammoth Mountain. The volcanic basalt (possibly andesite) flowed, cooled, and jointed into nature's most efficient geometric shapes: six-sided columns. Later, weathering and erosion revealed the columns. The top of the formation was subsequently planed off by late Pleistocene glaciers, while broken rock (talus) has accumulated at the base.* (Union Pacific Railroad)

ern edges of Lake Tahoe, in Donner Pass, and intermittently northward to the Cascades.

Over the next several million years, ash falls gradually gave way to outpourings of lava that mixed with water and earth materials to form steaming hot volcanic *mudflows*. The mudflows solidified as greyish-appearing andesite, which is much in evidence in the region immediately south of Tahoe, especially through Sonora Pass. The andesite flows covered upwards of 10,000 square miles of the northern Sierra at one time or another between 20 and 5 million B.P. Thereafter and into the ice ages, the focus of volcanism shifted southward to the Mammoth–Mono area, where a huge lava flow left Devil's Postpile (Figure 3-26) in its wake, and a series of nearby eruptions gave birth to an 11,034-foot mountain that would become a skiers' paradise, namely, Mammoth Mountain. There have been no eruptions in the twentieth century, but hot springs at the base of the Postpile, lying 25 miles east in Hot Creek and elsewhere, tell us that volcanism is far from dead in this part of the Sierra.

Block Faulting

Although uplift from faulting has been continuous in the Sierra since the earliest granitic intrusions over 200 million years ago, only in the last 10 million years has uplift of the batholith so exceeded its erosion that the highest crests of the range finally rose more than 10,000 feet above sea level. *Vertical displacement* (mountains rising and valleys dropping) of nearly 20,000 feet has occurred in the southern Sierra in the last 3 million years alone. Almost all of the recent faulting, including that associated with the 1872 Lone Pine earthquake, originated in the Sierra Nevada fault zone along the steep eastern escarpment of the Sierra. The 1980 and 1998 series of earthquakes near Mammoth, those in the northern Owens Valley in the 1980s and 1990s, and the Lee Vining earthquake of 1990 are further evidence of the Sierran dynamism.

In essence, the granitic batholith is still in the process of being block faulted so that it rises sharply on its eastern edge and tilts or dips downward ever so gradually on its western side, giving the Sierra an asymmetrical profile (see the cross-section in Figure 3-25). The long western slope is the one exposed to moisture-bearing storms coming from the Pacific. Consequently, numerous great rivers have incised steep, V-shaped valleys up and down the range's length. The eastern slopes, because of their abruptness and rain shadow location, handle only a small proportion of the Sierran drainage and thus exude short whitewater streams but no long rivers. Erosion by ice and wind, weathering of rock, and mass wasting are also wearing down the granite fortress; but uplift continues apace and keeps the Sierra a young and growing mountain range.

Figure 3-27 *East Sierra moraines along McGee Creek west of Crowley Lake. The McGee Creek moraine and till, seen here at an age of 2.5 million years, may be the world's oldest glacial deposit from the Quaternary period.* (Crane Miller)

Pleistocene Glaciation

The great alpine glaciers of the Pleistocene have come and gone on several occasions over the last 2.5 million years in the Sierra, and they may even return again in the not-so-distant future. Whatever happens, the valley glaciers have already made their mark more emphatically than any other agent of erosion or weathering affecting the Sierra range. Glacial landforms are ubiquitous today on both sides of the Sierra, from the 5,000-foot elevation all the way up to the summits of 14,000-foot peaks.

In sheer size, the U-shaped troughs and their tributary *hanging valleys,* now the conveyers of the nation's highest waterfalls, are the most impressive products of glacial scour. Glaciers thousands of feet thick and dozens of miles long carved Yosemite, Hetch Hetchy, Tenaya, and other previously V-shaped stream valleys ever more deeply and broadly into the granitic batholith. The best examples of valley glacier scour, including these three valleys, are found within the boundaries of Yosemite National Park.

The erosion potential of a valley glacier depends on a number of factors, not the least of which is the structure of the rock beneath and on the sides of it. The more closely jointed the granite a glacier is overriding, the more rock will be eroded. For instance, as Tenaya Glacier advanced down valley over and along closely spaced joints in granite, it scoured Tenaya Canyón some 2,000 feet deeper than it would have if all those fractures had not been there.

Interspersed between the glaciations, which lasted tens of thousands of years, were equally long interglacial periods when warming trends set in and glaciers retreated. As the glaciers wasted away, they left in their wake lateral and terminal moraines, *erratics* (boulders transported by glaciers), glacial lake beds, and other depositional landforms noted in Figure 3-27. The *moraines* are quite literally, "unconsolidated heaps of rubble" resulting from glaciers depositing rocks, gravel, sand, and other unsorted earth materials they had originally eroded up valley and then carried down valley. The till deposits accumulated along the sides of the glaciers as *lateral moraines,* at the point of farthest down-valley advance of the glaciers as *terminal moraines,* or at glacier retreat points as *recessional moraines.* The steepness of the eastern side of the Sierra and its drier climate, the latter contributing to *sublimation* (moisture converting from a solid to a vapor state) of ice, help explain the prevalence of moraines on the east rather than the west side of the Sierra. Like glacial erratics found any place in the world where glaciers have come and gone, the boulders here are plucked out of rock formations by the glacier and then carried varying distances by the moving ice. Once the ice thins out or the glacier stops advancing, the boulders are dropped. Wherever they come to rest, erratics usually appear somewhat out of place compared to the native rocks of the area.

By about 10,000 B.P., the last of the great Pleistocene glaciers had melted away in the Sierra. There have been minor glaciations since then, however, the most recent being the "Little Ice Age" barely 250 years ago. The five dozen or so tiny glaciers found in the Sierra today were born during that eighteenth-century cooling-off period. They owe their continued existence to high elevation—most are above 10,500 feet—and their exposure away from the sun, which diminishes ablation. Figure 3-28, shows the Palisade and other glaciers prospering under these geographical advantages.

These and some other Sierran glaciers appear to be experiencing additional effects from the cooling trend of the past 30 years. For instance, some of the heavy snowpack of the 1977–78 winter survived the following summer's melt season to become *firn,* and in turn glacial ice, because of the early arrival (in October 1978) of the next winter's snows. The 1986–92 drought resulted in a much reduced Sierran snowpack (in some winters about half of normal accumulation). Even so, drier but colder air minimized glacier wasting. There is no telling how the El Niño–driven storms of 1998 affected the existing glaciers, except to say that they helped nearly double the normal Sierran winter snowpack. It is too early to tell

A

B

Figure 3-28 *Eastern Sierran glaciers and Palisade glacier as seen from just north of Big Pine in Owens Valley. Despite early winter snow in the higher elevations, Palisade glacier still stands out to the left (south) of center in photo A and nearer center in photo B. It is California's largest existing glacier, yet it measures only a scant square mile in surface area. It is nestled in the cirque of 14,242-foot North Palisade Mountain, a cirque scoured out by much larger Pleistocene glaciers long ago.* (Crane Miller)

whether the glaciers are building noticeably more rapidly than they are ablating. However, we do know that the current cooling trend has resulted in substantial growth of hundreds of much larger glaciers only 600 miles away in the Northern Cascades of Washington.

Long before and long since the Pleistocene, a type of mechanical weathering of rock known as *exfoliation*, or the "peeling away" of granite, has characterized the wearing down of the Sierra almost as uniquely as has glaciation. The back side of Half Dome in Yosemite National Park, Moro Rock in Sequoia National Park, Tehipite Dome in Kings Canyon National Park, and other *exfoliation domes*, which look like giant onions with their skins peeling off, are almost as common a sight in the Sierra as are glacial landforms.

Why the domes weather the way they do is not fully understood, although alternate freezing and thawing of moisture in the rock joints, unloading or removal of overlying rock layers, and the crystalline structure of granite are all known to contribute to exfoliation. The weathered rock heaps up at the base of the formation into a *talus pile* or a more gentle *talus slope*.

Recreation and Retirement

It is obvious from the physiography thus far described that save for the western foothills, the Tahoe Basin, the Mammoth Lakes area, and Yosemite Valley, the 25,000-square-mile Sierra Nevada does not provide a terrain conducive to human settlement. Yet its vast dimensions and rugged mountainscape make the Sierra the premier alpine playground of California. People visit the Sierra by the millions the year round. The greatest influx of vacationers is during the summer, when the national parks and forests are overrun by campers, boaters, fishermen, sightseers, and the like. Even higher up on the popular back-country trails like the John Muir, the Pacific Crest, or the Mount Whitney, one is apt to encounter dozens of other hikers on any clear summer day. Of course, there are also still the less frequented trails, forests, wilderness areas, streams, lakes, meadows, and mountain summits where one is likely to see only wildlife.

Fortunately for its conservation, almost all of the Sierran high country is not accessible by motor vehicle. Most of the range's more than 2,000 glacial lakes, for instance, are at least a day's hike from a roadhead. In fact, from the southern end of the range at Sherman Pass (at 9,200 feet and therefore closed in winter) northward nearly 200 miles to Tioga Pass (at 9,945 feet), there is no way to cross the crest of the Sierra except by foot or on the back of an animal. Furthermore, Tioga, Sonora, Monitor, and Ebbetts passes are routinely closed in winter, seasonally extending the impassable crest of the Sierra another 100 miles northward almost to the Tahoe Basin.

Despite winter's severe restrictions on access to the High Sierra, the more snow there is, the more downhill and cross-country skiers there are. Ski resorts, often discouragingly crowded from autumn's first snows until the spring thaw, are scattered throughout the High Sierra in such areas as Donner-Tahoe, June Lakes (Grant, Gull, and June lakes), and Mammoth in the eastern part of the range, and Badger Pass (in Yosemite National Park), China Peak, and Mount Riba (Bear Valley) on the western slopes.

Environmentalists' concerns, however, over the negative impacts of more access roads, manicured slopes, waste disposal, wildlife depredation, and other modifications have mounted to the point of thwarting development of new winter, and even all-year, resort sites. Mineral King, just south of Sequoia National Park and the closest of any proposed Sierran sites for skiers from the Los Angeles metropolitan area, and Independence Lake, north of Donner Pass and convenient to vacationers from the Bay Area, Sacramento, and Reno, are notable casualties among recent year-round resort proposals.

The most significant constraint on growth and development in El Dorado and other California and Nevada counties bordering Lake Tahoe through the 1980s and 1990s has been a severe cutback in the number of sewer hookup permits issued for new construction, both residential and commercial. Until sewage treatment facilities, which are required to send their effluent outside of the Tahoe Basin, are expanded, development will likely proceed at a snail's pace compared to the boom decades of the 1960s and 1970s. Moreover, residents inside and outside of the basin are increasingly reluctant to support expansion of existing sewage treatment plants or construction of new ones. Of late, city, county, state, federal, and regional environmental and planning agencies (e.g., the Tahoe Regional Planning Agency or TRPA), all of which figure prominently in Tahoe Basin development, also tend to favor slower growth. On the other hand, when such matters come to a vote, real estate developers, chambers of commerce, and construction firms finance campaigns that sometimes lead to the defeat of slow-growth initiatives.

Other than the Tahoe Basin, the only significant growth area in the eastern Sierra is Mammoth Lakes. Although Mammoth Lakes is a year-round alpine play-ground, Mammoth Mountain is the area's main attraction, with 2 gondolas, 3 surface lifts, and 16 chairlifts. Mammoth is the nearest ski resort of this size to Southern California's estimated 500,000 skiers. Of the 10,000 skiers likely to be on the mountain on any Saturday during the season, most are from the Los Angeles metropolitan area. The skiers are weekend visitors, but their presence swells the semipermanent seasonal population by another few thousand—the people who work the lifts, restaurants, inns, motels, gas stations, and stores. To accommodate both weekenders and semipermanent residents and to beat the first snows of winter, the building of condominiums and other housing units has been going on at a fever pitch during many a recent offseason, especially since the local water district secured rights to additional water from Lake Mary that resulted in the lifting of a building moratorium in 1978.

This is the latest, and perhaps biggest, of several building booms to affect the area over the last quarter century. It is certain to keep Mammoth Lakes on the road to becoming one of the Sierra's largest and most congested cities. Lack of sufficient housing and parking facilities, inadequate access from U.S. 395, sewage disposal problems, and negative environmental impacts of all sorts are flies in the ointment of expansion that can be overcome with viable regional planning. However, another drought like that of the mid-1970s or 1986–92, when at times there was hardly enough snow to ski on, would slow the boom no matter what else was done.

Figure 3-29 *Empire Mine, Grass Valley. This northern Mother Lode mine produced nearly $1 billion worth of gold. Nearby Grass Valley lies along busy California 49; its population recently passed the 9,000 mark.* (Crane Miller)

Recent development at Mammoth and Tahoe notwithstanding, most of the Sierra's permanent population remains situated in numerous towns and small cities strung out along the length of the western foothills. From Greenville, Quincy, and other Feather River country towns in the north through Nevada City, Grass Valley, Auburn, Placerville (formerly Hangtown), and Sonora in the Mother Lode country to small, southern foothill recreational centers like Bass Lake, Shaver Lake, and Lake Isabella, no single community claims more than 11,000 residents, and most contain less than half that number. Many of the Mother Lode communities originated as mining camps during the Gold Rush, faded into obscurity once local placer deposits ceased yielding gold (Figure 3-29), and then revived in the present century as agricultural, lumbering, retail trade, retirement, and tourist centers. Away from the towns are increasing numbers of homes occupied by retirees and urban workers who commute to such cities as Sacramento. In fact, it was settlement in the unincorporated areas of the Sierran counties of Amador, Calaveras, and Nevada that helped render them the third, fourth, and fifth fastest-growing counties in the state between 1980 and 1990 (see Table 1-2).

Sierran Native American populations are small and scattered almost exclusively along the western foothills. Situated southwest of Sequoia National Park, the Tule River Indian Reservation, with nearly 800 residents and covering 55,356 acres, is the state's second largest reservation in area. Most of the rest of the native settlements of the region are relatively small rancherías. Two hundred years ago, upwards of 150,000 Native Americans lived in the Great Central Valley and the Sierran foothills. But in the wake of sweeping agricultural development, the hunting-and-gathering Maidu, Miwok, and Yokuts vacated the valley for the hills. Water, oak acorns, other seeds, fish, and game were plentiful in the foothills and became even more so as domesticated crops and livestock displaced the native flora and fauna.

One of the few remaining vestiges of aboriginal use of the natural landscape is at Chaw'se Indian Grinding Stone State Park in the Mother Lode country, where oak acorns are still ground by the Miwoks in holes hollowed out of limestone outcroppings. The 135-acre site is the only one in the state park system devoted to Native American culture.

THE GREAT CENTRAL VALLEY

Spanning 40–80 miles from east to west and 400 miles from north to south, the Great Central Valley's dimensions match those of the Sierra Nevada. But the two landform provinces bear no other resemblance. The Great Valley, Central Valley, or Sacramento–San Joaquin Valley(s), as this heartland of California is variously called, is flat as a pancake—on a typical hazy day not unlike Iowa or Nebraska, where mountains are nowhere in sight. The two great river systems that drain the Central Valley give it even more of a midwestern look, with the Sacramento from the north and the San Joaquin from the south meandering along somewhat like the mighty Mississippi or the wide Missouri.

Whatever the Great Valley lacks in scenery compared to the Sierra or in size compared to the Midwest, however, it more than makes up for in its agricultural landscape. No other single area of 25,000 square miles in the nation—or the world, for that matter—can come close to matching the Great Central Valley in terms of value of agricultural output, variety of crops grown, and crop yields per acre. And, beyond all that, the valley is rich in hydrocarbons—oil and natural gas.

Fold and Fill

The geologic explanation for the existence of this broad, productive valley plain in the midst of the Pacific mountain system begins some 200 million years ago with formation of a deep structural trough in the earth's crust. What is now the Great Central Valley was then a deepening geosyncline underlying the submarine western continental shelf of the North American Plate. As shown in Figure 3-25, the volcanic predecessor of the Sierra Nevada and the Klamath Mountains next began to rise from the sea, sending ever-increasing amounts of eroded materials into the down-fold. Marine sedimentation helped fill the trough to great depths—more than 40,000 feet below sea level in some places.

In time, the ancestral Coast Ranges were uplifted by the onslaught of the Pacific Plate; the Sierran granitic batholith tilted ever closer from the east; the last of the seas retreated (about 3 million B.P.); the great rivers covered the surface with alluvium and glacial outwash brought down from the mountains; and the sediment filled trough became the contemporary Great Central Valley depicted in map and cross-section in Figures 3-30 and 3-31. Today, the alluvial soils provide an essential basis for the world's most productive agriculture, and the older marine sandstones and shales yield oil and gas in amounts sufficient to keep California in the forefront of national production.

California's Cornucopia

The Great Valley is predominantly a land of prosperous, irrigated farms, as well as large and small cities that accommodate some 4 million residents. Urban settlement is strung out mostly along California 99 in the middle of the San Joaquin Valley and the newer I-5 and California 99E on both sides of the Sacramento Valley. Away from the beaten paths, small farm towns too numerous to mention dot the checkerboard farmscape seen from the air.

The parallel and perpendicular county and state roads, fence lines, windbreaks, irrigation channels, property lines, and utility line easements are all uniformly laid out in accordance with the grid pattern of the township, range, and section lines of the U.S. Rectangular Land Survey. The 160-acre *quarter-section* homestead (Figure 3-32 became the basic unit for the original agricultural subdivision of the valley. At present, legal title to virtually all the farmland in the Central Valley, as well as elsewhere in the state, is identified according to the rectangular survey (see Figure 3-21).

The quarter section is the basis for another, quite different form of federal regulation, namely, the 1902 statute that places a 160-acre limitation on the use of federal irrigation water. More recently, the limit was increased to 1.5 sections, or 960 acres, following decades of consolidation of farms into ever-larger, more productive units. To get around the limit and thereby continue to use relatively cheap, federally subsidized water, some large-farm operators divided their holdings into 960-acre units and put them in the names of relatives. However, in 1993, environmentalists won a lawsuit seeking stricter enforcement of limits on federal irrigation water use. The ruling will benefit small-farm operators, wildlife, and especially cities, which now pay dozens, and even hundreds, of times as much for water as do large agribusiness firms. Yet, if larger-farm operators are forced to break up their holdings and/or revert to greater use of privately owned well water, the rural character of the Central Valley will undoubtedly change. Moreover, food prices may climb as the economies of large-scale farm operation diminish. (A fuller discussion of the Central Valley project and the ramifications of greater groundwater use can be found in Chapter 5.)

The river systems, as well as the road and railroad networks, have profoundly influenced urban settlement

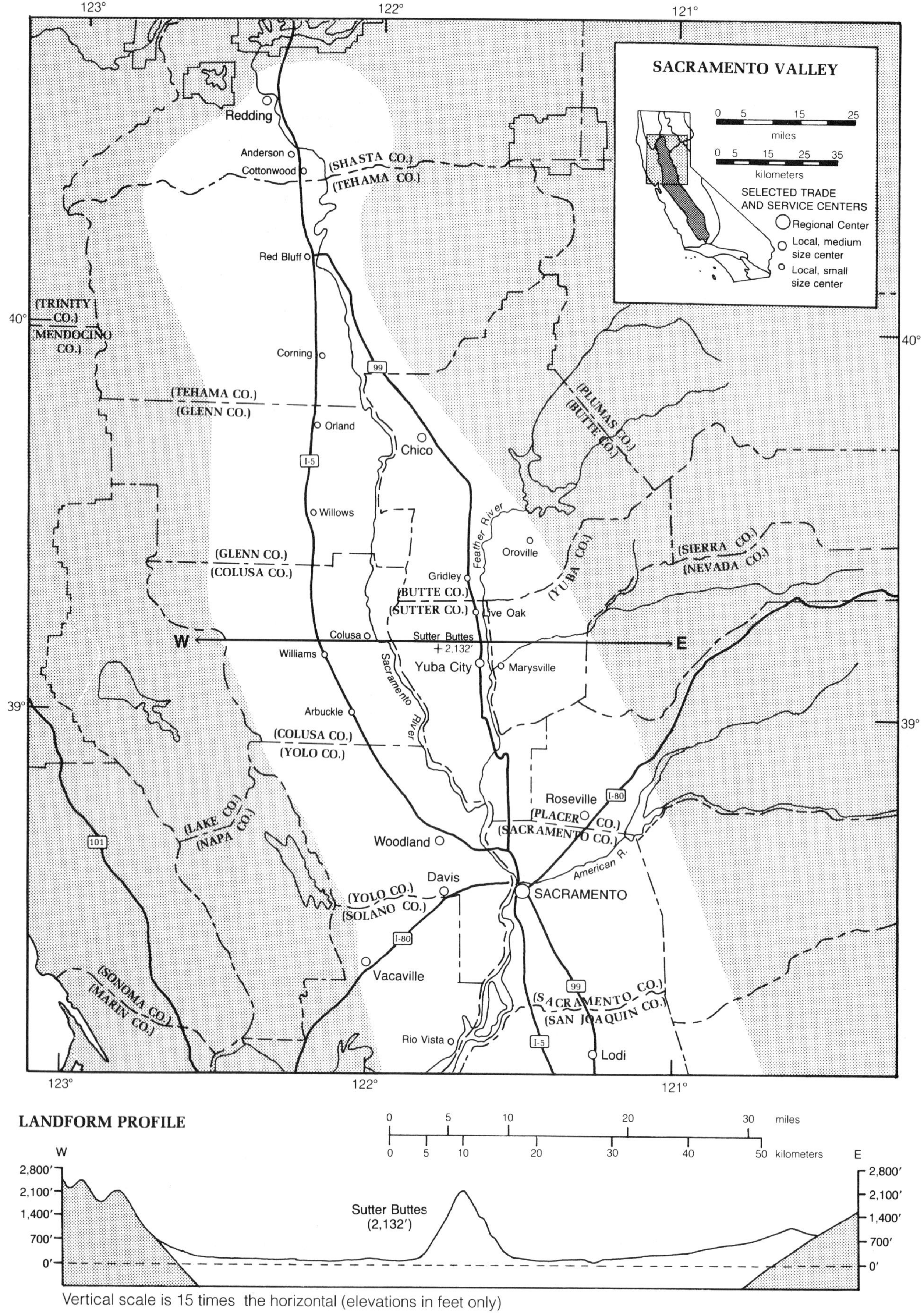

Figure 3-30 *Sacramento Valley and landform profile.* (Richard Crooker)

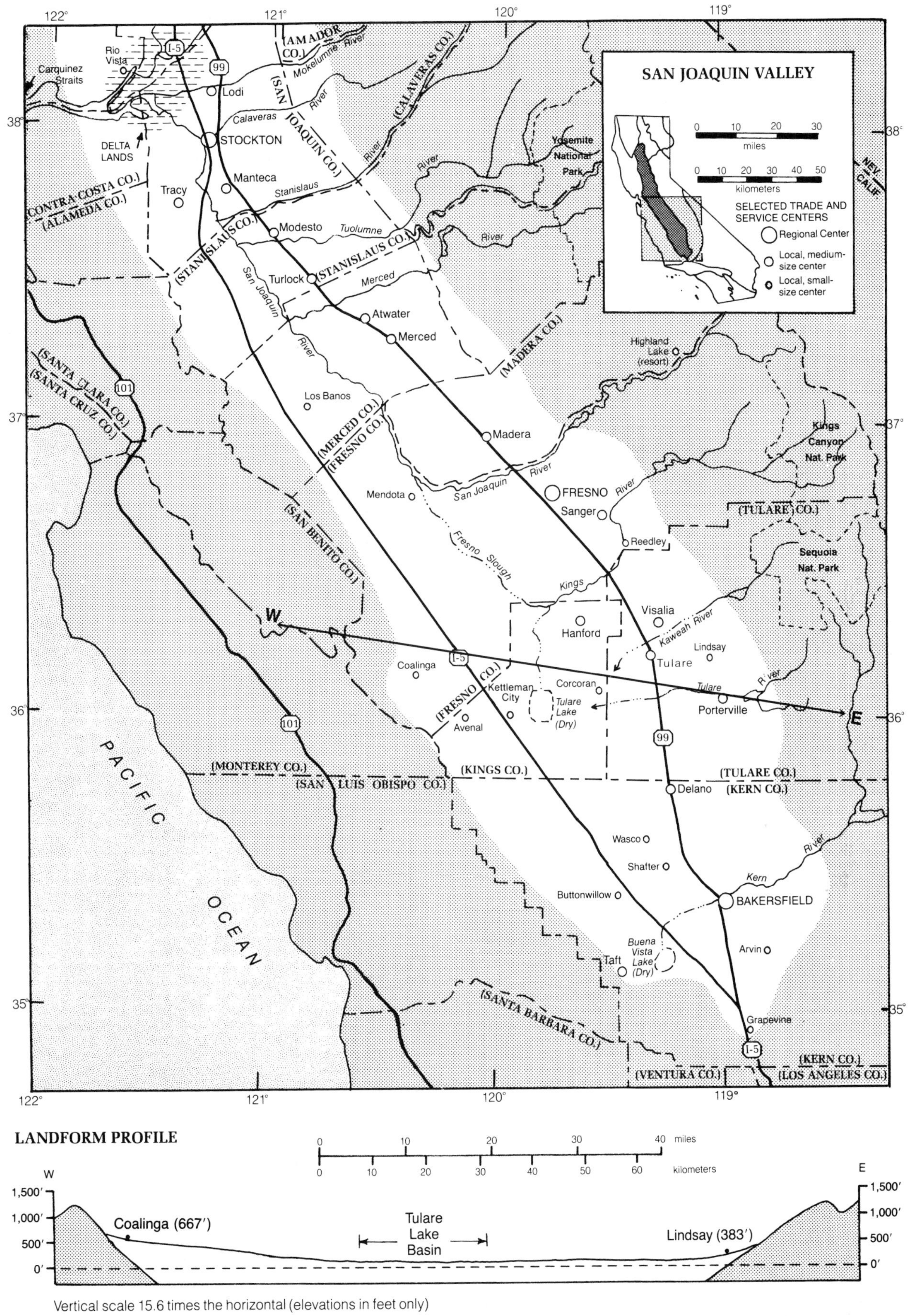

Figure 3-31 *San Joaquin Valley and landform profile.* (Richard Crooker)

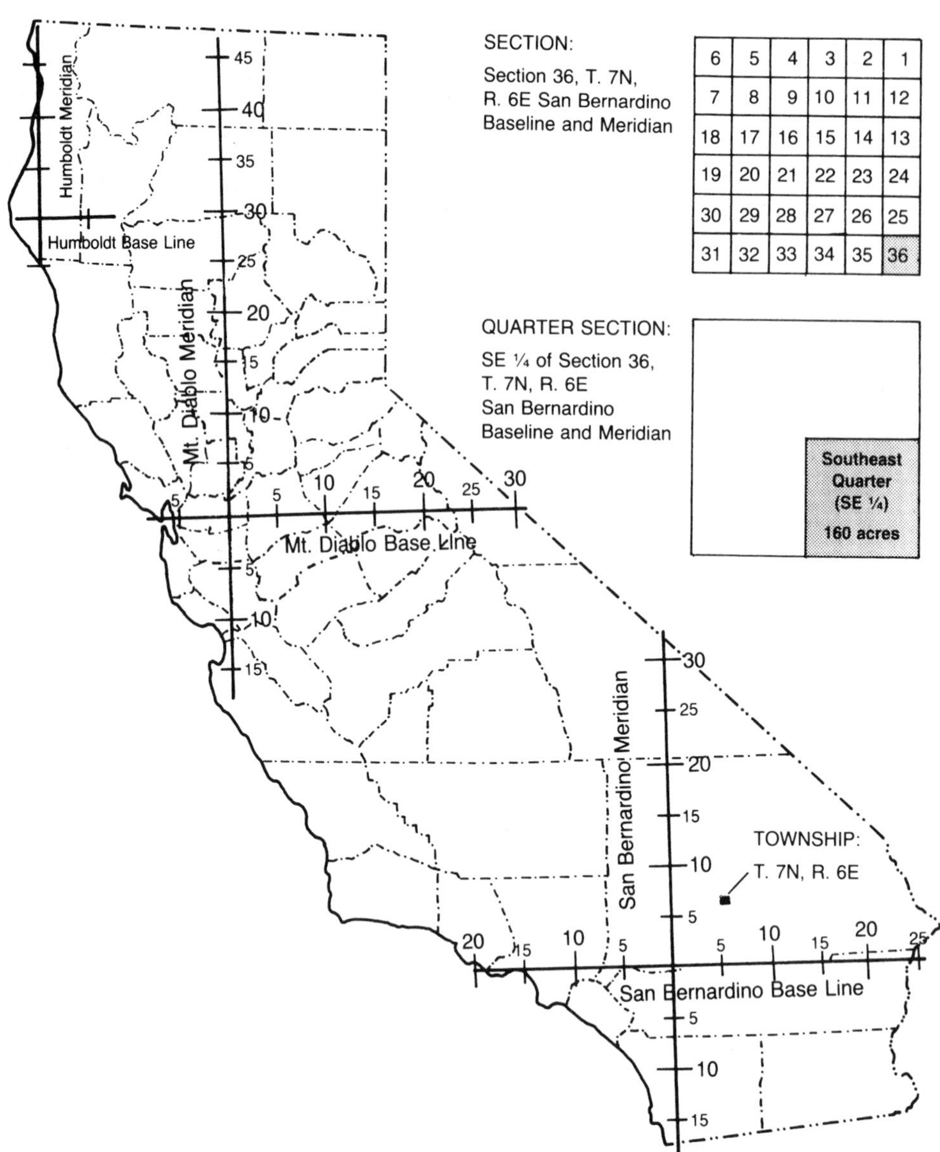

Figure 3-32 *U.S. Rectangular Land Survey.* (Richard Crooker)

patterns in the Great Valley. The preference for riverside sites is seen in cities from north to south in the valley: Redding and Red Bluff along the upper Sacramento River; Yuba City and Marysville opposite each other on the tributary Feather River; and Stockton on the navigable lower San Joaquin River—now including the Port of Stockton with the help of the manmade Stockton Deep Water Channel. Both the Sacramento and San Joaquin rivers drop only about 500 feet on their long, gentle courses through the valley and thus afford countless opportunities for navigable access to the sea.

Nowhere, though, are the advantages of riverside location better displayed than at the confluence of the Sacramento and American rivers. It is here that Sacramento, with a population now approaching 400,000 within its corporate limits and over 1 million in the greater metropolitan area, grew to become the Central Valley's largest city. Since its beginnings as the boat landing for Sutter's Fort, Sacramento has enjoyed access to San Pablo and San Francisco bays via steamers plying the Sacramento River. Navigable access to San Francisco Bay and the Pacific, coupled with the city's juxtaposition between the bay and the Mother Lode and its selection as the state capital during the Gold Rush, gave Sacramento quick and permanent ascendancy. Barely above sea level, riverfront Sacramento suffered the ravages of seasonal flooding until levees were built.

Old Town Sacramento eventually gave way as the central business district to an area several blocks east of the river, where the capitol building and a park were built. The river likewise relinquished commercial prominence as transcontinental railroad lines came on the scene starting in the late 1860s, followed by twentieth-century U.S. and interstate highways. Although the river indirectly experienced revival with development of the Sacramento Deep Water Ship Channel and the Port of Sacramento (Figure 3-33), both just west of the city and the river, the channel opened Sacramento up to

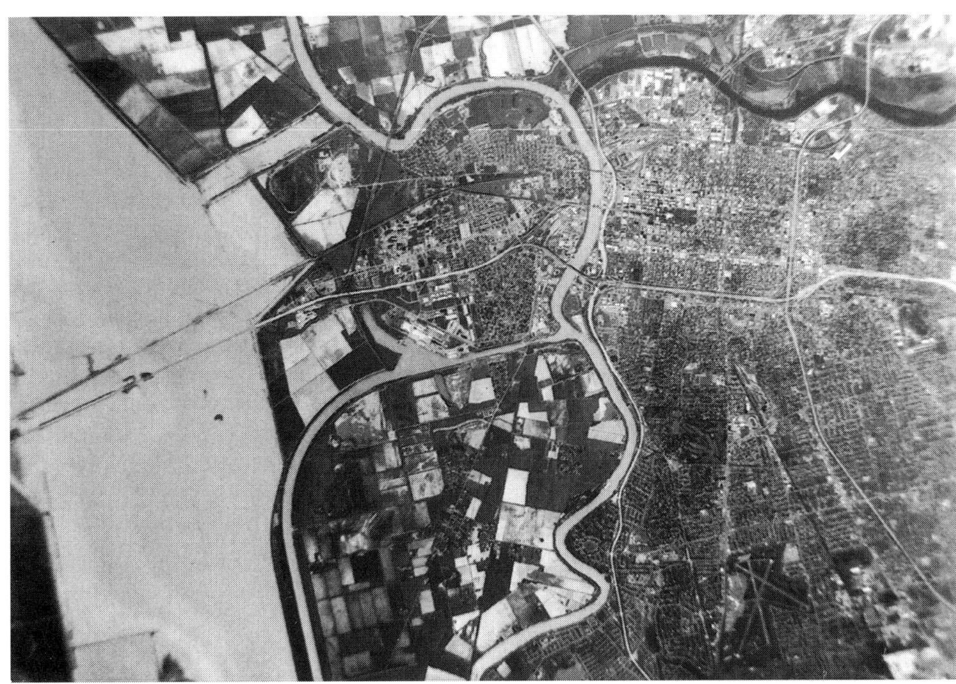

Figure 3-33 *Aerial view of the Port of Sacramento, the terminus of the Sacramento Deep Water Ship Channel (in the center of the photo). Rice, timber products, and other goods move by boat and barge out of the port. However, due to a relative lack of depth, the channel and harbor cannot hope to compete with California's other major ports in berthing container and other deep-draft vessels. Thus, traffic in both large, seagoing ships and barges has diminished significantly.* (NASA)

ocean-going vessels. As the transportation hub of California's prosperous heartland, the confluence of two great rivers, and the possessor of a wealth of developable land, Sacramento has earned a place of preeminence among the state's cities.

Except for Stockton, which claims third place in valley population ranking with 400,000 metropolitan area residents (including the cities of Lodi and Mauteca), San Joaquin Valley cities have developed with little or no dependence on river transportation. The Merced, Tuolumne, Stanislaus, and other Sierran tributaries of the San Joaquin River have figured prominently in the development of irrigation agriculture in the region. Fresno, for instance, emerged 12 miles south of the San Joaquin River as the food processing and distribution center of the San Joaquin Valley. Its metropolitan area, which includes all of Fresno County (the nation's most productive agricultural county by value), contains the Great Central Valley's second largest population, nearly 650,000. Fresno's central location in its own county, the San Joaquin Valley, and the state affords it many advantages, as well as agribusiness fame. The city is also prominent as a trucking, rail, and air terminal, and it is a convenient stopover and even residence for Californians who frequently travel north and south. And, as Californians who pay federal income taxes well know, Fresno is now an Internal Revenue Service center.

The city functions as a western gateway to Kings Canyon–Sequoia (California 180) and Yosemite (California 41) national parks, China Peak and Huntington Lake (California 168), and other Sierran resort areas. Unlike Sacramento, though, Fresno has no direct access through the Sierran landform barrier to the interior Great Basin states. Fresno's Highway 99 neighbors to the north, principally Modesto with 165,000 residents and the smaller cities of Merced (60,000), Madera (30,000), and Turlock (45,000) perform agribusiness, light industrial, retail, and other urban functions similar to those found in the larger city, but on a smaller scale.

None of the three counties—Kern, Kings, and Tulare—south of Fresno are naturally drained by the San Joaquin River system; rather, water spills out of the southwestern Sierra via the Kaweah, Kern, Kings, and Tule rivers into irrigation canals, groundwater basins, and basins of interior drainage. Buena Vista and Tulare lakes catch what is left of surface runoff, although some flow of Kings River is diverted away from Tulare Lake into the San Joaquin River via the Fresno Slough. In times of excessive runoff, as in the winters of 1969 and 1978, Kings River has surmounted manmade levees and temporarily made Tulare Lake the largest freshwater lake in the state. Though important to irrigated agriculture, these rivers were obviously never thought of as routes to the Pacific or even anywhere closer, and so their transportation significance was nil. Nevertheless, the southern San Joaquin Valley's largest city, Bakersfield, was founded over a century ago along one of these rivers—the Kern. Here, it joined with two pioneer wagon and cattle trails (from Los Angeles via Tejon Pass and from Mojave via Tehachapi Pass, both over the Tehachapi Mountains) and a railroad (also over Tehachapi Pass). The waters of the Kern, now diverted upstream into Kern River Canal, were then vital to Bakersfield's early growth as a cattle and railroad town.

In times of prolonged drought, such as during the 6-year spell that finally ended in the winter of 1992–93,

Buena Vista and Tulare lake beds dry out and become sources of deadly dust storms. Not only can the wind-whipped, blinding lake sands and silts cause chain-reaction vehicle pileups, as occurred late in 1991 when nearly 100 vehicles crashed and 17 people were killed on I-5, they can also spread the spores of the dreaded "valley fever." During the 1977 drought (see Chapter 4), for instance, winds stirred up 25 million tons of dust in the dry lake beds and carried some of it and the lung disease as far away as the San Francisco Bay Area. Equally blinding and lethal on the highways of the Great Central Valley are the tule fogs, which also are examined in the next chapter.

Late in the nineteenth century, the Kern River oil field began production, soon followed by others in the vicinity converting Bakersfield into an oil town. Agricultural development followed on the heels of petroleum discovery, and the city became a service center for both industries, reaching a present-day population of 175,000. Other southern San Joaquin Valley cities are numerous but modest in size, with Visalia (75,000) being the largest and fastest growing.

Anyone who has ever taken the long, boring drive on I-5 knows what to expect from the west side of the San Joaquin Valley: much newly irrigated cropland, a sprawling cattle feed lot (Harris Ranch), highway patrol cars, some oil fields, an aqueduct or two, a few scattered gas station–coffee shop–fast food clusters, and no cities. To some drivers, the flat land east of the highway and the low rolling hills of the Coast Ranges to the west add up to millions of acres of nothing. This lack of scenery and settlement continues up to the Delta lands, where the San Joaquin and Sacramento rivers meet.

If one leaves the dreariness of I-5 and ventures into the Delta lands, however, the sights improve: The *distributaries* (two or more rivers originating from a single river) of the two rivers have created myriad small, flat islands where truck farming takes place on reclaimed land and small boat landings serve fishermen, hunters, and water-skiers. Seasonal river flooding occasionally causes crop and property damage, and tidal bores increase ocean saltwater intrusion. Except for the narrow Carquinez Strait gap, however, the unique inland delta is protected by the Coast Ranges from direct exposure to the sometimes stormy Pacific.

Led in population by Rio Vista with some 3,500 inhabitants, towns in the Delta are few and small. A producing natural gas field at Rio Vista broadens the economic base of the town. Natural gas fields in the valley extend from the Delta lands northward to Sutter Buttes in the upper Sacramento Valley.

The prehistorical geography of the Great Central Valley, which is discussed in Chapter 7, differed markedly from what we see today. As a geomorphic province, the region contained the largest of the aboriginal language families in population, the Penutian (see Figure 3-19), and one of the largest tribes, the Yokuts, whose pre-1769 population may have numbered more than 15,000. Just as the valley is now the breadbasket of the state, its bunch grassland–herbivore–carnivore food chain was then the most abundant available to the hunting and gathering tribes of California. As mentioned earlier, however, the spread of agricultural settlement all but did away with the native landscape and its people. By 1910, the state's total Native American population had declined to a scant 17,000, and the Yokuts to barely over 500.[4] As for the San Joaquin and Sacramento valleys, there are no extensive Native American land-holdings left except for the western part of the Tule River Indian Reservation and a few rancherías.

Just as California's landforms influence human settlement patterns, so are they passive players in determining weather and climate. Be they the great granite walls of the Sierra Nevada, which capture copious amounts of rain and snow from incoming winter storms, or the low-lying coastal headlands and lagoons of the Peninsular Ranges, which allow cool Pacific breezes to penetrate well inland and ameliorate otherwise hot summer days, landforms passively but continually contribute to a variety of short- and long-term atmospheric conditions throughout the state. As the next chapter discusses, California's weather and climate are neither uniformly monotonous nor lacking in seasonal change. Indeed, the diversity of the land surface below ensures the dynamics of the atmosphere above.

[4] Today, Native Americans statewide number more than 250,000, but many of them are from outside of California and less than 10 percent reside on reservations and rancherías.

CHAPTER

From Droughts to Downpours

Anyone who lived in California during the 1990s could no doubt recall some of the many interactions among a fickle climate, fluctuating water supply, and uncertain energy supply. The start of the decade found California in its fourth straight year of drought, with winter precipitation dwindling to a mere fraction of normal and spring–summer–fall weather remaining as clear and dry as it had always been. By the winter of 1992–93, though, the rains had returned to the lowlands and the snows to the mountains, and once again, the state had persevered through a prolonged dry spell. There was plenty of water for producing hydroelectricity, irrigating crops, meeting industrial and recreational demands, watering lawns and gardens, washing cars, and even filling and later washing restaurant water glasses. Energy was cheap and seemingly abundant as hydroelectric output met about 20 percent of the state's power needs and fossil and nuclear fuels took care of most of the rest. And clean-burning natural gas, what little of it was required in California's normally mild winters, heated homes, offices, factories, and classrooms ever so economically. In all, California's climate, water supply, and energy production were performing about as expected, and the future appeared promising.

Although heavy precipitation early in the winter of 1997 brought floods to the Central Valley and heavy snowfall to the Sierra Nevada, much of California experienced a relatively dry winter with only scant precipitation later in the season. Yet, the recurrence of *El Niño* (an unusual warming of ocean waters) in the eastern Pacific later in 1997 and into 1998 was partially blamed for increased precipitation along the West Coast from British Columbia to California.

Through all the downpours, though, memories of the droughts of 1976–77 and 1986–92 lingered on. The earlier drought was short-lived, but its effect was compounded by an unprecedented rise in petroleum prices and accompanying shortages. Admittedly, the energy crisis was fashioned by international events, but the drought aggravated the problem in California, for as rivers and reservoirs dried up, so did hydroelectric production. Hydro's share of the state's total electrical energy production years ago was predicted to decline to about 10 percent by the year 2000, but certainly not to less than 5 percent as early as 1977. The unforeseen drought that caused hydro's demise could not have come at a more inopportune time. Petroleum imports and prices were skyrocketing; local reserves of natural gas were dwindling to near depletion; geothermal development was struggling along at a snail's pace; and advocates of coal, liquefied natural gas (LNG), and nuclear power seemed light years away from resolving their differences with environmentalists.

To make matters worse, the drought threatened to ruin California's number-one industry—agriculture—as well as to seriously debilitate most nonagricultural enterprise. Even the best-laid schemes of water resource experts offered little solace as communities resorted to importing water by truck when municipal reserves disappeared and the 400-mile California Aqueduct ceased carrying the precious liquid from a once water-surplus Northern to a usually water-deficit Southern California.

As the drought wore on, California's legendary self-contained water resources seemed precisely that.

But in the winters of 1978 and 1979, rain returned to the state and in unprecedented amounts. Almost as if to compensate for earlier shortcomings, weather conditions once again radically changed during the decade. The hydro picture brightened, water rationing ended, and the skiing industry revived. By 1983, for instance, hydroelectricity accounted for more than 40 percent of California's power generation (see Chapter 6). The record Sierra snowpack assured Californians that water would be available for a while, and residents once again returned to their former profligate use of the resource.[1]

During the early 1980s, precipitation and temperature conditions returned to normal. Later in the decade, though, the state once again experienced a drought, and as the 1980s ended, California was entering its fourth successive dry year. An increasingly rainless climate became all the more hostile when a record cold wave swept the state in the winter of 1990–91. California's $8-billion-a-year fruit and vegetable crop bore the brunt of the first blast of arctic air in late December when temperatures dipped far below freezing for several nights in succession throughout the Great Central Valley and other agricultural lowlands. Crop losses were unprecedented. For example, nearly 80 percent of the winter-harvested navel orange crop was destroyed, and varying degrees of damage befell other citrus and the strawberry, avocado, carrot, broccoli, cauliflower, celery, and lettuce crops. The protracted drought of 1986–91 had already diminished citrus and other tree crop production such that the 1990–91 winter freeze became the proverbial straw that could break the camel's back. Many farm owners were ruined financially, thousands of farm workers lost their jobs, and the retail prices of affected commodities rose sharply—all in the wake of hundreds of millions of dollars in crop damage from the freeze. Some farmers would reflect back on that dry, cold, ruinous winter by remarking that they had read about global warming but were now convinced that the next ice age had come to California. And to make matters even worse, a new Mideast crisis surfaced in 1990–91 and oil prices climbed back to the levels of the late 1970s.

Meanwhile, California's most abundant climatic resource—sunshine—came into sharper focus in the 1980s and 1990s as an alternative energy source. Collecting solar energy for water and space heating was nothing new to some California homeowners, but ways of directly converting the sun's energy into electricity for mass distribution were now at hand. What better place is there in the nation than California's sun-drenched deserts, after all, to locate solar power plants? And a solar energy fringe benefit exists in the atmospheric circulation patterns of the deserts as wind power.

Solar and wind power technologies ultimately may evolve to the point of competing commercially with coal and nuclear power as the principal surrogates for oil, gas, and hydro energy sources. Together, these last three met some 60 percent of the state's power needs, but now they provide a far smaller proportion. Given a rapidly changing energy picture, an environment long plagued by air pollution, a warm and sunny climate, a population outnumbering that of any other single state by millions, and an increasingly automated way of life, it is clear that California must develop its unique solar energy resource to the fullest potential.

THE LURE OF MILD CLIMATE

Placed against this mixed backdrop, the traditional lure of the Golden State has relied not upon availability of water and energy, but rather upon an irresistibly mild climate and a romantically embellished image. As Chapter 1 detailed, the romance of California has been promoted by many people for many reasons. A persistent element of this mystique, however, has remained the notion of "forever summer" weather. Rightly or wrongly, the climate has provided a consistent measure of value for immigrants to the state.

The Legacy of Horace Greeley

When Horace Greeley issued the now-famous dictum "Go west, young man, go west," he could not have hit upon a more appropriate theme for California.[2] In addition to economics and adventure, however, climate proved to be a dominating motive drawing tourists and immigrants to the state, and especially to the southern portion. To a native of New York or Minnesota, the weather was a source of amazement and delight, producing a sense of wonder that was conveyed in letters, books, advertisements, and lectures. Some went so far as to proclaim marvelous cures for all ailments that

[1] Here it is pertinent to note that a short animated film entitled *Water Follies: A Soak Opera* was produced by the Denver (Colorado) Water Department. It is suitable for all ages and gets the water conservation message across in a unique yet effective way.

[2] Greeley borrowed the phrase from J. B. L. Soule, who had first published that advice in the *Terre Haute Express* in 1851. Due to Greeley's greater prominence, however, his restatements of this advice became better known and credited, especially since he was able to publish it through the pages of the *New York Tribune*. Thus, Greeley is more widely credited with this directive than is Soule.

would soon accrue to pilgrims if they would but bask in the climate of the state.

Thus, by the early 1900s, Southern California had acquired a reputation as a Shangri La for the infirm, invalid, and afflicted, as well as the able-bodied and the ambitious. As Carey McWilliams recalled, the classic, if somewhat cynical, motto of the times became: "We sold them the climate and threw in the land." The draw of climate and claims of health benefits constituted part of the westward appeal of California to the rest of the nation. More than health considerations alone drew many immigrants to the state, however.

Midwestern Graffiti

Various scholars and observers have concluded that Southern California generally owes its particular ambience to the dominant influence of displaced midwesterners. Indeed, as early as the 1930s, a popular saying identified Southern California as the western seacoast of Iowa. The saying might well have substituted Illinois, Ohio, Nebraska, or any other of the midwestern states, since the number of native-born Californians seldom accounted for a very large percentage of the total population. Annual Iowa-day picnics or Ohio-day reunions were once common occurrences in the state, providing an opportunity to swap reminiscences of life "back home." Although such reunions became less common after the late 1950s, the immigration patterns certainly continued.

An interesting cultural effect derives from the fact that this midwestern migration has been rather constant. The continual flow of new residents into the state has created a layered form of cultural identity wherein the newcomers strive to develop characteristics that may be attributed to the "California personality" but in the process create an overall sense of never-ending growth and development with a midwestern twang. As is discussed more fully elsewhere, this national and regional cultural diffusion is rapidly being augmented by a liberal seasoning of international influences and cultures as well.

The Grapes of Wrath

California's role as a key agricultural state has also contributed to its steady appeal. Frequently, when poor weather or economic conditions created untenable situations in other parts of the country, California came to be viewed as an escape—a sunny, Mediterranean-like setting where food could be grown at least most of the year and the climate would be hospitable in the meantime. Having lived through brutal winter blizzards or scorching dust storms, was it any wonder that immigrants to the area saw a chance to improve their lives? The exaggerations of the fecundity of the soil and climate were misleading and often extreme, but there was just enough truth in the myths to encourage people to continue to see California as the fulfillment of unrealized dreams. John Steinbeck's *Grapes of Wrath* (movie and book) partially questioned the popular image of California, but for the most part the seekers of dreams persisted in chasing their dreams to the West.

In fact, the trend of American immigration into California had a curious twist. Initially seen as a resort for the rich, increasingly the state came to be considered a land of opportunity for the less affluent as well—an image promoted by land companies, railroads, and others. Throughout the early 1900s, this in-migration occurred at a steady rate, with occasional swells in the level of movement. The Depression years, for example, brought well over 350,000 persons across the country in old cars and trucks. As time passed, California increasingly lured working-class people from such diverse places as Ontario, West Texas, New York, and Oklahoma. Much of this continual flow, however, could still be traced to the small towns and farms that looked to California as an enchanted land where orange and palm trees grew in abundance and even country folk could share in some of the magic. And in modern times, California has also lured many more foreign and non-English-speaking poor. As demographic patterns demonstrate, Hispanic immigrants in particular look to California for a demonstrably better way of life.

The Resort Mentality: A Place in the Sun

California as a destination for permanent residence is one of the realities that has influenced the culture of the state. There is another facet to the lure of the state and its mild climate, however. With its ideal and varied weather, myriad tourist attractions, and wealth of geographic wonders, California has also been a prime setting for those who wish to escape the harsher climate of their native areas for a short time only.

Many parts of the state cater to a transitory population by providing the atmosphere, amusements, and diversion sought by this clientele. Palm Springs is widely known for its luxury hotels and golf courses and for its mild winter climate. San Diego has created a tourist haven of hotels, aquatic and wildlife parks, beaches, and golf courses; San Francisco is a mecca for visitors with its well-known Chinatown, Fisherman's Wharf, and Golden Gate Park; Lake Tahoe is a tremendous draw; and the growth of time-share condominiums at the seashore and on the ski slopes has brought a new meaning to the term *resort community* in California.

The appeal of the state as a goal for migration or a source of recreation persists. Such influence is clearly not a one-way phenomenon, however. With ever-increasing demands on its resources, power, and environment, the state faces the challenge of living with and understanding both the limits and the abundance of nature.

CLIMATES, MICROCLIMATES, AND CONTROLS

California is commonly said to have a Mediterranean climate. This reputation, which has lured millions of people to California, holds true only for the relatively small coastal portion of the state where most of these millions have settled. California actually has innumerable climates, varying from dry summer-subtropical and alpine-arctic *macroclimates* to highly localized *microclimates*, such as those found deep in a forested valley or high up on a north-facing mountain slope.

To add still more variety, California's climates have seasons—contrary to the belief of many outsiders—and they are not all "seasons in the sun." Every summer the northwest coast is fogbound and cool while the interior deserts, despite frequent thundershowers, swelter in the continent's hottest temperatures. Winters offer equally stark contrasts, with the Sierra and the Klamath Mountains often buried by record snowfalls while the lowlands never see snow. Spring and fall have their extremes as well, as when a hot, dry Santa Ana blows through the Los Angeles Basin on the same day that tule fog paralyzes the San Joaquin Valley.

Although Figure 4-1 depicts six climate zones (arid, semiarid, warm summer–Mediterranean, cool summer–Mediterranean, highland, and alpine), the true scope of California's climates and microclimates cannot be adequately depicted on a one-page map. Thus, a more

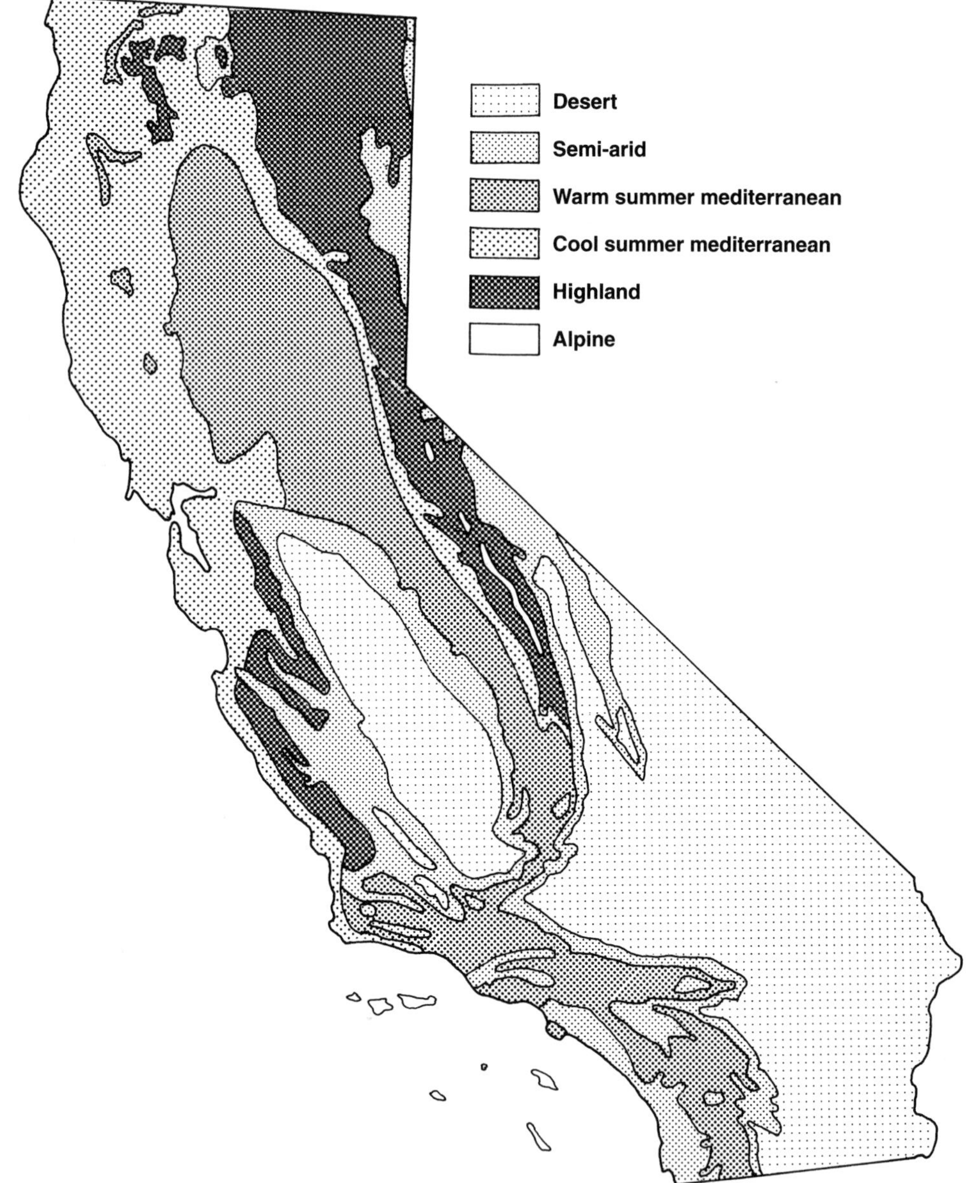

Figure 4-1 *California climates.* (Richard Crooker)

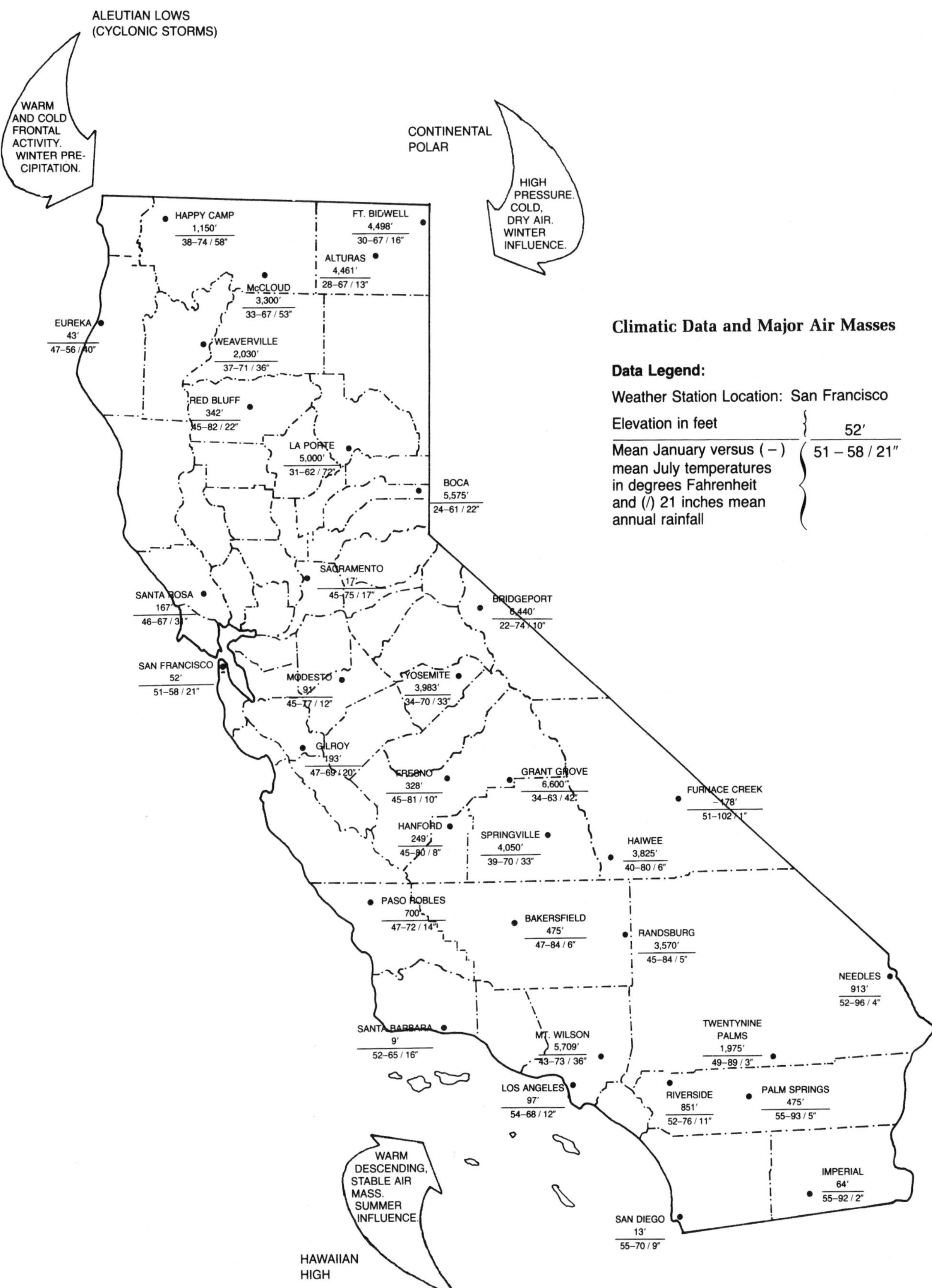

Figure 4-1 *continued*

Figure 4-2 *Orographic lifting over the White Mountains in eastern Mono County near the Nevada state line. Orographic precipitation can be seen occurring here on the western flanks of White Mountain (14,246 feet) and other peaks of elevations higher than 12,000 feet as a moisture-laden air mass moves from west to east (left to right) and must rise over the mountains. Note the increase in clouds and rainfall eastward as the air mass rises and cools and water vapor condenses. Figure 4-5 diagrams the orographic process.* (Crane Miller)

Figure 4-3 *Frontal lifting over the Pacific. In the satellite view over the northeastern Pacific (the clouds have parted over California and the snow-covered Sierra), two storms are identified by clouds swirling counterclockwise and inward into their low pressure centers, as into a vacuum. The northernmost cyclone is associated with the Aleutian Low (see Figure 4-6), which pumps out moist, cold air masses destined for the west coast of North America. Westerly winds push the storms southeastward, with many coming onshore in California in a normal winter. When such cold air invades an area of warm air and forces the warmer air to rise, cool, condense its moisture, and perhaps precipitate, a* cold front *occurs. Such a winter cold front (and the attendant wind-driven storm surf) intensify as they come onshore. Because the Hawaiian High (see Figure 4-6) blocks them from entry into California in summer and because the state is on a west coast and lies in the mid-latitudes, tropical* warm fronts, *where warm air invades zones of colder air, are relatively rare occurrences.* (NASA)

localized look at the state's climatic diversity is presented here in terms of the geographic features that control it. These *climate controls* include altitude, latitude, oceanic influences, air pressure systems, and winds.

Altitude and Air Masses

Whenever air changes altitude, it changes temperature. This in turn causes a change in the temporary state of the atmosphere, or what we call the "weather," climate being the long-term condition of the atmosphere at a particular location. Remembering that air has weight and therefore exerts *air pressure*, consider that as an air mass rises, the pressure upon it lessens and the air expands and cools. As air thins or cools, its ability to hold moisture or water vapor also diminishes, for the water vapor also has weight and therefore exerts *vapor pressure*. Put another way, colder air is drier air, much like the icy air mass that ushered in the cold but dry winter of 1990–91 mentioned earlier in the chapter. If an air mass loses altitude, on the other hand, the pressure around it increases and the air compresses and consequently heats up. The air mass we speak of may be *maritime polar* air (mP) from the Gulf of Alaska or *continental polar* air (cP) from Canada or *maritime tropical* air (mT) from the Pacific. Whatever its source, the incoming air will be forced to rise as it encounters California's mountainous terrain, and mountain-caused or *orographic lifting* (Figure 4-2) will take place. *Frontal lifting* (Figure 4-3), where lighter, warmer air is forced up and over heavier, colder air, and *convectional lifting* (Figure 4-4), where warm surfaces (usually deserts) radiate heat up into a cooler atmosphere, also cause air mass ascent. The latter two are more common in states with less varied topography than California's.

The temperature changes resulting from lifting occur without any infusion or subtraction of heat from the air mass and are termed *adiabatic*. When unsaturated or "dry" air (actually in a *vapor* or gaseous state) ascends or descends, it changes temperatures at a *dry adiabatic rate* (DAR) of 10° C per kilometer or 1,000 meters (5.5° F per 1,000 feet). Once rising air has cooled to its *dew point* (temperature at which the moisture in the air begins to *condense* or convert from a vapor to a liquid state), it will continue to cool at a reduced *wet adiabatic rate* (WAR) varying from 3° to 9° C per kilometer (2° to 5° F per 1,000 feet), assuming further ascent.[3]

Figure 4-4 *Convectional lifting. This occurs where a land surface is being intensely heated, such as in a desert in summer, and warm, relatively light air rises into an upper layer of moist, unstable air. The rising air* adiabatically *(by expanding) cools to its dew point temperature (see Figure 4-5), condenses its moisture, and often precipitates violently. Cloudbursts, sometimes causing flash floods, are most common in the southeastern deserts in summer.* (Crane Miller)

Condensation causes the formation of water droplets, which in turn accumulate to form clouds. Exhaling in chilly air produces the same effect; the moist breath is rapidly lowered in temperature by the cold air to its dew point and forms a cloud for an instant. Of course, the cloud produced by a human breath *evaporates* (from a liquid to a vapor state) almost as quickly as it formed.

The water droplets in the atmospheric cloud, however, may continue to grow and become heavy enough to fall to earth or *precipitate*. If the dew point of the air mass is above freezing, the precipitation will be in the form of rain; if it is at or below 0° C, the cloud will precipitate snow. Cloud formation or condensation gener-

[3] Although television, radio, newspapers, and other media in the United States still widely use the Fahrenheit (F) system of measurement, the metric Celsius or Centigrade (C) temperature scale is favored outside the United States, and the reader should be familiar with its use in studying climate. Thus, the Celsius scale is employed in this section, in Figure 4-2, and occasionally elsewhere in Chapter 4. Formulas for converting from one scale to the other are:

From °F to °C: °C = 5/9 (°F − 32°)
From °C to °F: °F = 9/5° C + 32°

The most used distance and area conversion equivalents are: 1 km = 0.62 mi, 1 mi = 1.61 km, 1 m = 3.28 ft, 1 ft = 0.3048 m, 1 cm = 0.39 in, 1 in = 2.54 cm, 1 sq km = 0.386 sq mi, 1 sq mi = 2.59 sq km, 1 acre = 0.4 hectare, and 1 ha = 2.471 a. The Federal Metric Conversion Act of 1975 calls for conversion to the metric system; but there is no mandatory deadline, so it may be a while yet before California "metrifies."

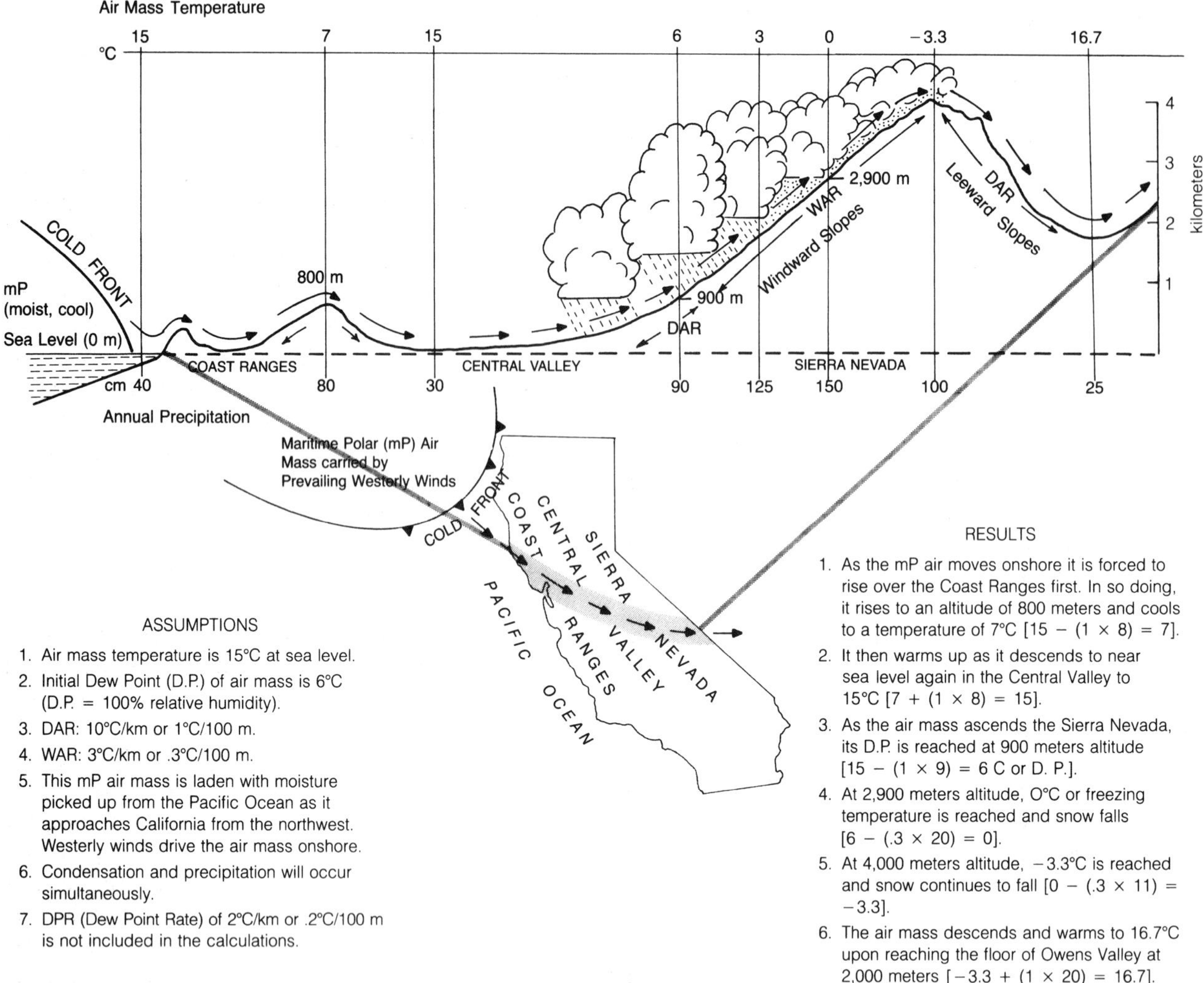

Figure 4-5 *Orographic lifting and adiabatic processes.* (Richard Crooker)

ally precedes precipitation, but the two can occur simultaneously.

Condensation also releases *latent heat of condensation* in the air mass, causing the slowdown of adiabatic cooling from the DAR to the WAR. Eventually, the air mass will begin a descent throughout which it will heat by compression at the DAR. Figure 4-5 illustrates these adiabatic processes at work as a winter storm tracks eastward over the Coast Ranges, Central Valley, Sierra Nevada, and Great Basin. The significance of the Sierra Nevada as a producer of *orographic precipitation* is evident from a comparison of the annual windward and leeward precipitation data given in Figure 4-5.

Mountains exert an extremely localized influence on precipitation wherever they abruptly border a broad lowland or a large valley. For instance, Mount Wilson at an elevation of 5,709 feet receives an average of 36 inches of precipitation annually, whereas at the base of the mountain and just a few hundred feet above sea level, Pasadena gets less than half that amount each year. Both places receives almost all of their precipitation in winter; at Mount Wilson, however, it often falls as snow, whereas in Pasadena and throughout the Los Angeles Basin, it is almost always as rain. Obviously, temperature differences are caused by elevation differences, and they also account for differences in the forms of precipitation emanating from a particular winter storm. It would be grossly misleading to say that it never snows in Southern California, but it would be close to climatic fact to state that it never snows in the Los Angeles Basin. The millions of people who live in the basin like the idea of no snow in their backyard but its availability less than an hour's drive away.

Latitude and West Coasts

California's latitude of from 32.5° to 42° north (N) and its position on the western coast of the continent place it in a unique climatic position in North America. Throughout the earth's continents, wherever there is a west coast between about 30° to 40° N or S, there will also be found a Mediterranean climate. This dry summer–wet winter climate occurs in two relatively small regions of the Northern Hemisphere: here in California and in the southern European and North African lands peripheral to the Mediterranean Sea. There are four even smaller and more widely scattered Mediterranean climate areas south of the equator in Chile, South Africa around Capetown, and Australia in the Adelaide and Perth regions.

In all, Mediterranean climates, including California's, occupy only a scant 2 percent of the world's land surface. It is noteworthy, too, that no such mild climate exists between 30° and 40° latitude on the east coasts of any of the continents, including North America. Why this type of climate exists solely on west coasts will be explained in the following sections.

A latitudinal range of nearly 10 degrees, coupled with a mountainous topography, carries California through a gamut of subtropical and mid-latitude climates and local variations. For example, coastal locations near the 42nd parallel (the Oregon border) are overcast most of the year, receive winter rainfall approaching 100 inches, and seemingly never experience heat or cold waves. By contrast, the San Diego metropolitan seashore area, at about 33° N, has clear days more than half of the year, less than 12 inches of winter rain, and an occasional hot spell. San Diego experiences the drier climate largely because it lies well to the south of the main Pacific winter storm track, whereas the northwest coast is usually right in the midst of it.

In the middle of the coastline, at about 38° N, is the San Francisco Bay Area, which experiences moderate winter precipitation and pleasant year-round temperature conditions. Yet the Bay Area also has microclimates that vary from foggy to dry because of the presence of several mountain ranges, a huge bay, and a large river. The general pattern from the southern to the northern end of the state is one of progressively wetter and cooler climates, with the Bay Area striking a happy medium between the extremes.

California's great latitudinal range also allows for some regional modification of the generally pervasive precipitation regime of dry warm seasons and wet cool seasons. The exception to the rule is the desert southeast, where meager annual precipitation tends to be concentrated in summer rather than winter. Tropical air masses from as far away as the Gulf of Mexico and the southeastern Pacific bring in the summer moisture. Because they usually track considerable distances over land, these tropical air masses tend to drop most of their moisture long before crossing the Colorado River or the Mexican border into California. There is usually enough unspent moisture in them, however, to be convected (by the rising desert surface heat) and condensed into towering cumulus clouds. The thunderheads in turn often produce thundershowers that are sometimes locally heavy enough to cause destructive flash flooding.

On occasion—typically only once every three or four years, and then in August or early September—a *chubasco* (squall) will come into southeastern California from the Pacific and the Gulf of California with the force of a hurricane. Such storms have been particularly devastating in the Imperial and lower Colorado River valleys, where they take lives, destroy crops, and wash away roads and bridges. *Sonoras*, named after the Mexican state over which they travel, track in from more southeasterly directions and more distant origins but can be as destructive as chubascos. Besides high winds, these storms bring brief but often drenching rains that can lead to flash flooding.

Oceanic Influences

More than any other single climatic control, the Pacific Ocean can be said to keep the California coast cool, but not cold, throughout the year. The marine air is cooled by one of the world's major cool ocean currents, the 400-mile-wide California Current, which taps the cold waters of the Arctic Ocean, Bering Sea, and Gulf of Alaska. Ocean currents, like the California, are among the great exchangers of heat and cold between the tropical and polar latitudes and are essential in sustaining heat balances in the global environment. In the case of California, the current literally imports coolness from colder environments and keeps coastal California from getting too hot.

Ocean currents result from a number of phenomena. They are set in motion by prevailing surface winds—the westerlies, in the case of California. The moving air pushes and pulls (drags) the surface water of the currents. Differences in the density (or weight) of water will also cause it to move in currents, as when colder, heavier polar seawater sinks to the ocean floor, spreads equatorward, and displaces warmer, lighter tropical water. The *Coriolis effect*, which on a counterclockwise-rotating earth involves the tendency of objects, winds, and ocean currents in the Northern Hemisphere to veer to the right of their path of movement, also contributes to the motion and direction of ocean and air (wind) currents.

Along continental west coasts like California's, winds and the Coriolis effect often combine to displace surface water seaward from a southeastward-flowing current. Colder water from the depths then moves up to the surface to replace the removed water in a process known as *upwelling*. The colder surface water in turn may drop the temperature of the air above it to its dew point, and condensation in the form of fog may result. Whenever northwesterly winds (blowing toward the land) prevail, which occurs off central and Southern California in late spring and along the northern coast throughout summer, fog is carried onshore.

It is noteworthy that upwelling sustains a significant nonclimatic resource for California as well: its ocean fishery. Upwelling brings nutrients from the cold depths of the sea to the surface, where they are vital to the growth of *phytoplankton* (microscopic plants) and *kelp* (seaweed). The sea plants are the foundation of a marine food chain that involves the phytoplankton being eaten by small fish, which in turn are consumed by larger fish.

Marine air buildup is sometimes amplified by an abrupt change in the coastline's shape or directional orientation, such as south of Point Conception in the Catalina Island embayment. Here the conformation of the coastline and islands cause an *eddy* (counterclockwise swirl) within the main air currents. This Catalina eddy can deepen the fog layer to several thousand feet and cause drizzle in the Los Angeles Basin in the normally dry months of May and June. Water-filled gaps in coastal landforms, such as the Golden Gate and the Carquinez Strait in the central Coast Ranges, allow marine air (sea breezes) to penetrate deep into the interior. Sacramento and Stockton, for example, benefit from their positions downwind from the Carquinez Strait by experiencing somewhat cooler summers than cities elsewhere in the Great Central Valley. Overall, though, the coastal mountain barrier prevents marine influences from affecting the climates of interior California.

The surface waters of the California Current generally range from 50° to 70° F throughout the year, and although never icy cold, they never seem warm enough to satisfy the swimmer who has experienced the 85° waters of Hawaii or Florida. The surfer, on the other hand, commonly wears a wet suit and couldn't care less about water temperature as long as the waves are right. For either person, the best beach weather is in August, when both air and water temperature hover around 70° F and the late spring fogs have long since gone their way, at least from along the Southern California coast.

The fact that the warmest weather starts about 6 weeks after the first day of summer (June 21 or the *June solstice*) is explained by the relatively long *seasonal temperature lag* produced by the Pacific Ocean. It takes large bodies of water much longer to absorb the sun's heat energy (*shortwave solar radiation*) and reradiate it (*longwave earth* or *terrestrial radiation*) than it does dry land surfaces. Consequently, this time lag is somewhat longer along the coast than inland from the sea, where maximum temperatures are reached only about 4 weeks after the maximum high-sun date of June 21 in the Northern Hemisphere.

The same lag occurs in winter, but in reverse, when maximum loss of earth radiation, and thus the coldest temperatures, occurs in late January, or about a month after the low-sun *December solstice* of December 21. Most of us are probably more familiar with *daily temperature lag*, when the warmest part of the day is at about 2 P.M., or a couple of hours after the sun has reached its highest angle in the sky at high noon.[4]

Air Pressure and Winds

Air exerts an average pressure of 14.7 pounds per square inch at sea level. An atmospheric physicist would refer to this air pressure as 1 atmosphere (atm); a meteorologist would say that it is the equivalent of either 29.92 inches of mercury or 1,013.2 millibars (mb) or 101.32 kilopascals (kpa). Air pressure varies from place to place with changes in temperature, altitude, and other atmospheric conditions. For instance, sea-level air pressure will usually vary from a *low pressure* of 980 mb to a *high pressure* of about 1,040 mb, depending on differences in air temperature. Along the equator, where temperatures are relatively high, the surface air is made lighter with warming and therefore rises, creating a zone of *equatorial low pressure*. As the air ascends, it cools off; as it becomes colder, it also becomes heavier. Once the air reaches higher altitudes, the atmosphere acts as a heat exchanger (much as ocean currents do in maintaining global heat balance) by forcing the tropical air poleward about 30°, where its heaviness causes it to sink down into regions of *subtropical high pressure*. In the Pacific between Hawaii and California, the mass of descending air is known as the *Hawaiian High* (Figure 4-6). It is this clockwise-swirling *high-pressure cell* that takes precedence over any of the controls heretofore mentioned in determining California's climates.

Because the subsiding air of high-pressure cells (circular-shaped isobars or lines of equal air pressure, spinning outward) and *ridges* (with isobars that are more elongated in shape) warms by compression and thus in a sense dries out, such high-pressure systems are associated with clear, stormless weather. When high pressure hovers over an area for several months every year, as the northeastern edge of the Hawaiian High does over California every summer and sometimes longer, mild weather prevails. In fall, the Hawaiian High usually begins to weaken and move away from California as it follows the sun southward—the Hawaiian High is normally strongest during the summer, when the sun is high, and weakest in the winter, when the sun is low. As winter approaches, the retreating ridge dissipates to the point of allowing low-pressure cells (circular-shaped isobars, spinning inward) and *troughs* (isobars more

---

[4] The terms *June* and *December solstice* are preferred to *summer* and *winter solstice* because the monthly designations are universal in application, whereas the seasonal ones apply only to the Northern Hemisphere. The principles of radiation not only govern the weather but also are the key to understanding solar energy, a subject that will be explored in Chapter 6.

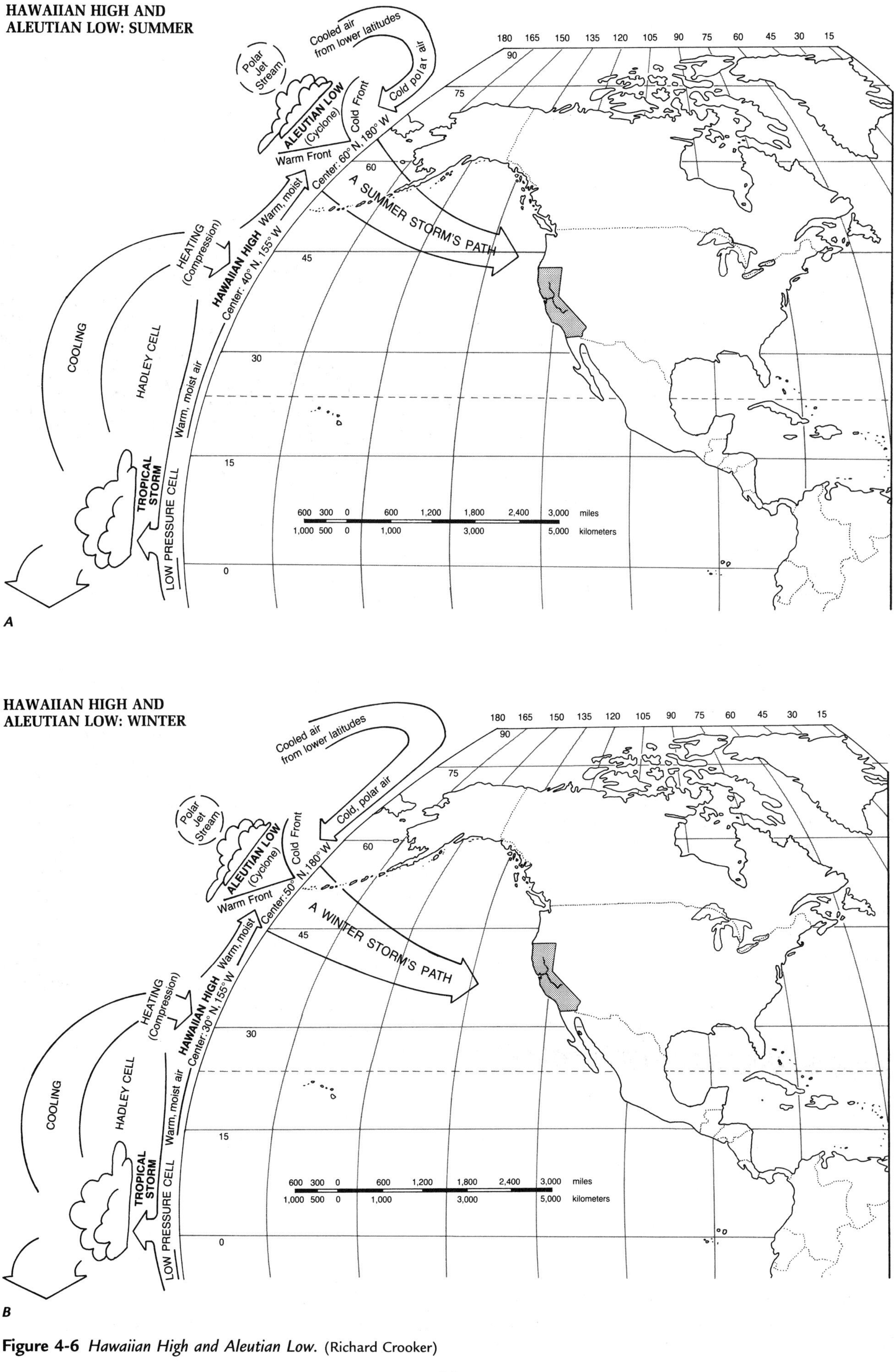

HAWAIIAN HIGH AND ALEUTIAN LOW: SUMMER

A

HAWAIIAN HIGH AND ALEUTIAN LOW: WINTER

B

Figure 4-6 *Hawaiian High and Aleutian Low.* (Richard Crooker)

elongated in shape) to migrate into California. These systems of rising air and condensing moisture usually bring rain if they originate in the central Pacific as mT air masses, or rain and snow if they are pumped out of the *Aleutian Low* (see Figures 4-3 and 4-6) as mP air masses. By spring, the Hawaiian High starts to reestablish itself in the North Pacific and in so doing pushes the storm tracks northward of California.

In some winters, the Hawaiian High happens not to leave the state's shores, and as a consequence California suffers through a drought. During the unusually dry winters of 1975 and 1976, the Hawaiian High not only did not weaken as expected but teamed up with a continental high-pressure cell. Together the two cells became a *blocking high*, rerouting Pacific-born storms far to the north and east of the western United States. The results were record snows back East and drought out West. But no sooner had the state begun to prepare for prolonged drought than the Hawaiian High again changed its movements, allowing countless storms accompanied by record precipitation to hit California in the waning winters of the decade.

Another explanation for the return of the rains in 1978, 1983, 1993, and 1997–98 concerns the advent of El Niño (Spanish for the infant Christ child) off the California coast during those relatively wet years. El Niño, which refers to the invasion of eastern Pacific cold currents (e.g., the California Current) by warm, tropical waters from the equatorial latitudes early in winter (about Christmas time), is often blamed for the abnormal storminess. A slackening of northeasterly trade winds, which blow southwest from the southwestward side of the Hawaiian High toward equatorial low-pressure zones (see Figure 4-6) at 15° to 20° N, apparently affords an eastward-flowing equatorial countercurrent greater opportunity for spreading warm water poleward. There is no question that warming ocean water increases the instability of the atmosphere above it as evaporation accelerates and air pressure diminishes, but whether a 1° or 2° C rise in water temperature will always trigger winter precipitation patterns is uncertain. In the fall of 1990, the ocean off Oregon warmed slightly, and the return of El Niño and a break in the long drought was anticipated, but the drought continued at least in California. True, the Pacific Northwest received drenching, flooding rains late that fall; however, the downpours soon turned to blizzards, and all the inclement weather brought to California was bone-chilling and crop-killing cold. As in previous cold, dry winters, the Hawaiian High had stubbornly held on again. During the winter of 1993, a dome of high pressure built up over British Columbia and Washington and served to divert storms from the Gulf of Alaska (mP air masses) westward away from the Pacific Northwest and then directly southeastward into central California. This time, dry, cold air sent temperatures plummeting in the Northwest while California experienced record-breaking precipitation. El Niño was probably not a factor, for sea temperatures off Oregon and California were about normal that winter. Late in the summer of 1997, though, El Niño returned to California's shores and persisted into the winter of 1998, bringing with it an unusually wet September in the Imperial and Coachella valleys and along the lower Colorado River, a December cloudburst that flooded Laguna Beach and Huntington Beach, copious amounts of January snow to the ski resorts of the Tahoe Basin in the Sierra Nevada, and record-breaking February rainfall for coastal Northern California.

The behavior of the Hawaiian High is difficult to predict, let alone explain. If some answers can be found—for example, through further study of the relationship of sunspot activity to changes in atmospheric pressure—we will be closer to more accurate prediction of long-term climatic change, as well as better forecasting of day-to-day weather.

Differences in air pressure from one place to another—as, say, from the Hawaiian High at 1,030 mb in the Pacific Ocean to a thermal low at 990 mb over the Great Basin—provide the basis for the horizontal surface circulation of air or winds. In the winter of 1975–76, the sinking air of the ocean high was spreading out toward the heated, rising air of a desert thermal low. The low-pressure cell acted like a vacuum as it pulled air counter-clockwise (in the Northern Hemisphere) into its center. The resultant wind, however, did not flow in a straight line directly from high to low, but instead moved roughly parallel to *isobars* (lines of equal barometric pressure such as that at 1,030 mb) because of the Coriolis effect's counterbalancing of the *pressure gradient*. California's prevailing resultant winds, the *westerlies* (winds are named for the direction from which they come), emanate throughout most of the year from the dry northeast side of the Hawaiian High.

The Hawaiian High and its westerly winds are augmented in their effort to keep the coastal California climate mild by the *sea breeze*. The sea breeze is especially welcome in summer, when it is often strong enough to penetrate well inland. A typical summer day will start out with the land heating up more rapidly than the ocean. A heat or thermal low deepens over the land while high pressure builds over the water. The cool California Current and the subsiding air of the Hawaiian High increase ocean air pressure all the more; by midmorning, the sea breeze is beginning to flow onshore, as illustrated in Figure 4-7. At night, the local pressure cells reverse position as the land cools off at a faster rate than the sea. The offshore flow is known as a *land breeze*.

The most significant deviation from the prevailing westerly wind and sea breeze circulation pattern occurs with the advent of the *Santa Ana*, a hot, dessicating *easterly* that sweeps down from the high deserts of the Great Basin into the Southern California coastal lowlands. Al-

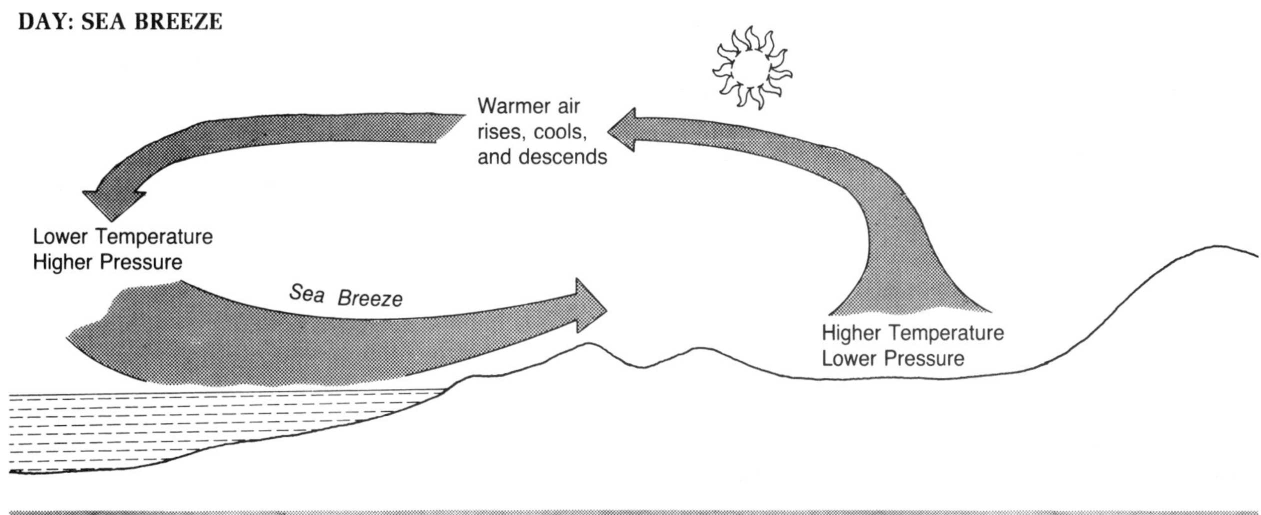

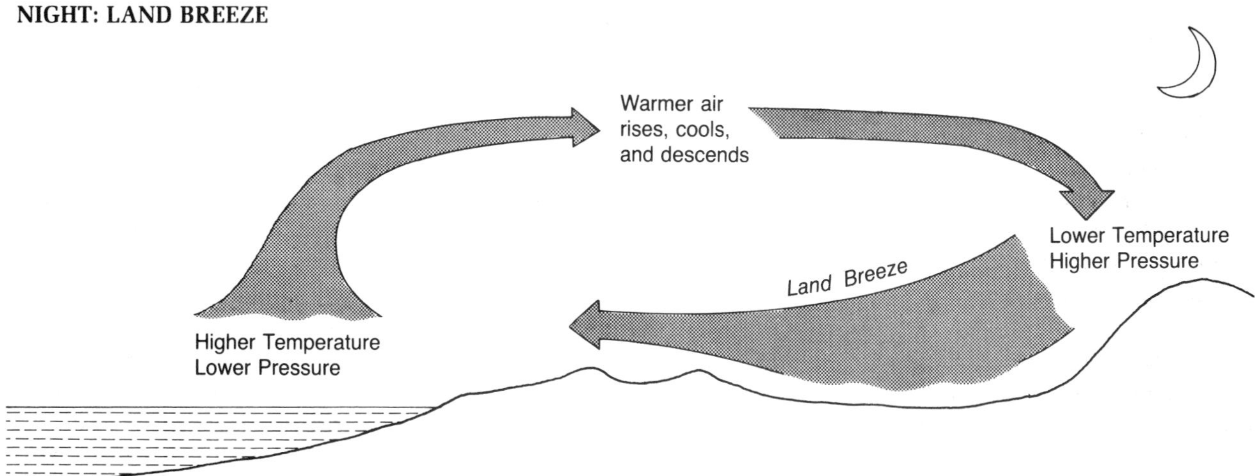

Figure 4-7 *Sea and land breezes.* (Richard Crooker)

though Santa Anas can occur any time of the year, they are most common in the fall, when the Great Basin begins cooling off and the Hawaiian High starts to lose its strength. These seasonal atmospheric changes allow high pressure to build in the dry western interior of the continent and low pressure to deepen offshore. Local mountainous terrain also plays a major role in bringing the Santa Anas to the south coast. The east-west orientation of the Transverse Ranges and the passes through and between these and the Peninsular Ranges literally let the Santa Anas into the lowlands, which are open to the sea. Beside flowing from higher to lower pressure zones, the winds also lose about 4,000 feet in altitude on their way to the sea. This means down-slope compressional heating at the DAR, as exemplified in Figure 4-8. It is typical during an autumn Santa Ana for beach city temperatures to be in the 90s while desert locales are in the 70s. In winter, colder Santa Ana–like winds usually follow the passage of storm fronts, but they occur anywhere in the state where high pressure follows departing low pressure.

Southern Californians regard the Santa Ana as anything from an ill wind to a dissipater of smog. Relative humidity in a Santa Ana is usually less than 20 percent, which brings on everything from the misery of dried-out sinuses to the problem of dried-out crops. The greatest hazard posed by Santa Anas, though, is for homes located in or near fireprone *chaparral* (tall, dense evergreen shrubs; see Chapter 7). Often gusting over 50 knots (nautical miles per hour), Santa Anas not only spread fires quickly but also blow down trees and power lines. On the brighter side, a Santa Ana usually blows the smog out to sea and affords unlimited visibility. Santa Anas generally last for two or three days before there is a return to the normal onshore flow of hazy marine air. In drier years, when high pressure seemingly stalls over Nevada or Utah, they occur in longer episodes and with greater regularity.

The antithesis of the steep pressure gradient and resulting high winds and clear skies of a Santa Ana is the stagnant air and near zero visibility of a *tule fog*. It seems a paradox that both weather phenomena happen

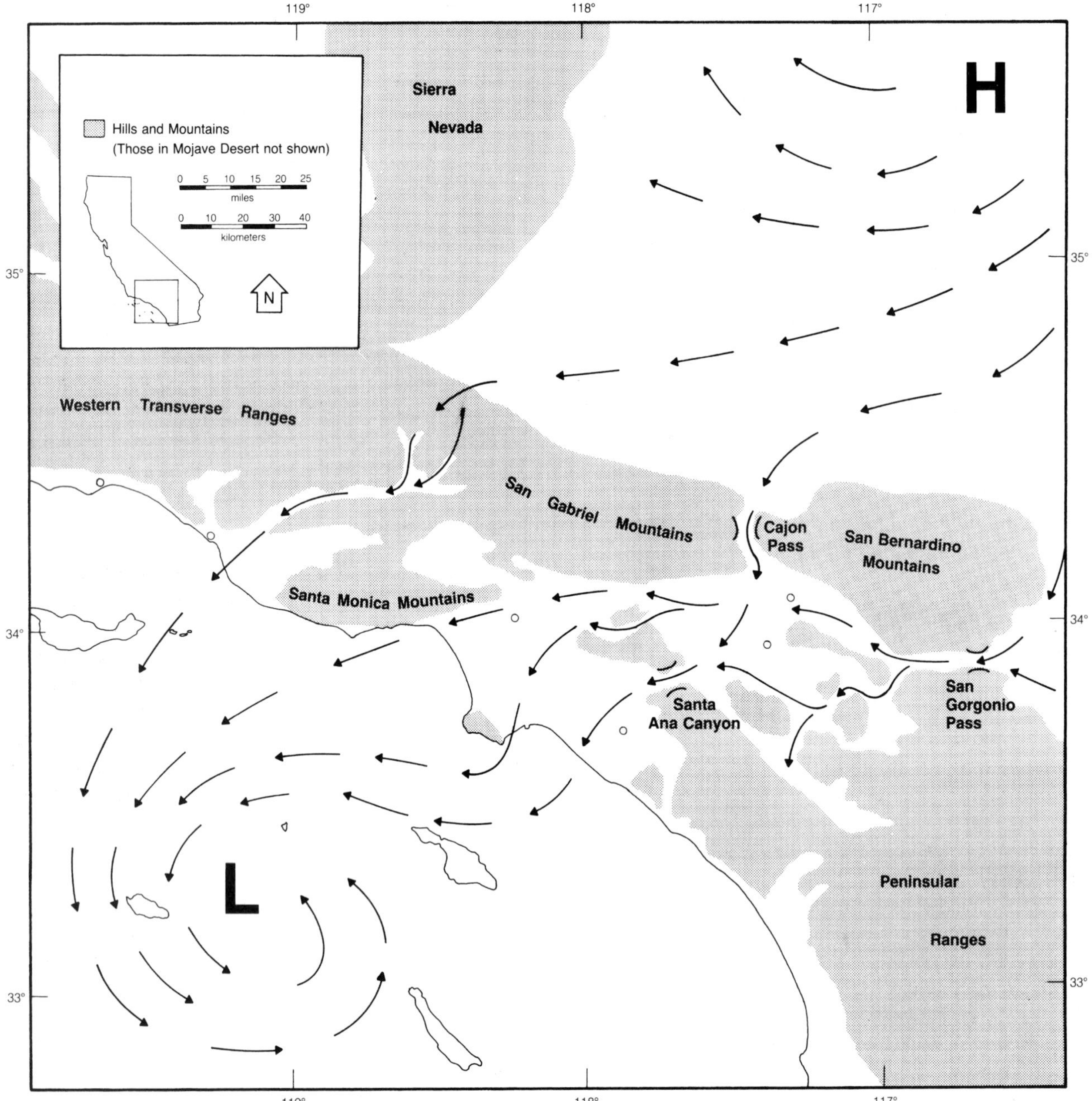

Figure 4-8 *Santa Ana winds.* (Richard Crooker)

about the same time of the year and close to each other. The tule fog season starts in the San Joaquin Valley a little later in fall than when the Santa Anas commence south of the Tehachapis, but the two have been known to be in full force on the same day.

Tule fogs derive their name from the tule reeds or cattails that grow in the swamps and deltalands of the Great Central Valley; early nineteenth-century Spanish explorers first dubbed the southern San Joaquin Valley the "Land of the Tulares" after the tule reeds that surrounded the then-huge Tulare Lake. The lake has since disappeared, but not the fogs that shroud cities like Bakersfield, Fresno, and Stockton anywhere from 20 to 40 days a year. The tule is a radiation fog that forms on cold nights when the ground is rapidly losing heat (radiation cooling) and there is little or no wind. Vertically, the fog may thicken to 2,000 feet in this way. Horizontally, it is so blinding that multivehicle crashes on the highways are inevitable.

The air pressure and wind conditions mentioned so far occur in the *troposphere* (lower atmosphere), which extends from the earth's surface up to the *tropopause* at an altitude of about 12,000 meters or nearly 40,000 feet; however, an upper atmospheric wind known as the *jet stream* occasionally has a profound effect on California's weather. The *polar jet stream* develops at about 50° N, where the tropopause drops sharply to some 7,000 meters in altitude and a steep pressure gradient is created by heavy polar air mixing with lighter tropical air.

Because the jet stream is far above the highest mountains and the friction of the lower troposphere, the Coriolis effect and pressure gradient forces balance out to produce a west-to-east airflow that sometimes exceeds 350 kilometers per hour in winter. Consequently, jet aircraft flying in the polar jet stream will gain or lose ground speed depending on which way they are going. In some winters, the rapidly moving stream of cold air oscillates in a wavelike motion far to the south of its normal course. When such a jet stream *wave* and its attendant upper troposphere ridge and trough dip down over California and the Southwest, dry, polar air rushes in and temperatures plummet to record lows.

Los Angeles was the victim of one of these arctic outbreaks late in 1978 when, for the first time in a December since 1897, subfreezing temperatures were recorded. The freeze played havoc with tropical garden plants and citrus and avocado crops, especially in interior valleys where the cold air remained trapped for several days, and even weeks in some locations, after a warming trend set in along the coast. Once in a great while, a lower-level storm will develop under the jet stream and snow will fall down to sea level—as it did from Palm Springs to Seal Beach late in January 1979. At no previous time, however, has the polar jet stream lived up to its reputation as the "Siberian Express" the way it did in the frigid blast of late December 1990, discussed early in the chapter.

Significant climatic change in California would obviously result from any increase in the wave pulsations of the polar jet stream or from a southward shift of its present course. The state's winters would become colder, drier, and longer. But there is another jet stream—the *subtropical jet stream*—to the south of California that could have quite the opposite climatic effect if it were to change its spatial pattern. A northward shifting of the subtropical jet stream would be likely to bring warmer, wetter, and longer summers to California. As it is, Southern California sometimes gets a taste of the tropics when an upper-level wave from the jet stream spurs a chubasco, a sonora, or merely several days of humid, muggy weather.

Colder or hotter, drier or wetter, California's climatic future in large part will be determined by how the jet streams behave. In the meantime, the upper atmosphere continues doing its part in exchanging the heat of the tropics and the cold of the polar latitudes, with California somewhere in between.

WEATHER MODIFICATION

Weather modification takes different forms, some accidental and others deliberate. Air pollution is perhaps the best, or should we say worst, example of the former; rainmaking, when it works, is a good example of the latter. We obviously don't want smog, but it seems an inevitable by-product of a high-tech society. Perhaps increasingly sophisticated technology will one day overcome this unwanted modification of the atmosphere, but in the meantime progress appears to be painfully slow at best. On the other hand, when precipitation results from the seeding of clouds, we are supposedly getting what we asked for from the heavens above. We say "supposedly" because sometimes induced precipitation gives us more than we bargained for—such as floods, destruction of property, and loss of life.

Unintentional: Smog

Unfortunately, California's climatic vernacular includes the word *smog,* which has been portrayed as everything from an animated monster to a toxic gas. Travelers disembarking from planes at Los Angeles International Airport say it brings tears to their eyes. Vacationers returning via I-5 say they get all choked up as they drive into the San Fernando Valley. Even 100 miles away in the desert, people say they can smell it coming. Even worse, smog destroys crops and forests and aggravates human respiratory ailments.

As for who or what causes smog, the blame rests on a combination of topographic, climatic, and cultural features existing in a given region. Such regions in California are predominantly urban and include the San Francisco Bay Area (particularly the East Bay and Santa Clara Valley cities), the larger cities of the Great Central Valley, and the Los Angeles Basin and neighboring coastal plains and valleys. We will focus on the interactions among the three principal elements of smog in the Los Angeles Basin. Perhaps it is a poetic injustice, but the basin is also known by the acronym SCAB (South Coast Air Basin), which includes some 6,500 square miles in Los Angeles, Orange, Riverside, and San Bernardino counties. Nevertheless, it is here more than anywhere else in the state that we find the optimum (or worst?) combination of natural and man-made elements that make smog a threat to life, limb, and property (as if Southern Californians didn't already have enough to worry about with earthquakes).

It has long been said with regard to the health hazards posed by air pollution that the basin topography of the Los Angeles region makes it the worst possible place to locate several million people. Except during the wind reversal of a Santa Ana or the high winds following a storm front passage, air pollutants are trapped by moun-

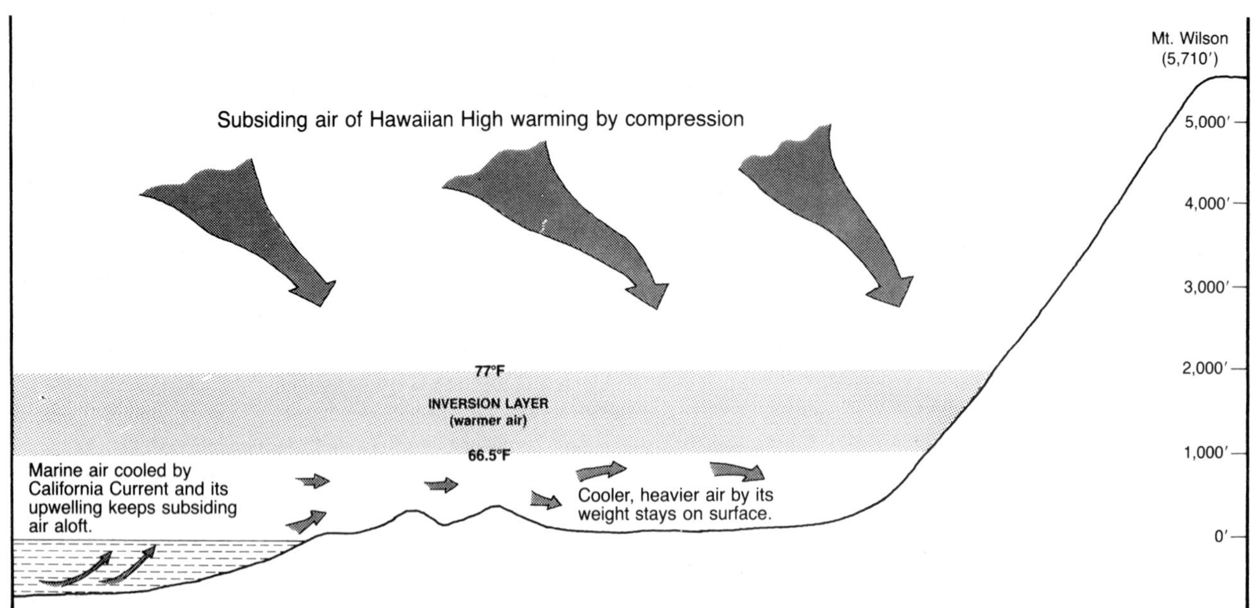

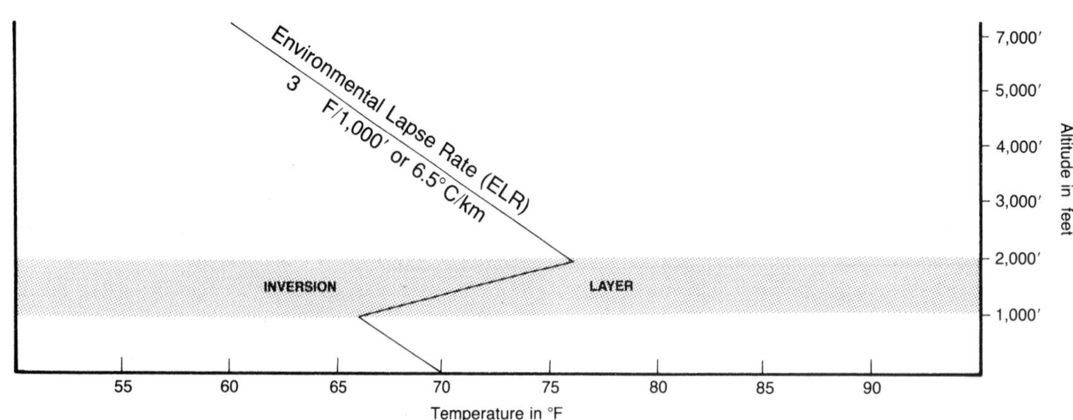

Figure 4-9 *Temperature inversion in Los Angeles.* (Richard Crooker)

tain barriers in the basin and the valleys that front on it. Another look at Figure 4-8 shows the three-sided containment posed by various sections of the Tranverse and Peninsular ranges. Three valleys—the San Fernando, San Gabriel, and San Bernardino—are almost completely surrounded by hills and mountains and as a partial consequence suffer more smog alerts than other parts of the basin. Sometimes smog builds up so heavily in the basin that it spills out through the mountain passes into the deserts. Fronting as it does on the ocean, the basin and valley topography is also conducive to temperature inversion, which is discussed next.

The purveyors of smog are the very same climatic conditions that give coastal Southern California its mild climate—namely, the Hawaiian High and the sea breeze. They team up to produce the *temperature inversion*, which involves warmer air aloft overriding cooler air nearer the ground. Normally, temperature decreases with an increase in altitude, but occasionally atmospheric conditions produce an inverse situation in which temperature increases as altitude increases. Such a temperature inversion occurs within an *inversion layer*, a layer of air that is warmer than the air beneath it.

Figure 4-9 illustrates how the warm subsiding air of the Hawaiian High meets with the cool marine air of the Pacific to form an inversion layer that on most summer days tops out at about 2,000 feet in altitude. The inversion layer then acts like a lid over the cooler, denser surface air layer, allowing no vertical movement of any pollutants. Horizontal movement of the contaminants is also thwarted: by mountains and hills to the north, east, and south and by the onshore sea breeze from the west. Besides keeping smog from escaping out to the ocean, the sea breeze, which rarely exceeds 11 or 12 knots and then only in the afternoon, carries pollutants far into the interior of the basin. Thus, millions of people in a sense become hermetically sealed in the SCAB pot to await their fate.

Of course, that fate will to a large degree rest on the source, kind, and amount of atmospheric pollutants that diffuse throughout the basin. Although the percentage has declined with wider use of emission control devices, about 65 percent of SCAB's smog is still produced by internal combustion engines in various types of vehicles. Actually, it is Southern California's abundant summer sunshine that photochemically catalyzes vehicle exhausts (oxides of nitrogen and hydrocarbons) to form the eye-smarting, gagging, foul-smelling main component of smog, *ozone* (O_3). People living anywhere in the South Coast Air Quality Management District (SCAQMD) are asked to restrict outdoor exercise and driving whenever ozone levels exceed two-tenths of 1 part per million (ppm) of air for 1 hour or more. Until recently, such first-stage smog alerts occurred on an average of 100 times a year, mostly in summer and mostly in the east San Fernando, San Gabriel, and San Bernardino valleys.[5] Although restrictions broaden to get more people and cars off the streets and some industries curtail emission of pollutants during stage 2 alerts, the former occur with much less frequency than stage 1 alerts.

Within the valleys suffering the most stage 1 and 2 alerts, there was a dramatic shift eastward of smog incident frequency in the early 1980s, from the Pasadena–Arcadia area of the San Gabriel Valley to the Upland–Fontana area of the San Bernardino Valley. Before the Kaiser Fontana steel mill shutdown in 1983, it and the sea breeze, as well as more cars and trucks, teamed up to render San Bernardino Valley smog perhaps the worst in the state.[6] Since the steel mill's demise, air quality in the Inland Empire has taken a turn for the better. No longer do Riverside and Upland fight it out to see which is the state's smoggiest city each year. Moreover, the Inland Empire's improving air quality continues despite unprecendented urban growth in the region.

Besides the component of smog we can smell and taste, there is one we can see. This is *nitrogen dioxide* (NO_2), which gives California smog its characteristic amber-beige appearance. As with ozone, NO_2 forms in sunlight from nitrogen oxide (NO) emitted mostly by motor vehicles. Incidents of unhealthful levels of NO_2 are fewer and farther apart in a normal year in the basin than with ozone alerts. Although both oxidants may contribute to emphysema and lung cancer, the relatively low levels of NO_2 normally in the air on smoggy days have been shown in laboratory experiments to destroy lung cells taken from humans and animals.

In the case of SCAB smog, that some 8 million motor vehicles are the worst offenders may be a blessing in disguise. Cars and the like don't burn coal, and it is the oxidants of coal burning that can bring on a killer fog such as the one in 1952 that contributed to the deaths of several thousand residents of London, England. The sources of SO_2, sulfates, H_2SO_4 (yes, sulfuric acid!), and other air effluents Londoners were breathing were largely industrial, although it was December and coal for domestic and commercial heating was being heavily used. The topographic contribution to that disaster was the shallow Thames River Valley, and the meteorological input came from a subsiding mass of cP air. In Southern California, meanwhile, the use of coal for heating buildings and firing thermal electric plants has been, and hopefully will continue to be, taboo. Even in heavy industry, with the exception of metals manufacturing, utilization of coal is practically nil in the basin.

There is, however, another source of SO_2 and sulfates in the basin that is cause for mounting concern: petroleum, which inherently contains sulfur. The energy industry had nominally increased the use of oil to fire its thermal electric power plants, expanding the use of oil to generate electricity from 4.7 percent in 1983 to 6.0 percent in 1989 (of the total of all power generated; see Chapter 6). There were fears that this trend would continue through the next decade, especially in light of shrinking local reserves of natural gas and abandonment of plans to build a liquid natural gas terminal (see Chapter 6); however, imports of natural gas from Alberta, New Mexico, and Texas, which now amount to more than 90 percent of the state's total annual supply, are now assured well into the next century. By 1995, electricity produced from natural gas and gas cogeneration (see Chapter 6) amounted to one-half of all power generated. When SO_2 and sulfates reach unhealthful levels in the basin, which is now an average of 30 days a year, power plants are asked to switch to natural gas, which produces about 1/100 the SO_2 of so-called low-sulfur fuel oil. On such "sulfate days," which may go on for a week or so if there is persistent heat, sunlight, humidity, and stagnant air, air conditioners and fans are

[5] The SCAQMD issues smog forecasts for various communities according to its Pollutant Standard Index (PSI), which lists five levels of air quality on a scale of 0 to 275+: good (0–50), moderate (51–100), unhealthful (101–200), very unhealthful (201–275), and hazardous (275+). Besides ozone, other pollutants considered in a PSI forecast include carbon monoxide, nitrogen dioxide, lead, sulfur dioxide, sulfates, and suspended particulate matter. A PSI of 100 is the federal standard; 200 is the threshold for a stage 1 alert, and 275 for a stage 2 episode. By the early 1990s, a diminishing number of stage 1 alerts indicated some progress in the war on smog in the SCAB, but victory was still nowhere in sight.

[6] In December 1981, Kaiser Steel Corporation announced the 1983 shutdown of its coke ovens, blast furnaces, and steelmaking furnaces due to competition from foreign steel manufacturers. Kaiser Fontana remained in operation but converted to finishing purchased steel slabs and becoming a steel supply and service center, relatively smog-free operations in comparison with steelmaking.

widely used and consequently the demand for electrical energy skyrockets to record levels for the year. In the afternoons of some summer days, demand can exceed 15,000 megawatts (1 MW = 1 million watts) in the basin.

By itself, natural gas is unable to support these levels of electricity production, but to avoid a potential air pollution disaster that might accompany increased fuel oil use, electricity is imported on lines from far outside the basin. Other stationary sources of sulfur oxidants and particulate matter, including refineries and petrochemical and metals plants, are also ordered by the SCAQMD to curtail operations on sulfate days. And the unloading of oil from tankers is altogether prohibited.

The long-term price for cleaner air will become apparent, in many cases painfully so, as compliance with the tougher, 1990 version of the Clean Air Act begins to take effect. Once all provisions of the new law are in place, which is estimated to be in about 2005, compliance costs are projected to range between $25 and $35 billion annually. The costs will be reflected in worker layoffs and retraining, higher power bills, less lavish lifestyles, higher prices for most manufactured goods and services, and a slower-growing economy. In Los Angeles, for instance, which is considered one of the nine smoggiest cities in the nation (the others are Baltimore, Chicago, Hartford, Houston, Milwaukee, New York, Philadelphia, and San Diego), pollution emissions from motor vehicles will be reduced by decreasing use of gasoline and increasing use of alternative fuels such as ethanol and methanol and of electricity. As required by law, in 1998 marketing of the first commercial electric cars started in California. By 2003, nonelectric cars must be emitting 60 percent less nitrogen oxide and 40 percent fewer hydrocarbon wastes than in 1990.

To cover the costs of achieving these goals, consumers will pay up to 10 percent more for fuel and new cars will cost upwards of $500 more than in 1990. The cost of cleaning up the air will cut across the entire motor vehicle industry, from the giant automakers, which will retool to produce electric cars, to the small auto body paintshop owner, who must reduce emissions of toxic fumes by 90 percent. Affected as well will be myriad other industries, ranging from the manufacturers of chloroflourocarbons (CFCs diminish the upper atmosphere ozone layer's ability to intercept deadly ultraviolet radiation before it strikes the earth), who must cease production altogether by 2000, to coal-burning power plants, which will spend *an additional $3 billion* each year to burn gasified or low-sulfur coal or install scrubbers. Although the utilities say it won't happen until after 2010, more power plants eventually will have to be built simply to accommodate increased use of electricity to recharge batteries in electric cars. One cannot help but wonder if the prospect of more electric cars, necessitating the construction of more power plants, is a trade-off that will prove to be a net benefit to overall atmospheric quality. However, the advent in the late 1990s of hybrid gasoline-electric cars (powered by both a combustion engine and batteries) and fuel-cell vehicles (which use hydrogen and oxygen to produce electricity) should diminish concerns about additional power plants.

Acknowledging that vegetation at the same time contributes to and helps alleviate pollution of the air,[7] certain species of both natural and cultivated plants are suffering from man-made smog. The principal natural vegetation affected is the commercially important ponderosa pine forest of the San Bernardino Mountains. Smog, mainly ozone, has caused a decline of nearly 50 percent in wood volume from yellow or ponderosa pine (*Pinus ponderosa*) trees in the San Bernardino National Forest since 1950. Conifers and shrubs that are relatively resistant to smog have replaced the dominant ponderosa in many parts of the forest, but they are mostly species of little or no value to the lumber industry, as well as ones that increase the fire hazard. Evidently, smog weakens yellow pines to the point of nonresistance to the ravages of the bark beetle and other pests.

Smog damage to farm crops poses a much more serious problem because it is more widespread in the state and the economic loss is greater. Losses averaged tens of millions of dollars annually by the mid-1970s in Southern California, Bay Area, and Central Valley agricultural counties alone. Ozone and peroxyacyl nitrates (PAN) do most of the damage, mainly to leafy vegetables such as spinach and lettuce, although SO_2 also "burns" leaves and eventually renders plants unfit to send to market. The sources of these crop and forest killers are the same mobile and stationary ones discussed earlier, which should make residents wonder how far out into rural California they have to go to escape smog.

Intentional: Rainmaking

As recent droughts wore on into successive winters, Californians increasingly gave thought to how the few storm clouds that now and then passed overhead could be made to give up their moisture. Clearly, it seemed, the time had come for somebody to do something more about the weather than merely talk about it—lest the state run out of water entirely. Yet neither drought nor rainmaking to combat it were new to California. Actually, the father of modern cloud-seeding methods perhaps was Charles M. Hatfield, just after the turn of the

[7] Plants exude hydrocarbons, such as pollens and terpenes, into the atmosphere to help form haze. Other natural air pollutants include volcanic dusts, blowing dust and sand, smoke from brush and forest fires, salt from breaking sea waves, and bacteria and viruses. On the other hand, *photosynthesis* in plants, whereby oxygen and carbohydrates are formed from chemical reactions among water, carbon dioxide, and light or solar energy, helps maintain the oxygen content of the atmosphere.

century in Southern California. Hatfield's technique involved filling evaporating tanks with a supersecret chemical formula that was supposed to induce rain. Nowadays, such formulas usually include silver iodide and/or dry ice that supercool ice crystals in clouds, causing them to gain weight and fall to earth. Twice, in 1904 near Los Angeles and in 1915–16 in San Diego, Hatfield experimented with his rainmaking formula (probably not silver iodide) and in both cases apparently met with rousing success—at least the downpours that followed would seem to indicate that he did. Nobody knows whether Hatfield's experiments actually produced rainfall, but a measure of his success is the fame he gained that later won him contracts in Canada, Honduras, and Texas.

Charles Hatfield and many more contemporary rainmakers have had to contend with the paradox of too much success in the form of floods following their work. The deluge that accompanied Hatfield's rainmaking in San Diego caused the city's Morena Reservoir to overflow; the resulting floodwaters washed out a dam, inundated city streets and buildings, and caused several drownings. Hatfield's original contract with the city called for his rainmaking to fill the reservoir and not much more. The city council refused payment of his $10,000 fee on the grounds that the results of his efforts far exceeded those stated in the contract. Hatfield sued the city, while the city demanded that he pay for all the flood damage. The whole mess eventually wound up a draw, with neither party realizing financial recovery.

There are economic and environmental benefits to be gained from intentional weather modification, however. For example, a recently completed 14-year cloud-seeding project over the San Gabriel Mountains produced nearly 100 billion gallons of precipitation over and above what normally would have fallen. The additional water, which encouraged a heartier watershed vegetation cover, was valued at more than $3 million above the total cost of the cloud-seeding operation. In another project, starting in the late 1970s and running into the following decade, research on winter cloud-seeding to augment the Sierra Nevada snowpack was conducted as part of the Sierra Cooperative Pilot Project. The SCPC's principal aim was to increase runoffs in both California and Nevada. If successful, the project would have been of inestimable value in furnishing additional hydroelectric energy and irrigation water for the two states. Unfortunately, results have been inconclusive.

The various experimental efforts to produce greater rainfall strongly underscore the importance of water resources to this state. As the following chapter illustrates, few issues have inspired more creativity or generated more debate or criticism than water needs and uses in California.

CHAPTER

5

Water: The Controversial Resource

Few, if any, of California's natural resources stir more innovative responses or cause more controversy among the state's citizens than water. Innovation is seen in the great aqueduct systems that crisscross the state, carrying water from where it is the most plentiful to where it is most needed. While such water-sharing schemes are of priceless benefit to recipient communities and farms, they often have an irreversible negative impact on the environments that give up the water in the first place.

Thus do the waters of controversy begin to flow, as the people losing the water and the environment it supports wage legal and political battles with the people who gain the precious liquid and all its benefits. On occasion, water rights disputes have even escalated into armed conflict. Works by such authors as Carey McWilliams[1] and Morrow Mayo,[2] as well as Remi Nadeau's *The Water Seekers* and even a full-length motion picture, *Chinatown*, have bubbled forth from the bottomless well of water conflict in California. But when all has been litigated, written, said, and done, the all-too-familiar doctrine of "the greatest good for the greatest number" seems to hold sway as large farms grow larger and big cities get bigger with water imported from somewhere else.

[1] Carey McWilliams, "Water! Water! Water!" in *Southern California: An Island upon the Land* (Santa Barbara and Salt Lake City: Peregrine Smith, 1973), pp. 183–204.
[2] Morrow Mayo, "The Rape of Owens Valley," in *Los Angeles* (New York: Knopf, 1933), pp. 220–246.

WATER RESOURCES: *An Uneven Hydrography*

California's surface streamflow averages 71 million acre-feet (MAF) of water annually (1 acre-foot, or AF, is the equivalent of 1 acre of land 1 foot deep in water, or 326,000 U.S. gallons), with another 6.2 MAF draining in from streams originating in Oregon (1.4 MAF) and from diversions of the lower Colorado River (4.8 MAF). But, as shown in Figure 5-1, nearly 75 percent of the natural streamflow occurs in the northern third of the state. This uneven distribution of natural freshwater resources assumes significance in light of the fact that about 80 percent of annual water consumption takes place in the southern two-thirds of the state. A review of Chapter 4 and the climatic controls that render the north wet and the south dry essentially explains the geographic disparity of water supply and demand in California.

When the hydrography depicted in Figure 5-1 is viewed in comparative terms, the Sacramento with its tributaries, including the Pit, Feather, Yuba, and American, stands out among individual river systems, supplying about one-third of all the state's streamflow. The river systems of the northwest, including the Smith, Klamath, Mad, Eel, Noyo, and Russian, combine to supply about 42 percent of total streamflow. Of the 25 percent of natural streamflow found in the southern two-thirds of the state, the San Joaquin River and its tributaries contribute about half of it.

Most of the south's remaining surface runoff comes from the interior-draining rivers of the southern San Joa-

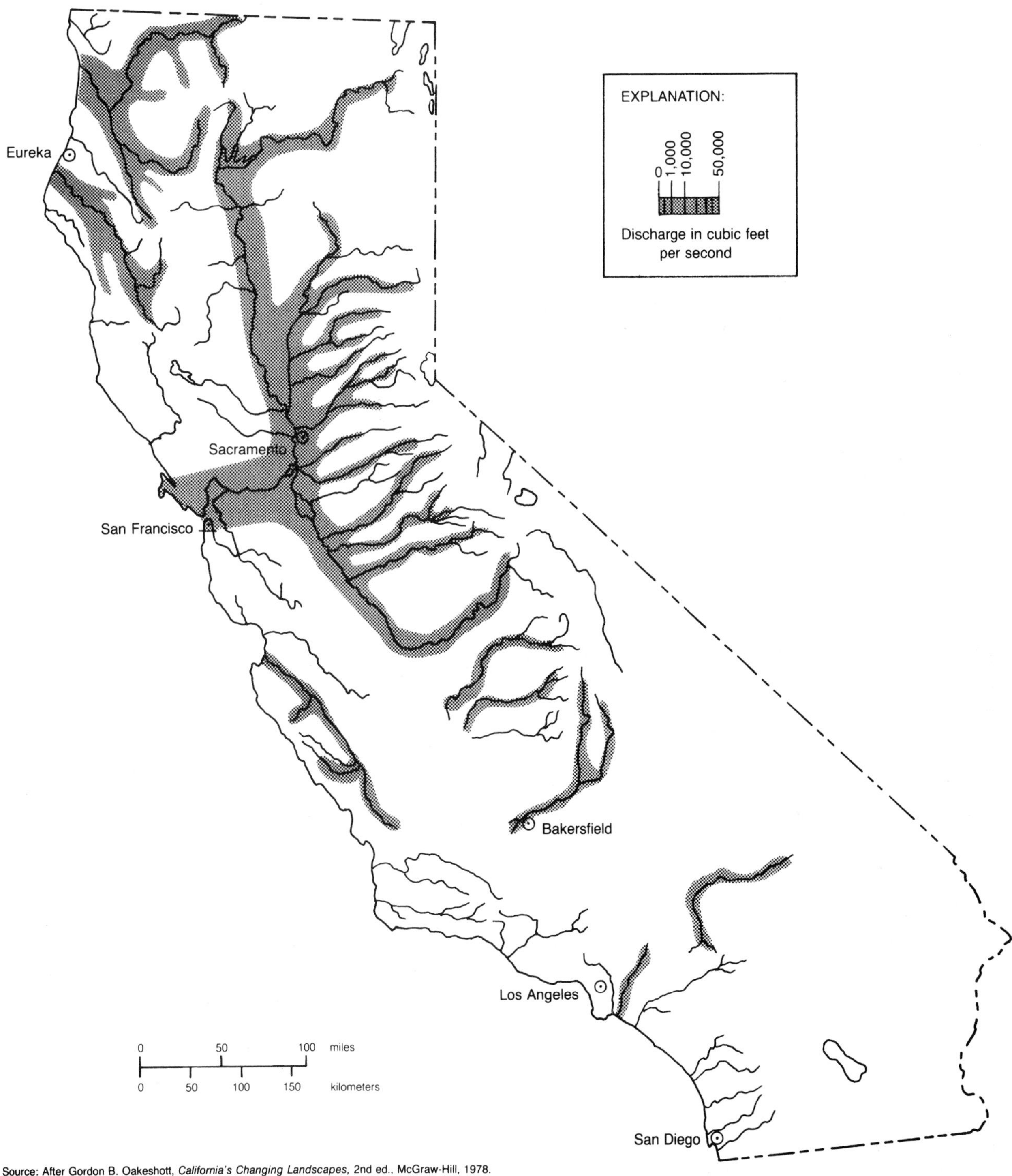

Source: After Gordon B. Oakeshott, *California's Changing Landscapes*, 2nd ed., McGraw-Hill, 1978.

Figure 5-1 *California hydrography. Hydrography or hydrology includes the study or science, mapping, and/or description of naturally occurring freshwater bodies (lakes) and courses (streams). Most of this natural resource is found in the northern third of the state, but its potential consumption is largely in the southern two-thirds.* (Richard Crooker)

quin Valley, the multitude of smaller streams rushing out of the eastern Sierra, and the "exotic" Colorado River, whose sources are entirely outside the state. There are numerous central and south coast streams emptying into the Pacific, as well as interior-draining water courses in the southern deserts, but the aridity of the south renders their flow *intermittent* rather than *perennial* and thus of minor consequence in the overall surface water supply picture of California. Fortunately for the total surface water supply, California's streams and rivers are relatively free of pollution compared to those in some eastern states, although heavy use of fertilizers and herbicides tends to obviate some of the benefits of minimal waterway pollution from mining and manufacturing industries in California.

WATER AND DROUGHTS

Net water use is the quantity of water delivered to farm headgates, city water system intakes, and other users in a region. It differs from *applied water* in that it takes into account the large amount of water reuse that commonly occurs in a region, *evapotranspiration* (water absorbed by soil, transpired by plants, and evaporated in the air), unrecoverable losses from water distribution systems, and outflow from a region. Net water use in California, as shown in Table 5-1, now averages about 34.2 MAF per year and is projected to rise by only 1.4 MAF (to 35.6 MAF) by 2010. Agriculture is far and away the biggest user of water, accounting for nearly 79 percent of all net usage. Although agricultural use (predominantly for irrigation) by 2010 will remain at a level of about 27 MAF annually, agriculture's share of net use will fall to 75 percent. The reason for this decrease lies in the projected urban population increase, which will boost water use by 1.6 MAF annually by 2010. Thus, urban users, who now account for little more than 16 percent of all consumption, will see their proportion rise to 20 percent. So-called other users, which include wildlife management districts, rural parks, power plants (for cooling), oil fields (for enhancing oil recovery), and water conveyance systems (for consumptive losses), will see little change over the next few decades. Their annual net usage will hold steady at about 1.7 MAF, or slightly under 5 percent of yearly totals, through 2010. As far as regional changes in net water use by 2010 are concerned, all the state's major hydrologic regions will show increases except for the southeastern or Colorado River region. Most of the drop in the latter region will be in the agricultural sector, where more efficient irrigation practices will aid water conservation and thereby reduce water use by 2010. Also affecting the region will be increasing withdrawals of water by the Central Arizona Project (CAP) from the lower Colorado River as the twenty-first century wears on, with that water destined mostly for the burgeoning metropolitan Phoenix area.

The net water use data for 1985 and projections for 2010 were provided at a time of plentiful precipitation. Unbeknown to the estimators, however, a long period of above-normal precipitation (1978–83) was drawing to a close and the state was on the brink of a historically unprecedented drought (1986–92).[3] Water resource peo-

[3] Although the drought began with the winter of 1986–87 and ended in the winter of 1992–93, periods are given in calendar years, that is, from January 1 to December 31. The "precipitation year" is from July 1 to June 30, and the "water year" is from October 1 to September 30.

TABLE 5-1 California net water use, in millions of acre-feet, 1985 and 2010.

	AGRICULTURE		URBAN		OTHER		TOTAL	
REGION	1985	2010	1985	2010	1985	2010	1985	2010
Central coast	1.01	0.98	1.31	1.53	0.13	0.13	2.45	2.64
South coast	0.75	0.57	2.82	3.59	0.19	0.20	3.76	4.36
Sacramento River	6.71	6.88	0.50	0.68	0.27	0.27	7.48	7.83
San Joaquin Valley	13.65	13.86	0.53	0.76	0.37	0.39	14.55	15.01
Colorado River	3.48	3.12	0.17	0.27	0.38	0.30	4.03	3.96
North coast and Lahontan*	1.35	1.34	0.26	0.36	0.34	0.39	1.95	2.09
Total	26.95	26.75	5.59	7.19	1.68	1.68	34.22	35.62

* Except for the Colorado River region, the Lahontan includes all of the Great Basin as mapped in Figure 3-1.

Source: California Department of Water Resources, Bulletin 160-87 (November 1987).

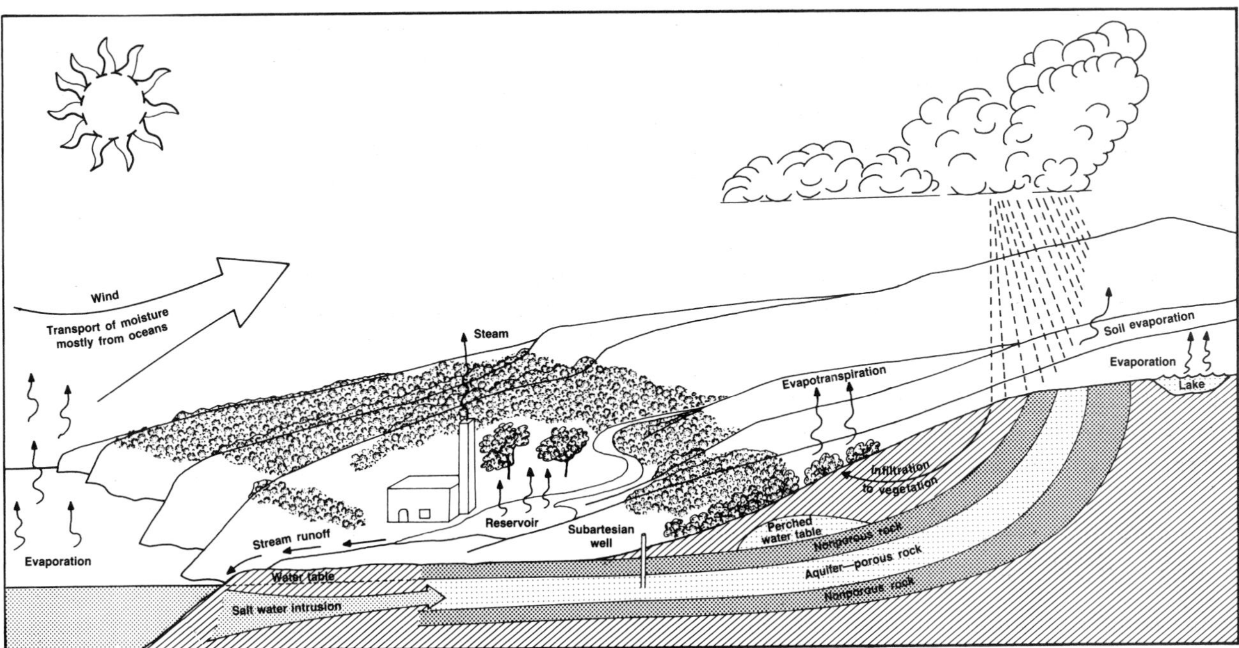

Figure 5-2 *Hydrologic cycle and groundwater.* (Richard Crooker)

ple in California always plan for periodic droughts, which they know to be an integral aspect of the state's geography. In 1986, however, no one anticipated that by 1992 the state would be entering the sixth year of the drought and that some would be labeling it California's "epic drought" of the twentieth century. Reservoirs throughout the state were at record low levels, with some so nearly dry that the eerie remains of wrecked airplanes, cars, and trucks and the ghostly hulks of buildings were exposed. From the plane that crashed in Calaveras Reservoir northeast of San Jose in the 1960s, to the Kern River town of Kernville that drowned decades ago in the impoundment of Lake Isabella, to the soggy ghost towns of Monticello in Lake Berryessa in Napa County and Mormon Island in Folsom Lake northeast of Sacramento, the drought brought back reminders of long-forgotten events and geographies.

By early 1991, short-term solutions to the drought crisis were being implemented throughout the state. In preparing 15 million Southern California residents for eventual rationing, for instance, the Metropolitan Water District (MWD) proposed a 17 percent cut (from the previous year's levels) in its deliveries to water agencies and a tripling of the price of water to agencies for amounts purchased in excess of 83 percent of what they consumed the year before. For long-term remedies to this and future droughts, Californians would look to tapping the state's undeveloped water resources. In the north coast hydrologic region alone, for example, there is a surface water resource development potential of several MAF a year. Here, as elsewhere, there are legal and natural constraints. The 1972 California Wild and Scenic Rivers Act precludes development of most of the state's northwestern streams. The Eel River, one of the region's largest streams, was reviewed for withdrawal from the Wild and Scenic Rivers System in 1985; but the Department of Water Resources recommended that the river remain protected from development, and so it did, "subject to future review."

Beyond any restraints imposed on California's hydrography to diminish the effects of drought, the state must also consider natural limitations in appraising its water resource development potential. Even if dams, reservoirs, diversion tunnels, and aqueducts are built to tap every ounce of this potential, California will still have precious little hedge against a prolonged drought. As illustrated in Figure 5-2, water in reservoirs doesn't simply lie around waiting for us to use it; instead, it evaporates into the atmosphere as an inextricable part of the *hydrologic* or *water cycle*. In fact, in the sunnier parts of California, even in years of normal precipitation, there exists a *water budget* deficit; evaporation from water bodies and evapotranspiration (see Table 5-1) from plants and soil far exceeds precipitation. In dry years, the deficit can extend statewide.

Drought or no drought, annual water consumption of 34 MAF is not likely to diminish unless a mass exodus of California's agricultural industry occurs. Irrigation agriculture alone accounts for almost 79 percent of all water consumption, and other agricultural and food processing activities account for another several percentage points. For instance, production of just 1 pound of beef from the time it starts out as irrigated alfalfa for cattle feed until it reaches the dinner table requires 2,500 gallons of water. And beef is only one of thousands of varieties of foods and fibers requiring varying amounts of water for growth and processing. Of the 16 percent or so of strictly nonagricultural consumption of water, residential and industrial users share most of it. A family of four uses an average of 360 gallons a day or 130,000 gallons a year for everything from drinking and dishwashing to flushing toilets and watering lawns.

By virtue of being the greatest water consumer, agriculture is one of the biggest losers whenever drought strikes. This has been true almost since the time large-scale commercial agriculture started in California. Barely a decade and a half after statehood, the drought of the mid-1860s all but wiped out a burgeoning beef industry by destroying more than 2 million head of cattle. In more recent droughts, such as the one that occurred in the 1970s, losses were sustained not from cattle perishing on the range from starvation and thirst, but rather from their being sold prematurely because pasturelands had dried up and feed prices were inflated. Their forced sales, in turn, put downward pressure on cattle prices, which contributed $500 million of an estimated $800 million total agricultural loss in the last year (1977) of the drought. Farmers growing *rain-fed* crops incurred much of the remaining $300 million loss because they depended on natural rainfall, which was less than 40 percent of normal in some *dry-farming* regions of the state. In all, the drought caused about a 17 percent drop in California's net agricultural income in 1977. Fourteen years later, as the 1986–92 drought dragged on and avocado and citrus orchards were increasingly stressed by the drier atmosphere, a record winter freeze (see Chapter 4) became the straw that broke the financial back of many growers. What some farmers dreaded even more than their crops becoming dessicated and frozen were proposals to reduce agriculture's share of the state's water supply by 10 percent, thereby freeing water to help the cities weather the drought. Such a move would no doubt delight city folk at first, but its appeal might wane as less irrigation water translated into higher prices and diminished quality for farm commodities in the supermarket.

Drought-induced losses suffered by nonagricultural sectors of the California economy were in some cases more severe than those experienced by the farming community. Hydropower losses alone, if quantified for the worst years of recent droughts, probably far exceeded those of agriculture, not only because relatively expensive fuel oil was imported to prevent an electricity shortage but also because of the inestimable costs to the environment and the health of residents posed by substituting a ''dirty'' energy source (fossil fuels) for a relatively ''clean'' one (hydropower).

The geography of the hydroelectricity shortage appears in Figure 5-3, where most of the Sierra Nevada and Klamath Mountains watersheds are shown receiving less than 40 percent of normal precipitation in 1976. These watersheds hold most of the state's hydroelectric dams, and by the fall of 1977, many of their reservoirs had gone dry. Imported hydroelectricity—for example, from the Columbia River via the Pacific Intertie line—was also in diminishing supply because the drought had spread throughout the western United States. Even more serious at the time was the drying up of municipal reservoirs—in affluent Marin County, for example, where water was trucked in so that residents had enough to drink and to maintain health and sanitation standards. During the 1986–92 drought, there was growing reluctance by local building departments and water districts to issue new building and water hookup permits. Of course, state and local governments can't prohibit people from moving to California, but persistent denial of permits for new residential construction eventually becomes a curb on growth. This practice came as a double blow to the home-building industry, which was already reeling from the spreading real estate recession of the early 1990s.

The effects of both recent droughts on California's natural vegetation and wildlife were serious and in some cases devastating. In the earlier drought's last year (1977), it was blamed for pushing commercial timber losses 250 million board feet over normal by making *conifers* (cone-bearing trees such as pines and firs) more susceptible to the ravages of pests and disease and by increasing the incidence of fire. Chaparral and other fire-prone plant communities became all the more menacing to suburbia, not simply in and of themselves but because of a heightened chance that there might not be enough water to control a fire.

Among the hardest-hit wildlife were salmon and steelhead trout, which on their return migration from the sea found their ancestral stream waters too low and/or too warm for successful spawning. Landlocked fish perished in countless numbers as lakes and streams disappeared, especially at elevations below 6,000 feet, where diversion projects and evaporation took their toll on existing water. Burros, mustangs, antelope, bighorn sheep, deer, and elk saw their browse vegetation vanish for lack of rain. Also faced with starvation were coyotes, foxes, hawks, and other predators that found their prey (rodents, for example) in declining supply. Because wetland habitats were disappearing for lack of rain and/or diversion of streams for irrigation, migratory waterfowl became hard-pressed to maintain their flyways across California. The dearth of water also promoted insect and disease infestation among most wildlife.

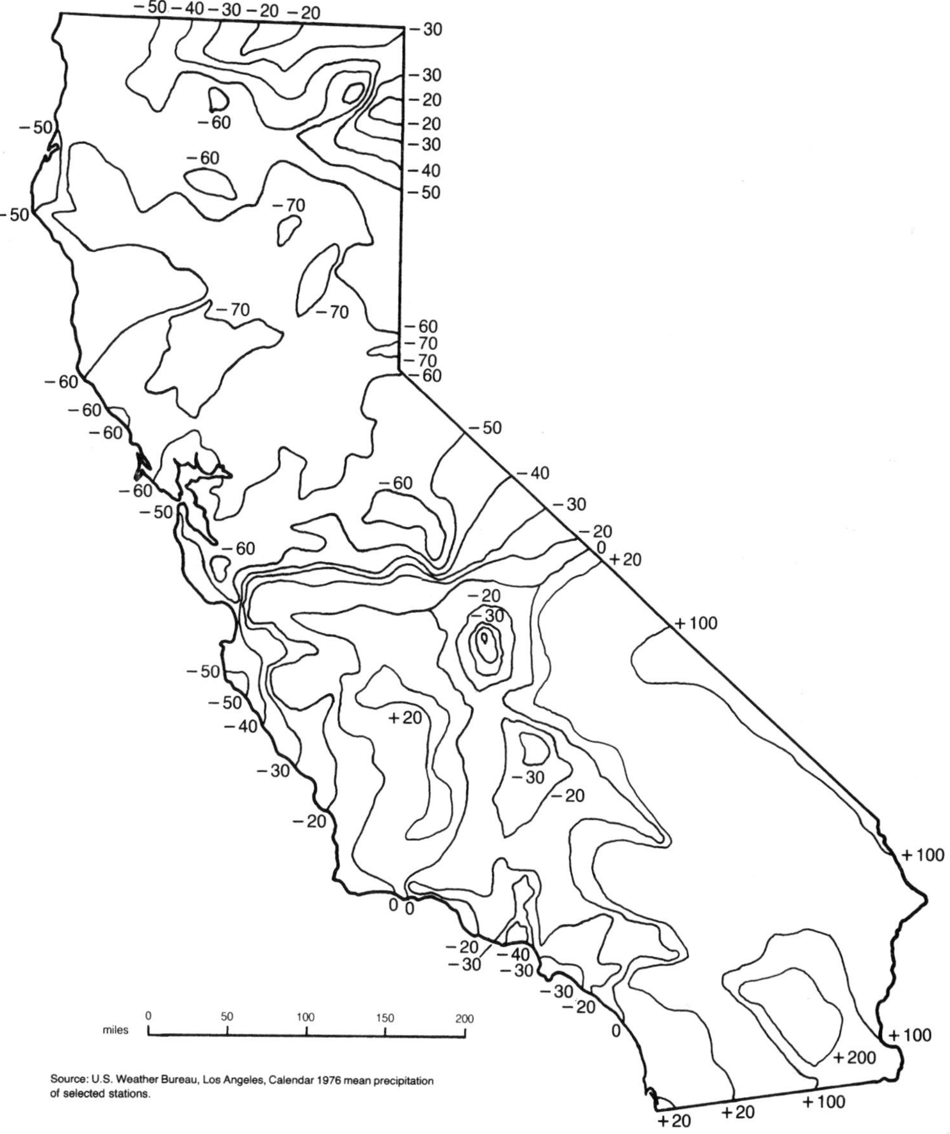

Figure 5-3 *Precipitation deviations from normal in California in 1976, by percentage. Below-normal rates are indicated with a minus sign, above-normal rates with a plus sign. Notice that above-normal precipitation occurred mainly in the southeastern part of the state.* (Robert Huson)

GROUNDWATER AND DROUGHT: *The Case of the Owens Valley*

California has a significant groundwater potential in its soils, subsoils, and certain rock formations, and it is this resource, properly developed and managed, that can stave off environmental and economic disaster in time of prolonged drought. The overall storage capacity of the state's groundwater basins totals about 850 MAF, almost 20 times the available active storage of all surface reservoirs in the state (their total capacity is 43 MAF). But far in excess of half of the groundwater capability is too deep to be pumped economically to the surface. During the 1970s' drought, *artesian* and *subartesian wells* (Figure 5-4), brought much of this naturally stored groundwater to the surface for use on the farm and in the city. The sandstone and other *permeable* (porous) rock strata that form the *aquifers* (see Figure 5-2) holding most of this capacity were extensively recharged by two years (1978 and 1979) of record precipitation following the mid-1970s' drought. Although *land subsidence, saltwater intrusion* (see Figure 5-2), and increased pumping costs often accompany groundwater utilization, they can be minimized if groundwater reserves are saved for dry years. Moreover, stored groundwater does not evaporate away like reservoir water does.

As with surface streamflow, most of the groundwater resources in California are remote from major population centers. The earliest major regional project aimed at rectifying the geographical inequities of water supply and demand in the state involved diversion of the eastern Sierra–Owens Valley drainage to the city of Los Angeles via the 233-mile Los Angeles Aqueduct. Even before the aqueduct began operating in 1913, and

A

B

Figure 5-4 *Owens Valley groundwater pumps and free-flowing wells. These pumps and wells are supplied with water by water-bearing rock formations (aquifers) that consist of alluvial deposits, lake bed sediments, and/or fractured volcanic flows, the latter typically yielding 10–20 cubic feet per second (cfs), or three to four times the yield of an alluvial aquifer. Water coming to the surface under its own hydrostatic pressure through a well is artesian (photo A); if it has to be pumped to the surface, it is subartesian (photo B). An artesian or capped pressure well usually taps deep, confined aquifers where relatively high pressure levels exist.* (Crane Miller)

continuing into the present, the Owens Valley Project of the Los Angeles Department of Water and Power (DWP) has been the subject of bitter controversy (the dispute is viewed in historical perspective later in the chapter). It is noteworthy that San Francisco actively sought far-off Sierran water resources before Los Angeles did, but political controversy over its Hetch Hetchy dam and 186-mile aqueduct projects delayed completion until some 15 years after the Los Angeles Aqueduct opened (the San Francisco water controversy is also discussed later in the chapter).

As seen in Figures 5-5 through 5-8, this diversion project, which enables Los Angeles to grow and satiates the thirst of most of the city's 3.5 million residents, now extends another 105 miles farther north to the Mono Basin and includes two aqueducts. Most of the water entering the 338-mile aqueduct system comes from surface streams, but recent droughts and the decline in Colorado River Aqueduct (see Figure 5-5) allotments (see the discussion of the Central Arizona Project later in the chapter) make utilization of the 38-MAF storage capacity of the Owens Valley Groundwater Basin very tempting to the DWP.

The actual amount of groundwater available in the basin literally depends on the weather. The Owens Valley proper averages only 6 inches of precipitation annually because of its rain shadow location; the orographically prolific eastern Sierra, however, usually forces several times this much precipitation from the heavens, except in dry years. The falling rain and melting snow move down the eastern slopes either by flowing in surface streams or by seeping into aquifers or underground water tables. The mountain runoff percolates into both the upper aquifer or *water table* and the lower or *confined aquifer*, with layers of nearly *impermeable* (nonporous) silt or clay separating the two zones.

Other sources of water supply to the aquifers include seepage from irrigation canals and ditches and groundwater flow from underground basins in Round Valley, Chalfant Valley, and the volcanic tableland between those two small valleys (see Figure 5-8). In wet years, the DWP also contributes to recharging the aquifers by spreading excess water via percolation in channels near Big Pine and Laws and on the alluvial fans of the western side of the valley.

Although diversion of surface runoff in the Owens Valley dates back to prehistoric times when native Paiutes irrigated some of their Round Valley lands, groundwater pumping did not commence until early in the twentieth century. Los Angeles drilled its first wells in 1908 to provide water for dredges used in constructing the aqueduct. The first drought-induced well drilling took place in the early 1920s; by 1931, pumping wells provided a then-record 142,630 AF of groundwater to

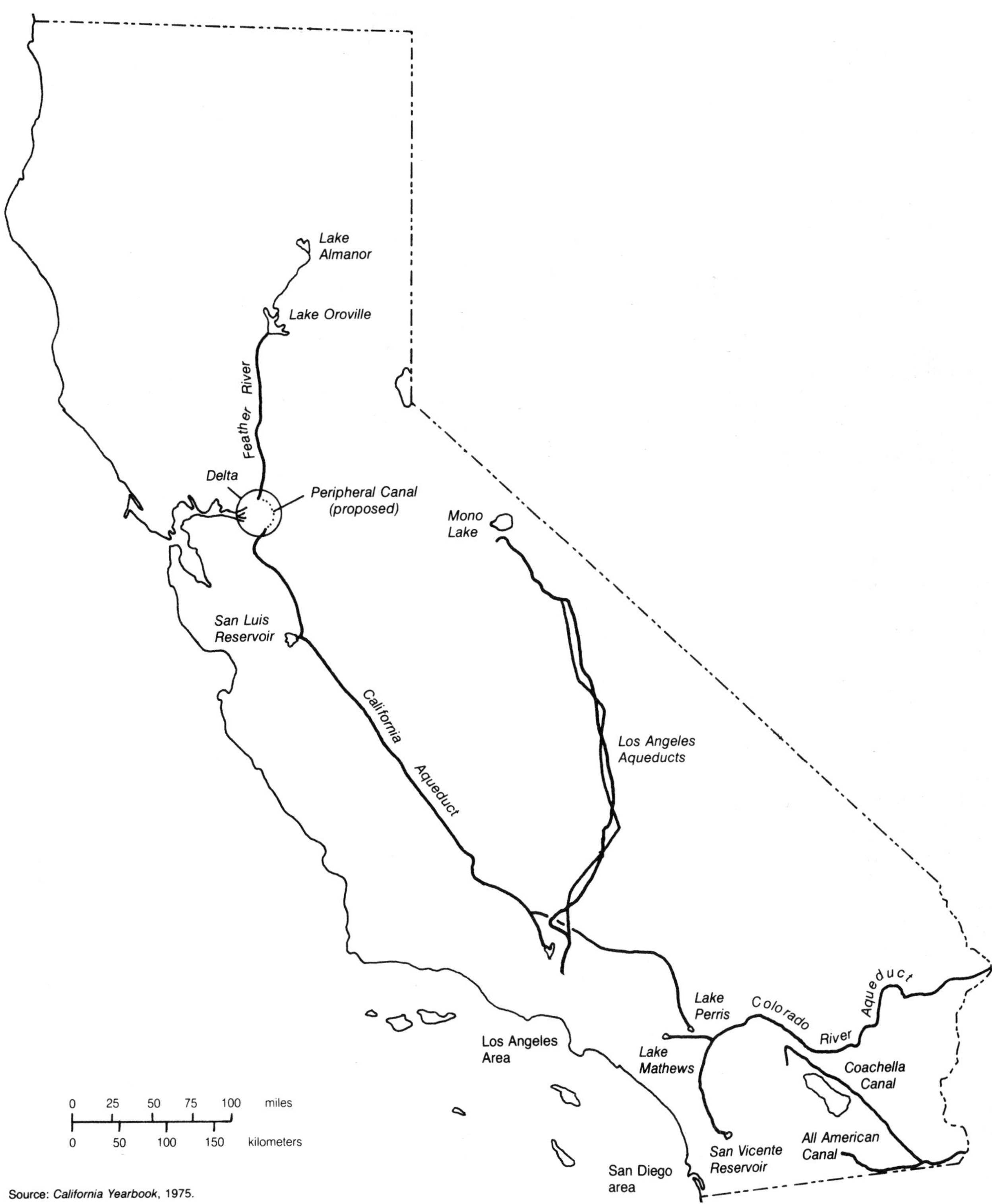

Figure 5-5 *Aqueducts serving Southern California. The "Feather River" in Northern California refers to the Feather River Project unit of the State Water Project. FRP water actually flows downstream from Oroville Dam through the natural courses of the Feather and Sacramento rivers on its way to the Sacramento Delta. Although originally "proposed" as a Delta unit of the SWP, plans for the Peripheral Canal have been withdrawn from the state project.* (Richard Crooker)

A

B

C

Figure 5-6 *Mono Basin, a 1,000-square-mile basin of interior drainage located at the edge of the northeastern escarpment of the Sierra Nevada and encompassing Mono Lake, 6,400 feet above sea level (photo A). If a court order in 1989 had not prohibited Los Angeles from continuing to divert water from streams feeding the lake, the water level would have continued to drop indefinitely. As the water level lowers, more of the unusual-looking, calcereous formations, emanating from springs and known as* tufa towers, *appear (photos B and C).* (A: NASA; B, C: Crane Miller)

be sent down the aqueduct to Los Angeles. The excessive 1931 pumping, however, slowed pumped well production to a scant 140 AF in 1932.

Throughout the rest of the 1930s, 1940s, and 1950s, pumped groundwater production was practically nil, and flowing wells averaged only about 10,000 AF annually. Then in 1960, about 2 years into a minor drought, pumped production revived anew with 40,460 AF bound for Los Angeles. Except for 1961 (111,880 AF), pumping during the rest of the 1960s averaged only about 11,000 AF each year. Then, spurred largely by the worst drought in decades, pumped production increased on average nearly tenfold each year, up to and including the last year (1977) of the dry period.

The DWP proposed in its June 1979 *Final Environmental Impact Report (EIR) on Increased Pumping of the Owens Valley Groundwater Basin* that pumping be increased to 112,000 AF in an average year and to 227,000

Figure 5-7 *Los Angeles Aqueduct in the Owens Valley near Lone Pine. The original 233-mile aqueduct, beginning just south of Tinemaha Reservoir at the Owens River intake and terminating in the northern San Fernando Valley at Los Angeles Reservoir, was completed in 1913. In 1940, a 105-mile extension of the aqueduct system to Lee Vining Creek in Mono Basin was completed, increasing the system's length to 338 miles. In June 1970, a second aqueduct from Haiwee Reservoir south to Los Angeles was put into operation. The second aqueduct increases water delivery capabilities from the eastern Sierra and Owens Valley to Los Angeles by nearly 50 percent.* (Crane Miller)

AF in a maximum year.[4] The former figure is about equal to the annual rate of the 1970s; the latter is nearly 50 percent of the projected total average annual export of surface water and groundwater from the valley.

The environmental impact of the proposed increase in groundwater pumping is much less a game of numbers, but much more a source of controversy, than the pumping itself. In 1972, Inyo County filed a lawsuit charging that groundwater pumping to supply the second aqueduct (completed in 1970; see Figure 5-7) would lower the water table and thus devastate the ecology of the valley. The next year, the state court of appeals supported the county's claim by ordering the city to prepare an EIR. In its 1978 Draft EIR, the DWP claimed that the water table would not lower sufficiently to cause any significant change in existing plant or animal life and that the ecology of the Owens Valley thus would remain much the same as it is today. "Quite the contrary," responded environmentalists, Inyo County, and even some state government agencies, who foresaw water tables dwindling below the rooting system of most plants once increased pumping was permanently implemented. This action, they stated, would doom animal life, as well as the vegetation it depends upon. The Owens Valley would become a barren desert filled with shifting sand and a dusty atmosphere polluted by blowing alkali. The few existing natural springs would disappear with increased pumping, leaving behind no natural oases around which some semblance of wildlife might otherwise survive. Many of these concerns were published in their original form, as received by the DWP, in the second volume of its 1979 *Final Environmental Impact Report*.

The 1979 EIR addressed geohydrology, flora and fauna, air quality, energy, and other issues related to increased groundwater pumping and impact mitigation. One engrossing section of the report dealt with "savings

---

[4] A maximum year refers to a dry year, one in which pumping would peak. Of the 112,000 AF pumped in an average year, 55,000 AF would be exported to Los Angeles and 57,000 AF would remain for in-valley use. The overall Los Angeles water supply picture looked like this before the 1986–92 drought:

Los Angeles Aqueduct		78%
Owens Valley	61%	
Mono Basin	17%	
Los Angeles area wells		17%
Colorado River MWD and California aqueducts		5%
		100%

By the end of the drought in 1992, the aqueduct was supplying only about half of the city's needs, with deficiencies being made up for by purchases from outside agencies. In 1989, for instance, the DWP purchased some 40 percent of its annual supply of water from the MWD. In the same year, the city was prohibited by court order from exporting water from Mono Basin until Mono Lake again attained a level of 6,377 feet above sea level.

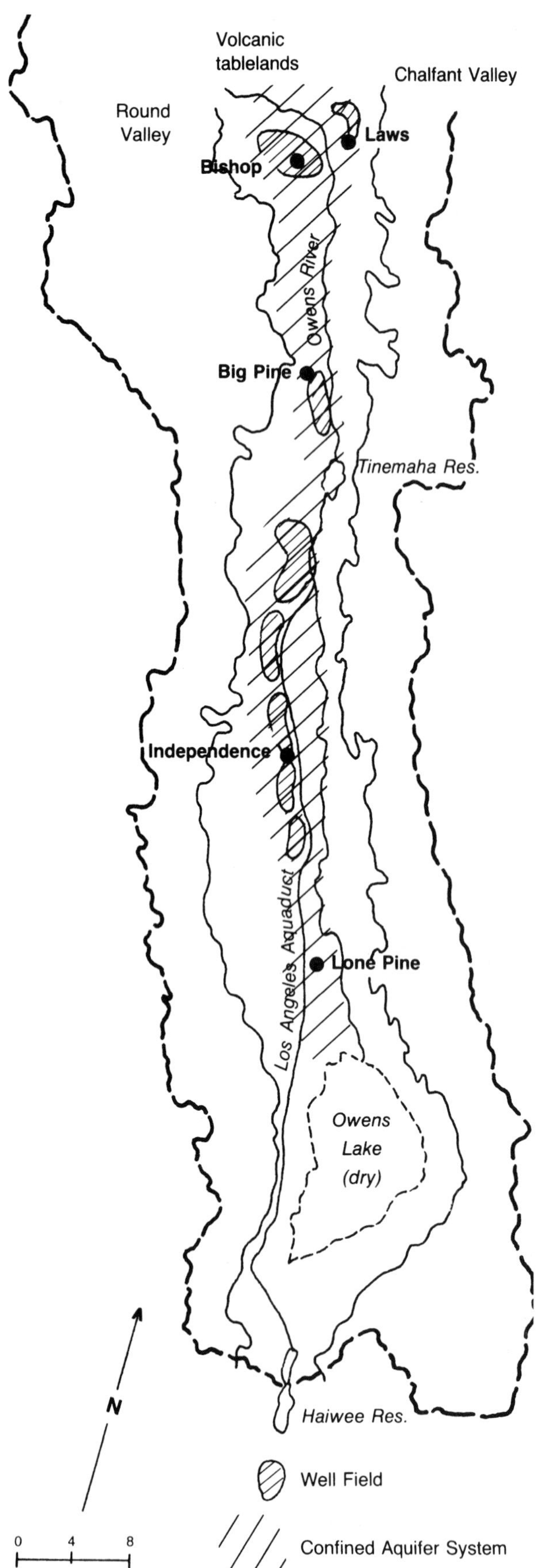

Figure 5-8 *Owens Valley Groundwater Basin well fields and confined aquifer system.* (Marc Blodgett)

of energy'' inherent in the Los Angeles Aqueduct system.[5] To clarify the claim, it was noted that *gravity flow* down the aqueducts from a 4,000-foot elevation in the Owens Valley to near sea level in the San Fernando Valley precluded energy use for pumping anywhere in the aqueduct system. As thorough as they were, though, the 1979 and a later 1980s report shared a shortcoming: Both were developed under the auspices of a single government agency, the city of Los Angeles. This could be likened to the mice guarding the cheese; however, as we will see, a third EIR was released in 1990 that would address this issue. Although it may not be intended as an impact mitigation per se, another benefit to the environment gained by the LADWP's presence in the Owens Valley is its prohibition of commercial signs and billboards on roadsides of U.S. 6 and 395 it owns.

The long-running legal wrangling between the city of Los Angeles and Inyo County over Owens Valley groundwater pumping finally was substantially resolved in September 1990 with the release of yet another EIR. Although this was the third such report issued since the court of appeals mandated an EIR in 1973, it was the first jointly prepared by the two governments and thereby perhaps the most objective report so far. The 1990 EIR pulled no punches in its assessment of the environmental damage done to the valley by increased groundwater pumping. It noted such negative impacts as severe lowering of some private wells because of pumping by nearby DWP wells, drying up of springs and wetlands, death to trees and shrubs on more than a thousand acres, failed regeneration of native vegetation on another thousand acres, and worsening air pollution from dust blowing off increasingly denuded land. The report also proposed remedies such as releasing water to create more pastureland and expanding cropland and native plant cover to reduce dust pollution—remedies that were regarded as largely experimental and therefore viewed with a certain degree of skepticism. To provide both an overview of the 1990 report and a means of comparison with the 1979 report, some excerpts from the 1990 EIR appear in Table 5-2. Lurking over all of this was the specter of a seventh year of an epic drought, which could have rendered all the report's recommendations moot had it continued for another few years. Fortunately, some ''natural mitigation'' of the enivironmental impacts of groundwater pumping may have come in the form of record rain and snow in the eastern

[5] Los Angeles Department of Water and Power, *Final Environmental Impact Report on Increased Pumping of the Owens Valley Groundwater Basin* (June 1979), pp. B1–3.

TABLE 5-2 Selected environmental effects of groundwater pumping in the Owens Valley from the Draft EIR, September 1990.

IMPACT	SIGNIFICANCE WITHOUT MITIGATION	MITIGATION	SIGNIFICANCE WITH MITIGATION
8-1: Groundwater pumping associated with the project has not and will not result in ground subsidence.	LS	8-1: None required.	LS
10-1: Vegetation in an area of approximately 300 acres near Five Bridges Road north of Bishop was significantly adversely affected during 1988 because of the operation of two wells, to supply water to enhancement/ mitigation projects.	S	10-12: Water has been spread over the affected area since 1988. By the summer of 1990, revegetation of native species had begun on approximately 80 percent of the affected area. LADWP and Inyo County are developing a plan to revegetate the entire affected area with riparian and meadow vegetation. This plan will be implemented when it has been completed.	LS
10-17: Meadow and riparian vegetation that were supplied by tailwater from formerly irrigated lands had been impacted.	S	10-17: The loss of meadow or riparian vegetation that was dependent on tailwater from formerly irrigated fields will be mitigated in the form of compensation by the restoration of meadow and riparian vegetation by the Lower Owens River Project.	LS
11-1: Changes of surface water management practices and increased groundwater pumping have altered the habitats on which wildlife depends. Vegetation changes have been significant in many locations throughout the Valley. Therefore, impacts to certain species of wildlife, which were entirely dependent upon the impacted habitat, can be presumed to be significant.	S	11-1: The importance of riparian, marsh and aquatic habitats is recognized for mitigation of the impacts to wildlife that occurred during the 1970 to 1990 period. Wetter habitats support many more species and greater populations of wildlife; therefore, water management to create wet habitats will be used to mitigate the significant adverse impacts of the project.	LS
11-2: The Agreement would protect native vegetation, improve fish and wildlife habitat, and result in beneficial impacts.	B	11-2: None required; however, LADWP would continue to conduct its program of on-going wildlife inventories, monthly wildlife censuses, raptor surveys, habitat assessments, breeding bird surveys, and other ecological studies.	LS
16-13: Air quality could be adversely affected by the construction and maintenance of new wells.	S	16-13: All areas disturbed during construction of the new wells would be wetted during construction to minimize generation of fugitive dust.	LS

Note: Section 8 addresses geology, soils, and seismicity; Section 10, vegetation; Section 11, wildlife; and Section 16, ancillary facilities. S = significant, LS = less than significant, B = beneficial.

Source: Los Angeles Department of Water and Power and Inyo County, Draft EIR, *Water from the Owens Valley to Supply the Second Los Angeles Aqueduct, 1970 to 1990 and 1990 Onward, Pursuant to a Long Term Groundwater Management Plan, Vol. 1, SCH 89080705* (September 1990) pp. 7-4 through 7-24.

Sierra and the Owens Valley in the winter of 1993. In the meantime, the sustained growth of the nation's second largest city and the environmental viability of the state's deepest valley hang in the balance.

IMPACTS OF STREAM DIVERSION ON MONO BASIN

Environmentally, the broader controversy involving both stream and groundwater diversion by the Los Angeles Department of Water and Power (DWP) may never be resolved. Decades ago, Owens Lake was sent down the drain, as it were, to Los Angeles; since 1941, the same thing has been happening to Mono Lake (see Figure 5-6).

Will Mono Basin one day resemble the alkali dust bowl that is now Owens Lake playa? Although the lake area is projected to shrink to 38 square miles (down from its present area of 60 square miles) in 100 years, will some future drought necessitate reducing even that area? The environmental impacts of increased water diversion in the Owens Valley and Mono Basin notwithstanding, another drought will leave no choice in the matter if the well-being of millions of Los Angeles residents is threatened.

Los Angeles's continued diversions (100,000 AF per year) of streams feeding Mono Lake may pose a grim if not deadly fate for certain basin wildlife. Specifically, it is the California gulls (*Larus californicus*) of Mono Lake (95 percent of the California population and 20–25 percent of the species population reside here) and the brine shrimp (*Artemia monica*) they feed on that are losing their lives and their breeding habitats. As the water level lowers by 1–2 feet per year, the gulls are increasingly brought into double jeopardy. First, land bridges form that connect their island nesting sites to the mainland, allowing coyotes (*Canis latrans*) and other terrestrial predators easy access. Second, salinity increases to fatal levels for newly hatching brine shrimp, the gull chicks' prime source of food, and brine flies, another major food source for basin bird life.

The first problem has been combatted by blasting the land bridges apart and erecting chain-link fences across them, but the gulls have continued their evacuation of the connected islands. The second problem has proved far more perplexing and may result in the abandonment of the Mono Basin as a rookery for the gulls and as a resting and feeding place for several species of migratory birds. The *Report of the Interagency Task Force on Mono Lake*, of which task force the LADWP is a member, notes that "experimental evidence indicates that populations of brine shrimp and flies are likely to decline with increasing salinities. It is unlikely that any of Mono Lake's major bird species, including the gulls, grebes, and phalaropes, will persist at the lake if populations of invertebrates disappear."[6] The distance of suitable alternative habitats for the birds only compounds the problem. The nearest viable environments are Abert Lake, 300 miles north in Oregon; the Salton Sea, 350 miles to the south; the Great Salt Lake, 400 miles to the east in Utah; and San Francisco Bay, 14,000 feet up and over the Sierra Nevada crest and then 175 miles to the west.

Matters worsened in early summer of 1981, when an abnormal percentage of gull chicks perished, the normal survival rate being about 50 percent. However, in a briefing document issued in May 1982 entitled "Los Angeles's Mono Basin Water Supply," the DWP stated that

> the exact causes are not known; however, salinity is not considered the cause as the lake level and conditions were approximately the same as in 1980. Research conducted by the City at ponds adjacent to Mono Lake under experimental conditions has shown that the brine shrimp can thrive in a salinity of approximately 15 percent salt. This salinity content will not be reached in Mono Lake for at least another 30 years. Other research also shows the brine shrimp can tolerate higher salinity. To date, there is no evidence that the brine shrimp could not adapt to an ultimate estimated salinity of 21 percent in Mono Lake, which will be reached very gradually over a period of 80 to 100 years. Although the brine shrimp populations for May and June 1981 were somewhat below the estimated average in previous years during the same period, the populations for July , August, and September 1981 were much higher compared to previous years during the same period. Latest research performed in summer 1981 by Dr. Joseph Jehl Jr., of Hubbs-Seaworld Research Institute, has determined that the migratory birds have visited the lake in their usual large numbers and have found abundant supplies of brine shrimp and brine flies on which to feed.

If one detects an unusual degree of contradiction between the task force report and the briefing document, one is reminded that controversy, especially over water, often breeds contradiction.

Over the remainder of the decade, things worsened with regard to Los Angeles's interests in Mono Basin. The gull population was sustained but made no dramatic comeback from the disaster of 1981. Trout fishermen got their way when the DWP was ordered to curtail its diversion of certain streams feeding into Mono Lake, lest those streams become inviable fisheries for lack of sufficient water. The drought commenced ever so subtly

[6] U.S. Forest Service, U.S. Bureau of Land Management, U.S. Fish and Wildlife Service, California Department of Water Resources, California Department of Fish and Game, Mono County, and the Los Angeles Department of Water and Power, *Report of the Interagency Task Force on Mono Lake* (December 1979), p. 20.

TABLE 5-3 Mono Lake: Benefit–cost comparison of selected benefits and alternatives.

ALTERNATIVE CONDITION OR LEVEL	WATER SUPPLY BENEFITS	POWER GENERATION BENEFITS	RECREATION BENEFITS	PRESERVATION BENEFITS	NET BENEFITS
No restriction	+5.1	+1.3	−2.9	−759.7	−753.0
6,372 feet	−10.8	−1.9	+0.4	0.0	−12.3
6,377 feet	−16.5	−2.7	+1.1	+22.6	+3.2
6,383.5 feet	−24.7	−4.2	+1.9	+63.0	+31.8
6,390 feet	−28.7	−5.0	+2.7	+85.9	+49.9
6,410 feet	−35.4	−6.7	+1.2	0.0	−43.4
No diversion	−41.2	−8.2	+1.2	0.0	−50.9

Source: Adapted from California Water Resources Control Board, "Summary" of draft *Mono Basin EIR* (May 1993), Table S-1.

in 1986, but by the winter of 1989, the eastern Sierran snowpack water yield was forecast at 50 percent of normal. Worst of all, though, was the lowering of the lake level below 6,377 feet (above sea level) in 1989 and the possible grave consequences this might have on nesting gulls and other wildlife. Environmental groups brought pressure to bear and in June 1989 obtained a superior court order prohibiting export of any more water from Mono Basin by the DWP until Mono Lake again reached a 6,377-foot level. That same year, a law was passed authorizing $60 million in state funds for projects to restore Mono Lake to its pre-1941 condition. Of those funds, $36 million was allocated to the DWP to find sources of water other than from Mono Basin, such alternative sources being subject to agreement by both the city and environmentalists. It took nearly 5 years, but by the beginning of 1994, the parties to the 1989 law had agreed to permanently stop diverting one-third of what Los Angeles had taken out of Mono Basin each year prior to the moratorium. Also in 1994, the state water board ruled in favor of the Mono Lake environmentalists by extending the moratorium on diversions. To compensate for the reduction, which amounts to about 35,000 AF (and serves 70,000 households) annually, Los Angeles will develop a $50-million water reclamation project in the San Fernando Valley. While this is a precedent-setting concession by the city, the eastern Sierra will have to experience normal precipitation for several years running before the city can tap into its eventual two-thirds allotment.

The aforementioned 6,377-foot level for Mono Lake is one of seven alternatives outlined in the State Water Resources Control Board's (SWRCB) May 1993 "Summary" of its draft *Mono Basin Environmental Impact Report.* Table 5-3 provides a benefit–cost comparison of water supply, power generation, recreation, and preservation benefits relative to the seven alternatives, ranging from "no restriction," which would allow the city to divert water based entirely on availability and need, to "no diversion," whereby diversions of the four streams would be entirely curtailed. It is noteworthy that under "Net Benefits," costs (also known as disbenefits, or minus or negative benefits) ranging from −$753 million to −$12 million were projected to occur under the no-restriction, 6,372-foot, 6,410-foot, and no-diversion levels, while benefits ranging from +$49 million to +$32 million were seen for the 6,377-foot 6,383.5-foot, and 6,390-foot alternatives. Most revealing, though, are the highest costs, −$753 million and −$50.9 million, appearing at the extremes of "no restriction" and "no diversion," respectively.

Remember, as we next examine water law and the history of Los Angeles's quest for water, that the city's other sources of water supply—the California Aqueduct, the Colorado River Aqueduct, and local groundwater and surface streams—are in some cases shared with communities inside and outside California and thus could become unavailable, as happened when the California Aqueduct was shut off in 1977. When precipitation returned to normal the following winter, the California Aqueduct was turned back on, and once again some of Northern California's water began flowing to Southern California. One wonders, though, if even a longer shutdown would really impose a lasting, or even short-term, limit to growth on Southern California. Voter rejection (Proposition 9, June 1982) of the California Aqueduct's proposed feeder aqueduct, the Peripheral Canal (see Figure 5-5), will have considerable bearing on the problem and will be discussed later in the chapter. In the meantime, the city of Los Angeles's decades-old ownership of property in, and legal rights to, the waters of the Owens Valley and Mono Basin are likely to perpetuate that city's continued growth.

THE WESTERN (CALIFORNIAN) REVOLUTION IN WATER LAW

Water is a necessity for humankind. It can be argued that water is *the* single most critical natural resource for human life. This is certainly true for people living in the arid West—and in much of California. Without water, the chain of life is broken, humans and animals are in jeopardy, and all fundamental elements of civilization are forestalled. Human location, transportation, recreation, and even spiritualization are short-circuited. Water provides life, crops, recreation, health, and protection. Control and distribution of water has thus become a major modern concern. In areas where water is in short supply, equitable uses are critical and basic to society's existence. California is one such area. As would be expected, laws dealing with water in California have reflected changing economic, environmental, technological, and social conditions. Since water has changed much of the face of California, making habitation, farming, and recreation possible in many places, legal control of this resource represents a significant area of legislation and litigation. It is therefore important to understand the fundamental legal doctrines that underlie much of the debate.

It has been suggested that two realities govern the "law" of water in the arid West: the basic question of availability of water and the related issue of how the available water is distributed, geographically, throughout the region. Much of early American water law was based on the premise that water was a ubiquitous resource (available virtually everywhere)—a concept that provided a "poor fit" for California and other western states.

Common Law Riparian Rights

The concept of *riparian rights* is the most common water doctrine in the United States. Simply stated, the riparian doctrine gives certain rights in the flow of water to the owner of land next to or overlying the water flow. These water rights are considered to be a part of the ownership interest of the land and a form of real property right. The extent of this right to use water has been debated for years, with two approaches emerging from the discussion. The "reasonable use" theory limits water use only to the extent that such use does not negatively affect other users. Thus, the premise is that all riparian users have equally protected rights vis-à-vis each other. The alternate theory, which might be called that of "domestic use," is that the riparian owner may take water for domestic purposes only (family, livestock, gardening) and that other uses of water may be infringements of other riparian owners' rights.

The ultimate thrust of the riparian doctrine is that (1) the water user must own land bordering the water flow and (2) the water user must limit use to those purposes deemed to be either natural or reasonable. Certain preferred uses, such as domestic or livestock over commercial or industrial, are recognized in riparian law. The doctrine of riparian rights has both advantages and disadvantages. Generally, diversion of water is not permitted under the doctrine, and this limits the use of water by nonriparian landholders. This may be a critical problem where there are few free-flowing streams. Because of the limitations of riparian doctrine, therefore, another major water allocation system was developed.

The Doctrine of Prior Appropriation (the Colorado Doctrine)

In the absence of abundant free-running water, the doctrine of riparian rights seems to be neither useful nor appropriate. In the dry western states, consequently, the doctrine met severe challenges. Due to extensive federal ownership of vast tracts of this land, early residents could not claim water use rights as arising from the ownership of land they did not legally possess. A practical solution to the problem of water rights evolved that created superior rights for those who first put the water to beneficial uses. This came to be known as the *doctrine of prior appropriation* and permitted the first person to appropriate water from a stream to continue using that amount of water against all subsequent users, as long as that original use was continuous. Appropriation rights, however, can be lost by failure to continue using the water acquired in this fashion.

In essence, this doctrine established a first-come, first-served legal process for water in the arid West. Miners and settlers were able to regulate legal control of water in accordance with this rule and establish some order and predictability in the process. It should be noted that the doctrine did not overturn riparian rules or take precedence over the rights of the federal government, which still owned the land, but only served to settle disputes between competing individuals, each of whom was claiming rights to water on land they did not own.

Later federal legislation attempted to clear up the question of riparian versus appropriation rights. Both the states and the federal government desired to respect the developing theory of appropriation while still preserving some measure of respect for the riparian doctrine insofar as it might apply to any particular future case. In its purest application (the Colorado Doctrine), however, prior appropriation separated water rights entirely from land rights and put the control of water in the hands of the state as a public trust. This often meant, in practice, that each western state could establish its own set of rules governing appropriation of water, thus undercutting riparian doctrine almost totally.

Most western states eventually developed a permit system whereby water users would apply for certain uses to the state and, upon compliance with permit requirements, would receive permission from the state to appropriate water for beneficial purposes. This placed significant power in the hands of the state agency, as well as heavy responsibility for monitoring competing uses and determining overall best interests of the public at large. Competing uses to be considered included personal, household, livestock, agricultural, recreation, fishing, and many others. Finally, the doctrine created a new approach to water use that tried to establish preferential uses of water and protect as many landowners as possible. The new system was better suited to the western states but needed further refinement to meet the specific needs of each state.

The California Doctrine: A Marriage of Convenience

Water distribution in California created a unique problem that called for a blending of ideas from both riparian and appropriation theories. The greatest difficulty was not the total absence or shortage of water, but rather its geographically imbalanced distribution. There is an abundance of water in the state as a whole, but most of it is located in the north where there is less need for it. The distribution of water is further imbalanced by its seasonality. Given this imbalance of location and occurrence, the state attempted from the outset to provide an orderly and consolidated water allocation plan that would serve the needs of all its residents equally.

To accomplish the goal of fair water use, the state of California has created a modified doctrine utilizing aspects of both the recognized approaches. Through various court cases, and eventually by constitutional enactment, a series of rules on water use has evolved.

First, the state declared that it would recognize and honor the water rights acquired under the prior appropriation process. This protected those water users who had established time priority.

Second, the state acknowledged that riparian rights did exist. The California courts noted that the original riparian rights had belonged to the original owner of the land—the U.S. government, which took title from Mexico—and that upon transfer to private parties these riparian rights were also transferred.

Third, riparian rights acquired by a recent transfer would not be absolute. They would be subject to those prior appropriation rights that existed at the time of transfer and would be further limited by a proviso that the riparian right would extend only to such amounts of water as were reasonably required for beneficial use of the land. This could prevent a riparian owner from selling water above the amount needed for the riparian land alone.

Fourth, the water resources of the state were declared to be so vital as to require such water to be used for the general welfare and for the public interest. This meant that water use would be subject to specific state control and regulation (although riparian and appropriation owners' rights were to be respected).

Fifth, all water above the amount needed for limited riparian and appropriation uses would be considered "excess water" subject to state control. This control would be exercised by a state water rights board.

Sixth, all uses of water that came about as a result of appropriation would be deemed to be "public uses" and thus subject to state control. In practice, this meant that the bulk of water available in the state was subject to governmental control. Once this fact was established by definition, it remained only for the state to determine priorities of use and provide a permit system for use and distribution.

The philosophical position of the state is reflected in the California Water Code, which places the highest priority on domestic use, followed by irrigation. Furthermore, applications by a municipality for use of water by its residents is given priority over most other competing uses. Beyond the priorities thus outlined, it is left to the judgment of the water board to determine allocations in such a way as to serve the broad public interest. All uses are handled by a permit application process, and the board acts on a case-by-case basis. Finally, the board is enjoined to give constant attention to state water plans for coordinated water use (see the discussions of the California Water Project and Central Valley Project later in this chapter).

The efforts of the state of California to develop a comprehensive approach to the use of water were not quickly or easily accomplished. The California Doctrine reflected an uneasy compromise of various rights and arrived at a position that would benefit the most people to the least detriment of any individual. Obviously, some individuals and localities were necessarily subordinated to the general welfare, a circumstance that created bitter and longstanding controversies—the conflict between Los Angeles and the Owens Valley being a case in point.

A CITY'S SEARCH FOR WATER: *Los Angeles and the Owens Valley War*

One of the enduring controversies in California, as well as a source of constant regional tension, was the appropriation by Los Angeles of the waters of the Owens Valley. Located in an arid part of the state, with an ever-expanding population, the city has required a dependable and extensive supply of water from outside its boundaries. Without water, Los Angeles would revert to its basic semidesert beginnings. The existence of a reliable source of water, however, has permitted the city and region to sustain a growing industrial, agricultural, and residential base. The development of such a water source by Los Angeles is, in large measure, a microcosm of the traditional conflict in California of urban south versus rural north—or more directly, the battle between city and country.

With a distinctly finite source of water for Los Angeles, coming from limited wells and the perennial (and unreliable) Los Angeles River, population growth was limited. Simply stated, in terms of water, the carrying capacity of the Los Angeles Basin had been reached by the end of the 1800s. Unless more water could be found, the city would be permanently frozen in population. The ultimate solution was an ambitious interbasin transfer of water.

Mulholland's Dream

As early as the turn of the century, many residents of Los Angeles began to realize the city's critical need for a reliable source of water. With a near drought in 1904 and the inability of the Los Angeles River and a few local wells to meet the demand, the city was ready to consider new methods of acquiring the needed water. Population growth figures for the area were astonishing, and so long-term planning was critical.

Based on a scheme developed by Fred Eaton, a former mayor and engineer for the city, it was proposed that the waters of the Owens River, 250 miles away, be brought to Los Angeles by means of a $25-million aqueduct. After passage of a bond issue by city voters, City Engineer William Mulholland undertook the amazing project in 1908. Pushing a pipe-and-flume, tunnel-and-trench system across the desert, he completed the job in 5 years. A remarkable feat of engineering even by today's standards, this system brought reliable water service to the city, as well as generating by gravity flow vast amounts of hydroelectric power along the route of the aqueduct.

Applying the theory of "the greatest good for the greatest number," the project did divert water from the Owens Valley and prevented a full-scale reclamation program from being completed there. Through a combination of agreements, legal theories, cooperative actions, and occasionally questionable dealings, the city had succeeded in acquiring rights to this water. Acquisition of excess water that normally would have gone into Owens Lake was accomplished by the process of appropriation. Furthermore, the city purchased riparian land in the valley to control these claims along the river's course. Under intense lobbying by the city, the federal government assisted by closing the remaining land in the valley to homesteading and by granting right-of-way on federal lands for the aqueduct. For several years after completion of the aqueduct, the city set about purchasing most of the rest of the riparian land of the Owens Valley, eventually gaining control of in excess of 90 percent.

During this period, and throughout the late 1920s, the city faced strong opposition from organized groups in the valley who contested the loss of control of the region and wished to preserve the valley for its residents. The city claimed that these opponents only wished to force the city to pay unfair prices for the land. Charges and countercharges flew back and forth and finally erupted in open violence.

The Owens Valley War

The sources of discontent over the project were many and varied. The harshest critics accused Los Angeles of "raping" the valley for the benefit of a few unscrupulous and greedy land speculators who wished to develop the San Fernando Valley with the water brought from the Owens River. The ranchers, businessmen, and farmers of the Owens Valley who had seen the possibility of reclaiming the land and prospering from its resultant fertility were understandably angry with and antagonistic toward the interlopers they believed were ruining their valley. Residents who were forced out or bought out under threat of eviction had lost their homes, even if the prices paid were fair—a fact not always conceded. Conservationists and naturalists who witnessed destruction of the valley with the removal of the water expressed their dissatisfaction. But as charges flew back and forth, the ultimate fate of the valley's water was already sealed. Various groups tried to hold out against the land purchases by Los Angeles but soon found themselves overwhelmed. Frustration eventually turned to violence, with dynamiting of the aqueduct occurring at least nine times, along with other forms of violence. These dangerous confrontations between armed ranchers and armed guards on the aqueduct only served to heighten bad feelings already at a fever pitch due to misunderstandings, mutual stubbornness, inept personal relations, and often unethical behavior.

Ultimately, the city triumphed, and some long-range benefits both to Los Angeles and to the Owens Valley were achieved. The emotionalism engendered by the conflict tended to obscure much of the true nature of the project and its role in the development of the state. Clearly, there were examples of greed, dishonesty, and self-serving financial manipulations on both sides. The city needed water, but its political machinations were often callous and heavy-handed. The Owens Valley residents wanted to preserve their homes, but if they could not, they frequently tried to bleed the city for every cent they could get above a fair and reasonable market value. Bitterness resulted on both sides, leaving a residue of ill feeling in the Owens Valley that has continued down to the present.

In historical perspective, although the dispute was extended, public, and acrimonious, it was only one small episode in the overall drama of water distribution in the state. As noted earlier in the chapter, this drama is far from completed. The controversy has raged on in the media and the courts and will continue to do so well into the twenty-first century. Increasingly, environmentalists and "friends of Mono Lake" have successfully blocked the Los Angeles Department of Water and Power from proposed water utilization plans in the Owens Valley. Using a combination of actions by political pressure groups, legal maneuvers, and administrative manipulations, valley environmentalists have thwarted further water diversions. The success of these efforts was underscored in 1990 when the California State Lands Commission supported their position in court hearings (arguably the first time in the state's history when a major state agency has done so). With ongoing drought problems, continued growth in the Los Angeles Basin, and the resulting demand for water, this battle will assume even greater significance in the future. Water may ultimately prove to be the "dictator" of lifestyles and population distribution in Southern California.

BUSINESSMEN VERSUS NATURALISTS: *The San Francisco Water Controversy*

The battles over water in California are not primarily north-versus-south controversies. Rather, they are a reflection of urban pressures brought to bear against rural lifestyles. This can be seen even more clearly in the case of the San Francisco Bay water controversy.

As with Los Angeles, the growth of San Francisco caused an increased demand for water. By the late 1800s, private water companies were unable to keep up with the burgeoning demand. Thus, Bay Area civic leaders began to cast about for a reliable and adequate water source. As a means of solving the water shortage, San Francisco proposed the then-novel expedient of transporting water from another region of the state by means of an aqueduct. The early 1900s saw the city attempt to purchase water rights along the Tuolomne River in the Yosemite Park area. Specifically, the city proposed to dam the Hetch Hetchy Gorge and from there build an aqueduct approximately 175 miles to San Francisco. The proposal was met with howls of dismay from naturalists and others, and the battle lines were drawn.

For 10 years the controversy raged between those who wanted to preserve and protect the natural beauty of the gorge and those who advocated the need to provide water in the populous Bay Area. Further complicating matters were various parties who saw a threat to the private water and power companies if the project were allowed to go forward. In environmental circles, John Muir, the Sierra Club, and other individuals and groups attempted to thwart the project. After wavering on the issue, however, the U.S. Department of the Interior finally granted approval in 1913. The project was completed in 1931, demonstrating the precedence of urban population needs over rural and environmental issues.

Clearly, this was not an easily resolved problem, and the final solution left a bitter taste in the mouths of many people. It did, however, provide a viable water source for San Francisco. The pattern was also followed by the East Bay cities, which collectively purchased water rights along the Mokelumne River and built the 95-mile East Bay Aqueduct to service their populations.

With continued population growth in the Bay Area, water demand also grew (as it had in the south). In an ironic parallel with Los Angeles's water problems, the San Francisco Hetch Hetchy debate has also shown remarkable persistence and further underscores more recent uncertainty over the ultimate wisdom of interbasin water transfers.

In what was perceived as an astonishing move, in the late 1980s, Ronald Reagan's prodevelopment secretary of the interior, Donald Hodel, proposed to dismantle the O'Shaughnessy Dam and reclaim Hetch Hetchy Valley. Environmentalists remained skeptical of hidden motives, while Hodel's prodevelopment friends were similarly nonplussed. The proposal raised significant questions: Where would the Bay Area communities find replacement water? Because the Hetch Hetchy project also produced substantial hydroelectric power, where would replacement power be found? What economic impacts would result? How would the actual physical process of removal of a massive dam be accomplished? How would a valley inundated with water since 1923 be restored to its "natural" state? How long would such a restoration take? Would substitute facilities ultimately have to be built? In short, the proposal raised many crucial questions going to the heart of California's water needs. Although Hodel (and Reagan) left office before this plan could be pushed any further, it did underscore the issues and problems of trying to find an appropriate balance between environmental concerns and the demands of a thirsty, increasingly urban California.

Thus, the transfer of water from one region to another is not unique to Los Angeles, San Francisco, or the East Bay. Water transfer from rural to urban locales merely reflects the realities of a constantly growing population in the state and related population clustering. One distinction, however, is that the needs of the south became increasingly more evident with the tremendous population rise in that region of California.

MORE WATER FOR SOUTHERN CALIFORNIA FROM SOMEWHERE ELSE

Just as imported water has enabled Los Angeles to grow into the second largest city in the nation, so has it helped provide the basis for Southern California as a whole to develop into the most populated and most prosperous part of the state. The 10-county region's total population stands at about 17 million, or nearly half of the California total, and annual personal income regularly surpasses $100 billion. Because the water that flows down the Los Angeles Aqueduct is intended for the exclusive use of the city (and even at that may be inadequate in dry years), neighboring cities have had to search elsewhere to ensure their own continued growth. In 1928, the California legislature created the Metropolitan Water District (MWD) of Southern California, which was charged with constructing the 242-mile Colorado River Aqueduct (see Figure 5-5) and using it to deliver water from "the River" to Los Angeles and other Southern California communities. In 1941, the first water from the Rocky Mountain states, via the Colorado River and Aqueduct, flowed into a reservoir in Pasadena.

Southland growth and development again boomed following World War II, and by 1960, the legislature and voters had approved what has become the most ambitious water redistribution project in the history of the world: the State Water Project (SWP). By the mid-1970s, the SWP began delivering water from Oroville Dam and Reservoir 600 miles southward to Perris Dam and Reservoir via a 200-mile stretch of the Feather and Sacramento rivers and the 400-mile California Aqueduct (see Figure 5-5). The last major link in the SWP, the Peripheral Canal, was voted down in 1982. The delivery capability of the State Water Project is many times that of either the Los Angeles Department of Water and Power (DWP) or MWD systems.

Although the DWP, MWD, and SWP all perform the same basic function of delivering water to Southern California from remote surplus source areas, some differences among the three systems are worthy of mention. The gravity flow of the Los Angeles Aqueduct precludes the need for pumping anywhere along the DWP line. In contrast, water must be lifted 1,617 feet by five pumping plants along the Colorado River Aqueduct and even higher on the California Aqueduct—for example, some 2,000 feet by the Edmonston Pumping Plant to surmount the Tehachapi Mountains. Each acre-foot of water delivered requires 2,000 kilowatt hours (kWh) of electricity for the MWD aqueduct and 3,170 kWh for the SWP aqueduct, and the cost of this energy is rising rapidly.

The DWP system enjoys another advantage: Its water is of higher quality, with a salt and other dissolved-solids content nearly one-fourth that of Colorado River water. SWP water, which is nearly equal to DWP water in quality, is often blended with MWD water to prevent the latter from corroding pipes and increasing the salinity of groundwater. The SWP serves a much more extensive area than either the DWP or MWD. Distributary aqueducts from the California Aqueduct serve the San Francisco and southern Coast Ranges regions as well as Southern California. Also, the Feather River Project unit (specifically Oroville Dam) helps control the flooding that once ravaged Yuba City and Marysville and provides hydropower to nearby communities.

Neither the MWD nor the SWP has been any more successful in escaping controversy than has the DWP. The main source of dispute over the Colorado River's waters has been the fact that they are shared by seven states and two nations (the United States and Mexico), most of which are water-short areas. Still other Southern California–based agencies, specifically the Imperial Valley Irrigation District and Coachella Valley Water District, use the Colorado's waters, diverting them into the valley for irrigation via the All-American and Coachella canals (see Figure 5-5). In the 1920s, the Colorado Compact and treaties with Mexico apportioned the river's waters among the several users. But Arizona and California continued to squabble over the Colorado until a 1963 court decision called for California users to lose a portion of their original allotment of river water starting in 1985. The Central Arizona Project (CAP) began operations on time and ultimately will take most, if not all, of what California users of Colorado River water must relinquish by law.

As mentioned at the outset of the chapter, California withdraws about 4.8 MAF annually from the lower Colorado River (net water use is about 4 MAF; see Table 5-1) whereas its basic apportionment is set at 4.4 MAF per year under the 1963 court ruling. Arizona is expected to be using all its annual apportionment of 2.8 MAF from the river later in the 1990s, once the CAP is fully functional. Because the recent epic drought did not spread to upper Colorado River Basin states, which through most of the 1980s and 1990s did not tap their full apportionments, California was able to use more than its legal allotment of Colorado River water over most of the 1980s. Moreover, conservation by and cooperation among the MWD, Imperial Valley Irrigation District, Coachella Valley Water District, and other Southern California users could diminish California's demands on Colorado River water. There is a fly in the conservation ointment, though, and that is that the "conserved" Colorado River water could simply be reallocated among California agencies and thereby do nothing to reduce the state's overall use of it. All this notwithstanding, urban population growth has far exceeded expectations in the service areas of these water agencies, and should it continue at the current pace (see Chapter

1), water shortages could result. Fortunately, by 1995, a decade after the CAP began to tap into the lower Colorado, the river was still satisfying all its users. Yet, as we will discuss next, a major drought in the upper Colorado River Basin would add to a potential shortfall anywhere downstream.

The MWD could suffer still further allotment cuts if possible litigation over the rights of Navajo Indian reservations to the Colorado's waters are resolved in the Navajos' favor or if all agencies that have legal rights to use of the waters of the Colorado suddenly exercise them. The latter event is not likely to transpire, but the overcommitment of river water rights should never have happened in the first place. At times, the Colorado River's bankruptcy has amounted to some 3 MAF more than normally flows in the river in a year. If a drought strikes the upper Colorado River states and they are forced to use their full allotments, the stream's bankruptcy will spread to Arizona and California.

Like the LADWP, the SWP has to contend only with opposition within the state. But, as the DWP well knows, this opposition can be formidable. Nevertheless, the development of the SWP progressed steadily through the 1960s and 1970s except for two setbacks: the halted development of perhaps the most important link in its system, the Peripheral Canal, and the 1972 California Wild and Scenic Rivers Act, which thwarted diversion of north-western rivers. Without the Peripheral Canal circumventing the Sacramento Delta to carry Sacramento–Feather River water directly into the California Aqueduct and Delta Mendota Canal, Southern California and the San Joaquin Valley are missing out on as much as 1 MAF of additional water a year.

Opponents of the Peripheral Canal contended that if this 1 MAF, or anything like it, were sent southward instead of into the Delta, saltwater intrusion (see Figure 5-2) from the bay would ruin farming and negatively modify the rest of the environment. Also, they said, the bay itself would not be as well flushed of saltwater. Farther inland, the Sacramento–San Joaquin Delta was targeted by the federal Environmental Protection Agency late in 1993 for tighter regulation of freshwater exports, which could profoundly decrease SWP exports from the Delta to Southern California. Proposition 13, the 1990s recession, and the accompanying fiscal conservatism of the state legislature also have had a long-term negative impact on financing of both state and joint state-federal water resource projects from special appropriations and/or the state's general fund. Moreover, as witnessed in 1979 by Santa Barbara County voters' rejection of a proposal to import SWP water and in 1991 by the city of Santa Barbara's decision not to build a $30-million desalination plant, the no-growth syndrome among Californians is apparently growing and taking its toll on water redistribution development proposals. Lastly, although in 1980 both the legislature and the governor approved construction of the 43-mile, $680-million Peripheral Canal, opponents gathered more than enough signatures to have the issue qualify for the June 1982 ballot.

The resounding defeat of Proposition 9 (the "Water Facilities Including a Peripheral Canal" referendum statute) by the voters in June 1982 could be attributed to taxpayers' fear of the facilities' costs as much as to anything else. The *California Ballot Pamphlet* for the primary election listed potential costs (from Senate Bill 200) in excess of $2.29 billion—$680 million for the Peripheral Canal, $139 million for relocation of Contra Costa Canal intake and construction and/or improvement of other Delta and Suisun Marsh facilities, $872 million for Los Vaqueros Reservoir, $493 million for Glenn Reservoir, $112 million for groundwater storage facilities in the San Joaquin Valley and Southern California, and an "unknown" amount for south San Francisco Bay Area groundwater storage facilities. Opponents predicted total costs of at least $3.68 billion and as much as $19.2 billion. They also argued that project goals could be achieved by far cheaper means.

Perhaps more significant than any other single reason given for the defeat of Proposition 9 was the existence in Northern California of an informed electorate anxious to get out and vote "no" versus a much less concerned electorate in Southern California. Northerners knew well what was at stake for their part of the state; equivalent knowledge seemed to have eluded the many Southern Californians. Otherwise, why was Proposition 9 soundly rejected in northern counties and only moderately successful in the south (see Table 5-4)? However the election results are interpreted, the environment of Northern California appears to have won a new lease on life, and Southern California may have imposed

TABLE 5-4 Selected counties' vote counts on Proposition 9, June 1982.

REGION AND COUNTY	YES	NO
Northern California		
Alameda	13,680	265,080
Butte	2,552	43,758
Marin	2,437	79,346
San Francisco	7,284	140,574
Santa Clara	29,525	249,616
Southern California		
Imperial	8,404	6,982
Los Angeles	895,716	572,721
Riverside	90,441	60,861
San Diego	306,670	111,013
Ventura	64,014	51,332

Source: Los Angeles Times, June 10, 1982, p. 19.

upon itself a limit to growth. Whichever side of the environmental and economic issues a Californian may be on, it pays to gain geographic insight into the issues and use that knowledge at the polling place.

THE CVP: *A Boon to Agriculture*

The Central Valley Project (CVP), portrayed in Figure 5-9, differs from California's other major regional water resources projects in two respects: (1) Its primary benefit is to agriculture rather than to municipalities and nonagricultural industry; and (2) it is funded and controlled by the federal government rather than by a city, as in the case of Los Angeles and its Department of Water and Power or by the state and its State Water Project. Actually, the CVP originated exclusively as a state-sponsored project with legislative, gubernatorial, and voter approval all given in 1933. But the Depression and the consequent reluctance of the public to buy $170 million in bonds led to the project's takeover by the U.S. Department of the Interior's Bureau of Reclamation in 1935. By 1938, contractors began construction of the project's first unit, Shasta Dam and Lake (Figure 5-10).

The CVP experienced delays caused by shortages of workers during World War II. Nevertheless, Shasta, Keswick, and Friant dams; a hydroelectric plant (at Shasta Dam); about 350 miles of main canals; some 200 miles of power transmission lines; and numerous pumping plants, bridges, and tunnels were essentially completed by the end of the 1940s. CVP developments since 1950 include impounding of the Trinity, American, and Stanislaus rivers to meet the Central Valley's growing demand for more irrigation water, municipal water supply, hydropower, flood control, and recreation facilities. A fringe benefit of the CVP system has been more water for operation of the Sacramento Ship Channel, completed in 1963, and the earlier-built (1933) Stockton Ship Channel.

The central aim of the CVP, which has been largely achieved with implementation of the projects just outlined, was to improve agriculture in the Central Valley as a whole by supplying water to the relatively arid San Joaquin Valley and by preventing saltwater intrusion in the Delta. Irrigated agriculture got its start in the San Joaquin Valley over a century ago and, with the help of surface stream diversion and groundwater pumping, steadily expanded in area through the early decades of the twentieth century. Blessed with two-thirds of the Central Valley's potential farmland but only one-third of its natural surface water, however, San Joaquin Valley irrigated agriculture increasingly turned to well water to sustain its growth. By the 1930s, overdrafting of groundwater resources, both in dry years and in years of normal runoff, threatened desertification of much of the newly won farmland. The solution lay in diverting some of the Sacramento Valley's two-thirds of all Central Valley surface water down to the San Joaquin Valley.

The CVP eventually accomplished this task with its Delta Cross Channel, Tracy Pumping Plant, and Delta–Mendota units carrying Sacramento River water into the heart of the San Joaquin Valley. This diversion freed up more San Joaquin River water for distribution from Friant Dam along the southeastern side of the valley via the Madera and Friant–Kern canals. But these marvels of the CVP notwithstanding, the desert once again threatens to reclaim hundreds of thousands of acres of San Joaquin Valley agricultural land if two problems are not soon resolved: (1) brine buildup, in part caused by the very water that irrigates the land, and (2) the implementation of a 960-acre limitation on federal irrigation water use. This latter move no doubt would result in farmers keeping their larger holdings by resorting to greater groundwater pumping and thus overdrafting (see Chapter 3 for more on the 960-acre limitation).

Besides transporting Sacramento River water across the Delta, CVP's Delta Cross Channel staves off tidal saltwater intrusion by releasing fresh water and thus helps maintain the agricultural productivity of Delta soils. But will still more valley land, this time in the western Delta, be lost to agriculture if and when an alternative to the Peripheral Canal becomes a reality?

Opponents point out that in low-runoff years, much of the Sacramento River's available flow would have been picked up by the Peripheral Canal and lost by the Delta Cross Channel, which in turn would have translated into less water available for tidal flushing and more farmland lost to saltwater encroachment in the western Delta. Proponents counter that the Peripheral Canal would have handled the release of water at several points along its course and improved water and land quality in the eastern or innermost Delta. But for these CVP and SWP Delta projects to perform their intended ecological tasks in times of drought, another massive input of fresh water will be needed in the system. Will it come from the proposed diversion of the Eel River in the far northwest when the moratorium imposed by the Wild and Scenic Rivers Act on damming of the Eel again comes under review? Or will one of the alternative water resources to be examined next come to the rescue?

WATER CONSERVATION AND REUSE

To most Californians, the drought of the 1970s is but a fading memory. The severity of the drought in 1977 prompted Los Angeles to adopt a water conservation program, but floods in 1978 and continued above-

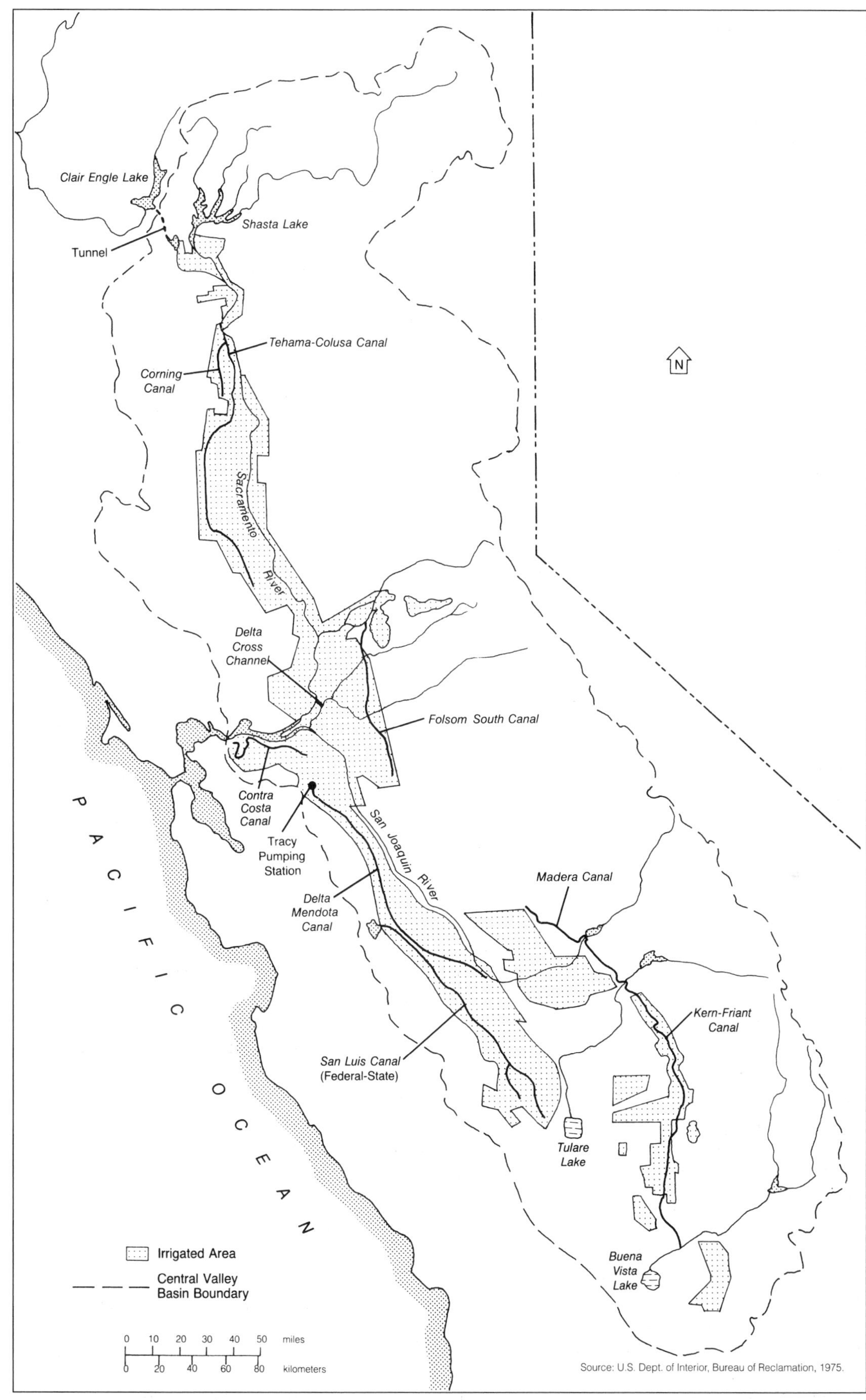

Figure 5-9 *Central Valley Project.* (Richard Crooker)

Figure 5-10 *Shasta Dam and Lake, which impound the state's largest river, the Sacramento, as part of the federal Central Valley Project. During the 1976–77 drought, the water level of the lake dropped dozens of feet. Going into the fifth year of the 1986–92 drought, lake level was down to the point of exposing roads not seen since this portion of the Sacramento River was impounded by completion of Shasta Dam in the early 1940s. Shasta Lake is the state's largest man-made reservoir, with a capacity of 4.5 MAF.* (Crane Miller)

normal precipitation through the early 1980s seemed to obscure the fact that this program was based on city ordinances that would be enforced in the event of another drought. And another drought was inevitable.

The ordinances prohibited certain water uses and called for water use cutbacks in one voluntary and four mandatory phases ranging up to 25 percent. Penalties for not complying with the mandatory conservation phases included fines, installation of flow restrictors, and even shutoffs. The prospect of such penalties might have prompted citizens to avoid them by practicing home water conservation techniques, such as replacing worn faucet washers, installing plastic inserts to reduce shower flow, and placing bricks or water-filled plastic bags in toilet tanks to displace flush water. But in January 1991, after a voluntary plan to cut consumption by 10 percent (from 1986 levels) in the waning months of the previous year had failed, the Los Angeles DWP sought a compulsory reduction of 10 percent in water consumption by all its customers, to go into effect March 1. Noncompliance would result in some of the stiff penalties just mentioned. But all restrictions, voluntary and otherwise, were lifted 2 years later as a result of above-normal winter precipitation.

The costs of administering mandatory water rationing programs could be staggering and make *reclaimed water* projects appear all the more necessary despite the relatively high cost of the product they produce. For example, the DWP's 1978 Draft EIR estimated that reclaimed wastewater could cost between $144 and $170 per AF, whereas local groundwater cost only $29 per AF, when available, and Los Angeles Aqueduct water cost $39 per AF. And reclaimed water is not getting any cheaper. In its 1990 Draft EIR, the DWP projected reclaimed water costs at four different plants at between $400 and $900 per AF for implementation dates from 1991 through 1995.

Figure 5-11 illustrates a plan for improvements to Los Angeles's wastewater collection and disposal system submitted by the city's Department of Public Works pursuant to the federal Clean Water Act. Note that the plan suggests only possible groundwater recharge, irrigation, recreational facility, and industrial uses for reclaimed water. Wastewater quality criteria still need to be developed for groundwater recharge and industrial use, and the California Department of Health expressly prohibits the use of reclaimed water in domestic drinking water systems.

Recharging and recycling groundwater have met with considerable success in the city of Fresno, which, like most other San Joaquin Valley cities, depends heavily on groundwater to supply municipal and industrial needs. To resolve the problem of a falling water level directly beneath Fresno, the city and the Fresno Irrigation District have entered into agreements whereby the district delivers water from the CVP's Friant–Kern Canal (see Figure 5-9) into the "Leaky Acres" recharge basin. In the several years since the recharge operation be-

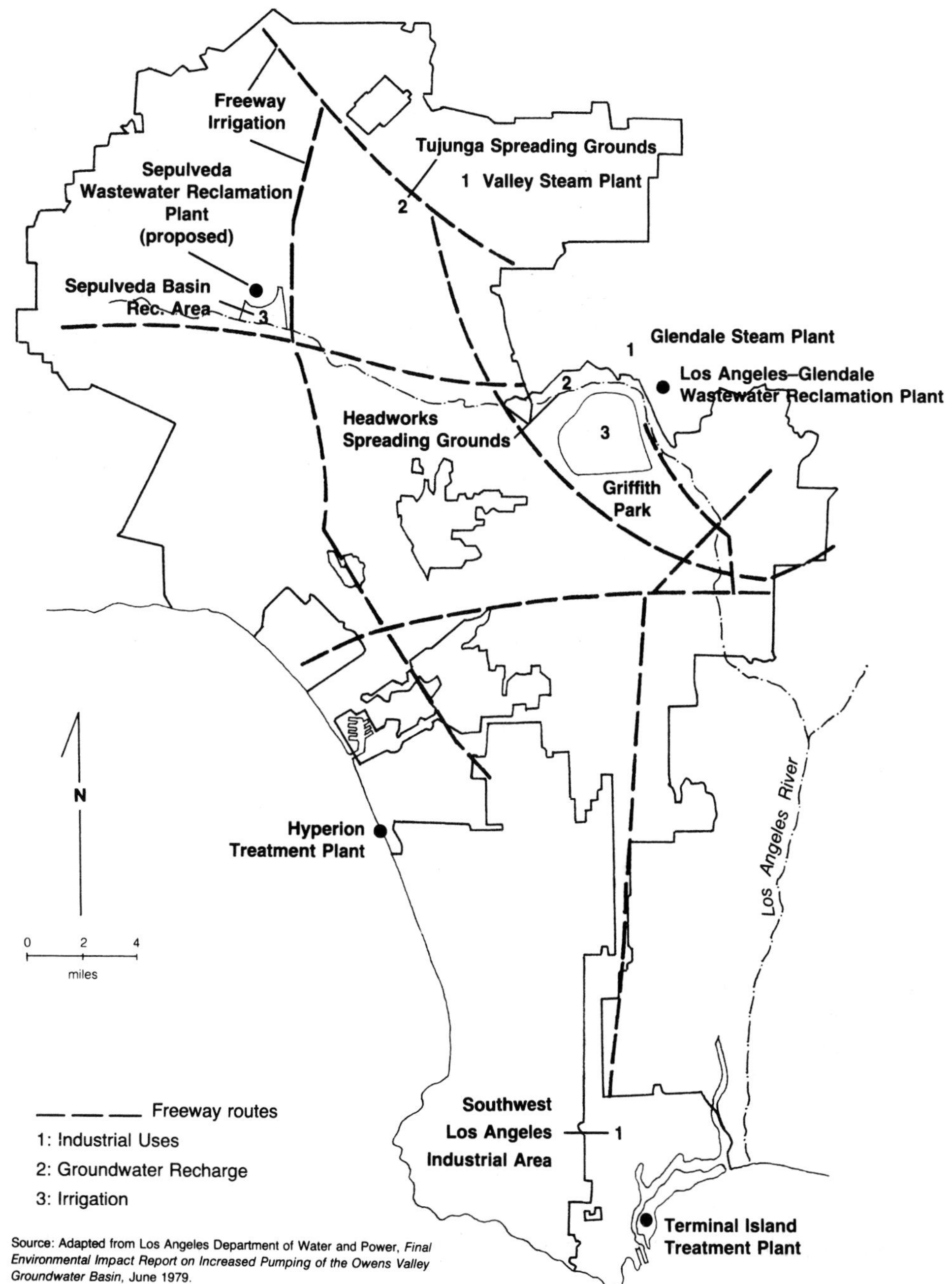

Figure 5-11 *Potential uses of reclaimed water within the city of Los Angeles.* (Marc Blodgett)

gan, the quantity and quality of groundwater beneath Fresno have steadily improved. Another facet of the agreement involves keeping the water table beneath the city's wastewater disposal plant from rising too high by pumping it and recycling the water through crop irrigation. The Fresno Irrigation District in return allows the city additional water from the Kings River equivalent to 46 percent of the recycled water it receives. This water is then used either for recharge or crop irrigation east of Fresno.

There are some 200 wastewater reclamation plants in California "conserving" about 200,000 AF of water for reuse each year. Irrigated agriculture is far and away the main user of reclaimed water in California, and the industry considers its use a major step toward water conservation. Other water-conserving practices employed by agriculture include drip irrigation, use of antitranspirants to cut down on evapotranspiration from leaves, control of water-wasting weeds in canals, improvements in dry farming methods, coating of reservoir water surfaces with a harmless chemical "film" to reduce evaporation losses, and implementation of special systems to produce greenhouse crops with a minimum of water use.

THE PACIFIC AND THE COLUMBIA: *Prolific but Unlikely Water Supply Alternatives*

In times of drought especially, many a Californian might cast envious eyes upon the waters of the Pacific Ocean and/or Columbia River and be overheard to mutter, "There it is, let's take it"—much as the water seekers of Los Angeles did decades ago when "discovering" the Owens Valley. Nobody denies the seemingly infinite water supply these two sources have to offer. But an unfavorably changing energy picture and mounting environmentalist and regional opposition make the prospects for development of either source appear dimmer with each passing day.

Seawater can be desalinized either by evaporation under a transparent cover, using the sun as an energy source, or by evaporation after bringing the water to a boil by heat energy derived from fossil fuel or nuclear sources. The former method is relatively cheap, but it is also slow, cumbersome, and generally inadequate for large-scale use. The latter process, on the other hand, is capable of meeting the demands of large populations but has become prohibitively energy expensive. It takes some 20,000 kWh of electricity to render 1 AF of seawater *potable* (drinkable), whereas one-tenth that much energy delivers 1 AF of Colorado River fresh water to Lake Mathews and an actual net gain in electrical energy is realized with each AF of Owens Valley fresh water delivered to the San Fernando Valley.

There are ecological constraints on desalination as well. They range from the environmental impacts of the buildup of salts from a nearby desalination plant on land and/or at sea to the closing of a plant because of an oil spill. The latter was brought to mind in 1990 with the accidental oil spill off of Huntington Beach in Southern California and again in 1991 with the far-off deliberate oil spill growing out of the Persian Gulf War. The Persian Gulf spill threatened not only to become the ecological disaster of the decade but also to shut down Saudi Arabia's desalination plants, from which that nation gets most of its fresh water.

Steam desalinization plants not only use copious amounts of energy and pose inherent environmental problems but also are expensive to build and operate. Inflation was a major factor in halting development of two such facilities that came closer to realization than most of the others proposed for California. One was the Bolsa Island Nuclear Desalting Plant, with a freshwater production capability of 150 million gallons a day, which was to be located on a 40-acre artificial island two-thirds of a mile off the Orange County coast. The other was a San Luis Obispo County plant, with a 40-million-gallon daily production capacity, which was to be built adjacent to the existing Diablo Canyon nuclear plant. Antinuclear sentiment, the California Coastal Plan, concern for disposal and/or sale of accumulated salt by-products, and a declining population growth rate were other factors that undermined the proposal. Could it be that desalinization is an idea whose time has come and gone—at least until a cheap and safe source of energy, such as fusion power, is developed?

Finally, in 1991, with no end to the drought in sight and local reservoirs all but dried up, Avalon (on Catalina Island) and Santa Barbara became the first California cities to venture into the desalination business. And costly ventures they were, with the larger, more economical Santa Barbara plant alone costing $30 million to build and able to produce 10,000 AF of fresh water annually only at a staggering $1,900 per AF. By comparison, once Santa Barbara's Gibraltar Reservoir and the nearby U.S. Bureau of Reclamation's Cachuma Lake began to fill with the drought-breaking runoff of 1992–93, reservoir water was once again available at $230 per AF. And if the local reservoirs could not meet the demand and Santa Barbara and San Luis Obispo counties were ready to pay for it, the State Water Project (SWP) stood ready to build a $320-million link with the California Aqueduct and supply imported water at $500 per AF. Nevertheless, Santa Barbara voters overwhelmingly (82 percent) supported a citywide referendum on the advisability of developing the emergency desalination facility. Not only would the plant be there if another drought came along, but Santa Barbarans would not have to wait 5 years from the date of the start of construction on an SWP pipeline to ease their water supply situation.

The other "water dream" shared by many Californians concerns the Columbia River. Any resident of British Columbia, Washington, or Oregon who has traveled in the Owens Valley, seen the movie *Chinatown*, or read *The Water Seekers* would no doubt turn red with rage at the merest suggestion of exporting Columbia River water to California. Yet schemes aimed at accomplishing wider distribution of the Pacific Northwest's vast water resources, such as those suggested by the formation of the North American Water and Power Alliance (NAWAPA), are still bandied about. There is no question that Canada could reap considerable income from the sale of surplus water—not water *rights*—to the United States or that Washington and Oregon would derive benefits from a federal Pacific state water transfer project. As it is, Southern California has been receiving hydroelectricity for years from the Columbia River via the Pacific Intertie system.

From 1986 through 1992, as during previous droughts, elected politicians from Los Angeles to San Francisco exhorted their counterparts in the Pacific Northwest and back in Washington, DC, to legislate an aqueduct into existence that would export just some of the Columbia River Basin's vast surpluses of fresh water to California rather than have it flow out to sea. But opposition by northwesterners to exporting water to California appears to be stiffening, and with good reason when Portland's "brownouts" and other rigors of the 1970s drought in the Pacific Northwest are recalled. There is also mounting resistance among many states to spending more federal money on California's water problems. In any event, water could well be the next major natural resource to limit California's economic development.

Just when the drought that started so subtly in the winter of 1986–87 was about to spawn the driest winter on record 4 years later in 1990–91, the rains and snows returned in unprecedented amounts to all of California, in what was hailed as the "March miracle." Despite warnings from water resource experts that the epic drought was far from over, record precipitation for the month quickly dampened talk of building a desalination plant in nearby Baja California or importing water by tanker from British Columbia. The euphoria was not without substance, for runoff from the March storms was so great that the SWP resumed maximum pumping of Sacramento–San Joaquin Delta waters into the California Aqueduct and its San Luis holding reservoir to the south (see Figure 5-5). As it turned out, though, the March miracle was not such a drought-buster after all. That would come two years later with the record precipitation of the winter of 1992–93. Whether that winter's generous rain and snow or the El Niño–induced precipitation of 1997–98 signaled the arrival of a string of wet years, it seems inevitable that people and droughts will continue to come to California. Thus, plans to accommodate the former in the face of the latter must be vigorously pursued. Alternatives to planning for more equitable distribution of water resources are marginally acceptable at best, for they range from pricing inequities, such as farmers paying $2 per AF for irrigation water while suburbanites spend $2,000 per AF for desalinated water, to outright abandonment of farms and cities and the return of much of California to the desert that it once was.

As we reflect back on water and look ahead to energy, the focus of Chapter 6, we are reminded of how inexorably linked these two resources are. During a drought, as water supply and hydroelectricity output simultaneously dwindle, we increasingly turn to alternative sources of power, some of them, such as nuclear and fossil fuels, being potentially more harmful to the environment than others we would prefer to use but are still learning about. On the other hand, when there is an abundance of precipitation accompanied by significant gains in hydropower production, we gain additional time to research and develop environmentally friendlier "green power" sources, such as the sun and the wind.

CHAPTER

Energy: The Assumed Resource

Like most Americans, Californians assumed for many years that the sources of energy were unlimited. Within the past few decades, however, this assumption has undergone serious challenge. As a consumer of one-quarter of the total energy resources of the world (with less than one-fifteenth of the total population), America has been forced to reexamine its future. The United States no longer has a monopoly on petrochemical energy sources. Alternate sources are needed, and alternate lifestyles may be necessary. The question is, Can Americans in general, and Californians in particular, meet this challenge?

LIFESTYLE CHANGE: *Can California Maintain Its Mystique?*

In many respects, California epitomizes the Western world in its attitude toward and use of nature and energy. The commitment of Western civilization to science and technology has permitted great improvements in living standards for humankind. Much of this progress, however, has been accomplished by following a basic social and religious premise: Subdue the earth. Subduing the earth has, in the short term, brought about a lifestyle of unheard-of luxury, nowhere more evident than in California.

Part of the great mystique of California has involved the use of the land for pleasure. No place is inaccessible; if there is no way to get there, build a road! Freeways and roads have thus become an important feature of the state. Because the beaches and shoreline are beautiful, everyone should be able to enjoy their natural beauty; thus, the beaches and shoreline have been "improved" with vast concrete parking lots, artificial fishing jetties, and sand groins to prevent subsequent erosion. To this way of thinking, mountains may be striking evidence of nature's majesty, but ski lifts and lodges, artificial snow, mountain-forest condos, and improved roads make them even better. Similarly, the vast California deserts that are among the last American frontiers possess a fragile, balanced ecology, so dirt bikes and off-road and recreational vehicles must be brought in, the better to see this wilderness before it is destroyed. And, of course, in each instance nature is brought to heel for the enjoyment of humans and the "toys"—boom boxes, beer coolers, jet skis, snowmobiles, and so on—they bring along to desert, beach, or mountain as inevitably as trash and garbage. Thus, democracy, religion, technology, capitalism, urbanization, and mobility bring new meaning to the concept "natural environment."

Is the Mobile Culture Affordable?

More than any other single artifact, the automobile is the icon of California, a state that has relied heavily on individualized transportation. Although mass-transit systems exist, these are the exception; for the most part, the typical pattern is single-passenger transport. Even in the face of reduced gasoline supplies, most Californians have been reluctant to give up the convenience of the automobile. The vast numbers of commuters have shifted approach only to the extent of moving toward smaller, more fuel-efficient cars.

Although some attempts have been made to get Californians out of their cars, these efforts have had only limited success. Los Angeles's much-vaunted light rail system suffered setbacks from the beginning, and efforts to establish a "subway" in the City of Angels were plagued with problems with fire, tunnel collapse, and questionable geologic siting. The Bay Area's BART system has enjoyed some success, but the total volume of single-occupant automobile commuters remains astronomical. Even San Diego's experiment with a novel "trolley" system has not met the most optimistic expectations. In short, Californians seem blissfully wedded to their automobiles as a means of maintaining their sense of independence and privacy—at least while commuting to work.

In terms of recreation in the state, Californians have seldom been deterred by distance in pursuing their entertainment, either as spectators or participants, be it theater, sports, hiking, or sightseeing. And it is not just a matter of the family van, sedan, or wagon, either; the proliferation of recreational vehicles, off-road vehicles, motorcycles, and sports cars has not slackened perceptibly into the new century, even though the availability and cost of gas have become important factors.

As the new century evolves, the question of mobility will become more vital. Clearly, Californians cannot continue in their present pattern. If mobility is to remain a part of the state's identity, forms of transportation must change, because the costs and impracticality of the current mobile culture cannot be borne much longer. This becomes even more evident as gridlock spreads from one metropolitan area to another, as state and local governments struggle to maintain roads, and as air quality continues to deteriorate.

Is the Poolside Goddess a Thing of the Past?

The aquatic image is also a part of the California mystique. Southern California in particular is a vast checkerboard tract filled with backyard swimming pools and Jacuzzis. The lifestyle promoted in fashionable magazines, decorators' drawings, newspaper ads, and "house beautiful" publications includes the swimming pool and spa as vital elements of the good life. The social and sexual revolution in suburbia focused upon this symbol, and common wisdom dictates that a home's value jumps with such an addition. No apartment complex or condominium development could survive without the accompanying water sports playground, and cost traditionally has not been a consideration.

Heating water, however, is not an inexpensive proposition. Rising energy costs have mandated a reexamination of this luxury symbol. The stereotypic golden-tanned poolside goddess relaxing by bubbling, heated water may soon become an anachronism. With both water and energy becoming critical resources in the state, certain priorities must be considered. Swimming may well continue, but in winter months, at least, Californians may have to develop hardier constitutions.

Can Californians Become Energy Spartans?

The United States traditionally has been one of the world's largest consumers of energy, accounting for over one-quarter of the world's total energy use. Californians generally hold their own in this energy usage, in terms of both amount and variety.

Californians, like Americans elsewhere, continue to insist on some relatively marginal uses of energy. The extent of these uses can easily be seen during the Christmas season with the prolific and prodigious displays of lights and electrical decorations. Throughout the rest of the year, Californians insist on widespread outdoor lighting of houses, patios, tennis courts, pools, and driveways. Likewise, luxury and convenience appliances remain big sellers in department and discount stores. Californians seem unable to function without garage door openers, trash compactors, electric toothbrushes and can openers, televisions, and electric knives. Yet in the face of rising costs and limited supplies, Californians may have to become reacquainted with their wrists, shoulders, and backs as energy sources.

The California penchant for air conditioning may undergo a severe test. Although hermetically sealed skyscrapers and schools, enclosed shopping malls, and modern tract houses may not have enough design flexibility to permit natural air circulation, thermostatically controlled indoor climates may soon be an expensive luxury. The assumptions of unlimited energy that have made such design and construction possible in the past can no longer be indulged. Californians must reluctantly face the harsh reality that their energy uses are often wasteful and that the old assumptions do not serve. Petroleum and water, electrical, and other forms of energy are valuable assets to the state. Simple belief in unlimited existing sources cannot take the place of responsible searches for alternate forms of energy.

ENERGY AND ELECTRICITY: *Consumption and Supply*

Among the 50 states, only Texas has consistently outranked California in energy consumption. But when it comes to transportation, the Golden State leads the Lone Star State and all others in the United States. In 1994, for instance, transportation accounted for 50 percent of total energy use in California, leaving the other

TABLE 6-1 California electricity generation sources, by percentage, 1983–1994.

SOURCE	1983	1986	1989	1994
Coal	12.2	12.1	11.3	10.0
Geothermal	4.4	7.0	5.8	5.0
Hydropower	43.2	31.6	20.4	22.0
Natural gas	30.7	27.9	34.3	50.0*
Nuclear	4.8	19.3	22.2	11.0
Oil	4.7	2.2	6.0	1.0
Solar, wind, biomass	0.001	0.01	0.01	1.0

* In 1994, for natural gas, the 50 percent includes *cogeneration* (8 percent), which is the simultaneous production of electricity and thermal (heat) energy from the same plant. Natural gas is also produced at landfills and other sites from *biomass*, or organic wastes that, in the absence of oxygen, decompose to produce methane (natural gas) and other gases, which in turn can fuel the production of electricity. Such biomass-produced methane is sometimes referred to as "garbage gas."

Source: California Energy Commission, QFER Form 1 (October 18, 1990), and *Energy and the Economy* (1994).

major categories far behind: industrial and other users, 28 percent; residential, 13 percent; and commercial, 9 percent. Transportation's dominance as an energy consumer is attributed largely to Californians' love affair with the automobile and widespread aversion to public transit. Cars alone in California consume about one-fifth of available energy each year. Simply observing the number of people in each car—usually only the driver—and then the sheer numbers of automobiles involved in a typical rush hour should be enough to convince anyone that the car is the king of energy consumption in California. Even all the new car pool lanes added to metropolitan-area freeways appear woefully underused during the rush hour commute. Thus, not surprisingly, California's proportional consumption of petroleum (50 percent of the state's total energy supply in 1994) is several percentage points above the national average. California's other suppliers of energy in 1994 were natural gas (29 percent), hydropower (3 percent), nuclear power (6 percent), geothermal power (4 percent), and coal (6 percent). Even though solar power plants and wind farms collectively formed the smallest energy producer category, barely 2 percent by the mid-1990s, they are expected to at least double that percentage within a decade.

In terms of electrical energy supply and demand, the proportions change noticeably. Transportation all but drops out of the picture, while commercial, industrial, and residential consumers share almost equal thirds of the user pie. The annual demand for electrical energy has skyrocketed in California over the past 40 years, increasing from 56 billion kilowatt hours (kWh) in 1960 by nearly ninefold heading into the twenty-first century. As we turn to an examination of each of the sources of electrical energy (Table 6-1) available from inside and outside California, bear in mind our earlier discussions of the influence of climate, weather, smog, and water resources on electrical energy supply and demand. Later in the chapter, we will also consider some of the impacts of deregulation of California's electricity industry, which commenced in January 1998.

HYDROPOWER

In theory, hydroelectric energy can be generated wherever a controlled water supply can be dropped to a lower elevation. Given California's abundance of high mountain ranges and its orographic (mountain-related) precipitation patterns, such hydropower dam sites are plentiful in the state. Locally, rugged topography also facilitates *pumped-storage* hydro projects such as at Castaic (Figure 6-1). Operated by the Los Angeles Department of Water and Power (DWP) in cooperation with the California Department of Water Resources (DWR), the pumped storage facility has an installed capacity of 1.25 million kilowatts (kW) or 1,250 megawatts (MW), enough power to supply several hundred thousand Los Angeles Basin customers with their electricity needs during daily peak demand periods. When needed, additional water originates from the west branch of the DWR's California Aqueduct as its waters are being pumped up and over the Tehachapi Mountains. Where water supply is unreliable or relatively small, however, which includes most of the southern third of the state, hydropower development is not practical. And a long drought can all but eliminate hydroelectricity production within the state.

Currently, hydropower meets about 22 percent of California's electrical energy needs, but in another 20 years, it is projected to decline to less than 10 percent, droughts or lack of them notwithstanding. The drop of hydro's proportional input is attributed to two factors: (1) the predicted doubling of electricity consumption in the next 15 years, and (2) an already almost fully developed hydropower potential.

California has several major developed hydropower source areas, both inside and outside the state. The Sacramento and its tributaries make up the premier hydro river system in the state, boasting such hydropower facilities as Oroville Dam and Shasta Dam (see Figures 5-5 and 5-10). The San Joaquin River system is a distant second but claims one of the state's unique hydro diversion projects in Southern California Edison's (SCE) Big Creek Project (Figure 6-2), developed by SCE early in the twentieth century to provide up to 700 MW of hydropower to the Los Angeles Basin. The unde-

A

B

Figure 6-1 *Castaic Power Project.* Penstocks *(photo A) both deliver water from Pyramid Reservoir to reversible generator-pumps and carry it back uphill to Pyramid, which is the* upper forebay. *The reverse pumping is done during normally low power demand periods, with power coming from steam electric plants in the basin and water coming from the* lower, *or* pumping, forebay *(photo B). The* surge chamber *or* tank *seen above the penstocks acts as a giant shock absorber to relieve excess pressure in the tunnel and penstocks during a sudden shutdown.* (Crane Miller)

Figure 6-2 *Big Creek Project in the southwestern Sierra Nevada. The waters of the San Joaquin River are diverted at Florence Lake (7,328 feet) and from Mono Creek at Lake Edison 13 miles west to Huntington Lake (6,950 feet). Tunnels, penstocks, powerhouses, and dams downstream from Huntington Lake combine to provide the hydroelectric output. In this aerial view eastward with the snow-capped High Sierra in the background, Shaver Lake (5,370 feet) appears in the foreground while Huntington Lake can be seen to the northeast (upper left).* (Crane Miller)

veloped rivers of the northwest augment the state's hydropotential, but with the exception of the Eel River after 1985, the Scenic and Wild Rivers Act protects them from development. As mentioned in the previous chapter, the Eel River's protected status was extended after review in 1985. Environmentalists, who object to the inundation of wildlife habitats, damage to fisheries, changes in microclimate, displacement of people, and elimination of free-flowing streams caused by dam building, may also put a damper on development of what's left of California's hydropower potential. However, they now seem to have their hands full fighting the proponents of coal and nuclear power development, which pose ecological hazards far greater than those of hydropower.

California's two main outside sources of hydropower are power plants along the lower Colorado River and the Pacific Northwest/Southwest Intertie, which imports Columbia River energy. The Bureau of Reclamation's Hoover Dam is the showpiece of Colorado River Basin hydro development. Its power plant has a rated capacity of 1,345 megawatts (MW), which is enough electricity to air condition and light up Las Vegas, Nevada, around the clock and still have 513 MW left over to send to Los Angeles—assuming there is plenty of water flowing down the Colorado.

Southern California gets another massive input of interstate hydroelectricity from the Pacific Intertie via direct current (DC) and alternating current (AC) transmission lines. The 845-mile DC line is the longest-distance high-capacity line of its kind in the world, stretching from a converter station near the Dalles and other Federal Columbia River Power System dams to the Sylmar DC-AC converter terminal. DC lines use fewer cables (usually two per circuit), fewer conductors, cheaper towers, and less *easements* (rights of way) than AC lines and thus are about 35 percent less costly to build than AC lines. Yet DC lines have the same transmission capacity as AC lines, which in the case of the Pacific Intertie is 1,400 MW. The Southern California utilities sharing the Intertie's output from the Sylmar station are SCE, the LADWP, and the local power agencies for the cities of Burbank, Glendale, and Pasadena.

Adding still more hydropower to the southland's hydro pool are the DWP's Los Angeles Aqueduct power plants in Owens Gorge (110 MW, with storage in Crowley Lake)and San Francisquito Canyon (74 MW, including power generated at the Los Angeles and Franklin Canyon reservoirs in the Santa Monica Mountains). Indeed, Los Angeles and neighboring communities are well off hydroelectrically despite the region's lack of a dependable local water supply.

The central purpose of intertie systems is to ensure a supply of electricity during emergencies and during times of routine power plant maintenance. In essence, the Pacific Intertie is insurance for Los Angeles against the type of blackouts that have occurred in New York and other major cities. But an Intertie is reciprocal as well. For instance, when a severe blizzard in the winter of 1968 caused hydroelectric outages in the Pacific Northwest, SCE sent power generated at local steam electric plants to Oregon and Washington via Pacific Intertie AC lines, thus averting what could have been a major disaster. The intertie system also links Northern California's Pacific Gas and Electric Company (PG&E) with both SCE and the Federal Columbia River power projects.

Actually, Pacific Intertie is international in scope by virtue of its use of some of the so-called *Canadian entitlement power*. Three dams built along the upper Columbia River in British Columbia provide "stored capacity" of some 2,800 MW, which by international treaty and subsequent agreements can be used by U.S. utilities until such time as Canadian utilities need it. The DWP, for example, has contracted to purchase a substantial proportion of Canadian entitlement power for transmission to Los Angeles over Pacific Intertie DC lines.

The ultimate in intertie exchange developments will be a *national grid* connecting all major utilities from coast to coast. With such a nationwide system, a *peak power demand* at 6 P.M. in New York, for example, could be met with surplus power from the Pacific Northwest or California, where it would be only 3 P.M. and peak demands would not be occurring. Three hours later, *peaking power* supply could be shifted, with New York augmenting the West Coast's electrical energy demands. Ownership, control, and dissimilar power markets in the four time zones are the biggest obstacles to development of a national grid. Power loss over long-distance transmission lines is another deterrent. In the meantime, though, Pacific Intertie already affords California participation in a *regional grid* system that goes a long way toward preventing blackouts of the sort that have plagued the East Coast.

OIL AND NATURAL GAS

Since the late nineteenth century, petroleum and (later) natural gas have been California's most valuable mineral products, and for a few decades in the twentieth

Figure 6-3 *Moss Landing Plant, on Monterey Bay. This and other oil- and gas-fired steam electric plants along the California coast and inland rivers meet most of the state's electricity demands. Steam electric plants, whether fossil fueled or nuclear powered, consume much water for cooling purposes, which explains the preference for waterside location. As mentioned in Chapter 4, natural gas is much preferred over oil as a fuel because it is far less harmful to the atmosphere and to humans.* (Pacific Gas and Electric Company)

TABLE 6-2 California natural gas sources and requirements, in million cubic feet of gas per day, 1990–2010.

SOURCES	1990	1995	2000	2005	2010
California sources	466	455	459	448	393
California percentage	9.2%	7.8%	7.2%	6.7%	5.5%
Out-of-state sources*	4,594	5,415	5,945	6,248	6,733
Out-of-state percentage	90.9%	92.3%	92.7%	93.7%	94.1%
Withdrawal/EOR†	−6	−3	8	−30	29
Total	5,054	5,867	6,412	6,666	7,155
REQUIREMENTS					
Utility electricity generation	1,458	1,501	1,753	1,804	2,120
Other‡	3,815	4,366	4,659	4,862	5,055
Total	5,273	5,867	6,412	6,666	7,175

* Includes utility purchases, customer gas and out-of-state transport, and other.
† EOR (enhanced oil recovery) is injection by steam into oil-bearing geologic zones to improve the ability to extract oil by lowering its viscosity, or thinning it.
‡ Includes residential, commercial, and industrial space heating, other nonelectric cogeneration, and other.
Source: California Gas and Electric Utilities, *1990 California Gas Report*.

Figure 6-4 *"Texas towers" in the Santa Barbara Channel. Although unsightly, these facilities pump oil from California's prolific offshore reserves. On January 28, 1969, a well drilled from one platform blew out, leading to an oil spill that spread along 20 miles of seashore and 40 miles out to sea, befouling beaches and killing birds by the thousands. GOO (Get Oil Out) and other groups opposed to offshore drilling were quick to organize and warn of the potential for another such disaster.* (Standard Oil Co. of California)

century, California led all states in production. But in the 1960s, the mounting energy demands of an expanding population and an increasingly automated society forced California into importing these hydrocarbons. Today, California ranks fourth behind Texas, Alaska, and Louisiana in petroleum production and much farther down the list in natural gas production. In the last two decades of the twentieth century, though, daily crude oil production in California steadily slipped from more to less than a million barrels (bbl).

In-state and imported oil and natural gas presently provide about 50 percent (see Table 6-1) of California's electrical energy through their use as the fossil fuels that fire steam electric plants (Figure 6-3). The significance of fossil fuel power is exemplified by the LADWP, which can generate more than 3,300 MW from its steam electric plants versus about one-third that amount from its hydroelectric sources.

California's own natural gas supply is all but exhausted. As Table 6-2 shows, California imported 93 percent of its natural gas in 2000 and projects 94 percent imports by 2010. In-state petroleum reserves are dwindling as well, causing oil companies to periodically pressure the government to increase offshore drilling leases (Figure 6-4). Even though in the 1990s onshore oil reserves, as mapped in Figure 6-5, were estimated at 3.7 billion barrels (bbl; 1 bbl = 42 gallons) and those offshore only at 776 million bbl, offshore oil reserves may one day surpass all those on dry land in California. Their

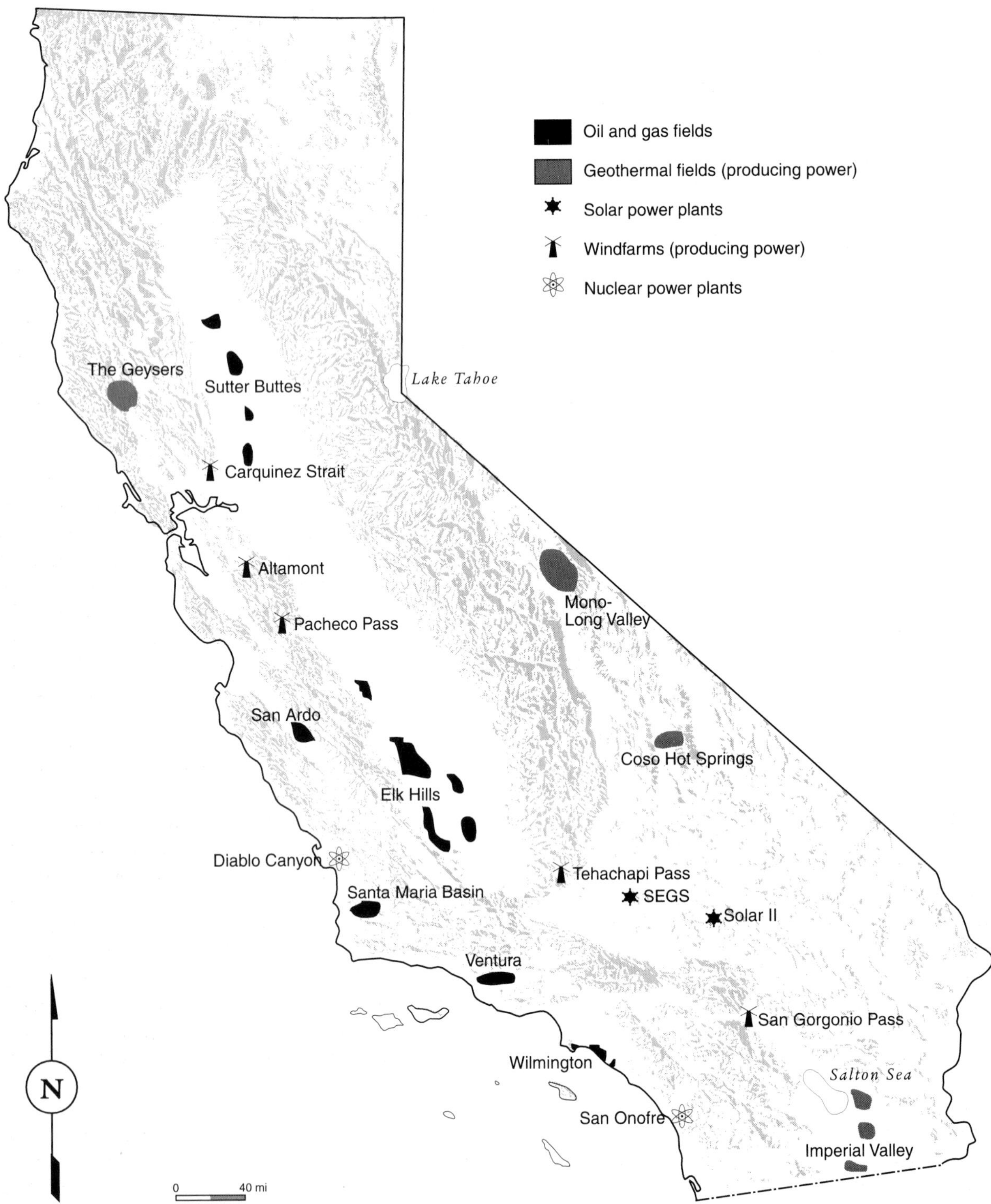

Figure 6-5 *Energy map of California. Oil fields are largely found in sedimentary basins in the southern Coast Ranges and western Transverse Ranges, and offshore of Santa Barbara, Ventura, Los Angeles, and Orange counties, while gas output focuses on the sedimentary formations of the Sacramento Valley. Power-producing geothermal fields are found at the Geysers in northern Sonoma County, in Mono and Inyo Counties, and in the Imperial Valley. SEGS (solar energy generating systems) and Solar Two solar power plants are all located in San Bernardino County. The three major windfarms are in Altamont in Northern California, Tehachapi in Kern County, and the San Gorgonio Pass in San Bernardino County.* (Crane Miller)

exploitation may never be fully realized, though, because of constraints evoked by memories of the devastating Santa Barbara Channel oil spill of 1969 and by the passing into law of the California Coastal Plan. Although 1.3 million acres offshore from Humboldt, Mendocino, Sonoma, Marin, San Mateo, Santa Cruz, San Luis Obispo, and Santa Barbara counties are under study by the federal government for its proposed oil and gas leasing program, environmentalists are certain to fight tooth and nail to prevent oil rigs from invading this as yet untouched and spectacularly scenic coastline. Meanwhile, environmentalists a decade ago won an important victory in the government's decision to delete from proposed oil lease sale all offshore tracts from Dana Point in Orange County southward along the San Diego County coastline. The California contribution of natural gas to creating electrical energy declined to 7 percent by the end of the century (see Table 6-2). But supplemental supplies via pipeline from Mexico and Canada, by ship as liquefied natural gas (LNG) from Alaska and Indonesia, and as coal gas from coal gasification plants may forestall the demise of natural gas from steam electric plant use and also maintain adequate supplies to meet winter space heating needs. The LNG supplement appears to hold the most promise, for it will about double the amount of both the already contracted Canadian supplement and that of coal gasification plants.

Location of an LNG terminal poses some thorny problems. Wherever LNG is unloaded from tankers, it must be reconverted to gas, with the unloading of the low-temperature liquid and its reconversion requiring the utmost care to prevent explosions. State law expressly prohibits such a facility operating anywhere near populated areas. Thus, with the public safety in mind, the California Public Utilities Commission (PUC) long favored siting the state's first LNG terminal at Cojo Bay near Point Conception, which is centrally located with regard to San Francisco, Los Angeles, and San Diego, but some 40 miles away from the closest city of any appreciable size—Santa Barbara.

The Point Conception site also presents problems, however, including some of the roughest seas found anywhere along the California coast, the usual seismic hazards, Native American claims to the land, and the federal government's coolness toward LNG as a desirable energy source. To prevent Point Conception's becoming a graveyard for tankers, a half-billion-dollar breakwater would have to be built. Another extra cost in developing the terminal would be incurred in building structures that could withstand a 7.5-magnitude earthquake. Native American claims, which were expressed by an encampment at the site of up to 50 Native Americans in 1978–79, are not likely to be resolved until they at least are granted some kind of permanent access for religious purposes. In any case, the federal government would rather see greater development of domestic reserves of natural gas, even though it gave California the go-ahead to import LNG from Indonesia. In the final analysis, each of these problems seems capable of solution. If and when they are ever solved and the joint PG&E-SCG Point Conception terminal developed, it satisfy about 10 percent of California's natural gas wants.

For more than three decades, Northern California has increasingly relied on pipeline imports of natural gas from Alberta, Canada; and this foreign source apparently will continue to be reliable for the next several decades. PG&E's Canadian gas comes from the Western Canadian Sedimentary Basin (WCSB), which underlies much of Alberta and portions of British Columbia and Saskatchewan. The WCSB is often referred to as the "Alberta gas bubble" and is reputed to be one of the most prolific natural gas fields in the world. Current annual output from "the bubble" is 3.5 trillion cubic feet (TCF), with established reserves of 70 TCF and estimates of at least another 70 TCF in reserves yet to be discovered. The WCSB thereby claims a known reserve life of two decades and an estimated life of two additional decades or more. Confidence in this long-term source of natural gas is best substantiated by PG&E's wholly-owned subsidiary, Alberta and Southern Gas Company, Ltd., recently gaining approval from the Canadian National Energy Board to extend its export licenses at full current volumes through the year 2005.

Assurances of long-term Canadian gas imports and the possibility of LNG imports notwithstanding, the import of overseas crude oil will no doubt continue to increase both in amount and in environmental impact. Until the mid-1970s, most tankers were small enough to be unloaded inside California's refinery ports. Now a growing number of the vessels are supertankers too large to enter the state's ports. This means that the large tankers must lie outside and have their petroleum carried into port by smaller ships in a process known as *lightering*. The negative impact on the environment comes in the form of fumes escaping into the atmosphere when the oil is being transferred from the larger to the smaller vessel.

If what has been said so far seems frustrating as regards the best efforts of some Californians to establish more fossil fuel terminals, consider the Sohio controversy. In the mid-1970s, Sohio (Standard Oil of Ohio, which is partly owned by British Petroleum) proposed construction of a $1-billion terminal in Long Beach to transfer Alaskan crude oil to an unused gas pipeline and pump it on its way to Midland, Texas, and from there to refineries in the Midwest. The purpose of the project was not to supply California with more oil, but rather to save about a dollar a barrel in extra shipping costs incurred in shipping Alaskan oil through the Panama Canal and to increase Alaskan production. The president of the United States, the governor of California, and

A

B

Figure 6-6 *Los Angeles–Long Beach harbor, which in 1997 was the nation's busiest cargo port. That same year, a new, $200-million coal-loading facility was completed. The view in photo A is northward with the Vincent Thomas Bridge in the background, the Los Angeles harbor cruise ship center to the left (west), and container-loading facilities to the right (east); the latter handle about 3 million containers annually. The new coal transfer facility, seen in photo B, handles several million tons of coal annually.* (Crane Miller)

the voters of Long Beach all approved the proposed terminal, and the principal air pollution issues had essentially been worked out with some tradeoffs.[1]

In March 1979, however, Sohio suddenly canceled the proposed project; apparently it could not countenance any further delays in gaining final regulatory approval from local and state agencies. As it was, Sohio had already spent $50 million and filed more than 700 permits and applications in trying to get the project underway. There also existed uncertainty over threatened citizen suits and the possibility that Sohio might not gain access to the pipeline it planned to use.

Although the import of hydrocarbons through California seaports suffered from the shelving of the LNG and Sohio projects, and despite not being a coal-mining state, California witnessed a dramatic expansion of its coal-shipping facilities before the end of the century. Strip-mined in Colorado, Utah, and other Rocky Mountain and Great Basin states, coal travels by train to Northern and Southern California seaports like Stockton and Los Angeles–Long Beach (Figure 6-6), where it is loaded on ships bound for Far Eastern ports. For the past quarter century, power plant operators in Japan, South Korea, and elsewhere in eastern Asia have been converting from oil to gas, thereby creating an increased demand for American coal. One cannot help but wonder what impact the international treaty on cutting emissions of carbon dioxide and other greenhouse gases agreed upon at the December 1997 global warming conference held in Kyoto, Japan, will have on California coal exports. Whatever the outcome, it will be well into the twenty-first century before a plurality of nations has ratified the treaty and its impact is felt. Another issue the treaty raises is how it will affect a rapidly growing coal-fired power base here in California.

COAL

California is not a coal state, in terms either of minable reserves or imports. But several nearby western states lay claim to what may be collectively the world's richest reserves of "strippable" coal—coal that is accessible to surface or *strip mining* as opposed to tunnel or *shaft mining*. California utilities presently use a portion of this coal, mostly in power plants outside of the state. Previously of no commercial significance in California, coal-fired power plants finally began to come into their own in the 1980s, but only on a relatively small scale. By mid-1988, there were seven operational coal-fired plants in the state, generating a meager 315 MW; by contrast, the DWP and SCE alone were receiving some 2,000 MW of coal power from plants in Arizona, Nevada, New Mexico, and Utah. The largest in-state coal-fired plant is SCE's Coolwater facility at Daggett in the Mojave Desert, with a current capacity of 100 MW (it was rated at 631 MW in 1982). Farther north, PG&E listed small coal-fired plants as operational in 1988 in Amador, Kern, and San Joaquin counties, and more recently has proposed many other sites. In September 1990, PG&E suffered a major setback in its efforts to base coal power production in California when the state supreme court let stand

[1] Under the original Clean Air Act, no new sources of pollution were to be allowed where the air was already dirtier (as in SCAB) than federal regulations allowed. As was the case with lightering, unloading of tankers at the terminal would have involved release of hydrocarbon fumes, but Sohio agreed to suppress other sources of smog as a tradeoff. There also exists the chance of an in-harbor oil spill, but this hazard is present at oil tanker terminals everywhere. The Los Angeles Basin is a major refinery center, although it would never have refined the Sohio project oil.

an appellate court decision denying operation of a nearly completed $70-million coal-fired cogeneration plant near Hanford in the San Joaquin Valley. The Hanford facility, which was built by GWF Power Systems, was intended to cogenerate steam and electricity for the Armstrong Tire Company and PG&E. In 1988, the city of Hanford won a Kings County Superior Court test of its EIR report stating that building and operating the plant would have no significant negative environmental impacts, and so its development proceeded. However, environmentalists and farmers appealed the decision, contending that there would be a significant rise in air pollution and crop damage and a drop in groundwater table levels if the plant were built and operated. The appelate court ruled that the city had failed to obtain enough information to proceed with the project, notably in omitting study of the cumulative impacts on air quality of this and more than a hundred other cogeneration plants planned for the valley.

Another case of coal-caused air pollution, though far removed from California, also threatens coal power's reputation. This time, the foul air is in Grand Canyon National Park, where the atmosphere is being hit from two sides. One is from the west in the form of Los Angeles smog (see Chapter 4); the other is from 80 miles east at the coal-fired Navajo Generating Station, a 2,250-MW facility in which the LADWP claims 22 percent ownership. Researchers are trying to identify the worst offender, which will have a bearing on the Environmental Protection Agency's plan to force $1 billion in air pollution controls on the plant.

Coal, whether it is burned inside or outside California, poses monumental economic, environmental, and social headaches for the entire. Far West. Some remedies are being developed, however. Coal may be regionally plentiful and therefore a relatively cheap form of energy raw material, but it is dirty, bulky, and thus costly to transport, even by *unit train* carrying the single product on a regular schedule from strip mine to market.

One answer to the hauling problems is the mixing of pulverized coal with water into a *slurry*, which is then pumped through a pipeline to its ultimate destination. A coal slurry pipeline now operates between the Black Mesa coal mine in northeastern Arizona and the jointly operated (by the DWP, SCE, and four other agencies) Mohave Power Plant in Nevada south of Las Vegas.

Slurry may provide a solution to another, more serious problem—that of air pollution from coal burning, which was touched on in Chapter 4. As the Germans demonstrated during World War II, product slurry can be distilled into a liquid fuel that is rid of much of the sulfur inherent in solid coal and thus may not necessitate the use of *scrubbers* to clean up the exhausted fumes and smoke following its burning. At present, though, we likely are a few years away from a *coal liquefaction* process that will render coal liquids competitive with solid coal or oil in price.

Another type of coal-caused air pollution, one that is less amenable to solution, is that caused by coal dust kicked up at strip mines sites. Between more strip mines with their coal dust and more power plants with their coal fumes, the Great Basin, once one of the clearest regions in the country, may one day become nearly unbreathable.

As for the socioeconomic benefits and costs of increased coal exploitation in the Far West, they range from more jobs and income for some residents to the disruption of Native American lifestyles and the social disorganization of fly-by-night towns following strip mines from site to site. It seems inevitable that the "cowboys and Indians" are destined to be replaced by strip miners and coal power plant operators. When that happens, what will become of the magnificent western landscape that has drawn so many tourists and retirees? The strip-mining people claim they can make it look like new again when they are finished with a site. Ranchers and environmentalists say it will never be the same.

Perhaps the coup de grace with respect to coal for California comes when other western states are regarded as energy colonies for California. The death of the giant Kaiparowits coal power project proposal in southern Utah in 1976 was undoubtedly a signal from nearby states that California had better look in its own backyard for coal-fired steam electric plant sites. To outsiders, California's 1,200 miles of coastline seemed ideal for such sites, especially since steam electric plants need copious amounts of water for cooling purposes, and the prospect of future droughts acts as a limiting factor to the development of interior Great Basin sites. To most Californians, though, there were already more gas-and oil-fired steam electric plants along the coast than was desirable. Furthermore, the state's Air Resources Board looks dimly upon coal power plant development in urban areas, either along the coast or in the interior, and the federal Clean Air Act would seem to preclude such development in any event. California may have to await the perfection of liquefied and/or gasified coal, which can be burned in already existing oil- or gas-fired steam plant boilers, before coal will supply appreciably more than the present 10 percent of the state's electrical energy needs.

GEOTHERMAL POWER

Mention of geothermal energy development does not trigger the same negative feedback that oil and coal do. Maybe this is because most Californians are unfamiliar with its environmental impacts. Or perhaps the apparent lack of furor can be traced to the fact that only 5 percent of California's electrical energy demands are

Figure 6-7 *Geothermal electricity at the Geysers, 70 miles north of San Francisco in Sonoma County. The Geysers' maximum capacity is forecasted to remain at 2,000 MW through the year 2030.* (Pacific Gas and Electric Company)

presently satisfied by geothermal power. Yet California is the sole major national producer of commercial geothermal power and holds the greatest potential for its future development.

PG&E's Geysers facility in Northern California's Sonoma County (Figure 6-7) is the largest geothermal power plant in the contiguous United States. Presently, it has an electrical generating capacity of 1,000 MW, but the Geysers' estimated potential capacity is double that, a power output sufficient to serve one-half the present needs of the nearby San Francisco metropolitan area. In South Geysers, near Calistoga, the California Department of Water Resources closed down a 7-MW geothermal power plant in September 1990 and subsequently got out of the geothermal electricity business. Unocal and the Northern California Power Agency, a 14-city consortium, still operate in South Geysers, however.

Other than in Sonoma, geothermal fields have been tapped for power production in five other counties in California—Imperial, Inyo, Lake, Lassen, and Mono. Among the newer geothermal power facilities are the California Energy Company's 240-MW Coso plant at China Lake in Inyo County (see Figure 6-5) and Mammoth Pacific's development of the Casa Diablo Hot Springs geothermal field several miles east of Mammoth Mountain in Mono County. The geothermal field with perhaps the greatest potential in the whole state is in the Salton Trough portion of Imperial County (See Figure 6-5)—an area many thought was largely unusable. After San Diego Gas and Electric shut down a $188-million experimental geothermal power plant in 1987, Magma Power entered the Imperial Valley, overcame "impossible difficulties" (discussed later in this section), and in January 1989 started generating 126 MW of geothermal power. Magma's three *online* (transmitting electricity to customers) geothermal power plants serve the electricity needs of some 100,000 households in Southern California. The company's success draws attention to a 1983 federal law mandating that electric utilities sign power purchase agreements with independent alternative energy contractors like Magma Power. Such nonhydrocarbon energy projects are obviously in harmony with both the 1990 version of the Clean Air Act (see Chapter 4) and society's broader goal of diminishing its dependence on fossil fuels. Although seemingly overly optimistic to many observers of the geothermal scene, some foresee Imperial eventually outpacing Sonoma as the leading county in geothermal power production, not just in the state but in the nation. With improved technology, the Brawley, East Mesa, Heber, and Salton Sea sites alone in Imperial County could be producing more than 4,600 MW of geothermal power by the year 2030.

As with projections on the future use of coal in California, there is considerable variance as regards geothermal energy. Most estimates peg geothermal sources at supplying 5–10 percent of the state's electricity coming into the twenty-first century. In fact, as Table 6-1 shows for 1983–94, the low end of the range (5 percent in 1994) is the most reliable figure. Advancements in geothermal recovery technology, as discussed next, will point the way to the future.

In undergoing radioactive decay, magma and semimolten rocks deep within the earth release heat energy, which convects toward the surface. Where this heat concentrates in rocks near the surface and mixes with either naturally occurring groundwater or artificially injected water from the surface, it constitutes a geothermal field that can be tapped as steam, hot water, or hot rocks.

Hot-water or *flashed-steam* systems are the most commonly found in California—throughout the Salton Trough, for example, from the Imperial Valley southward to Cerro Prieto in Baja California. Unfortunately, flashed-steam systems are much more costly and difficult to develop than dry-steam systems because the hot water usually contains dissolved minerals that corrode pipes and other equipment. The brine buildup problem has slowed development of the Salton Trough geothermal field to a snail's pace, and potential negative environmental impacts, such as noxious hydrogen sulfide odors, hot wastewater and brine disposal, and the atmospheric effects of cooling towers, portend still more delays. Yet Magma Power's success indicates that some of these problems are being solved, and geothermal power production is on the upwing in the Imperial Valley.

On the positive side, geothermal power plants have a distinct advantage over fossil fuel and nuclear power plants in that almost all the activities involved in producing geothermal power take place at or adjacent to the plants. For example, a geothermal power plant's raw material—steam or hot water—is extracted from the earth right at the plant site, whereas oil, coal, or uranium are often mined hundreds or even thousands of miles away from where they will eventually be used to generate power. Not only are these fuels costly and often risky to transport (tanker oil spills, for example), but in some cases they must also be refined and stored before use. Note, too, that hydro, solar, wind, and biomass plants offer on-site advantages similar to those of geothermal power plants. These non-nuclear, nonhydrocarbon forms of electricity production collectively constitute what is increasingly being referred to as *green power*.

NUCLEAR POWER

The films *Chinatown* and *The China Syndrome* are similar not only in title but also in the fact that they added fuel to the fires of controversy over the development of certain resources: the former focusing on a big city's water schemes, the latter coming down hard on the nuclear power industry. *Chinatown* was released decades after Los Angeles had firmly established its water storage and distribution system, and so the movie was not likely to cause the undoing of anything. *The China Syndrome*, on the other hand, came out right before nuclear power plant proliferation was underway and at a time (March 1979) when the nuclear industry was already taking its lumps. There were five nuclear power plant closings in Maine, New York, Pennsylvania, and Virginia; antinuclear protests in New Hampshire; a cooling system malfunction causing dangerous radioactive steam leakage at the Three Mile Island nuclear power plant in Pennsylvania; and the discovery of possibly hazardous levels of radioactivity from old buried radium mine tailings in Denver, Colorado.

In the same month, however, the nuclear industry got some good news, too. A U.S. District Court judge in San Diego voided a 1976 California law requiring the state Energy Commission to determine that provisions for nuclear waste disposal exist before the construction of a nuclear plant. In essence, the ruling confirmed federal rather than state control over such matters and lifted what had effectively been a moratorium on nuclear plant development in California.

The nuclear reactor accident most detrimental to the nuclear power industry worldwide was the April 26, 1986, thermal explosion and fire at Chernobyl in the then Soviet Ukraine. Soviet government officials attributed 32 deaths to the accident, but environmentalists claim as many as 10,000 may have died by the fifth anniversary of the disaster. Since then, tens of thousands more have fallen victim to radioactive fallout, and the death toll may rise even more dramatically. For instance, of the 2.2 million people in the fallout area in the Ukraine and Byelorussia (now Belarus), 400,000 were children, and some 15,000 of them have been found to have enlarged thyroid glands, a condition that may affect the gland's ability to produce thyroicin, which is vital to metabolism and body growth. Despite the scope of the Chernobyl disaster, the number of operational nuclear reactors in the former Soviet Union dropped only from 51 to 45 between 1985 and 1990, and 25 new ones were being built in 1990 according to the International Atomic Energy Agency.

All told, the foregoing events seem symptomatic of the many troubles afflicting the nation's most suspect energy source, *nuclear fission*, whereby heat energy is released by the chain reaction splitting of uranium (U-235) atoms, which leads eventually to the generation of electricity. At the root of these troubles are vital public concerns, such as (1) the disposal of radioactive wastes, (2) the location of plants in earthquake-prone regions, and (3) the possibility, albeit remote, of a reactor meltdown in a plant. A near meltdown was effectively dramatized in *The China Syndrome*, and for a few days in March 1979, there was a chance that the Three Mile Island accident might lead to a meltdown.

The suggestion that nuclear power's time may have come and gone in California gains credence when both the present and future status of nuclear power plants is pondered. Nuclear power's share of California electricity production soared from 4.8 percent in 1983 to 22.2 percent in 1989 (see Table 6-1). During that 6-year period, the San Onofre nuclear facility increased its operational capacity to 2,200 MW; the Diablo Canyon plant near San Luis Obispo came online at an eventual capacity of 2,160 MW, but only after nearly two decades of vehement environmentalist protests; the Rancho Seco plant at Elk Grove near Sacramento continued operational through much of the period at a rated capacity of 875 MW; and the LADWP began tapping into its 9 percent share of Arizona's Palo Verde Nuclear Generating Station near Phoenix, the nation's largest with a capacity of 3,810 MW. These four sources alone served the electricity needs of several million California households. However, since mid-1989, when Sacramento voters—no doubt with the Chernobyl accident, as well as more localized concerns, in mind—recommended the closing of the Rancho Seco plant, nuclear power's share of electricity generation in the state has been put on hold perhaps for some time to come. It could even decline, unless the city of Los Angeles increases its import of nuclear power from Arizona. A fifth source of nuclear electricity generation, the Humboldt Bay Power Plant (65-MW capacity), was *decommissioned* (taken offline and deactivated) back in the 1970s and is unlikely to be reactivated.

Nuclear power's problems don't end at the power plant, but stretch far beyond in the form of disposal of radioactive waste. A recent case in point is the Ward Valley waste site in the eastern Mojave Desert, some 20 miles west of the lower Colorado River. Although antinuclear activists succeeded in 1994 in convincing the federal government to postpone transfer of the site from federal to state ownership, the issue of how to safely bury everything from radioactive power plant debris to radioactive waste from California's booming biotech industry is far from resolution. These and other industries may face severe curtailment if waste sites are not found.

In the meantime, nuclear innovations, including development of breeder reactors and fusion power, could one day prove a boon to the industry. Different types of *breeder reactors*, any one of which "breeds" enough fuel to replenish itself and drive another reactor over a number of years without the assistance of uranium enrichment facilities, are presently in operation at several locations in Europe and the former Soviet Union. The specter of breeder reactors overproducing plutonium and thus fostering the runaway buildup of atomic weaponry has caused many government officials in this country to take a dim view of breeder reactor development. So far, no commercial breeder reactor power plants are operating in the United States, although a demonstration plant along the Clinch River in Tennessee was proposed by Westinghouse and a demonstration core has been installed at a nuclear power plant in Shippingport, Pennsylvania, as part of the Naval Reactors Program.

By comparison, nuclear fusion technology is far behind that of breeder reactors. *Nuclear fusion* is the opposite of fission in that atoms combine rather than split to produce energy. Before fusion can be useful, however, it must yield more energy than is invested in heating it. Temperatures of about 100,000,000°C must be reached for fusion to occur, and so far this has been impossible to attain. Laser fusion may provide an answer, but this research, too, is in its infancy.

Nuclear power received a boost at the 1997 Kyoto global warming conference when it was reported that 442 nuclear power plants in 30 countries were supplying 17 percent of the world's electricity and thereby were holding carbon dioxide emmission levels (principally from hydrocarbon-burning power plants) at least 8 percent lower than they would otherwise be. For instance, in France, where nuclear plants supply 75 percent of the nation's electricity, emissions of carbon dioxide were cut by 26 percent on a per-capita basis in the 1990s. Yet this benefit of expanded nuclear-generating capacity in California seems to be outweighed by environmental and cost concerns, not the least of which is the $4-billion price tag for a third nuclear power plant in the state. Moreover, combatting hydrocarbon emissions by expanding nuclear power potential is not envisioned as part of national policy on global warming.

The ultimate solution to the nuclear power puzzle may lie not in economic considerations, or even in finding viable radioactive waste sites, but in alternative energy sources or green power—a solution that has been staring us in the face all along. Given California's diverse physical geography, which promotes a broad range of energy alternatives, there is far less urgency for nuclear power plant proliferation than almost anywhere else in the United States. Except in time of drought, interaction between the normal precipitation regime (see Chapter 4) and the extensive mountainous terrain (see Chapter 3) assures abundant hydroelectricity for California. The state leads the nation in geothermal power production, and by some estimates it may be producing 27,000 MW of geothermal electricity by the year 2030. And, as we will see next, climate and terrain facilitate a solar and wind energy wave of the future that has already arrived.

SOLAR ENERGY

Solar energy derives ultimately from the nuclear fusion process that takes place deep within our sun. The sun's nuclear energy rises to the surface and is emitted there in the form of energy known variously as *solar radiation, electromagnetic radiation,* or *shortwave radiation.* Solar energy travels out into space at a constant speed of 186,000 miles (300,000 km) per second. The earth, an average of 93 million miles (150 million km) or about 8 minutes and 20 seconds away, intercepts only a tiny fraction of the sun's total energy output. The intercepted energy, or *solar insolation,* then changes form as it penetrates deeper into the earth's atmosphere, losing much of its dangerous shortwave component long before reaching the troposphere. X-rays are absorbed 50–60 miles above the earth and ultraviolet rays are mostly screened out in the ozone layer 10–30 miles above the earth, although the latter process may have been weakened by the release of fluorocarbons. Once the incoming solar radiation reaches and passes through the troposphere, it loses much of its energy to diffusion, reflection, and absorption caused by the presence of water, vapor, dust, chemical molecules and clouds. Finally, the approximately 50 percent of remaining solar radiation is first absorbed by the ground and water surfaces of the earth and then reradiated as *ground radiation* or *longwave radiation* (principally *infrared* energy) that heats and maintains the air we breathe.

This transformation to longwave radiation and its retention by any barrier to escape, be it a cloud layer in the sky or the well-insulated walls and roof of a house, provides us with our most basic use of solar energy—*solar heating.* Southern Californians started amplifying this *greenhouse effect* in 1909 with the building and installation of *solar collectors* to heat water. By the late 1920s, natural gas was economically outcompeting the sun, resulting in some 15,000 solar collectors in the greater Los Angeles area falling into disuse. Now, though, solar energy use appears on the threshhold of an unprecedented renaissance as natural gas grows scarcer, oil becomes more expensive, coal pollutes more of the environment, and nuclear fuels seem more dangerous than ever before.

Other signs of a renewed solar surge can be seen in the 1978 Energy Tax Act granting a federal tax credit to those homeowners who installed solar equipment during a limited period that expired later in the 1980s; the involvement of Arco Solar, Exxon, General Electric, General Motors, the Jet Propulsion Laboratory (JPL) of Cal Tech, Luz International, Siemens, and others in solar research and development; federal, state, and city government promotion of solar development; and increased competition from green power providers caused by the 1998 deregulation of the state's electricity industry (discussed in the last section of the chapter). For solar energy's resurgence to last and prosper, though, inadequate financial incentives from government and private moneylenders and lack of consumer confidence in the solar industry must be overcome.

Figure 6-8 illustrates the fortunate position California occupies in relation to the rest of the nation as regards the development of solar power plants. It is one of only four states that has an appreciable amount of undeveloped land lying inside of the 500-langley-maximum solar radiation *isoline* (line along which a given value is equal). (A *langley* is a unit of illumination used to measure temperature equal to 1 gram calorie per square centimeter of irradiated surface.) The Great Basin, Mojave Desert, and Colorado Desert portions of California have ample area for space-consuming arrays of solar collectors and concentrators (mirror-lens systems) that would provide the heat energy necessary to run the turbine generators of each *solar thermal electric* power plant developed. Several such plants were in operation in the region before 1990.

In October 1980, the site of the nation's first solar electric plant, Solar One (see Figure 6-5), was dedicated near Daggett in the Mojave Desert. Solar One, a joint venture of the LADWP, SCE, the U.S. Department of Energy, and the California Energy Commission, was capable of generating 10 MW from 1,818 *heliostats* (mirrors) reflecting the sun's rays onto a collecting "power tower" containing a steam boiler. Steam then drives a turbine and generator.

Although successful, Solar One was purely a pilot or experimental solar power venture, with a planned shutdown taking place in 1987. Well before the closure, however, the privately owned Luz International Ltd. (LIL) and its California subsidiaries were rapidly filling the gap with more than a dozen developed and/or planned Solar Energy Generating System (SEGS) sites in the northern Mojave Desert (see Figure 6-5) capable of generating more than 700 MW of electricity. However, late in 1991, citing problems ranging from inconsistent government energy policies to falling oil and gas prices (oil and gas intended for fossil fuel power plants, making them more competitive), Luz filed for bankruptcy. Despite the filing, nine Luz SEGS continued sending electricity over SCE lines, and the SEGS plants came under the control of the KJC Operating Company, a wholly owned subsidiary of Kramer Junction Company, which was incorporated in 1991. Also in 1991, SCE decided against investing in the troubled Luz facilities, instead opting to build Solar Two at the former Solar One site. A $39-million facility, Solar Two is jointly funded by SCE, LADWP, the U.S. Department of Energy, and the Sacramento Municipal Utility District (SMUD).

The negative environmental impacts of large areas of desert covered by unsightly solar collector farms (130 acres at Solar One) and possible microclimatic thermal imbalances seem mild in comparison with the potential impacts of fossil fuel and nuclear power plants. Development of *photovoltaics* or *solar cells,* which convert solar energy directly into electricity and which have already proven reliable in supplying on-board power in all man-

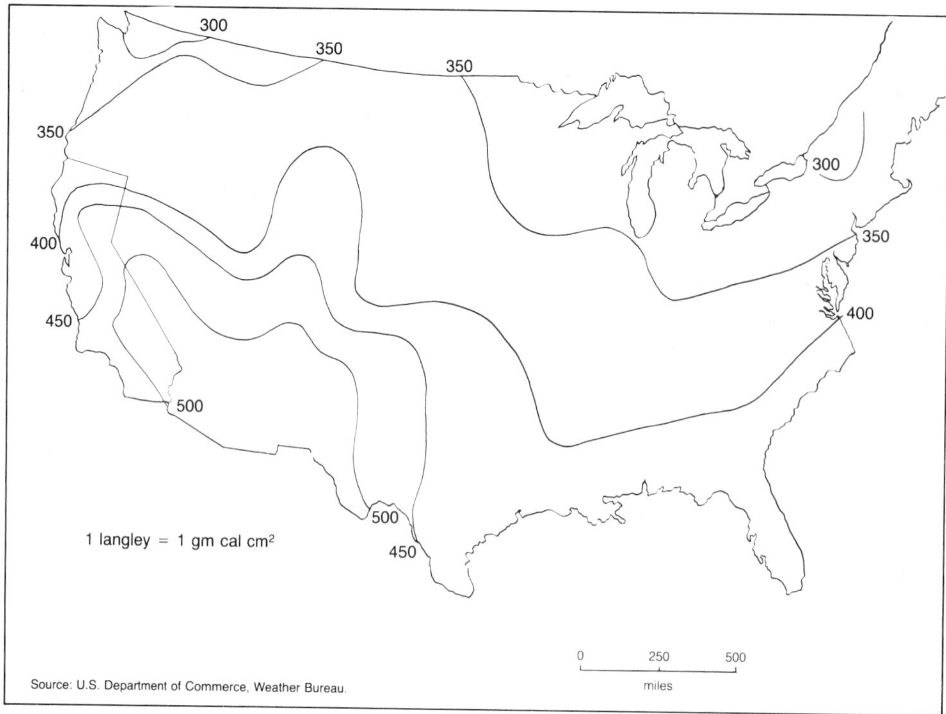

Figure 6-8 *Annual mean daily solar radiation (in langleys) in the United States.* (Karen Geissler)

ner of satellites, is expected to progress to the point in the next several years of rendering solar cells economically competitive with conventional electrical energy sources. Electricity from photovoltaics currently costs 30–100 times more than from conventional power plants. But solar electric plants energized by photovoltaics consume relatively little desert land by eliminating the need for solar collectors. A third desert solar energy innovation, already tested by the oil-short Israelis, is the *solar pond*. Plans were underway to develop solar pond power plants by the DWP at Owens Lake and by SCE at the Salton Sea in the 1980s.

Given the intermittency of solar energy—that is, there is none of it at night and greatly reduced amounts on cloudy days—the storage problem is perhaps the biggest single drawback faced by the solar industry. For space and water heating, heat can be stored in well-insulated water tanks or rock beds. For commercial solar electric plants, the storage problem is much more difficult, although some solutions may lie in utilizing electrochemical storage (storage batteries and fuel cells), pumped storage and production of hydrogen for solar electricity (with the hydrogen burned as a fuel when needed).

If solar cells accompanied by any one of these storage facilities could be adapted for use in individual structures, such as a single-family house, that building's occupants could receive enough solar energy to meet all their heating and electrical needs around the clock and be free of any energy input from utilities or other outside sources. For example, taking into account such geographical variables as location, average cloud cover, and hours of darkness, 1 square foot of area anywhere in the United States receives, on average, the equivalent of 17 watts of electrical energy per hour from the sun. Using this 17-watt amount and applying only a 10 percent efficiency factor to the use of solar cells, a 20- by 30-foot panel of solar cells could produce about 25 kWh of electricity per day in one year's time. This is sufficient output to satisfy the electricity needs of a single-family household of four. For larger numbers of people and/or bigger buildings, photovoltaic panel size would simply be increased. Another technique, involving either home solar cells or wind turbines, would tie homes into a reciprocal system whereby surplus electricity generated at homes during the day or during high-wind periods would be transmitted to nearby power plants. At night or during windless periods, power plants would return electricity to the homes.

WIND POWER

Harnessing *wind energy* to generate power, *bioconversion* of agricultural and municipal wastes into fuel, and generating electricity from *ocean thermal* differences (temperature change from ocean depths to surface) represent still more ways in which we can utilize solar energy. Again, as stressed earlier, California's sunny climate, mountainous topography, and location next to the Pacific Ocean place it in an enviable position in North America for utilizing all forms of solar energy. Although contributing only 1 percent of all electricity generated in the state in 1994 (see Table 6-1), forecasters see solar,

Figure 6-9 *Wind farm at San Gorgonio Pass near Palm Springs. In 1990, the wind farm had the third-highest-rated capacity in the state, with 212 MW from 3,292 turbines. First was Altamont Pass east of San Francisco Bay near Livermore, with 661 MW from 6,242 turbines; and second was Tehachapi Pass east of Bakersfield and the southern San Joaquin Valley, with 393 MW from 4,361 turbines. A scant 2 MW from 62 wind turbines came from three additional wind farms in Northern California (Boulevard, Carquinez Strait, and Pacheco Pass).* (Claudio Ribas)

wind, and biomass more than doubling the percentage within a decade.

Given California's unique atmosphere–land–water relationships, the development of wind power appears to hold the most promise among the more unconventional forms of energy. The high, continuous ridges of the Sierra Nevada, lying almost perpendicular to the prevailing westerlies from the Pacific, provide the windiest, albeit perhaps the most hazardous, wind power sites in the state. Environmentalists are not likely to be enamored by the prospect of windmills along the John Muir or Pacific Crest trails, but it is such high-altitude locales that offer the strongest and most constant winds in the state. Yet they are remote from user markets compared to several recently developed wind farm sites (see Figure 6-5) that in 1990 had a cumulative capacity of nearly 1,300 MW from some 14,000 turbines. A decade earlier, there were only a handful of wind turbines generating less than 10 MW of electricity. By 1995, some 17,000 turbines were generating nearly 2,000 MW of electricity. This presently accounts for about 1 percent of state electricity production, but it can rise to as much as 8 percent when the winds are howling simultaneously at the state's three principal wind-farming sites: Altamont, San Gorgonio, and Tehachapi passes (see Figure 6-5). California alone generates 85 percent of the world's wind-produced electricity. It is distributed by PG&E and SCE from privately and cooperatively operated wind farms. Much of the technology comes from Denmark, which seems destined to become the leader in wind power development in Europe.

San Gorgonio Pass (Figure 6-9) exemplifies the mix of climatic and geomorphic conditions conductive to large-scale wind farming in many mountain passes in California. This is especially noticeable heading into summer each year as the westerly flow (west to east) of air begins to pick up speed. As discussed in Chapter 4, the surface of the desert interior heats up, the warmed air rises, and thermal low pressure prevails. In turn, the low-pressure area acts like a vacuum, pulling in the descending air of the Pacific subtropical high-pressure cell onshore. Once the westerly winds have reached their gateway to the desert, San Gorgonio Pass, they are intensified, in a kind of "Venturi effect," as they are funneled through the 2,616-foot pass with 10,000 plus-foot mountains lining its north and south sides. At the eastern end of the pass, thousands of turbines await the winds.

DEREGULATION, CONSERVATION, AND THE ENVIRONMENT

When deregulation of California's electric industry took effect January 1, 1998, consumers realized an instant benefit: a 10 percent rate reduction that applied whether they continued with their existing electricity service provider or switched to a competitor. The longer-range impacts of deregulation, which is intended to lower electricity prices by allowing some 200 in-state and out-of-state power providers to compete with the existing (prior to 1998) 3 regional utilities (PG&E, SCE, SDG&E) and 30 municipal power companies (including the LADWP and SMUD), will be varied and sweeping. Ranging from increased consumption of cheaper electricity to the growth of the green power industry, deregulation will significantly influence energy conservation and related environmental initiatives in twenty-first-century California.

Conservation is one of the most promising sources of energy, but it has had trouble getting much past the hypothetical stage. Energy-efficient appliances, the requirement that zero-emission vehicles constitute 10 percent of the new car market by the 2003 model year, the now defunct 55-mile-per-hour speed limit, better gas mileage from more compact cars, utility company consumer conservation and education programs, buildings designed to be energy efficient—all have had some effect, but energy consumption continues to increase at an alarming rate. If the present pace does not abate, electricity use will double in the nation by 2010. Even now, Americans use twice as much energy per capita as citizens of the most developed countries in western Europe.

Californians are a bit better off than residents of most other states as regards the cost of energy. The state's mild winters largely preclude the use of heating oil, the outlandish bills for which have driven many easterners to California. In the summer, Los Angelenos and San Franciscans utilize air conditioners, but considerably less than Chicagoans or New Yorkers do. And when there are oppressive heat waves and smog attacks, most California urbanites have the option of quick escape to the beaches or mountains. But getting there, or even back and forth from work, is usually accomplished with a motor vehicle, often occupied by only one or two people. California needs more and better rapid-transit systems if it is going to to put a dent in both the nation's highest gasoline consumption levels and its worst smog conditions.

The premise that conservation is the best energy source appears to be gaining support. One study concluded that $200 billion spent in conservation would save more than $200 billion put into exploration for new fossil fuel reserves.[2] And Americans would not have to become energy spartans to the point of sacrificing lifestyle and economic growth and prosperity. The study also supported commitment to the expanded use of all forms of solar energy. Even hydropower's future appears brighter as higher-capacity turbines are developed for both existing and new hydroelectric dams. It is conceivable that green power, coupled with adequate conservation measures, could supply a quarter of the nation's total energy needs by 2010.

The most important benefit to be reaped from energy conservation is giving the environment we depend on a longer lease on life. The bottom line is, the more energy we conserve, the less we place pressure on the environment—be it emissions from automobiles and fossil fuel power plants; dust from strip mines; thermal pollution of the ocean, lakes, and streams by warm water discharged from steam electric plants; or radiation above normal background levels wherever a nuclear accident has occurred.

As we shift focus from the atmospheric to the biotic component of the California ecosystem in Chapter 7, remember that energy is the basis for everything that functions in any environment, be it human or natural. When energy is plentiful, nonpolluting, and readily distributed throughout an ecosystem, whether by power line or food chain, that environment will prosper. When energy becomes scarce, polluting, and unevenly distributed, however, the environment in question will deteriorate. At first glance, the latter scenario unfortunately seems to fit California more closely than the former. But when California's most abundant energy source—the sun—is brought to mind, it is not difficult to perceive the state's human population and natural resources on the brink of a new prosperity.

[2] Roger Stobaugh and Daniel Yergin (eds.), *Energy Futures* (New York: Random House, 1979).

CHAPTER

From Redwoods to Sagebrush

California's majestic landforms, innumerable climates, more than 1000-mile ocean front, and great latitudinal range provide an unrivaled habitat for plant and animal life. This complex of natural influences has produced a native California flora that is unique and yet at the same time mirror almost any other part of North America, from the Arctic to the tropics or between the coasts.

Stately redwood groves are all but exclusively Californian but at the same time represent southward invasions of the coniferous rainforests of the maritime Pacific Northwest. Ancient bristlecone pines are perched high atop several mountain ranges west of the Rockies, but it is in the alpine tundra of California's White Mountains that they are most numerous and were first discovered to be the oldest of living trees. Douglas fir and yellow pine forests by the millions of acres underscore California's high ranking among lumber-producing regions in the West. Oak woodland–grassland pervades the Sierran foothills and Coast Ranges and in the green of spring resembles the deciduous woodlands of the East. Chaparral—a fire-prone shrubland characteristic of Mediterranean climates—belongs almost exclusively to Southern California (although many a hillside resident would sooner give it away). Sage-covered high deserts and cactus-dotted low deserts give eastern California the appearance of the arid Southwest. It is here in the parched Mojave Desert that the world's longest-living plant, a creosote bush thousands of years older than the most ancient of the bristlecone pines, ekes out an existence. And native fan palms in oases scattered about the Salton Trough and southern Great Basin add a final tropical touch to California's natural landscape.

Given such diversity of plants, and remembering that wildlife variety, population, and distribution are regulated by the availability of food and shelter within natural communities, it can be assumed that California's fauna is no less diverse than its flora. Whether one ponders reptiles and rodents forging an existence in the arid desert, seabirds and sea mammals prospering from a bountiful ocean, remnant herds of antelope and elk browsing on Great Basin vegetation, bears and coyotes rummaging through Forest Service garbage cans, or salmon and steelhead trying to swim past hordes of Labor Day fishermen upriver to their ancestral homes, California's wealth of wildlife is everywhere to be found. Most of these and a multitude of other animals can be seen in their natural habitats even by the typical carbound Californian, if he or she is but willing to get away from the city.

Human interaction with California's flora and fauna began perhaps 50,000 years ago; but only in the last few centuries, with the advent of European settlement, has human modification of natural communities in a comparatively short time taken on a whole new dimension. In some cases, entire native landscapes have been altered beyond recognition, as witnessed in the speedy displacing by agriculture of the once-pervasive bunchgrasses of the Great Central Valley or the cutting away of most of the virgin stands of coastal redwood. In other instances, individual species either have been eliminated altogether from the California scene, as in the case of the ill-fated grizzly bear, or have come dangerously close to local extinction, as with the condor, sea otter, tule elk, and other fauna too numerous to mention. Yet countless species of exotic plants and animals, ranging from Australia's eucalyptus trees to Europe's wild hogs, have been introduced and prospered as successfully in the California wilds as they would have in their ancestral homelands. With a human population of 33 million so recently attained, a quickened pace of ecological change in California is inevitable.

What these modifications forebode for the future of the biotic elements of the environment is uncertain. Knowledge of California's contemporary biogeography, however, is a step toward intelligent participation in the conservation of the state's rich biotic resources.

PRINCIPAL BIOMES

Biomes or *ecosystems,* those biogeographical units that contain both particular physical environments (landforms, water bodies, climates, and soils) and the energy-producing plants and energy-consuming animals found there, are generally classified at several different areal levels.[1] The sun is the ultimate source of biome energy, and our solar system could be considered the largest ecosystem of which we have any appreciable knowledge. The hierarchy of ecosystem sizes ranges from the earth down through the continents and regional biotic communities to units as small as a local stream or a pond.

California's diverse physical geography has made for literally dozens of different biotic communities named for characteristic plant assemblages that appear dominant, such as the pine-oak woodland of the western Sierran foothills or the coastal sage association bordering the seashore of central and Southern California. Because this chapter presents a condensed biogeography of California rather than a complete naturalist's guide to the state, we will systematically examine only some of the more representative biotic communities in each of the broadly based *principal biomes*—those biomes consisting of California's coniferous forests, woodlands, grasslands and marshlands, desert shrublands, chaparral and coastal shrublands, and littoral (tidal shoreline).

Coniferous Forests

California's coniferous forests, generally described as areas nearly or completely covered by needle-leaved, cone-bearing trees, occupy nearly one-fifth of its 158,693 square miles. The natural distribution of this 17 million acres of commercially accessible redwoods, firs, pines, and other evergreen softwoods is determined largely by climate and topography, as a comparison of Figures 3-1, 4-1, and 7-1 demonstrates. Wherever orographic (mountain-related) precipitation is heaviest, which includes the west flanks of the northern Coast Ranges, Klamaths, Cascades, and Sierra, verdant forests of conifers are likely to be the prevailing form of natural vegetation. Despite rain- and snowfall mostly in winter and rarely in other seasons, humidity, latitude, nutrient supply, *aspect* (direction of slope exposure), soil moisture, and other local conditions also favor forest growth along the northern coast and windward mountain slopes.

Adaptation by distinct associations of plants and animals to these localized environmental influences has led to the formation of a myriad of biotic community types throughout California's coniferous forest biome. Most are *climax communities:* They have developed to a point of stability or equilibrium with their environment throughout several stages of *ecological succession*. Succession may take hundreds of years, as when a red fir (*Abies magnifica*) forest slowly establishes itself on a former meadow, or it may occur in a matter of decades, as when lodgepole pines (*Pinus murrayana* or *contorta*) quickly invade a recently burned-over area.

In terms of special adaptation and long succession, and even world renown for that matter, California's redwoods are unique among climax forests. The distinction actually is shared by two spatially separated species of redwoods, both of which are almost exclusively Californian and variations of the single genus *Sequoia*: (1) *Sequoiadendron giganteum* (formerly *Sequoia gigantea*), which are commonly known as sequoias or simply the "big trees" and are found only in small groves scattered along the western slopes of the Sierra Nevada; and (2) *Sequoia sempervirens*, usually called redwoods, which extend in a coastal-oriented band from several miles above the Oregon border southward through Del Norte, Humboldt, and Mendocino counties and still farther south in isolated coastal mountain groves well into Monterey County. Both species enjoy unparalleled stature—with the tallest coastal redwood claiming a world's record of nearly 400 feet and the General Sherman tree in Sequoia National Park measuring 27.5 feet in diameter from a dozen feet above its base. The species are not short on longevity, either; many a sequoia and coastal redwood are in excess of 2,000 years old. The majestic General Grant sequoia is an estimated 3,500 years of age, following by only a few hundred years some scraggly-looking bristlecone pines (*Pinus aristata*) as the oldest of living trees (Figure 7-2). Indeed, older sequoias and redwoods are often spoken of as living fossils in deference to an ancestry that dates back perhaps a hundred million years when *Sequoia* and similar genera were distributed

[1] Biomes are classified by the dominant vegetation found within them, whereas ecosystems are likely to be determined by climate and soil conditions within their boundaries. Both can be thought of as systems.

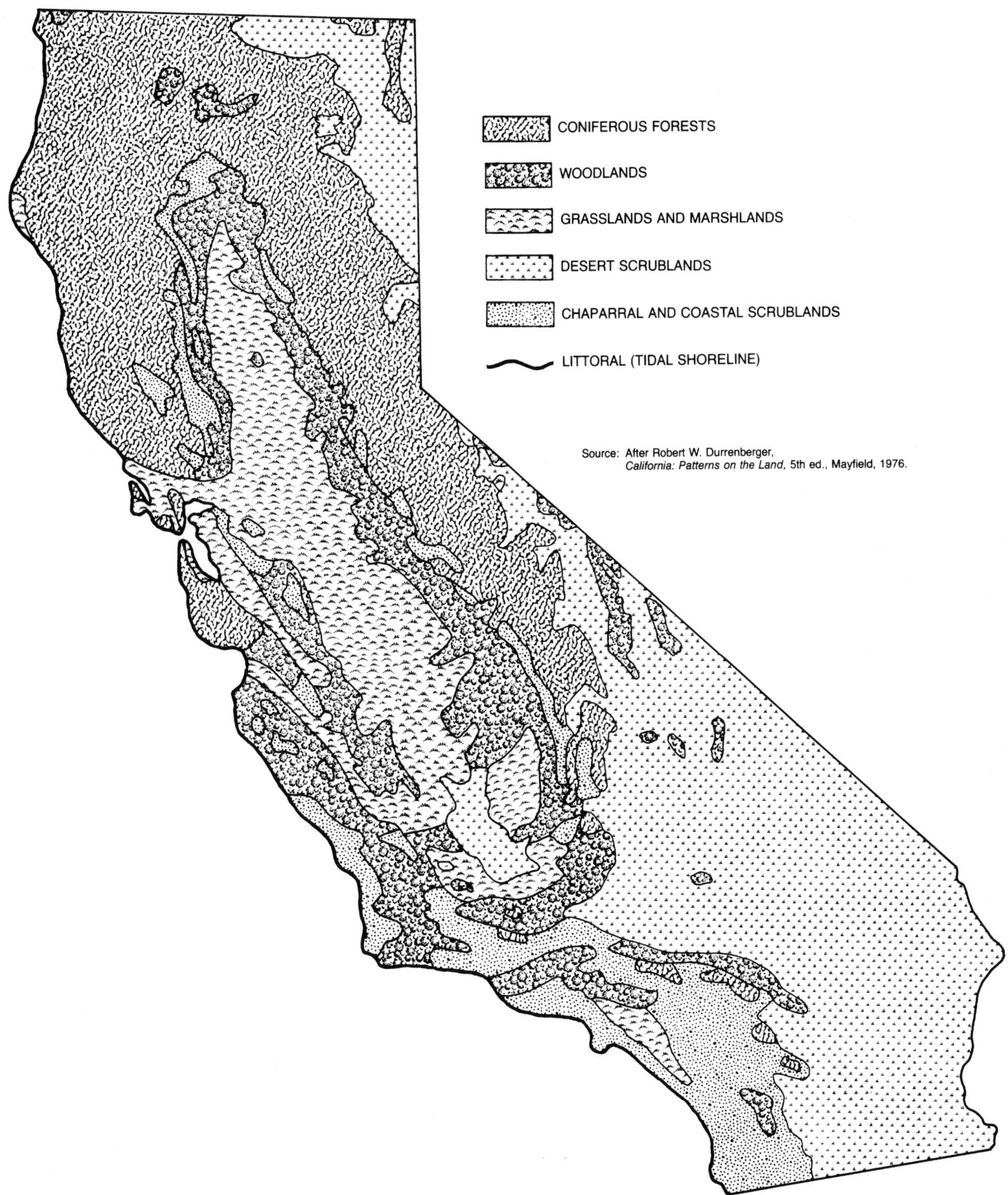

Figure 7-1 *California's principal biomes.* (Richard Crooker)

far beyond the boundaries of California as we know it. Then, as now, these trees endured because of long-acquired survival features including fire-resistant, fibrous, nonresinous bark; moist and decay-resistant wood; lack of lower limbs, which deters the spread of fire over a forest floor; abundance of small cones, each filled with hundreds of seeds; ability to seed-root or stump-sprout; and certain flood tolerance capabilities.

For all their environmental adaptations, the coastal redwoods' accommodation with California's wet-winter/dry-summer precipitation regime is most remarkable (Figure 7-3). Copious amounts of winter rain-

Figure 7-2 *Bristlecone pines in the White Mountains east of the Owens Valley. These are the world's longest-living trees, with some individuals dated at more than 4,600 years of age. Their longevity owes to a seeming paradox of harsh environmental conditions including the presence of thin, alkaline soils; highly reflective, heat-avoiding soils and rock outcrops; a mean elevation of the groves of 11,500 feet above sea level; meager annual precipitation of 15 inches; and unobstructed gale-force westerlies. The bristlecones have adjusted to these timberline hardships by growing slowly; having a minimal amount of live foliage, which reduces moisture loss through transpiration; having dense, decay-resistant wood; and clustering on shady slopes, where snow stays on the ground longer.* (Crane Miller)

Figure 7-3 *Redwood rainforest in foggy Armstrong Redwoods State Reserve in northwestern Sonoma County. The Parson Jones redwood is 310 feet high and nearly 1,500 years old. A moist environment, lack of lower limbs, and thick bark help preclude the spread of fire through a redwood forest; if burned down, however, the forest will regenerate itself as seeds flourish in barren soil and sprouts issue forth from stumps and roots.* (Crane Miller)

fall coupled with mild temperatures contribute to the redwoods' presence along the central and northern coasts, but it is the lingering fog of summer in this region that ensures their success. Year-round high relative humidity and the attendant cool air in California's fog belt help satisfy the trees' water requirements by diminishing moisture loss from leaves and soils caused by evapotranspiration. Were it not for fogs drizzling upwards of 20 inches of moisture on the forests in a normal rainless summer, most of the state's 1.5 million acres of coastal redwoods would simply not exist.

During the Pleistocene, when these cool, cloudy summers extended farther south and inland, redwoods populated the mountains and coasts of Southern California. Today, although *endemics* (species originating naturally and exclusively in a given region) that are *relict* in their area (endemic in a different time and sometimes different climate), such as big-cone spruce (*Psuedotsuga macrocarpa*) and Torrey pine (*Pinus torreyana*), still barely manage a natural existence in the southland, redwoods are likely to be found doing well only in the artificial environments of irrigated ornamental plots and experimental forests.

Few, if any, North American trees command the deep respect and high value accorded the redwood. In the Sierra, its groves are named after generals; in the northwest, after churchly architectural forms. Among environmental activists, "Save the redwoods!" is a commonly heard cry. And in retail lumberyards, the price of redwood, like oil or any other rapidly disappearing resource, is skyrocketing. Coastal redwood is a durable wood, highly prized by architects, builders, and other users. The increasing demand for the wood over the last century or so has resulted in the cutting of more than 90 percent of the original old-growth redwood stands. Today, demand persists as the resource shrinks. California has less than 30 billion board feet (1 BF is equal to the volume of a board 12 inches by 12 inches by 1 inch) of commercially recoverable redwood saw timber in reserve, but with an average annual cut rate of 1 billion BF, double the average yearly replenishment rate of new or second-growth redwood, the resource may run out more rapidly than anyone realizes.

As shown in Figure 7-4, the 58,000-acre Redwood National Park, adjacent expansions of the park, Muir Woods National Monument, and state parks preserve some of what is left of the virgin stands from the chain

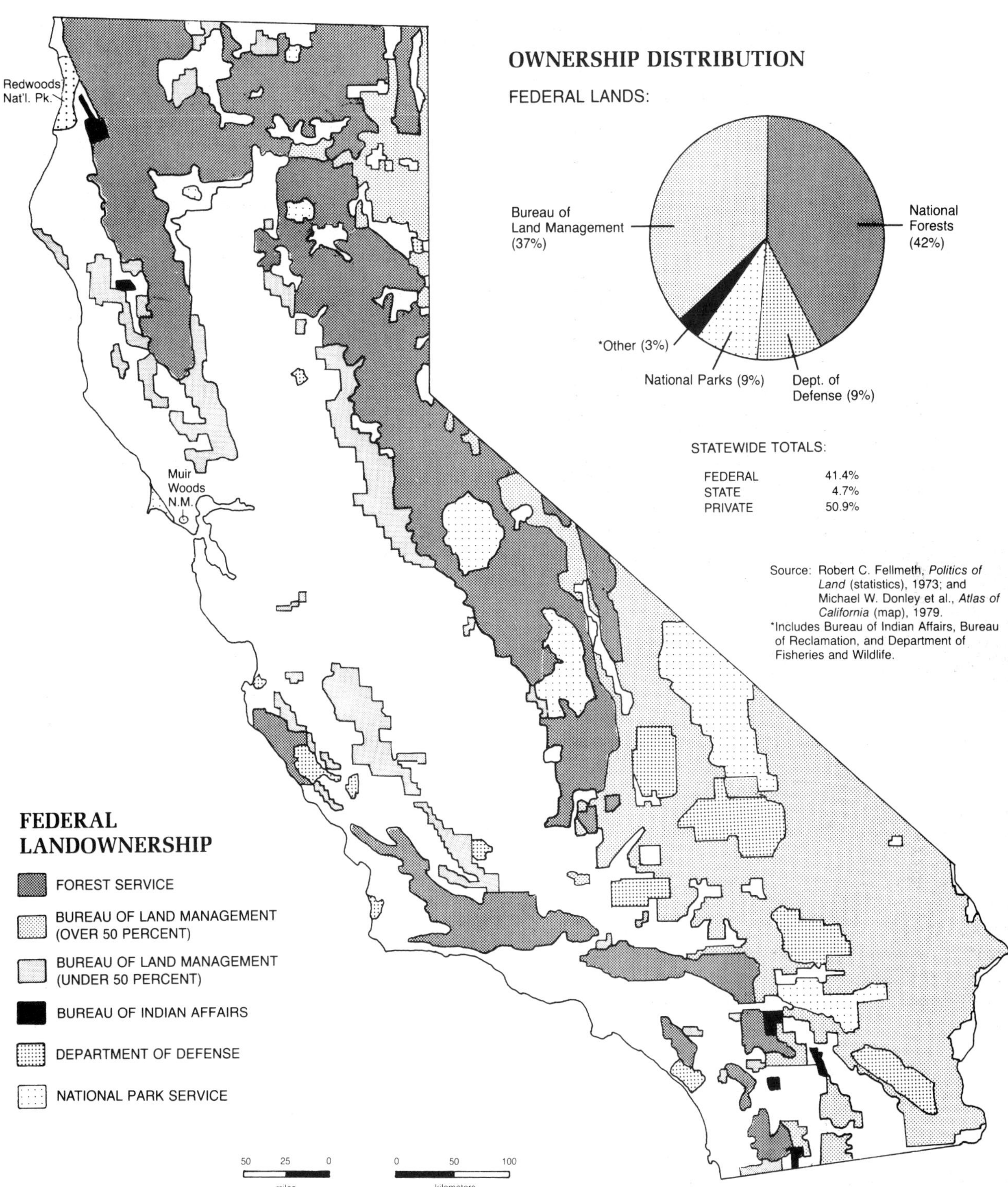

Figure 7-4 *Federal landownership and distribution in California. The Bureau of Indian Affairs administers lands in trust, much of the land not being federally owned. The Department of Defense slice of the pie chart does not reflect 1990s military base closures and possible transfers of ownership, the latter sometimes not occuring until many years later, if at all.* (Richard Crooker)

saw and even some of the ravages of nature. Some foresters contend, however, that further major expansion of protected areas for coastal redwoods will eventually weaken the forest's ability to stave off fire, floods, disease, and other natural disasters. This seems a reasonable argument as far as it goes. But the fact remains that the national park and other existing preserves are really so meager, so spatially disjunct, and so pressured by

TABLE 7-1 California conifer harvest, value, and acreage, 1988 and 1994.

SPECIES	YEAR	BOARD FEET IN BILLIONS	PERCENTAGE	MILLION $	PERCENTAGE	MILLION ACRES	PERCENTAGE
Yellow Pine	1988	0.94	20.3%	$195	29.2%	2.65	11.5%
	1994	0.44	18.9	230	20.9	—	—
Douglas Fir	1988	1.19	25.8	128	19.1	1.77	7.7
	1994	0.53	22.7	258	23.4	—	—
Redwood	1988	0.78	16.8	175	26.2	1.57	6.8
	1994	0.49	21.5	313	28.4	—	—
Other*	1988	1.72	37.1	171	25.5	17.02	74.0
	1994	1.50	36.9	101	27.3	—	—
Totals†	1988	4.63	—	669	—	23.01	—
	1994	2.96	—	902	—	—	—

* Other commercial conifers mainly include red and other firs, sugar pine, and incense cedar.

† In 1988, there were 23,013,000 acres in all species of conifers, or 23 percent of California's total land area of 100 million acres. Employment in 1988 in California forest products industries totaled 107,500, with 66,400 in lumber and wood products and 41,100 in pulp, paper, and associated products. Forest products employment increased from 86,000 in 1971 to 107,800 in 1979, and then declined to 83,700 in 1982, a year in which high mortgage interest rates depressed the home-building industry and thereby diminished demand for lumber. Relatively high mortgage interest rates, while they last, obviously can forestall the cutting of conifers.

Sources: California State Board of Equalization, Timber Tax Division; California Department of Forestry, FRRAP Information and Analysis System; and California Department of Employment, Report to the Governor on Labor Market Conditions.

recreational activity that they represent anything but a viable natural ecosystem. Redwood National Park boundaries and jurisdiction, for instance, do not extend inland far enough to directly influence logging activities in upstream portions of the Eel River, Smith River, and other major regional watersheds where greatly accelerated erosion eventually affects drainages all the way to the sea.

The related problem of cutting more timber than is being replaced occurs not only on private lands bordering the national park but throughout the state in privately owned forests that make up almost half of the entire 17-million-acre commercial timberland resource. And, although many companies strive for *sustained yield* (maintaining the yield of a renewable resource, such as trees, at a given level into the foreseeable future) through tree farming and other sound forestry practices, the fact persists that the bulk of redwood forest remains in private hands and the species continues to lose ground because of it. What's more, the public's national forests (see Figure 7-4) are subject to multiple-use management, which includes lumbering, grazing, mining, recreation, and utility and transportation line easements. About 1 million acres of wilderness areas in national forests are exempted from these uses (except for some recreational uses), but most of the wilderness land is far inland of the coastal redwood range.

Fortunately for the conservationist, logger, and tourist alike, redwoods do not stand alone in northwestern California. Intermingling with redwoods near the coast and forming nearly pure stands inland on the windward slopes of the Klamaths, Coast Ranges, Cascades, and northern Sierra, Douglas fir forests constitute the most prolific raw material source for the state's forest products industry. As shown in Table 7-1, in 1988, for example, Douglas fir led all other species in timber harvest with nearly 1.2 billion BF cut. Although still the leader in 1994, harvests of Douglas fir and all other species declined steeply. Yet the value of all conifer harvests increased from $669 million in 1988 to $902 million in 1994. Because the main stands of *Pseudotsuga menziesii*—not a true fir as its "false hemlock" scientific name implies—lie in the less accessible interior mountains of the northwest part of the state, they were not exploited by the timber industry as early nor as readily as were coastal redwoods. But a post–World War II construction boom, modern road-building equipment, and fleets of hauling trucks thrust Douglas fir from obscurity into prominence almost overnight. The flexible, knot-free wood soon became the chief raw material for everything from plywood to rough carpentry.

In recent years, Douglas fir has been the wood most responsible for these effects: (1) California places second, after Oregon and before Washington, among the nation's leading lumber producers; (2) the forest products industry as a whole (but mostly lumber products) has captured about 5 percent of the California economy; (3) some 100,000 Californians are employed in various forest products industries; and (4) the main source of income in the northwestern part of the state derives from timber. As regards the latter and the sawtimber and milling industries in particular, automation and a fluctuating demand for new housing have contributed to chronic high unemployment in the region. Conservationist pressures to exclude more forests from cutting have aggravated the situation, especially as far as jobless loggers are concerned.

If jobs and a whole industry are to be saved, sustained yield must soon be achieved in the Northwest's Douglas fir forests. But there is a loser in such an undertaking: the already beleaguered coastal redwood. Douglas fir grows faster than redwood and other second-growth conifers, progressing from seedling to saw timber size in 30–60 years. Also, Douglas fir prospers in both rainforest and drier alpine climates. Nevertheless, the removal rate of the species still more than doubles the growth rate, and extensive cutting of remaining virgin stands continues for the most part to sustain the timber industry. To narrow the disparity between the two rates and save something of the old-growth Douglas fir forests at the same time, timber firms generally do not reseed cut redwood areas with redwood but rather with quicker-maturing Douglas fir. Both species are subject to *clear-cutting*, which can leave a hillside barren except for stumps. And with regard to reseeding a logged-over area, both require the drenching sunlight of a clear cut. Unlike redwoods, however, Douglas firs do not defend well against fire, flood, and pest, nor can they grow back from stumps. As paradoxical as it may seem, the redwood's natural competitive advantage has furthered its undoing. As foresters correctly point out, redwoods can regenerate on their own, whereas Douglas firs need help. And because Douglas firs can be harvested more often, they get that help. Apparently, if sustained yield is one day to be realized by Douglas firs, it will continue to be at the expense of coastal redwoods.

Of California's 23 million acres of conifers, 16.4 million acres are considered commercial forestland, with 47 percent of this privately held, 52 percent in federal hands, and 1 percent owned by Native Americans and others. Obviously, the private forests, where there is the least government regulation, are of the greatest concern to environmentalists. One of their goals, which is to promote outright state or federal purchase of privately-owned old-growth forests, was realized in 1999 when the federal government purchased and created the 2,738-acre Headwaters Forest Reserve. The reserve, which lies to the southeast of Redwood National Park in Humboldt County along the headwaters of the south fork of the Elk River and which is part of a larger $480-million purchase from Pacific Lumber Company, is administered by the U.S. Bureau of Reclamation. Environmentalist also are seeking greater representation on the California Board of Forestry, which oversees the logging industry, and increased power vested in the state Department of Fish and Game to control logging. Earlier in the 1990s, in an unrelated move, the federal government created the 305,337-acre Smith River National Recreation Area in the northwestern corner of California, thereby preserving much of the old-growth habitat of the spotted owl (*Strix occientalis*) and a small seabird called the marbled murrelet (*Brachyramphus marmoratus*).

Ascending the mighty Sierra Nevada anywhere along its 400-mile western front, one cannot help but be struck by the immensity and variety of coniferous forests. Here, indeed, are forests that rival those of northwestern California by almost any measure. The first to be encountered in an eastbound ascent of the Sierra are the oak–pine woodlands, but as foothills give way to longer and steeper mountain slopes, California's greatest gathering of conifers soon prevails. Averaging from 2,500 to 7,500 feet in altitude range, this is the Sierran yellow pine belt (Figure 7-5). Here, tall stands of *Pinus ponderosa* and sugar pine (*Pinus lambertiana*) reach for the sky on sunny slopes and ridges, while mixed forests of Douglas fir, white fir (*Abies concolor*), and aromatic

Figure 7-5 *Yellow pine forests, here seen at about 5,000 feet. These forests dominate the western slopes of the Sierra.* Selective cutting *of mature or rotation-age trees is practiced in these forests rather than clear-cutting, the latter not being essential for the growth of seedling yellow pines. The cleared areas in the background are scars left along the South Yuba River by the Alpha and Omega gold-mining operations of decades ago.* (Crane Miller)

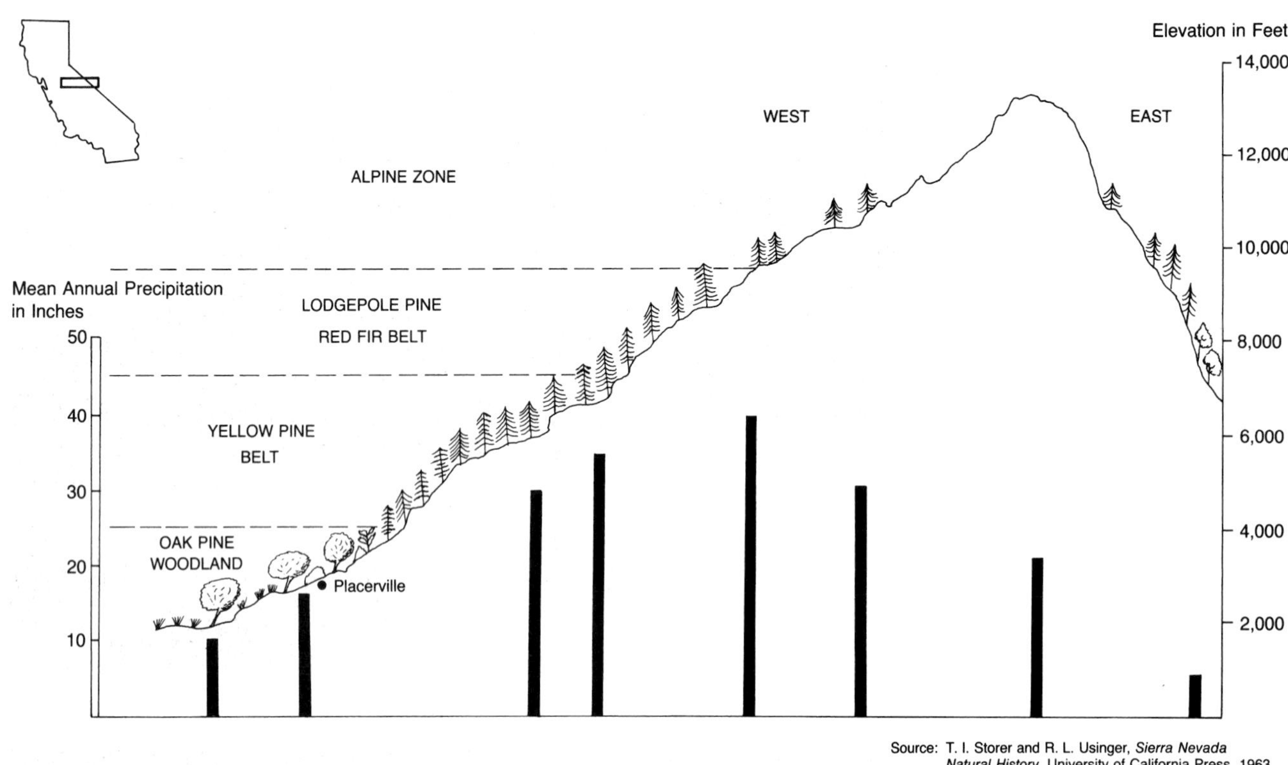

Figure 7-6 *Vertical zonation of precipitation and vegetation in the central Sierra Nevada.* (David Vechik)

incense cedar (*Libocedrus decurrens*) take cover on shaded mountainsides beneath groves of giant sequoia. All these species provide sawtimber of varying grades; together, they comprise a several-million-acre forest resource, the largest single area of its kind in California. Yellow pine alone (mostly from the Sierra, but elsewhere as well, as noted in Figure 7-5) is second to Douglas fir in volume of lumber produced statewide. In 1994, for instance, nearly a half billion board feet of yellow pine was harvested (see Table 7-1).

Above the yellow pine belt on up to 9,500 feet, the lodgepole pine–red fir belt prevails. Besides its namesake species, this second long wall of conifers contains the ponderosa lookalike Jeffrey pine (*Pinus jefferyi*) and the bluish green mountain hemlock (*Tsuga mertensiana*). The four species add still more to the Sierran timber resource.

From 9,500 feet to timberline, forests thin out into scattered groupings of trees nestled on granite flats or clinging to precipitous ridges. Commonly seen in the thin environment of this high alpine belt are spectacularly deformed Sierra juniper (*Juniperus occidentalis*), bulky-looking limber pine (*Pinus flexilis*), wind-blasted foxtail pine (*Pinus balfouriana*), and dwarflike whitebark pine (*Pinus albicaulis*). Of no commercial worth, these loftiest of California's conifers nevertheless add the final living touch to the priceless solitude of the High Sierra.

The role of orographic precipitation in accounting for the vertical zonation of different climax communities is brought into noticeably sharp focus in the Sierra Nevada. In this relationship, profiled in Figure 7-6, average annual precipitation attains its maximum in the upper yellow pine and lower lodgepole pine–red fir belts of the central Sierra. As expected, these wettest forests are also the richest forests. They contain not only the many conifers previously described but also small groves of aspen (*Populus tremuloides*; see Figure 7-7) and other broadleaf deciduous trees and shrubs, wildflowers and other flowering plants, flowerless and seedless ferns, and all manner of lichens and mosses.

Limited largely to winter and occurring at elevations above 5,000 feet, precipitation over these forests falls mostly as snow. Hundreds of inches of snow accumulate annually, forming both a protective blanket of insulation for vegetation against windchill and a slow-melting reservoir of water to see plants and animals through the long summer's drought. The cold, snowy winter, combined with a warm, dry growing season (3–7 months, depending on altitude and latitude), also favors predominance of evergreen over deciduous trees, for the former keep their needle-shaped leaves the year round and thus have no need to expend energy producing new foliage every spring.

In essence, California's Mediterranean climate, as discussed in Chapter 4, has largely precluded the nat-

Figure 7-7 *Aspen and similar deciduous trees in the Sierra. These trees prefer the higher and more northerly reaches of the ranges, as seen here along lower Lee Vining Creek near the eastern entrance (Tioga Pass) to Yosemite National Park. These hardwoods are more prolific in the American and Canadian Rockies where, unlike in California, wetter summers and colder winters prevail.* (Crane Miller)

ural existence of large deciduous forests anywhere in the mountainous regions of the state. Myriads of microclimates, such as in the shade of a large tree or on a sun-drenched south-facing slope, have further fostered a broad variety of native and endemic species of evergreen conifers throughout California's forests.

In terms of photosynthetic plants and the food chains they energize, California's coniferous forests provide the best of worlds for wildlife. One such world and its food chain can be found at forests' edge in Sierra streams, where plentiful sunlight activates green chlorophyll in algae, which is fed upon by insect larvae and water snails. These small aquatic animals are then eaten by endemic rainbow trout (*Salmo gairdnerii*; see Figure 7-8), which, in turn, are sometimes taken by nonendemic *Homo sapiens*. (Fishermen often use the term "native" to refer to a trout that was born in a stream rather than in a hatchery.)

Another such producing and consuming ecosystem lies deeper within forests on the western slopes of the Sierra. Here, chipmunks (*Eutamias* species, or spp.) and squirrels (*Citellus*) scurry about harvesting pinecone seeds while at the same time keeping a wary eye out for predators such as the grey fox (*Urocyon cinereoargenteus*) and mountain coyote (*Canis latrans*), which rarely deviate from their carnivorous instincts, or the black bear (*Ursus americanus*) and raccoon (*Procyon lotor*), which as omnivores will eat just about anything they can get their claws on.

Figure 7-9 shows yet another forest food chain, but this time in the hypothetical construct of an *ecological pyramid* representing avian as well as aquatic and terrestrial fauna involved in the transfer of energy through a biotic community. Of course, these are only three examples of countless intricately different food chains that make California's coniferous forests a self-regulating system. They should, however, suffice to emphasize that every ecosystem has certain components that are vital to its success as a climax community. Should one of these components fail or be forced to fail, as with the

Figure 7-8 *Rainbow trout, the dominant trout of all cool streams in the western Sierra. A cousin of the rainbow, the steelhead trout, and its traveling companion, the Pacific or Chinook salmon, are born in freshwater streams, migrate to sea when 4 to 18 months old, remain in the salty Pacific for 2 to 5 years, and then return to their ancestral streams to spawn and die. With the building of dams on most western Sierra streams, their return was forever blocked, so steelhead and salmon have largely abandoned the Sierra for still-wild northwestern streams.* (Crane Miller)

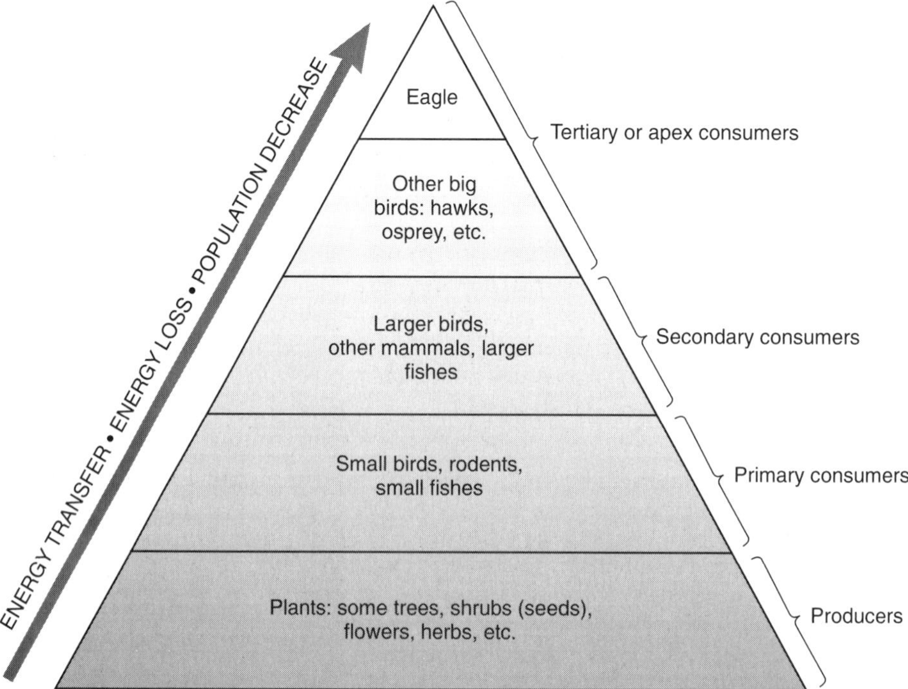

Figure 7-9 *Energy pyramid.*

loss of a predator species, the ecosystem as a whole may eventually fail.

Humans as hunters and fishers often represent the ultimate consumer or *apex species* (see Figure 7-9) in an ecological pyramid. Native Americans fulfilled this role for millennia as they sought fish and game in California's forests, lakes, and streams. Because their numbers were small and they took only what was required for food, clothing, and shelter, their overall impact on native wildlife populations was not destructive. But European and American settlement, especially after the Gold Rush when loggers, miners, and ranchers began exploiting the forest regions of northwestern California and the Sierra, brought disturbing new dimensions to hunting. Hunting solely as a subsistence activity soon gave way to hunting for sport and for the purpose of ridding the landscape of unwanted animals. Weaponry changed as well, from snare and bow and arrow to chemical poison, rifle, and shotgun. No animal, however small or large, could long escape the new hunter bent on its eradication. And escape they did not, as witnessed by the extinction by the mid-1920s of the California grizzly bear (*Ursus arctos* or *horriblis*) and the near extinction of other such forest and woodland predators as the weasel-like marten (*Martes americana*), the bearlike wolverine (*Gulo luscus*), the red fox (*Vulpes fulva*), the mountain lion (*Felis concolor*), and the bald eagle (*Haliaetus leucocephalus*).

Populations of these predators must have seemed infinitely large to hunters at first. But as is the case with any food chain excluding man, the quantity of apex consumers is always relatively small. The reduction of small predator populations by overhunting brought on a subsequent rise in the already larger numbers of their prey. Forest and woodland communities from the Sierra to the Klamaths were thrown out of ecological balance as herbivore populations boomed, local vegetation was overbrowsed, and animals invaded foreign territory seeking food. Drought and disease checked some population explosions, but ironically, it was humans with their guns and poisons who stepped into the void created by the predators' forced absence. Today, for instance, it is hunters more than natural predators that control mule deer (*Odocoileus hemionus*) populations and it is animal exterminators by some other name that poison ground squirrels (*Citellus* species) in order to save a golf course from having more than 18 holes.

Nowhere in the Sierra, though, has human overpopulation and overdevelopment posed a greater threat to a strictly coniferous forest–alpine lake ecosystem than at Lake Tahoe. One of the clearest bodies of fresh water found on earth, the lake itself is in danger of being turned pea soup green from excessive algae buildup caused by bulldozed forest organic material (soil and vegetation, basically) that drains into the lake basin. The air above the lake is likely to become more polluted as the numbers of motor vehicles increase and the prospects for a mass-transit system decrease despite the $7 million provided by a state referendum to begin purchasing land for a light-rail transit system from the South Lake Tahoe airport to the casinos at Stateline, Nevada. The forest around the lake seems destined to lose more ground, not simply from being cleared away from the path of oncoming buildings and roads, but from more extensive depredations, such as housing and commercial developments. Much of the forest will be left intact to furnish evergreen amenity and to camouflage structures, but it will—or already has—become the habitat mostly of man and domesticated animals rather than of native wildlife. Moreover, motor vehicle exhausts, winter salting of the roads, bark beetle infestations, and the inability to conduct prescribed burning (because of the existence of so many structures) have turned much of the Tahoe region forest brown.

These and other modifications of the Tahoe ecosystem appear inevitable in view of the fact that the winter population alone—and Tahoe is a place for all seasons—now pushes 150,000. Twenty years ago, the flood of new residents and the building of new homes seemed the principal threat to Tahoe's serenity. But during the 1980s and 1990s, new construction slowed to a snail's pace as sewer hookup (sewage and solid wastes have been exported from the Tahoe Basin for several decades) permits became scarcer and the bistate Tahoe Regional Planning Agency (TRPA), as part of its general plan, declared some 9,000 lots unsuitable for improvement. About 3,000 lots still remain for development, but only 300 permits were issued annually through the 1990s. The TRPA action has spoiled the plans of many retirees and thereby slowed the immigration of new residents to a trickle, except for an influx of mostly Latino service workers employed in the casino-hotels and living in a barrio near Stateline. Tahoe's only incorporated city, South Lake Tahoe, barely budged beyond 20,000 population in the 1980s and 1990s. Thus, it is the burgeoning tourist population that poses the biggest threat the Tahoe Basin.

Woodlands

If there is any single landscape that best typifies past and present California, it is that of seemingly infinite numbers of oaks and other stately trees randomly scattered over grassy, rolling terrain as far as the eye can see. Of little worth to the lumber industry and thus spared from the axe and the saw, these woodlands appear today much as they did 150 years ago, when a pastoral life reigned over California, and for hundreds of years before that, when oak acorns and pine nuts were staples in a hunting-gathering and trading economy.

Now, as then, woodlands pervade the foothills of the Cascades and the Sierra, the stream courses of the Central Valley, the central and southern Coast Ranges, and the inner coastal valleys of Southern California. In some places, such as the Los Angeles Basin and the Santa Clara Valley, native trees have given way to housing tracts and shopping centers. But these days many an oak and even an occasional parkland are left in the midst of an expanding suburbia to serve as living reminders of a once natural landscape.

The distribution of woodlands, as noted earlier and mapped in Figure 7-1, is based on their ecological role as a transition zone or *ecotone* between the well-watered forests of higher elevations and the drier grasslands and shrublands of lower-lying locales. As expected where such an intermediate environment exists, there is a mixture of conifers down from the forest and broadleaf trees up from the hills and valleys, all adapted to a respectable 10–30 inches of annual precipitation (almost all of it as winter rain), a long and warm dry season, and a variety of soil conditions and slope exposures.

The genera most often represented in these woodland ecotones are oak (*Quercus*) and pine (*Pinus*), which include several endemic species noted for their special adaptations to California's unique Mediterranean woodland biomes. The deciduous California black oak (*Quercus kelloggii*), for example, mixes with coulter pine (*Pinus coulteri*) from the central Coast Ranges southward into the Peninsular Ranges of San Diego County and with yellow pine in the upper foothills of the Sierra Nevada. In both associations, yearly rainfall is at least 25 inches, winter frosts occur regularly, elevations are above 1,500 feet, and soils vary from rocky to loamy.

The evergreen coast live oak (*Quercus agrifolia*), on the other hand, occasionally consorts with black and other oaks, but almost exclusively in outer coastal shrublands and grasslands and rarely with pines of any sort. Its Sierran cousin, interior live oak (*Quercus wislizenii*), is also evergreen and associates with chaparral, but appears more gregarious in that it is often found with digger pine (*Pinus sabiniana*) and blue oak (*Quercus douglasii*). The latter two species prefer the fine-grained alluvium of the lower reaches of the foothills and consequently are among the most drought-resistant trees (requiring only 10–20 inches of rain a year) in the state.

Last, but most majestic of all woodland trees, is the valley oak (*Quercus lobata*; see Figure 7-10), sometimes standing more than 125 feet tall and claiming several

Figure 7-10 *Valley oaks in the eastern Sacramento Valley. These trees and* Quercus *species elsewhere in California usually congregate in open woodlands rather than in thickly treed forests. The lack of dense stands and commercial species of oaks and other hardwood trees such as found in the eastern United States has helped spare California oaks from use as raw material in furniture and cooperage (wine vats) manufacturing. Actually, oak trees were far more significant to the aboriginal economy than today's, for their acorns were both staple food and trade items among native peoples.* (Crane Miller)

hundred years' seniority. It is basically a *riparian* species, most often congregating along stream courses that meander through broad, alluvial valleys. Nineteen oak species survive to this day in California, but some, such as the mesa oak (Englemann or Pasadena oak; *Quercus englemanii*), are on the brink of extinction, mainly because of urbanization.

California's woodlands are also the meeting place of wildlife. Most of the forest fauna described earlier frequent the woodlands as well, with many of the larger herbivores and carnivores migrating seasonally between the two principal biomes. Mule deer are among the most plentiful of the transients, regularly exiting the yellow pine belt of the Sierra or higher forests of other mountain ranges at the first sign of winter—usually a 6-inch or heavier snowfall in mid-November. As the deer come down into the woodlands to browse on grasses, herbs, and shrubs, they are both met and followed by the ubiquitous coyote and the rare mountain lion. Occasionally, bobcats (*Lynx rufus*) will attack deer bogged down in an unusually heavy, early snow, but squirrels, other small rodents, California quail (*Lophortyx californica*), and other birds are more their cup of tea.

Another woodland predator that rodents, birds, and even lightning-fast black-tailed jackrabbits (*Lepus californicus*) must contend with is the poisonous western rattlesnake (*Crotalus viridis*). Because the rattler's principal adversary is the overfearful human, it has been hunted into near extinction in forest recreation areas and consequently enjoys more peace and freedom in the woodland biome. Predators of all sorts, especially Cooper's hawks (*Accipter cooperi*) and red-tailed hawks (*Buteo jamaicensis*), benefit, when perched high atop oak or pine vantage points, from the openness of woodlands in seeking their prey. Overall, the woodland ecotone appears to offer an unrivaled variety of food chains involving avian and terrestrial fauna.

In the riparian woodlands of the Great Central Valley, the beaver (*Castor canadensis*) is undoubtedly one of the most controversial of native inhabitants. In times past, beaver were quite numerous and seemed in no danger of being eliminated from the local ecosystem by fur hunters. But in recent decades, their industrious dam-building activities have made them the scourge of irrigated agriculture. Apparently, beaver chose irrigation ditches for dam sites as often as they would slow-moving streams, at least until these large rodents (weighing up to 50 pounds) were reduced in number and distribution to their present predicament of extreme rarity. Beaver dam a ditch or stream with gnawed-down riparian (streamside) trees, brush, and mud in order to form a pond in which they and their young can avoid predators. Ironically, beaver ponds flooding cropland understandably invite the wrath of farmers, against whom they have no effective stronghold. Beaver have been introduced along streams in the High Sierra, but the shortage of aspens (see Figure 7-7) in forests dominated by conifers limits population growth in the new colonies. The inner bark or cambium layer of some poplars (aspens and cottonwoods of the genus *Populus*) and willows (*Salix* spp.) is a staple item in the beaver's diet.

If the beaver is to survive in the long run, it likely will be along foothill streams that provide its kind of vegetation and that lie eastward of irrigated farmland but below the yellow pine belt. Even here, though, the beaver will no doubt have to contend with the pesky little muskrat (*Ondatra zibethica*), introduced a half century ago, which has also turned out to be a destroyer of irrigation systems.

Grasslands and Marshlands: The Central Valley as an Aboriginal Environment

One would be hard-pressed to find another landscape anywhere that has changed as dramatically and rapidly as the Great Valley. For thousands of years since the last ice age, the valley harbored many of California's native peoples, wildlife, grasslands, marshlands, and riparian woodlands. But in barely more than a century, this natural landscape has essentially vanished. What was it like, and how did it lose its aboriginal character so quickly?

Before the mid-nineteenth century and the first signs of ranching and farming, the Central Valley was carpeted nearly end to end with perennial bunchgrasses that were golden brown from summer through winter and reborn as a sea of green early every spring (if there had been a wet winter). Ribboning the landscape, as they do today, were the wandering, tree-lined courses of the Sacramento and San Joaquin rivers, their tributaries, and the great in-valley Delta into which they converged before flowing on toward the Pacific. Far to the south were still more rivers and marshy deltas, but with no exit to the sea. Here, as described in previous chapters, Sierran streams came to rest in tule swamps that stretched over hundreds of thousands of acres of bottomland.

This was a valley teeming with wildlife. Chinook salmon (*Oncorhynchus tshawytscha*) ran the streams from the mountains to the sea and back. Beaver and their dams abounded but rarely got in the fishes' way. Warm-water fish, amphibians, and resident and migratory waterfowl populated the tules. Coyotes, grizzly bears, and wildcats stalked herds of deer grazing shrubbery in river woodlands.

Of all the valley's indigenous fauna, however, none was as dependent for sustenance on the grasses of that day as the pronghorn antelope (*Antilocapra americana*) and tule elk (*Cervus canadensis nannodes*). These animals numbered in the thousands and gave the valley the appearance of one of the great natural rangelands of the world. Today, although a few dozen tule elk survive in a state preserve west of Bakersfield, antelope, elk, and bunchgrass no longer grace the valley landscape. About 6,000 pronghorn remain in northeastern California, mostly in Modoc County. A second controlled herd of tule elk, numbering about 400, occupies the central part of the Owens Valley. Limited hunting of both *ungulates* (hoofed animals) is permitted, but serious hunters are likely to go elsewhere—to Wyoming, for example, where bigger, better, and more numerous elk and pronghorn antelope await them.

Once the most numerous of valley endemics, the winter-run Chinook salmon suddenly faces extinction in the ecosystem. For millenia, more than 100,000 Chinook would return from the Pacific to their ancestral home in the Sacramento River and its tributaries, there to spawn and live out the few remaining days of their 4-year life cycle. The last year their population was known to reach such proportions was 1969 when 117,000 adult Chinook were counted. However, by 1985, the count was down to less than 4,000, and by 1990, it was under 450. What has caused the almost overnight demise of the fish? Some point to agriculture and its modification of the ecosystem; others suggest the massive rearrangement of water resources brought on by such projects as the Central Valley Project. But large-scale farming was active in the valley, as was the CVP (see Chapter 5), for decades before the demise of the Chinook was first noticed. Another theory tentatively identifies the Red Bluff Diversion Dam. Often during the 1986–92 drought, more water than usual was diverted to meet the increased demands of municipal and agricultural users. This caused a lowering and warming of water in the Sacramento River upstream from Red Bluff Diversion Dam to the base of Shasta Dam, with the latter location the principal spawning grounds of the winter-run salmon. Shallower and warmer spawning grounds often spell doom for the salmon eggs therein. To mitigate these matters and allow the salmon to bypass often deadly fish ladders, diversion of water was curtailed at Red Bluff and dam gates were opened indefinitely in 1991. But lasting remedies involve a return to normal precipitation, be it anything but downpours like those that caused the Modesto-to-Marysville floods of January 1997; appearance offshore of the 1997–98 El Niño, whose warmer waters carry a diminished nutrient base and thereby deter salmon migrations; possible declaration of the winter-run as endangered; and month-long bans on commercial salmon fishing from Mexico to Canada such as those recommended by the Pacific Fishery Management Council in 1997. Spring-, summer-, and fall-run salmon have been far less adversely affected.

Like the flora and fauna, the Great Valley's original cultural landscape prospered, both in numbers of people and ways of life. Though estimates of aboriginal population vary, the fact that the valley accommodated the largest pre-1769 population of any of California's landform provinces seems substantiated by archaeological settlement sites, data collected by ethnographers from living informants early in this century, and the written accounts of Spanish, Mexican, and American explorers and pioneer settlers. Missionaries' vital records would have further verified the relatively large size of the valley's aboriginal population, but the mission system never penetrated this region. It is important to emphasize that rather than being evenly distributed throughout the valley, native Californian subsistence areas and thus population concentrations focused on the lake and marshes of the southern valley, the central Delta region, and the riparian environments of the rivers and sloughs. These locales, rather than the drier, open grasslands, contained the bulk of biotic resources from which these people derived their relative economic stability.

The southern half of the Great Valley was predominantly Yokuts country, with perhaps as many as 50 tribes of the linguistic stock represented there. Lakeside Yokuts ate fish and waterfowl caught from the shallow lakes of the Buena Vista and Tulare basins and fashioned rafts and huts with tule or bulrush (*Scirpus acutus*) taken from basin swamps. They no doubt evacuated the lowland swamps as the humidity and insects of summer approached and headed for the cooler foothill woodlands some 20–40 miles away. Here, they harvested oak acorns on their own and traded for them with other Yokuts tribes. Once leached of its poisonous tannin, the acorn meal provided a staple item in their diet.

Although seasonal variations in the availability of plant foods and the need for animal protein rendered hunting indispensable, Yokuts who preferred it to fishing and gathering almost never found a scarcity of ungulates and smaller game browsing the grasslands or drinking from countless streams and waterholes. In any case, the valley food resources were so varied and the inhabitants so skilled at utilizing it that drought and other climatic catastrophes may have produced shortages but rarely, if ever, starvation.

From the Delta region northward through the Sacramento Valley, Yokuts gave way to Maidu, Miwok, and Wintun as the principal languages spoken, but the level of subsistence resources and activities remained every bit as varied as in the south. Both sections of the Great Valley had their riparian woodlands, poorly drained wetlands, and multitude of streams from which the tribes drew their sustenance. The middle Sacramento probably did afford a more dependable salmon fishery than the lower San Joaquin, because the former normally receives more runoff and is less likely to run dangerously low in time of drought.

Unfortunately, the Sacramento Valley's wetter environment may have accelerated the spread of an epidemic (possibly malaria) that decimated native populations during the 1830s. Some years later, when John Sutter started grain farming on his rancho New Helvetia, most of the native peoples of the Sacramento Valley—perhaps numbering as many as 50,000 a few decades before—were gone. Native American communities here and elsewhere in California were devastated by introduced diseases that accompanied white settlement. Lack of natural immunity among aboriginal peoples often caused a disease to reach epidemic proportions shortly after its introduction.

Just as the coming of the white man was fatal to the aboriginal culture, so it was to the flora and fauna of the Great Central Valley. Some seemingly indestructible oak woodlands survive in riparian habitats and in the foothill margins of the valley, and a few salmon still swim the Sacramento and its tributaries, but the domesticated plants and animals of agriculture have all but swept away any semblance of the aboriginal landscape. And with it went California's only extensive tracts of native prairie grassland. A few patches of perennial bunchgrass remain in the valley and on distant hilltops, but they are a pale reflection of the way it was. To the Native Americans, bunchgrasses were important in that they sustained vast herds of game. The rancheros of the mid-1800s looked on the grasses similarly, but as forage for their cattle. In leading the initial assault on the bunchgrasses, the livestock not only overbrowsed and trampled them to death but, in so doing, also outcompeted antelope and elk for the tasty perennials.

The next wave of invaders onto the grassland were fescues (*Festuca* spp.) and other annual grasses, first those native to California and then those from as far away as Europe. In a sense, domesticated livestock cleared the way for the annuals by laying more ground bare than already was—bunchgrasses, as their name implies and by nature, do not provide a complete ground cover even when dominant and, once destroyed, cannot revive the way a logged-over redwood forest can. The seeds of the annual grasses arrived by every means imaginable: by wind, by wildlife and livestock, and by wagon train. Consequently, annuals quickly took over as the wild grasses of the Central Valley. Even they, however, have long since relinquished the bulk of valley land to crops, fences, irrigation system, roads, and cities. What has become California's horn of plenty now affords precious few reminders of its flourishing nonagricultural past.

Desert Shrublands

Another name for the desert shrubland ecosystem might be "rain shadow biome," at least insofar as it fits California's biogeography, for prevailing westerly winds and the towering mountain ranges that stand in their way combine to guarantee aridity throughout the lee side of California. Here, mapped as desert shrubland in Figure 7-1, natural vegetation and wildlife have adjusted to getting along with less than 10 inches of precipitation in most years and with only a trace of precipitation, or even none at all, in some years. Compounding the dryness are temperatures that regularly soar above 110° F in summer in the low deserts (at lower elevation, principally in the Colorado and Mojave deserts) and dip below freezing on winter nights in the high deserts of the Great Basin. The desert climate relents a bit in early spring and late fall with pleasantly mild temperatures and in late summer with the occasional thundershower, but through all the seasons desert flora and fauna are prepared to survive the worst of climatic extremes a mid-latitude rain shadow environment has to offer.

The key to survival is making do with what little moisture is or becomes available, and it is desert plant life, because it underpins the food chain, that has made the most significant ecological adaptations. These adaptations are viewed in various ways by different naturalists, but in general they can be envisaged in terms of a desert plant's ability either to gather water with its rooting system, store water in its tissue, or come to life only when, where, and while there is sufficient moisture. In other words, desert plants can be categorized as (1) *water gatherers*, (2) *water storers*, and (3) *opportunists*. But this is an admittedly arbitrary and vastly simplified classification of *xerophytes* (drought-adapted plants) and thus merits closer examination.

Throughout the deserts of California and North America, water-gathering perennial shrubs, like those pictured in Figure 7-11, monotonously dominate the landscape. The most conspicuous feature of this evergreen scrub cover is the amount of open space between individual plants. This openness is attributed to each plant's rooting system extending laterally far beyond the

Figure 7-11 *Owens Valley desert shrubland. Great Basin sagebrush and rabbitbrush are mixed with piñon pines in this view from the upper edge of the alluvial* piedmont *(coalesced alluvial fans) flanking the eastern Sierra. Porous, granitic soils derive from the adjacent mountains. Such drought-adapted plants are called* xerophytes. (Crane Miller)

diameter of its visible or above-ground parts so it can absorb soil moisture from as wide an area as possible. The roots spread out horizontally so as to remain close to the surface and thereby maximize the benefit of whatever meager, shallow-penetrating precipitation occurs. In effect, competition for rooting territory keeps water-gathering shrubs apart. The extensive rooting systems of big sagebrush (*Artemisia tridentata*), burroweed (*Franseria* spp.), creosote bush (*Larrea divaricata*), rabbitbrush (*Chrysothamnus nauseosus*), shadscale (*Atriplex confertifolia*), and other desert dominants also reduce soil erosion and protect against uprooting brought on by high winds or flash floods. The latter function is especially significant to all manner of small mammals, reptiles, birds, and insects that seek out the shrubs for shelter and other habitat amenities. The microclimate of shade supplied by a shrub helps ensure the survival of smaller plants as well. Testimony to the survivability and thus longevity of these desert plants rests with a lowly creosote bush in the Soggy Dry Lake region of the Mojave Desert. This plant, named the "King Clone" for its ability to clone itself, is more than twice the age of the most venerable of the bristlecone pines (see Figure 7-2).

Even humans have sought food and shelter from the water-gathering plants. The Cahuilla tribes of the Salton Trough region, for instance, ate seeds from the Chia sage (*Salvia columbariae*) and beans from the mesquite tree (*Prosopis* spp.). Mesquite with its probing taproots can act as a reliable indicator of groundwater, usually within 30 feet or less of the desert surface. Joshua trees (*Yucca brevifolia*), when concentrated in their natural alluvial slope habitat, also tell us that groundwater may not be far underfoot.

When things really become grim in the desert, such as during the 32 months of no precipitation that occurred between 1909 and 1912 in the south central Mojave, the water gatherers resort to other means of survival. In this instance, creosote bushes lost nearly all of their leaves but did not die; evidently, they were able to continue limited photosynthesis in stems or in bark. Leaf loss also reduced transpiration and thereby conserved moisture in the plants. Even when these evergreen desert shrubs have a full complement of foliage, the small size, reflective color, waxiness, and hairyness of their leaves help keep water loss and heat gain to a minimum. Some of the same mechanisms also see the plants through unusually cold or long winters.

Whatever the season, in a prolonged dry spell, the plants enter a period of dormancy during which only enough live tissue remains to handle the scant amount of moisture available. Longevity and the slow growth needed to attain it—for example, many a big sagebrush more than 200 years old exists today in the Owens Valley and Inyo Mountains—is yet another reason the evergreen desert shrub is perennial in every sense of the word. These shrubs, bristlecone pines, and many other native California plants literally take life slow and easy, especially during dry times.

Cacti and other water-holding *succulents* make up the second category of desert plants, the water storers. From the giant saguaro (*Cereus gigantea*) of the Colorado Desert to the smaller, more ubiquitous opuntias (*Opuntia* spp.), the cacti, with their comparatively shallow rooting systems, are notably unique in their ability to soak up surface moisture created by a cloudburst and then store the water in the internal tissue of their stems. A cactus becomes bloated after rain and then shrinks during drought as it uses its own water to survive. The leathery skin and needles of a cactus not only protect the inner reservoir against overheating and consequent water loss but also repel browsing animals.

Although mountainside thermal belts often accommodate opuntias and the like, lengthy winter frosts largely preclude any extensive distribution of water storers in the high deserts of the Great Basin. Rather, it is in the warm, low deserts of southeastern California's where native cactus gardens are to be found. Other tropical genera, such as succulent, swordlike century plants (*Agave* spp.) growing out of rock piles and native fan palms (*Washingtonia filifera*) populating faulted canyon oases, also grace the state's warm deserts.

Finally, we come to the opportunists or *ephemerals*—those desert plants that seem to be here one day and gone the next. Wild buckwheats (*Eriogonum* spp.), sand verbena (*Abronia villosa*), goldfields (*Baeria chyrsos-toma*), and other annual wildflowers so varied that they defy easy identification essentially make up this category. Despite the implication of their classification as annuals, wildflower seeds and bulbs sprout forth only following sufficient precipitation, which may not occur for years on end in the desert. For dormant seeds to germinate, precipitation should be spread out intermittently over several weeks or more rather than coming all at once. This explains why the best wildflower displays can be expected to follow a good winter's worth of rain and snow, not a summer's single tropical downpour, even though the amount of moisture produced in both cases may be roughly equivalent.

Although wildflower seeds generally await their opportunity to come to life in the shelter of large perennial shrubs, they can be inadvertently carried just about anywhere by animals, flash floods, winds, or even humans. Consequently, wildflower crops may be found in such diverse places as alluvial slopes and dry washes and along the edges of alkali sinks, sand dunes, and highways. Any of these locations from the Salton Trough north to the Great Basin are good bets for finding the desert in bloom following a normal or wet winter, but the Mojave and the Colorado in April are rarely surpassed in beauty. The "desert day" or "wild Easter" lily (*Hesperocallis undulata*), sprouting from a bulb to heights sometimes exceeding 5 feet, could be said to be the most appropriately named of the ephemerals that herald the blooming of the California deserts in the springtime.

Hiding out from searing solar radiation in summer, avoiding excessive heat loss in winter, and conserving water in every season are challenges that obviously must be met by the fauna as well as the flora of the desert. What is amazing is that the desert teems with animal life despite the aridity.

Both the rich variability and the survival adaptations of the fauna become apparent when studying any one of a number of desert food chains, but these features are perhaps best typified by the members of a seed–rodent–reptile–bird food chain. Many desert rodents, including the bush-nesting harvest mouse (*Reithrodontomys raviventris*) and the rock-dwelling pocket mouse. (*Perognanthus* spp.), rely on seeds as the staple of their diet. Pocket mice enjoy an advantage over other seed harvesters in that they are outfitted with a pair of cheek pouches that enable them to carry a relatively large number of seeds home after a single outing. Long-tailed kangaroo rats (*Dipodomys* spp.) also have cheek pouches, but their ability to produce water chemically from plant foliage and seeds is undoubtedly one of the most significant survival capabilities of any desert animal.

Like kangaroo rats, grasshopper mice (*Onychomys* species) dwell underground, but their diet includes insects as well as seeds. Living in burrows by day and harvesting and hunting by night enable many rodent species to escape from the worst of summer heat, although some burrowers, like the round-tailed ground squirrel (*Citellus tereticaudus*) and antelope ground squirrel (*Ammospermophilus leucurus*), are diurnal foragers. Ground squirrels never stay in the hot sun for too long, and their light sandy color may minimize heat absorption by maximizing reflection of the sun's rays.

But in spite of desert rodents' quickness afoot, defensive coloration, and other survival adaptations, they are constantly in jeopardy from a host of predators. Perhaps most effective in keeping rodent populations from exploding are various species of the venomous rattlesnake (*Crotalus* spp.). Once their winter hibernation is behind them, many rattlers hunt both nocturnally, seeking the likes of kangaroo rats and mice, and diurnally, going after ground squirrels, lizards, and small rabbits. Because they are cold-blooded animals, rattlers need air temperatures consistently above 70° F before they'll emerge from their homes in burrows or rock crevices on a regular basis. Yet daytime temperatures exceeding 100° F will find them coiled in the shade of shrubs or rocky overhangs and unlikely to expose themselves to direct sunlight, where they would lose coordination, lapse into heat prostration, and die within a few minutes. Like the desert wildflowers, rattlesnakes like to be out and about during the milder days of spring, and this is the time of the year to exercise the most caution when walking in the desert.

Other than humans, rattlesnakes fear few predators save the hawk. Hawks, owls, and other large birds of prey, however, more commonly hunt smaller reptiles and mammals. California's largest native lizard, the chuckawalla (*Sauromalus obesus*), is among their favorite prey. Besides predatory birds, coyotes, bobcats, and the like are also present in the desert and play important roles as apex species in the food chain.

Quite skillful at negotiating the steepest of terrain, bighorn sheep (*Ovis canadensis*) once roamed the mountains of California from the Pacific to the Great Basin and from the Klamath Mountains down to the Peninsular Ranges. But hunters, ranchers, livestock, and feral burros have greatly reduced their numbers. Today, Peninsular and Sierra varieties of bighorn are legally protected from being hunted, and their populations have been revived by animals imported from as far away as the Canadian Rockies. Bighorn are found at higher elevations than mule deer and are most likely to be seen along rugged mountain crests such as in the California Bighorn Sheep Zoological Area just east of Kings Canyon–Sequoia national parks. Protected from hunters since 1873 in California, the desert bighorn or Nelson variety was removed from total protection from hunting in 1987; that is, limited hunts supervised by the California Department of Fish and Game (DFG) were to be permitted. The rationale for reducing the protected status of the desert bighorn included the increase in their DFG-estimated population from some 3,700 in 1972 to nearly 4,800 in 1986. Because only two dozen or so desert bighorn hunting tags are issued annually and therefore fetch a high price individually, increased revenue to maintain the animals was another consideration. Prior to California's joining 13 other western states and 2 Canadian provinces (Alberta and British Columbia) in legalizing hunting of the bighorn, as much as $79,000 (in Oregon) had been paid for a single bighorn tag.

Petroglyphs (Figure 7-12) and other archaeological evidence indicate human settlement in California deserts dating back at least 8,000 years. The aboriginal peoples were basically hunter-gatherers and, as would be expected in an arid environment, their populations never approached the densities of those in milder climatic regions. But those peoples who settled along the lower Colorado River and far to the north in the Owens Valley were the only ones known to practice irrigated agriculture in aboriginal California. Their basketry was examplary, too, especially their pitch-covered baskets used to carry precious water from spring or stream to settlement site.

The advent of white settlement in the mid-nineteenth century signaled the beginning of hard times for both native desert peoples and their environment. As noted in Chapter 3, surviving Native American popu-

Figure 7-12 *Petroglyphs. Engraved in dolomite centuries ago by predecessors of the present-day Paiutes in the Inyo Mountains, these depict mountain or bighorn sheep, humans, and other figures.* (Crane Miller)

lations have become concentrated on a few widely scattered reservations, rancherías, and in some desert cities and towns, and their lot is generally a poor one. The plants and animals of the California deserts have also lost much ground in the past century, giving way not so much to the spatial expansion of a few small cities as to transient human activity.

The development of dirt roads, paved highways, interstate freeways, power lines, microwave towers, railroads, and fences has been harmful enough in itself, but add to this the destruction of desert habitat by off-road vehicles (ORV) and one wonders if much of anything will be left in time. Motorcyclists running down rabbits, dune buggies running over desert iguanas (*Dipsosaurus dorsalis*) and tortoises (*Gopherus agassizi*), and ORVs of all sorts crushing vegetation, accelerating surface erosion, and carving up *desert pavement* (the wind-abraded and -polished surface of closely packed, sometimes cemented-together pebbles) and *intaglios* (large figures carved in the sand-and-gravel landscape by human hands)—all threaten to turn the desert into a lifeless dust bowl.

If the Mojave and Colorado deserts are to remain a viable recreational and wilderness resource to millions of nearby urbanites not bent on their destruction, greater support must be given to maximizing the number of roadless areas (each normally to be in excess of 5,000 acres) under the California Desert Plan as it was completed in 1980 and updated thereafter. The U.S. Bureau of Land Management (BLM), which controls 12.5 million acres of desert, started the inventory phase of the plan late in the 1970s. This California Desert Conservation Area (CDCA), as it is officially designated, may indeed become a bastion for the permanent preservation of California's desert environment. Before we become overly optimistic about the desert's future, however, we should again ponder the environmental impact, especially on the bioclimate, of the possible development of coal-fired and solar thermal power plants (see Chapter 6) in the state's deserts. In the meantime, 1.38 million acres of the eastern Mojave has been designated a National Scenic Area to be managed by the BLM.

Chaparral and Coastal Shrublands

In pinpointing the distribution of this principal biome in Figure 7-1, we should note that although it is concentrated in a mixed fashion on the windward and southwest-facing slopes of Southern California's coastal mountain ranges, dense patches of "pure" chaparral are also found under the same conditions in the Sierran foothills and farther north. By the same token, along the coast north of Monterey, the more open coastal shrubland exists almost exclusive of any chaparral.

Despite differences in the ground cover density of the two plant communities, however, they share many of the same dominant evergreen species, including laurel sumac (*Rhus larina*), different varieties of ceonothus (*Ceonothus* spp.), purple sage (*Salvia leucophylla*), and scrub oak (*Quercus dumosa*). Scrub oak translates into

Figure 7-13 *Fire in Swall Meadows in the eastern Sierra. Whether in suburban chaparral scrubland or rural coniferous forest, fire is both destructive and beneficial to the habitats involved.* (Crane Miller)

Spanish as *chaparro*, from which the word *chaparral* was derived. Chaparral is, of course, California's version of Mediterranean vegetation, and in most ways it is the same thick, evergreen scrub that circles the Mediterranean Sea and goes by the name *garigue* or *maquis*, but with different species. Chaparral plants serve as reliable sources of shelter, food, and protection from invaders for a great variety of wildlife. Indeed, chaparral animals are not likely ever to see many Sunday picknickers or intrepid hikers in their territory.

Although chaparral may be singularly unattractive for recreation, it serves humankind by preventing excessive storm runoff and therefore functions as watershed vegetation that minimizes slope erosion. At the same time, though, chaparral poses a greater hazard to Californians than any other biome because it is a *fire climax community*. Put another way, chaparral should burn every so often or it will degenerate. There are a number of climatic and botanical explanations for this hot-tempered behavior (Figure 7-13). In adapting to California's long warm-season droughts, many chaparral species produce volatile oils that, when kindled, bring about the quick destruction of the plants by fire. Deadwood and debris also contribute to the rapid spread of fire. On the other hand, fire clears away the debris and produces ash that provides nutrients to the soils. After being burned beyond recognition, many a chaparral species (scrub oak, chamise, and toyon, for example) will stump-sprout to live another day. In the case of some manzanita and ceonothus species, their seeds germinate only after a burn.

Painful familiarity with chaparral's propensity to burn has, it seems, done little to prevent home builders from developing the Santa Monica Mountains, Verdugo Hills, San Gabriel Mountains, and other chaparral locales overlooking the Los Angeles Basin. Perhaps the tangible and intangible rewards gained from a view site in Bel Air or a secluded sanctuary in Mandeville Canyon cancel out memories of the devastating chaparral fires of 1961 and 1978.

Miraculously, no deaths resulted directly from the Bel Air conflagration. But 30 years later, the most destructive suburban wildfire in California history would claim dozens of lives when it ravaged the hills overlooking Oakland, Berkeley, and San Francisco Bay. By the time the October 20–23, 1991, Berkeley–Oakland Hills fire was declared controlled, there were 24 confirmed deaths, 25 people listed as missing and feared dead, 148 injured, and 5,000 evacuated. Moreover, 1,800 acres were burned, 2,700 structures were destroyed, and financial losses were estimated at $5 billion. The fire's origin notwithstanding, fire weather more typical of Southern than Northern California was ostensibly to blame for the blaze raging unchecked for three days. High, dry winds, likened by many victims to Southern California's Santa Anas, blew in from the dry, eastern interior and whipped up uncontrollable firestorms amid stands of native conifers and introduced eucalyptus trees dessicated by 6 months of normally dry summer weather and 5 years of abnormally long drought. Clearing brush and replacing wood shingle roofs in suburban California will reduce the ravages of future fires, but the dry summer climate guarantees their occurrence.

The Littoral

Life along the more than 1,000 miles of California shoreline goes and comes with the ebb and flow of the Pacific's tides. The regular rise and fall of sea level created by the gravitational forces of sun and moon acting on a rotating earth help carve out and build various coastal habitats for myriad tidal communities of plants and animals.

When the tide is high, there is little to see, but ebbtide can reveal all kinds of worlds. On broad, sandy beaches, long-billed curlews (*Numenius americanus*) can be seen digging for tiny crustaceans while hordes of flies circle about seaweed (kelp of various genera) that has washed ashore. On the outer reaches of rocky headlands, countless tidepools appear as aquaria, each filled variously with abalone (*Haliotis* spp.), anemones (*Anthopleura* spp.), barnacles (*Balanus* spp.), hermit crabs (*Pagurus* spp.), mussels (*Mytilus* spp.), starfish (*Pisaster ochraceaus*), sea snails and worms, and tiny fish of all descriptions. On rocky crags safely above pounding surf and incoming tides, ashy petrels (*Oceanodroma homochroa*) and cormorants (*Phalacrocorax* spp.) are sometimes visible in their nests. In the isolation of offshore rocks, glimpses of seals (*Callorhinus* spp.) and southern sea otters (*Enhydralutris* spp.) may be caught. In tidal marshes and lagoons, *halophytic* (salt-tolerant) grasses and other plants contain everything from lowly mud-flat crabs (*Hemigrapsus* spp.) to high-flying kingfishers (*Megaceryle alcyon*). And California seagulls (*Larus californicus*) crowd the sky in an endless search for prey.

Sadly, much of the natural littoral has disappeared, having succumbed to the understandable desire of millions of Californians to reside there. The Santa Barbara oil spill of 1968 and other unnatural events have compounded destruction of the seashore. Aimed at protecting the environment, including coastlines, from such abuse, the National Environmental Policy Act of 1969, the California Environmental Quality Act of 1970, and the California Coastal Act later in the 1970s have saved much of what remains on the California littoral. Yet it sometimes seems futile to strive for what is ecologically sound when the nation's hunger for energy may dictate more seaside oil refineries, ports, and offshore drilling rigs for the state.

There are, however, comeback stories, that of the sea otter being among the more remarkable. Two centuries ago, there were an estimated 16,000 otters along the coast. By one century ago, they had been hunted to near extinction by American, English, Mexican, Russian, and Aleut and Northwest native fur hunters. But in 1937, a small band of the little mammals was seen swimming off Carmel. Since then, their population has increased to some 2,000 and their range has extended to as far south as Point Loma. Pismo Beach clam diggers, Morro Bay abalone fishermen, and the like would just as soon see otters disappear once again, for man and animal in this case compete for a diminishing supply of shellfish. But sea otters, now fully protected, are likely to continue to increase their numbers, according to Department of Fish and Game estimates.

With the sea otter eliminated early in the nineteenth century as the principal prize of those seeking commercial gain from California's offshore fauna, the marine fishery assumed the limelight. The California fishery is prolific in everything from abalone to yellowtail, in 1994 producing a catch weighing nearly 330 million pounds and valued at about $145 million. Albacore, bluefin, skipjack, and yellowfin tuna (*Thunnus* spp.) are the most valuable of the commercial fish, but their numbers and value have declined dramatically in recent years. From 1987 to 1994, for instance, the value of all tuna caught declined from nearly $74 million to $15 million. The tuna sport fishery has declined as well. For example, the San Diego sport fishing fleet reported a drop in the number of tuna caught from some 90,000 in 1990 to only a few thousand in 1991. Much of the tuna brought into San Diego is caught in Mexican waters up to 100 miles south of the border, but in the summer of 1991, the northward migration of tuna was stalled 50 miles farther south. The tuna's reluctance to venture into California waters made warmer (possibly by El Niño), and more turbid (perhaps by pollution) may be at the root of the problem.

THE SUCCESS OF EXOTICS

Since the Franciscans first brought the Old World grape (*Vitis vinifera*) and domesticated livestock to California in 1769, plant and animal species by the thousands and from all over the earth have been introduced to California. Much of what the Spanish missionaries and others grew and raised in California for the first time would eventually become the basis of the state's most important industry: agriculture (see Chapter 10). Non-native, wild plants and animals have found environmental riches in California as well. And, unlike crops and livestock, they have done so without human help, although some wild creatures of today, such as desert burros, mustangs, and other *feral* animals, are descended from domesticated lines. In any case, introduced plants and animals have benefited from the state's diversity of climates, landforms, and soils. There seems to be an ecological niche in California for just about every living thing, no matter its origins.

Figure 7-14 *Eucalyptus and other introduced and native trees, overlooking Brentwood, the UCLA campus, Westwood, and Century City. Eucalyptus was first introduced to this western part of the Los Angeles Basin in the 1870s, at a time when few people and still fewer trees graced the lowland landscape. Today, the basin is jammed with people, buildings, freeways, and other human artifacts, yet it is greener than ever before, partly because of the succesful introduction of plants from elsewhere in the world.* (Crane Miller)

Figure 7-14 shows what many consider to be the most successful plant, at least in terms of surviving on its own, ever introduced into the state: the blue gum or eucalyptus tree (*Eucalyptus* spp.). A native of Australia, where there are some 400 species of the tree, the eucalyptus was probably first brought into California in the 1850s. During that decade, nursery people in San Francisco began advertising the availability of seedling gum trees; before long, the trees were competing with native species throughout much of the Bay Area. Planting of eucalyptus trees in Southern California began in the following decade, probably under the supervision of citriculturist, viticulturist, and former fur trapper William Wolfskill. Cultivation continued through the 1880s as eucalyptus became a familiar part of the Los Angeles

Basin landscape. Growers foresaw use of eucalyptus in the manufacture of rail ties and other hardwood items, but softer woods that were easier to work with were preferred. Eucalyptus oil was another product thought to have a great sales potential, but it, too, never caught on.

With a market never really developing for eucalyptus products, plantings of the trees tailed off by the end of the century. Thereafter, about the only cultivation involved occasional use of eucalyptus as an orchard windbreak. Nevertheless, the eucalypts have prospered on their own and today seem as much a part of the California scene as any native tree, especially in cities. In Santa Monica and Santa Barbara, for example, at least three dozen different species of eucalyptus line the streets.

Another successful foreign tree, the tamarisk (*Tamarix gallica* and *Tamarix aphylla*), was introduced to the Mojave and Colorado deserts by ranchers and the Southern Pacific Company early in the twentieth century. Tamarisks originated in the arid eastern Mediterranean region and thus prospered in California's deserts. In fact, they have become a nuisance—their seeds readily dispersed by both wind and birds to places where they're not wanted. Along with palm trees (Figure 7-15), however, they have served their original purpose of providing shade and windbreak protection against blowing sand.

Bearing little resemblance to its barnyard cousin the hog (*Sus* spp.), the California wild boar or pig has become the state's most hunted non-native big-game animal. The domesticated hog was introduced to California by Father Serra and the Franciscans in the eighteenth century, but the wild Old World boar came somewhat later. Nearly as many wild hogs (32,000 in 1978) as mule deer (35,000 in 1978) are killed legally by hunters each year. California's other major big-game animal, the black bear, looms a distant third in number killed annually (a record 935 in 1967).

Wild hog populations are concentrated in the Coast Ranges from Mendocino through Santa Barbara counties and on Catalina and San Clemente islands. In these locales, boar subsist on insects, roots, wild barley, and oak acorns. Oak woodlands, coastal shrub, and chaparral also provide them with shelter and refuge. Although some California wild hogs weigh more than 600 pounds, a 200-pounder is considered the norm. When cornered, a wild hog of these proportions is a dangerous animal. On the run from hunters and their dogs in open country, however, a hog rarely will turn and go on the attack.

Most geographies of introduced fauna would ignore insects, especially the tiny Argentine ant (*Irodomyrmex humilis*). But for many Californians, notably those living in the suburbs, no creature of nature, large or small, can match the Argentine ant as a nuisance.

Figure 7-15 *Palm trees in the heart of downtown San Diego. Fan palms* (Washingtonia filifera) *are native to the southeastern deserts of California but have been successfully introduced throughout much of the rest of the state.* (Crane Miller)

Fleas once outnumbered ants, but Argentine ants are now the most numerous of household pests in California. The 1986–92 drought rendered the Argentine ant more of an unwanted companion than ever before, for it invaded structures ranging from single-family homes to high-rise office buildings in search of food and water that was in diminishing supply outdoors. Furthermore, the bulldozing of land for development is dislodging the wily, six-legged insects from their natural habitats and making them more a part of the human environment. For Argentine ant populations to be controlled, torrential rains must fall for several consecutive winters, so that the ants are lured outdoors and then washed away. The advent of El Niño in 1998 certainly helped.

We have examined only a tiny sample of the many exotic plants and animals in California. Perhaps we could learn as much about the proliferation of introduced flora and fauna by simply gazing out over the artificial landscape of almost any city. Seen in this light, the greatest significance of the massive introduction of plants to California is in the improvement of the quality of urban and suburban environments. Many of the most densely developed sections of the state are all the more pleasant to work and live in because so much exotic vegetation prospers in California.

The density of development in such high-rise urban areas notwithstanding, there is quantitatively and qualitatively more vegetation today than there was before urbanization. Most of the west side of the Los Angeles Basin, for instance, was a sparsely treed grassland in prehistoric times, which was barely two centuries ago (the city of Los Angeles was founded in 1781).

Increases in *biomass* (the actual weight of living matter), significantly augmented by introduced vegetation, have improved the looks of suburbia as well, be it frost-sensitive bougainvillea (*Bougainvillea* spp.), hedgerows replacing coastal sagebrush near the ocean in Southern California, or all-weather oleander (*Nerium oleander*) lining a freeway in Northern California. Because they carry on photosynthesis, exotic flora have also made the state's urban air all the better to breathe.

Faunal introductions, from English sparrows (*Passer domesticus*) filling suburban skies to mosquito fish (*Gambusia affinis*) controlling insects in flood channels, have made city life more attractive as well. Unwelcome exotics, such as lawnseed-eating starlings (*Sturnus* spp.) and dozens of weed species, detract from the picture, but in general, the impact on the California landscape of the introduction of thousands of species of plants and animals has been a successful one.

Biogeography, as we have applied it to California in this chapter, and the historical geography of the region, as we will examine it in the next chapter, can both be viewed as landscape studies. We tend to describe a landscape in terms of a particular feature that appears to dominate it. If a redwood forest blankets hilly terrain for as far as we can see, we are likely to describe the landscape in terms of its natural vegetation rather than its topography or some other less obvious physical feature. Hence, "redwood forest" and "hilly terrain" are descriptions of natural or physical landscapes. Similarly, as we next venture back in time and ponder the impact of humans on the landscape over time, we will consider the succession of cultural landscapes, or the *sequent occupance*, of California that originated thousands, or perhaps tens of thousands, of years ago when the first humans came on the scene. From the aboriginal landscapes of the first Californians, who for millennia minimally modified their environments, through the mission- and rancho-dominated Spanish and Mexican periods, to the first signs of farming, mining, and urban landscapes that marked the beginning of American occupance, the prehistorical and historical geography of California is long, varied, and colorful. But cultural landscape change can be brutally quick and unforgiving as well. We examined this dynamic earlier with regard to the rapid demise of the Central Valley aboriginal environment, and we will see it again with regard to the rapid decline of the native population.

CHAPTER

8

The Historical Geography of California

Scholars have long debated the issue of how much influence the physical environment has had upon the unique development of cultures throughout the world. According to *geographic determinism,* the human race, its creations, and its evolution are essentially conditioned by the natural environment. At the other extreme, scholars deny that nature and geography alone are so omnipotent in the unique development of a culture or people.

Perhaps the most balanced view is to recognize the interrelationships between people and the environment. Certainly, many factors influence the evolution of a particular culture, and geography is one of these important factors. California provides an excellent example of a situation in which the physical environment has played a significant role in historical events and cultural development. Geography has constantly shaped people and events at each point in California's evolution. Aboriginal lifestyle was intimately related to environment; European settlement patterns were conditioned by distance, weather, and nature. The offerings of the land, from fertile soil and climate to gold and other minerals, drew Spaniards, Mexicans, and Americans alike. The size of the state, its natural resources, and its generally mild weather have continued to play key roles in the economic, social, and cultural life of the state, especially as manifest in agriculture, transportation, oil, and moviemaking. The same factors that brought early immigrants continue to attract modern-day newcomers to the state. Thus, the continuity of California's development can, in part, be attributed to the realities of climate and geography, with economics playing an increasingly major role.

THE ORIGINAL CALIFORNIANS

Evidence of human presence in California is among the oldest reported in the United States. We know that as early as 29,000–34,000 years ago, primitive people began a Southern California tradition on Santa Rosa Island by barbecuing a dwarf mammoth. There is some additional evidence of human presence in California as early as 50,000 years ago. Evidence of established societal or tribal existence, however, does not appear until much later.

The aboriginal population of California constituted the first wave of immigration to the state. Anthropologists conclude that this immigration into the area came about as a natural result of nomadic wandering through the Bering Strait, across a land bridge, and onward down the North American continent. By the time of the first Spanish forays into California, the Native American population had grown significantly. Although the estimates vary widely, it is generally agreed that at least 133,000 and perhaps as many as 300,000 tribal peoples populated California at the time of the first Spanish contact. It is widely accepted as well that California at that time reflected the greatest population density of any *nonagricultural* area in the world and had the greatest population density in North America.

The nature of the California Indians can be de-

Figure 8-1 *Main aboriginal groups in California.*

scribed as diverse and varied. In language alone, there were at least 135 dialects in no less than 6 different parent language stocks, ranging from the Penutian and Hokan to the Shoshonean and Athabascan (Figure 8-1). Similarly, dress, habitation, and physical stature all varied widely. The only common element, in fact, has been the way California Indians have been characterized historically. Perhaps the most widely circulated image is that derived from the term gratuitously placed on them by the American Forty-Niners—"Diggers," which implied root grubbers and insect eaters of low intelligence and primitive demeanor. Applied indiscriminately as a label of contempt for all California aboriginals, it was a gross oversimplification.

Individual tribal groups in California exhibited vast creativity and frequent sophistication in certain areas of their lives. The Mojave and Yuman tribes were far more than casual agriculturalists, with an economy strongly reliant upon flood farming. The variety of crops cultivated by these groups included maize, squash, tepary beans, riverine plants, and marsh grasses. The artistic ability of some tribes was especially evident in their intricate basketry, which has subsequently been described as perhaps the best in the world. Coastal tribes became quite adept at harvesting resources from the ocean using tar-caulked canoes, harpoons, and shell fishhooks. Various groups valued religious ceremony, introspection, and spiritual pursuits. Telling stories, playing with games and toys (tops, dolls, and so on), gambling, and other recreational activities were not uncommon in most California tribes. Indeed, the notion that California Indians were simple and dull-witted was highly inaccurate.

Although generalizations about such a diverse group are difficult to make, certain common traits are evident. Certainly, measured against European and American standards of aggression, technology, and materialism, California Indians were less "developed." They were peaceful rather than warlike. They were well adjusted to their environment, living in harmony with it rather than destroying it. By modern standards, they lived an uncomplicated, simple life instead of a hectic and competitive existence. Their religious life was generally highly developed, and their social life apparently was much more complex than previously believed.

The aboriginal inhabitants of California were attuned to their environment and reflected many of the characteristics we now associate with the state. Isolated by mountain and desert from the rest of North America, they developed a culture of appreciation for and reliance on nature. Largely individualistic, living in villages or rancherías of approximately 130 people, they built dwellings suited to their surroundings. Planks were used in the northwest, bark in the central regions, earth on the coast, and brush in the deserts. Dress was, for the most part, "early suntan" attire, consisting of topless two-piece apron skirts for the women and *au naturel* for the men. In cooler weather, a blanket or cloak of hides would be added. Food was provided by nature in the form of acorns, game, fish, and other natural bounty. Only with the coming of the Spanish friars did organized agricultural production become common with non–Colorado River tribes.

However, the absence of European-style farming did not mean that these tribes were uninvolved with plant resources. Increasingly, experts are recognizing that California Indians engaged in effective indigenous horticulture. Methods of plant management included coppicing (pruning), weeding, annual burning, selective harvesting, soil manipulation, and tillage. These methods required intimate knowledge of their consequences, because plant resources were critical to tribal diets, weapons, tools, shelters, medicines, and other vital lifestyle items. One of the reasons that European immigrants (and scholars) erroneously concluded that the California Indians were backward and unintelligent was that their methods involved less overt modification of (or violence to) the natural environment.

Singing, dancing, and chanting were familiar pastimes, as were gambling games, athletic contests, and storytelling. Public health and sanitation were accomplished by use of sweathouses and periodic burning of the living structures, and herbal medicines were in common use. The dominant form of religion was *shamanism*—the belief that supernatural spirits work for the benefit or to the detriment of humankind through the intervention of the priest or shaman.

In retrospect, the California Indian appears much more well adjusted, efficient, harmonious, and admirable than generally depicted in the past. If environment does help form social styles, then it may be said that the society of these peoples accurately reflected their surroundings. The mild California environment demanded no excesses from its inhabitants, and they did not seek unnecessary complications for their lives—although, in all fairness, some California tribes (e.g., the Mojaves) did not live in gentle environments.

The Spanish, Mexican, and American waves of immigration brought major changes, and their collective and progressive impact on the California Indians was destructive. Disease, starvation, and violence such as organized slaughters (with bounties paid for each dead native) caused a drastic reduction in the native population, with various children's diseases, pulmonary ailments, venereal diseases, and other similar maladies taking a huge toll. Similarly, imposed changes in living patterns disrupted the flow of tribal life, whether it was the forced relocations of tribes into mission dormitories by the Spaniards or the forcible ejection of tribes from whole areas by the Americans. Victimized in turn by each incoming group, the California Indian was, as the respected California historian W. H. Hutchinson put it, "vital to Spain, useful to Mexico, and an annoyance to the United States." The discovery and colonization of California by Europeans truly spelled the end to the simple Native American culture, which could not hope to resist "civilization."

Ultimately, and ironically, these once despised native peoples may have had a better sense of social balance and environmental responsibility than the groups that were to follow. As modern society increasingly recognizes the problems associated with environmental degradation, the California Indians and their ecological practices must be seen in a much more positive light. The respect and reverence of these "Native Californians" for the earth have encouraged practices just now being adopted by a concerned population. Whereas European settlers tended to reduce the populations of plants and animals, the California Indians generally adhered to the concept of sustained yield and multiple uses for both—an approach currently receiving attention in California and much of the rest of the world.

Thus, although the California Indians became victims of their times, many of their ideas and concepts (especially as they relate to stewardship of the earth) have reemerged as sensible and coherent approaches to life in California.

EUROPEAN EXPLORATION

Spanish Dominance

The discovery, exploration, and eventual settlement of California by the Spaniards came about in the aftermath of initial explorations of Mexico by Hernando Cortés and others in the early 1500s. Curious about the nature and extent of this new territory, Cortés, Coronado, Ulloa, and others. Explorers undertook several expeditions to the north. It was Juan Rodríguez Cabrillo who first entered what would be called Alta California when he sailed into San Diego Bay and on up the coast in 1542. Prophetically enough, he named the coastal area near Los Angeles and Santa Monica "the Bay of Smokes" for the profusion of campfires in the vicinity. Cabrillo died during the course of the voyage, and the reports of the expedition generated little interest in the region among Spanish officials.

Although Spain provided the initial impetus for exploration, California did have some other European visitors. Reportedly, Britain's salty sea dog Sir Francis Drake first set foot on California's shoreline in 1579. He spent a month on shore establishing England's sovereignty (so he claimed), engaging in trade and communication with the native peoples, and repairing his ship, the *Golden Hinde*. Having made his ship seaworthy enough to carry the treasure he had stolen from Spanish towns and galleons in the New World, he set sail for England, undoubtedly filled with tales of his adventures and romances in sunny Nova Albion.

Concerned by Drake's activities, the Spaniards sent another expedition up the California coast to seek "harbors in which galleons might take refuge" from sea rovers such as Drake. Thus, between 1584 and 1602, Francisco Gali explored the coast, Sebastián Cermeno sailed along the shoreline noting likely harbors, and Sebastián Vizcaíno made an extended excursion, stopping in San Diego, Catalina, Santa Barbara, and Monterey Bay. These expeditions increased Spain's knowledge of the area, but not its immediate interest. Thus, Alta California remained largely an ignored outpost until the 1760s.

One other notable foreign incursion involved the Russians at a somewhat later date. Fort Ross in present-day Sonoma County was founded by the Russian-American Company in 1812. Although primarily concerned with the fur trade in Alaska, operating in concert with Americans, the company did extend southward in search of both food and other fur sources (primarily sea otter). With the decline of the sea otter and the fur seal populations, however, the Russians sold out to John Sutter in 1841 and withdrew permanently from California.

Mission Settlement Patterns

As with many colonizations, the settlement of California was largely inspired by international competition. Spurred on by the commercial activities of the Russians and the presence of English and Dutch privateers along the Pacific coast, the Spaniards recognized the need to protect their holdings from foreign rivalry. The most effective way to accomplish this goal was to initiate the tripartite policy of mission, presidio, and pueblo.

According to the Spanish concept of empire in the region of Alta California, the missions were intended to "Christianize" the native populations while, at the same time, establishing and reinforcing loyalty to the Spaniards. The presidios were the frontier forts designed both to control the native populations and to protect Spanish interests from foreigners. Thus, they were located in strategic sites to enable them to protect the missions and to permit control of critical ports or trade centers. The pueblos were the civil arm of the colonization effort, designed to draw civilian settlers into the region and further reinforce Spanish rule. The pueblo residents interacted with the presidios by selling them food products and by acting as a form of "reserve militia."

Under the energetic leadership of José de Gálvez, the inspector-general of New Spain, expansion was initiated. With the guidance of Captain Gaspar de Portolá and Father Junipero Serra, the first presidio and the first mission in Alta California were established in San Diego in 1769. Over the course of the next 50 years, 20 more missions were founded, some accompanied by presidios, some by pueblos, some by both. Approximately 30 miles apart (one day's travel by horseback), they formed a chain strung from San Diego to Sonoma, connected only by a dusty path known as El Camino Real (the royal road). These missions were remarkable institutions that assumed the key role in California life. From the viewpoint of the civil authorities, they were inexpensive extensions of the empire, requiring only a couple of padres, a handful of soldiers, and infrequent supplies.

Each mission reflected a certain adaptation to California realities. Structurally, the missions evolved an architecture of red tile roofs, massive buttresses, and thick walls as protection against brush fires and earthquakes, the heat and sun of summer, and the cold of winter (Figure 8-2). Agriculturally, each mission operated on a huge tract of around 100,000 acres located in fertile coastal areas supplied with both ample water and native population. The dedicated but paternalistic Franciscan friars intended to bring civilization and progress to the California Indians. Unfortunately, more often, they brought dependence bordering on slavery, diseases for which the native population had no immunity, and early death for many.

Figure 8-2 *Santa Barbara Mission, a classic jewel in the California chain of missions.* (Union Pacific Railroad)

Economically, the missions were probably highly successful. The missions developed a crude form of industrial-manufacturing enterprise, in which weaving, blacksmithing, cattle raising, farming, tanning, masonry, and carpentry became common. Thus, the California Indians learned from the friars how to toil long hours for their food and livelihood, whereas before the padres' arrival they had merely lived successfully off the natural bounty of the land.

Considerable historical debate has focused on the ultimate "morality" of the mission and its impact on the native California population. As a convenient personification of the Spanish church, Father Junipero Serra has been proposed for canonization by some and reviled as an imperialist exploiter by others. Regardless of the characterization, the *impact* of the missions on California culture and population was substantial, permanently transforming the face of California.

The presidios and pueblos played a less vital role in the settlement of California, although they were important in extending secular authority to the frontier. The presidios at San Francisco, Monterey, Santa Barbara, and San Diego served to guard against foreign invasion and native uprising. With their limited capabilities, however, it was fortunate they were not put to any severe test. In spite of inducements of land, stock, and implements, the establishment of pueblos was not a major success. The pueblos of San Jose and Los Angeles are the only enduring evidence of the civil townships created by Spanish authority (Figure 8-3).

Figure 8-3 *Pico House in the old Pueblo de Los Angeles, one of the first hotels in Southern California and a favorite of early California cowboys.* (Richard Hyslop)

THE MEXICAN PERIOD

The End of the Mission

The mission period ended suddenly with the disintegration of the Spanish empire in the Americas in the early 1820s. The geography of the mission colonization effort inevitably resulted in its separation from the Spanish empire. All of Spain's American colonies were largely independent of the mother country. Thus, in the early years of the nineteenth century, Spain's American colonies began seceding en masse. After 10 years of sporadic struggle, the Republic of Mexico was formed in 1821. Always a distant outpost at best, California had been little affected by these independence efforts; Mexico, however, asserted its natural trade and political rights as heir of Spain, and California became an official part of the republic.

During the mission period, the establishment of private ranchos had been successfully opposed by church authorities, with fewer than 20 land grants being made to private persons. After the Mexican revolution, however, the number of private land grants increased dramatically.

The turbulent Mexican period brought several changes to California life. Isolated as it was from Mexico, California did not respond well to governors sent to rule from distant Mexico. The *Californios*[1] had been accustomed to being left alone, and a series of unpopular and incompetent governors only helped formalize a tradition of conspiracy, insurrection, and rebellion, as well as bring about major socioeconomic upheavals.

Although initiated earlier, perhaps the single most dramatic event was the secularization of the missions by order of the Mexican government in 1834. Reflecting a concern over the traditional power held by the church and its historical ties to Spain, the revolutionary Mexican government decided that the vast holdings and influence of the church should be reduced. Under government edict, missions were to become civil pueblos, and mission lands were to be divided up among tribal families. The actual mission buildings were to be reduced to the role of parish churches, with the mission friars converted to parish priests. Ostensibly, such changes would have brought treatment of the California Indians in line with the new Mexican government's republican principles. However, corruption, avarice, ineptitude, and fraud deprived the California Indians of their intended legacies. The vast mission holdings were divided up, eventually becoming private ranchos, and for a brief time, the colorful and romantic rancho characterized California life.

The Romance of the Rancho

Of all historical periods, that of the California rancho is perhaps the most romanticized. The image of the cool, tile-roofed rancho, with *ollas* of water dripping peacefully in languorous breezes blowing through enclosed patios, is indeed compelling. The colorfully dressed, devil-may-care *caballero* riding or dancing with reckless abandon also provides a dramatic historical figure. Undoubtedly, these images have been overdone, reflecting as they do only a small percentage of the population during a very limited period. Nevertheless, they represent a colorful evocation of a way of life that held sway for some people. Cattle ranching *was* the predominant economic underpinning of Mexican California. The entire population owed its social and economic existence to this activity, which had replaced the mission almost entirely by the late 1830s. The secularization of the missions brought about an increase in private rancho grants from 20 in 1821 to over 600 by 1846. This covered most of the appropriated lands up to and including the Sacramento area.

The rancho was a natural outgrowth of several economic and geographic factors in California. Perhaps most important was the fact that the California land and climate were ideally suited to raising cattle and horses. With little or no attention, the California ranchero could let the herds roam the open range, multiplying in wild profusion. *Vaqueros*, or cowboys, engaged in lively cattle roundups twice a year, once for branding and once for slaughter. The best of the wild horses were culled from the herds to provide the predominant means of work, play, and transportation. Because the ranchos ranged in size from 4,500 to over 100,000 acres, there was plenty of land for this form of enterprise.

By relying simply upon beneficial climate, abundant land, and biological reproduction, the ranchero could prosper without artificial or unreasonable efforts. Indeed, with cheap or free land grants and with plentiful Indian and vaquero labor, many of the ranchos were hugely successful. This prosperity has produced the nostalgic image now associated with the period of the rancho. Dashing horses and handsome vaqueros, bright costumes and joyous fiestas, dances and songs, lively rodeos and lovely señoritas—all have come to characterize this period in the minds of many people. If the reality of life in that period was not always exactly as imagined, it nevertheless has provided a fascinating legendary background for the state.

FOREIGN INCURSIONS AND EARLY STATEHOOD

Mountain Men, Sailors, Pioneers, and Heroes

A long-standing consequence of California's geographic isolation from Spain and Mexico was the constant influx of foreign visitors. Early restrictions on commerce with outsiders were never well enforced in California, especially given that Spain itself could seldom provide the trade items, news, and exchange desired by the Califor-

[1] This was the name given to the non-Indian settlers and rancheros who made their homes in the region of modern-day California.

Figure 8-4 *Overland and sea routes into California.*

nios. Thus, early American incursions into California took the form of Yankee ships that entered its ports for trade and commerce.

American fur traders and explorers also blazed trails into California in the early 1800s, beginning with Jedediah Smith and followed by others such as James Pattie, William Wolfskill, and John Fremont (Figure 8-4). Although these trappers, mountainmen, and sailors were not welcomed by the government of California, they began a tide of American immigration that would

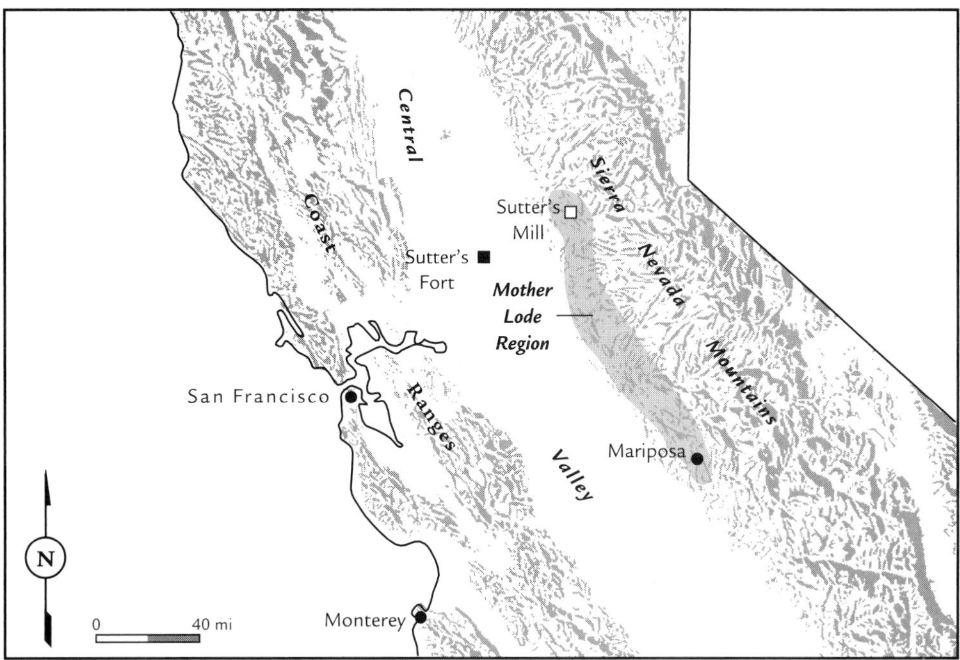

Figure 8-5 *Mother Lode region: gold country.*

soon lead to a significant shift in demographic and political patterns in the region.

Attracted to the relaxed lifestyle, pleasant weather, business opportunities, and pretty señoritas, American sailors and trappers stayed in California to become influential rancheros themselves. Through letters and by word of mouth, California soon became known as an idyllic and profitable place for settlers. A steadily growing number of Americans began to arrive in the area, and American influence increased far in excess of actual numbers settled in the area. By the early 1840s, hundreds of Americans were entering California annually by various overland trails, with the Bidwell and Donner parties being only two of the more famous groups. With growing immigration, the groundwork was laid for the formal political control of California by Americans.

There is strong evidence that even prior to the Mexican-American War, the administration of President Polk had its eye on eventual annexation of California. With the outbreak of the war in 1846, American forces seized various points in the region. The American desire to annex and the general enthusiasm of the American settlers in California for this move were aided by the general indifference of the California population to the American takeover. Although there were a few skirmishes in the Los Angeles area, the American occupation was for the most part peaceful, and California formally became a part of the United States in 1848 with the signing of the Treaty of Guadalupe Hidalgo. Spurred on by the influence of the new mining population, leaders adopted a constitution in 1849, and the state of California was admitted to the Union in 1850.

The Gold Rush

If any anomaly of geography or nature helped California become a vital part of the United States, it was the presence of a malleable yellow metal in its streams and mountains. The discovery in 1848 of gold precipitated monumental changes in the area. Population, travel patterns, socioeconomic shifts, racial balances, and myriad other changes took place. The impact of the Gold Rush was not lost on America; *Harper's Weekly* observed as early as 1859 that it was "the most significant, if not the most important event of the present century connected with America." This is an observation widely shared by scholars up to the present.

John Sutter did not consciously or directly seek gold. He was interested in the wealth to be gained through another medium: commerce. Convinced that lumber would be a valuable commodity to immigrants, Sutter commissioned James Marshall to survey and construct a sawmill on the American River at modern-day Coloma. In the process of building, Marshall happened to spot gold flecks in the tailrace (a canal in which water flows to and from a mill wheel). Slowly at first and then like lightning, the news spread, bringing would-be miners from all parts of the world. In a year's time, over 10,000 treasure seekers were scattered over the gold country, with hundreds of thousands yet to come.

The peak of gold production was reached by 1852, but the effects of gold on the state endured. By the 1850s, California had become the most populous region in the western United States. Also, gold drew people away

Figure 8-6 *Mono County Courthouse in historic Bridgeport is the oldest continuously operating courthouse in California.* (Richard Hyslop)

from the coast and into the interior of the state, a demographic pattern not previously encountered.

Although the lode deposits were found in a strip running from San Diego to Siskiyou County, the bulk of the gold was located in the Sierra foothills, in a region that became known as the Mother Lode region (Figure 8-5). The tenor of the times is captured in the names given to the camps and towns: Rough and Ready, Whiskeytown, Angels Camp, Hangtown, Fiddletown, French Gulch, Chinese Camp, Mormon Bar, Drytown, Oroville, Lazy Man's Canyon, Chile Gulch, and Poverty Flat, to name just a few (Figure 8-6). These names reflected ethnic influences as well as cultural humor; many of the towns were later to achieve fame through the works of Bret Harte and Mark Twain.

More significantly in the long run, San Francisco, Stockton, Eureka, and Sacramento experienced expansion of agricultural and mercantile activity in response to the needs of the mining regions (Figure 8-7). The Gold Rush thus helped to awaken California to more than mineral wealth alone. It acted as a magnet to population. It provided a stimulus to service industries such as agriculture, cattle, shipping, and trade. Finally, it drew to California the kind of Yankee aggressiveness needed to develop the state. As Californio Mariano Vallejo observed, "The Yankees are a wonderful people. . . . If they emigrated to hell itself, they would somehow manage to change the climate." If the physical climate was not changed in California as a result of gold, certainly the political, economic, and social climates were affected in permanent and dramatic fashion (Figure 8-8).

Figure 8-7 *Restored buildings and boardwalks of "Old Sacramento." The area represents the city's desire to re-create some of the atmosphere and spirit of the 1850s, when this approximately 28-acres waterfront section was the launching point for many Forty-Niners.* (Crane Miller)

Figure 8-8 *California state capitol. Sacramento was a population center of 12,000 when the state government was finally settled there in 1854 after being moved between San Jose, Vallejo, and Benicia. This magnificent building was completed in 1874 and has recently been restored to its original appearance.* (Sacramento Convention and Visitors Bureau)

The Decline of the California Indian Population

The first years of California statehood were turbulent, dramatic, fast-moving, tumultuous, and violent. The sleepy legacy of Spain was forgotten as the state forged ahead to face the challenges of a modern world.

The American impact on the tribal peoples of California was devastating and permanent. If the Spanish and Mexican periods represented earlier phases in the brutal treatment of the California Indian population, the Gold Rush was almost the final blow. With their simple lifestyles, the California Indians could not hope to cope with avaricious gold seekers who tore up their lands and societies looking for the elusive metal.

There is no history of savage warfare here, as with the Plains Indians, but the genocidal effects were just as pronounced. Alcohol, measles, bullets, and culture shock all helped bring about the decline of the California Indian population. The few instances of "warfare" were cruel, one-sided events that included wholesale massacre of tribal women and children by the white settlers. Gradually, the natives were killed off or pushed farther into high mountain or stark desert country. This sad chapter in California history is partially captured in Helen Hunt Jackson's *Century of Dishonor*.

One of the few instances of physical resistance by the Indians was known as the Modoc Wars. Initially, the Modoc tribe had lived in the northeastern part of the state, near Mount Shasta, in a region of timber, lakes, rivers, and lava beds. As more white emigrants passed through the region, the native inhabitants became increasingly hostile. In 1852, after tribesmen massacred an emigrant party, local ranchers and miners immediately retaliated. Sporadic battles were waged for the next 12 years; finally, in 1864, the remaining Modoc Indians were moved to a reservation following a tenuous peace agreement. Placed on the reservation with their traditional enemies the Klamaths, a group of Modocs under a chief named Captain Jack made their way back to the Lost River area. Talked into returning to the reservation once, the group finally decided to make a stand on the ancestral homelands. When an army force was sent to capture them, the group took to the lava beds and, relying upon their intimate knowledge of the terrain, fought the army to a standstill. Eventually, the shortage of water led to the surrender or capture of Captain Jack and his people in 1873, ending the only notable "Indian war" of California history (Figure 8-9).

The harsh attitudes of the Americans and the policy of eliminating California Indian opposition resulted in a tragic decline in the native population. Estimated at around 100,000 in 1850, by 1916 this population had dropped below 20,000. Deprived of land, dignity, spirit, and lifestyle, the California Indian became a victim of the state's "progress."

Figure 8-9 *Modoc Lava Plateau and Mount Shasta in the Lassen region. This was the general area of the Modoc tribe of Captain Jack.* (Crane Miller)

In many respects, the fate of the California Indian is epitomized by the story of Ishi, the "last wild Indian" in America. Discovered in a corral in Oroville in August 1911, Ishi was the last of a small band of Yahi Indians who had fled from contact with civilization and had lived in the Mount Lassen region. As the last survivor of this small band, Ishi eventually was able to reveal much about the last days of the California Indian to his protector and friend, Alfred Kroeber, an anthropologist at the University of California. Ishi lived and worked with Kroeber until he died of a white man's disease, tuberculosis. His true story is a touching and sad indictment of the treatment of the California Indians, encapsulating much of the tragedy of the decline of the native inhabitants of the Golden State.[2]

It should be added that recent legal actions have successfully underscored the fact of the mistreatment of the California Indians. Beginning in the late 1930s, various court decisions have held that they are entitled to both land and financial compensation. Somewhat belatedly, California Indians are receiving part of the bounty of the land that was taken from their ancestors.

As noted earlier in the chapter, the recent appreciation for the contributions of the early native Californians is a final irony. Although the disastrous decline of the native tribes cannot be corrected, misperceptions about their culture can. One example of this can be found in recent scholarship concerning the relative technological sophistication of certain of the tribes. Where once these groups were thought to have developed little technological capabilities, it is now known that some groups, such as the Nicolinos, created an aqueduct system to collect and dispense water. Likewise, anthropologists have pointed out that the Colorado River tribes, such as the Chemehuevi, used irrigation systems as a basic part of their agricultural practices. Thus, the modern water transfer systems that serve cities like San Francisco and Los Angeles (see Chapter 5) may be seen as the technological "descendants" of the "water works" of native Californians, completed long before whites set foot in the state.

Similarly, research by such authorities as John P. Harrington and Thomas C. Blackburn reveals rich, multifaceted tribal cultures. Ranging from tribal stories to pictographs and petroglyphs to gigantic geoglyphs in the Mojave, the spiritual, cultural, and artistic contributions of these peoples are just now beginning to be fully acknowledged and appreciated—further underscoring the tragedy of their decline.

The Rise of the Beef Industry

One benefit of the Gold Rush was a diversification of the economic base of the state. With the initial need to feed growing hordes of miners, the cattle industry was able to expand significantly. Although cattle, raised for their hides, had always been a staple of the rancheros in the southern part of the state, the Gold Rush provided a source of eager customers for beef, which by the 1860s numbered more than 3 million head. The decline of the Gold Rush, a serious drought, and high interest rates brought problems to the industry, and other agricultural pursuits assumed equal or greater importance.

Eventually, the economic and geographic realities of ranching in the state caused the evolution of certain forms of cattle production. Large cattle operations became the rule, as the less successful small ranchers were phased out. Fenced-in ranges began to replace free-roaming range cattle operations. Cross-breeding and improvement of the meat and dairy aspects of cattle raising became more critical and widespread. Like the state itself, changing conditions led the cattle industry to diversify, broaden, and improve itself. The beef industry evolved into a key agricultural activity, largely avoiding competition with irrigation farming by grazing cattle on the nonirrigated portions of agricultural lands.

In the context of cattle, agriculture, and landholding, the impact of the Land Act of 1851 was critical. Following California's admission to the Union in 1850, the status of the Spanish and Mexican land grants was placed in doubt. Although the Treaty of Guadalupe Hidalgo had guaranteed protection of the rancheros and land grants, many new American immigrants felt justified in claiming parts of these lands by squatter's rights.

Because of imprecise boundaries, confusion over historical sources of titles, conflicting claims, and financial difficulties, Congress passed the Land Act of 1851. Over the next several years, the rancheros and other claimants were required to present to land commissions and courts proof of title before their titles would be

[2] Kroeber's wife, Theodora chronicled this story in her book *Ishi in Two Worlds*. Ishi, she wrote, "was the last wild Indian of North America."

recognized. This placed a tremendous burden on many rancheros, whose proofs of title often required extensive and costly searches of records in Mexican and Spanish archives. The prolonged legal expenses often led to the loss of ranchos by debt-ridden Californios.

The whole story of the land title debacle is one in which legality and morality were largely absent. The long-term impact was to shift much of the California agricultural and ranch lands from the hands of the legitimate Californio owners to the hands of the recently arrived American squatters and others who benefited greatly from the questionable largesse of the state and federal governments.

TRANSPORTATION AND CALIFORNIA'S EVOLUTION

One of the most important factors in the development of modern-day California has been the transportation revolution. Just as transportation, or lack of it, was instrumental in shaping early Californian customs, culture, and society, it was to have the same impact in the American period. The creation of new means of transportation brought the state into significant contact with the rest of the nation, thus affecting all aspects of life in the Golden State.

The Gold Rush acted as a tremendous spur to the evolving transportation system. It soon became evident that mule trains and stagecoaches, clipper ships and ferryboats, and dispatch riders could not meet the growing needs of the state. Although the names of Wells Fargo and Butterfield flash large in the imagination, those stagecoach lines were essentially primitive stopgaps for the real transportation links that were to come. Indeed, the overland stage operations enjoyed only a brief ascendancy, flourishing during the 1850s and 1860s, although some stage operations continued in the Bodie region into the 1880s. Meanwhile, a real transportation revolution loomed on the horizon.

The story of California's development is, to a large degree, a tale of railroad domination. Politically, economically, and geographically, the Central Pacific Railroad Company and its successor, the Southern Pacific, helped mold the state more than any other single entity. The extent of this influence is described in Frank Norris's *The Octopus*, a novel whose title captures the company's pervasive presence.

Although local lines had begun to develop as early as 1855, the critical event was the connection of these local lines to a transcontinental system. Such a system had been discussed in the 1850s, but the the Civil War and the immensity of the task delayed its achievement. However, when the federal government sweetened the stakes with legislation granting 20 sections (square miles) of public land for every mile of track laid, in addition to loans and other encouragements, the project was a reality. The climax of the story is the now-famous meeting of the Central Pacific and the Union Pacific at Promontory Point, Utah, on May 10, 1869, sealed with the driving of the golden spike connecting the two lines.

The California involvement revolved around the efforts the Central Pacific under the leadership of the "Big Four"—Mark Hopkins, Leland Stanford, Collis P. Huntington, and Charles Crocker. These four Sacramento businessmen established the Southern Pacific as a holding company to acquire the Central Pacific's assets and develop rail lines within the state (Figure 8-10). Through their sometimes unscrupulous practices, the Big Four fixed rates and built a rail transport monopoly that extended to every railroad line coming into the Golden State. The exorbitant rates charged by Southern Pacific led to occasional violence on the part of farmers and businessmen. Because the railroad barons also dominated the political scene, however, little real relief occurred until the rise of the reformist Progressive movement in the early 1900s.

The lasting effects of this railroad revolution on the development of California are immense. The whole complexion of population growth and settlement was colored by railroad activity. Transportation and growth patterns largely mirrored desires, expanding where Southern Pacific wanted, or was paid, to go, and bypassing those areas the railroad wished to ignore. Thus, because Southern Pacific had little stake in real estate in the south, development of that part of the state had to wait until competitor lines, such as Santa Fe, were able to break the monopoly and penetrate the state.

In terms of landownership alone, Southern Pacific Railroad was (and still is) one of the single largest landowners, with over 10 million acres, in the state. Because its business activities, including construction of additional lines and land sales, depended upon supplies of people, the railroad also influenced ethnic balances with its recruiting of immigrants from both China and Europe. Its efforts in encouraging immigration resulted in increased labor supplies and more customers for business, farmers for the ample land, and tradespeople for the ever-expanding population. On a positive note, Southern Pacific helped link the state to the rest of the nation and the world through trade and population mobility. On a negative note, it provided a focal point for racial hatred, greed, and questionable business and political dealings. Its excesses led businessmen and farmers alike to organize against it and provided reformer writers like Frank Norris fertile grounds for their works.

The railroad has played a significant role in the evolution of modern California. Without the efforts of Southern Pacific and the Big Four, California would be a different place than it is today. In an era of "robber barons," the Big Four acted as many others around them did and at no time apologized for their methods. They

Figure 8-10 *Southern Pacific. Southern Pacific was acquired by Union Pacific in September 1996. Trains bearing the familiar Southern Pacific markings, although officially a part of Union Pacific, continue to be major freight carriers, with efficient and powerful diesel locomotives such as these pulling 90- to 100-car loads.* (Union Pacific Railroad)

brought progress to the state and placed a lasting imprint on its social, political, economic, and geographic landscape. One need only examine the persistence of the names of these four men to realize how deeply this influence has been felt.

DRY FARMING AND IRRIGATION COLONIES

Agriculture remains one of the primary economic activities in California. If any one word describes the agricultural pattern of the state, it is "diverse." With the wide range of climatic and soil conditions throughout California, diversity is almost mandated in crop production. The hundreds of crops grown in the state have solidified its reputation as the country's leading agricultural producer.

The nature of California's cultivation of the land has evolved through two stages—dry farming and extensive irrigation. This evolution has brought with it many changes in the economic and human elements of agriculture in the state.

The early period of California established dry farming as the dominant agricultural mode. Certainly, the native Californian inhabitants of the region engaged in little agricultural innovation, relying primarily upon hunting and gathering rather than organized cultivation. With the arrival of the Spaniards, farming assumed greater importance, especially through the efforts of the missions. For the most part, this agricultural development took the form of basic grain, vegetable, and fruit crops that could be grown without extensive irrigation, although the Franciscan friars did introduce some irrigation projects in the southern part of the state. The Mexican period did not further crop growing as a major economic activity, because most emphasis was placed upon livestock.

With the impetus of the Gold Rush and the American presence, organized agriculture assumed a key role. Grain became the dominant crop in the state, with California ranking second in the nation in wheat production by the early 1870s. The success of these efforts led farmers to more diversification of crops, expansion of orchard cultivation, and experimentation with vineyards.

Additional stimulus for California agriculture was provided by both the transportation revolution and the federal government. Railroad development permitted widespread marketing of the expanding volume of crops, while government policies provided further encouragement and assistance for farm diversification and growth.

The increasing demand for California food crops brought with it a necessity to free agriculture from the whims of nature. Although dry farming had been a highly profitable endeavor, irrigation proved to be a more dependable and efficient means of production. And although the missions had tried to develop irrigation systems, it was left largely to growers in the American period to capitalize upon the potential of large-scale irrigation agriculture.

The first irrigation colony of any significance in California was developed in the mid-1850s. This was a part of a Mormon settlement in San Bernardino where over 4,000 acres were brought under direct irrigation. Other irrigation colonies followed, including some in the San Joaquin Valley and in Riverside County. From this beginning, irrigation techniques spread throughout the state, fostering fruit crops, grapes, and other high-value agricultural products.

The legal foundation for irrigation, the Wright Act of 1887, established the concept of "appropriation and beneficial use" of waters and led to the development of recognized irrigation districts. This, in turn, led to massive efforts to cultivate huge tracts of land such as the Imperial and Coachella valleys. The eventual success of those efforts underscores the impact of irrigation on California's history, economy, and geography.

With the growth of tensions between residential users and growers, and with the reevaluation of water use priorities, *techniques* in irrigation farming are undergoing serious scrutiny. Extended drought conditions in the 1980s and 1990s brought the era of low-cost, government-subsidized, unlimited-use agricultural water to an end. However, the basic *concept* of irrigation farming appears to be a reality for the foreseeable future. California remains dependent on such irrigation farming to retain its position as the leading agricultural state in the nation. Thus, with the growth in the population and the gradual preemption of agricultural lands for residential purposes, high-yield irrigation farming has become even more vital. A trend toward consolidation and large-scale production, use of efficient machine labor, and high-density growing will undoubtedly continue, as will efforts toward further irrigation of formerly unusable acreage. Although the manufacturing and industrial character of the state has assumed greater importance, agriculture will continue to play a key role in the life of the state.

THE "BLACK GOLD" RUSH: *The Rise of the Petroleum Industry*

When Gaspar de Portolá made his first exploratory march up the coast toward Monterey in 1769, he happened upon a spring with large marshes of pitch and tar. This was the spring of the Alders of San Estevan—now known as La Brea Tar Pits. Although the presence of oil was recognized quite early in California, it was not until the late 1800s that any organized effort was expended to recover and use this resource. Certainly, both the native Californians and the early Spanish settlers had used the crude tar from oil seeps to caulk canoes, baskets, and the roofs of adobes, but it remained for American entrepreneurs to put this California resource to significant use and turn a profit.

In its own unique way, the petroleum boom was every bit as romantic and significant as the Gold Rush in California. Based on a steady, if undramatic, growth, commercial oil production began around the 1860s. Pico Canyon is frequently cited as the location of the first *commercial* oil well in California. It was not until the 1890s, however, that the first boom period began. Measured against later production, the state's first "boom" in oil was more of a snap-crackle-pop. Nevertheless, this initial period was similar to its earlier gold counterpart, ranking as a "black gold" rush. Through the late 1800s, independent oil companies engaged in a variety of activities, mostly in the south. By 1895, the combined output had risen to over 1 million barrels a year. With the further stimulus of the developing automobile, as well as increased rail and manufacturing usages, oil production became even more important.

The second major oil boom began in the 1920s with the discovery of vast fields in Huntington Beach, Signal Hill, Santa Fe Springs, Torrance, and Dominguez. Certain factors urged this development along, including improved drilling technology, increased industrial and heating uses, and the growing popularity of the automobile. With the easy availability of petroleum, the automobile became a sort of unofficial state emblem, a position it retains today. The boom of the 1920s was sustained by these developments and further nurtured by the military requirements of the nation in World War II.

The latest chapter in the oil saga is the modern period, dating from approximately the early 1950s. Exploration for new wells has continued on land, primarily in the southern portion of the state. The Tidelands Act of 1955, however, ushered in a new era of offshore exploration, drilling, and development (Figures 8-11 and 8-12).

Figure 8-11 *Richmond refinery wetland. Reflecting the environmental concern of Chevron employee Peter Duda, a former dried-up wastewater treatment pond was converted into a wetland. It is now home to more than 100 species of migratory birds, as well as many varieties of insects and animals.* (Courtesy of Chevron Corporation)

Figure 8-12 *Offshore oil development. Some efforts are being made to disguise the oil rigs as islands, as in Long Beach (photo A). Other wells, such as those off Huntington Beach (photo B), are far enough off the coast so that camouflage is unnecessary.* (Richard Hyslop)

With this new activity has come rising controversy. Environmentalists have pointed to the dangers of oil leakage and spills, such as the Santa Barbara oil spill, and have engaged in an ongoing battle with both the state and the oil companies to prevent development. For their part, the oil companies have increasingly attempted to reduce friction through aesthetic masking of offshore sites, public relations campaigns, and safety precautions against environmental pollution. Meanwhile, oil production in the state has remained a key part of the economic picture and promises to remain so as long as Californians continue their long-standing love affair with the automobile and the oil supply holds out!

THE ASCENT OF THE WESTERN STARS: *The Making of the Movie Capital*

As California began to mature, many businesses and industries were also growing in the state. One of the most widely recognized symbols of the Golden State, the

Figure 8-13 *Movie Studios. Movie studios still abound in Culver City and Hollywood, ranging from picturesque Paramount (photo A) to smaller Chartoff/Winkler Productions (photo b).* (A: Greater Los Angeles Visitors and Convention Bureau; B: Richard Hyslop)

glamorous movie industry, emerged at the turn of the century. In many ways, the geography of the region contributed heavily to this development. Several factors made Southern California extremely attractive to the newly expanding independent moviemakers. The mild climate meant that less money had to be spent filming on location, open space was readily available in such sleepy towns as Hollywood, and the Southern California region was about as far away from New York and other major Eastern cities as these producers could get. This last factor was crucial, because many of these enterprising filmmakers were engaged in running battles with patent holders and the so-called movie trust back East. California offered both distance and the Mexican border if the legal climate grew too warm.

Soon after the filmmakers arrived, the Hollywood area achieved dramatic success, in terms of both economics and publicity. By the early 1900s, California had come to be known as movieland, and by 1915, Hollywood was the self-proclaimed "film capital of the world." With the rest of the country ripe for this type of entertainment, the state had acquired a new major industry. The movies employed a wide variety of persons, including actors, carpenters, designers, researchers, writers, and publicists. Indeed, drawn by the popularity of early westerns, such famous (or infamous) characters as Emmett Dalton (formerly of the train- and bank-robbing Dalton Gang) and Wyatt Earp (formerly marshall of Tombstone, Arizona, and principal player in the gunfight at OK Corral) tried their hands with the film industry in Hollywood.

With the development of sophisticated equipment and sound and color technology, the industry brought even more attention to the state. To some people, Hollywood came to represent California; to others, California came to represent America. With a system that rewarded stars such as Mary Pickford and Charlie Chaplin with weekly salaries in excess of $10,000, the "good life" of California became even more appealing.

Through the peak period of the 1930s and 1940s, the movie industry significantly affected the state, as well as the rest of the world. Internally, at least three key factors were in evidence. First, the tourist attractions of Hollywood drew millions of people and their money to the state. Second, a major center of industry was created in the state, spinning off secondary economic activities such as cosmetics, electronic technology, broadcast enterprises, and clothing. Third, the magic aura of the Golden State was artificially enhanced, luring still more people to the West Coast.

Ultimately, the television industry reduced the role of Hollywood as prime dispenser of dreams and entertainment for America, and many movie companies disappeared. Although the filmmaking industry has subsided substantially from its peak period of the 1930s and 1940s, it retains a preeminent role in the life of the state. The majority of films made in America still originate in Southern California. West Los Angeles, Century City, Universal City, Culver City, Studio City, and Burbank have joined Hollywood as movie entertainment cities (Figure 8-13). Even though television has surpassed movies as the single most important dispenser of popular arts, television itself has become *the* major customer of the movie industry. Furthermore, the lure of Hollywood continues, as evidenced by the throngs of tourists who annually trek to view Hollywood Boulevard's "walk of the stars," Mann's (Grauman's) Chinese Theater, and Buena Park's Movieland Wax Museum.

From the first commercial film produced in California in 1908 (*The Count of Monte Cristo*), to the first "talkie" (*The Jazz Singer*), to the most recent thriller, comedy, action-adventure, or dramatic spectacular, the California movie industry has molded attitudes, economics, and social styles in an astonishing fashion. Like its host state, the movie industry has built upon a golden legend. In the process, California has added one more romantic attribute to its identity.

WORLD WAR II: *Enter Defense Plants, Exit Japanese Americans*

California enjoyed glamor, success, expansion, romance, excitement, and action during the course of its development. The Gold Rush, oil boom, Hollywood, transportation revolution, and agricultural bonanza all brought progress and change to the state.

The Great Depression of the 1930s, however, brought a different tone to California, the rest of the nation, and the world. Unemployment in the state rose drastically, while crops rotted in the fields and orchards for lack of markets. At the same time, lured by its climate and its movie-created reputation, thousands of unemployed migrated to California seeking the golden dream. Sadly, the state could not live up to its legend and the economic situation remained bleak, as did the lives of hundreds of thousands of Californians, new and old.

The agricultural industry was particularly hard hit. Conditions of life for farm workers were deplorable at best. John Steinbeck's anguished *Grapes of Wrath* gave accurate voice to the despair of these workers, underscoring the depth of the depression in both spirit and social reality. Although efforts were made to provide relief, it was not until America's entry into World War II that California and the nation climbed out of economic depression.

In many ways, World War II had a profound and overwhelming influence on the modern development of the state. The war brought with it a significant shift to urbanized, industrialized lifestyles, as well as an increasing racial ambivalence reflected in the roles of Japanese Americans, Mexican Americans, and African Americans in the life of the state.

Well before the attack on Pearl Harbor in 1941, relations with Japan had been seriously strained, a fact reflected in California in both legislation and popular attitudes. This reaction in California had its roots in a variety of motives, both genuine and contrived. Certainly, patriotism and legitimate concern over security played roles in the hostility and suspicion that was generated toward Japanese Americans. Equally important, if less noble, was the historical sense of jealousy, greed, and competition that the successful Japanese businessmen and farmers engendered in the minds of "native" Californians.

With a tradition of hostility toward Asians and other foreigners, the California response to the Japanese bombing of Pearl Harbor was predictable. Foreigners in general, and Asians in particular, had been forced out of the state before. Here was a natural reason to gather, incarcerate, and reject the Japanese again. (It should be noted that although Hawaii was also under martial law, Japanese Americans there were not rounded up and quarantined from the rest of the population.) Under pressure from Californians and military zealots, President Franklin Roosevelt on February 19, 1942, authorized military control of "enemy aliens" by the War Department. Thereafter, citing military necessity, the head of the Western Defense Command, General John L. De Witt, issued relocation orders resulting in the forced "voluntary" internment of approximately 110,000 Japanese Americans from California, Washington, Oregon, and Arizona, two-thirds of whom were native-born Americans.

Figure 8-14 *Stone guard/entry houses, bleak reminders of Manzanar detention camp. The snow-covered Sierra in background give testimony to the cold winters in the camp, located between Lone Pine and Independence off U.S. 395 in the Owens Valley.* (Richard Hyslop)

Treated as untrustworthy, ordered to assembly centers, and then transferred to "relocation centers," Japanese Americans were sheared of their basic civil liberties and forced to live in stark barracks encircled by barbed wire and patrolled by armed guards in barren locations such as the Owens Valley (Figure 8-14). Remarkably, the majority Japanese Americans retained their sense of patriotism to the United States—a patriotism that saw many of the young men join Nisei units of the U.S. Army and serve with unusual valor.

The costs of this shameful policy were immense. Economically, Japanese Americans suffered losses of at least $365 million in the forced sales of homes, businesses, and land. Psychologically, the humiliation and hatred they suffered was a bitter indictment of American attitudes in general, and Californian racial attitudes in particular. Politically, this suspension of constitutional guarantees was a frightening precedent inspired largely by hysteria and racial hatred. Finally, in a period of intense demand, some of the best businessmen and agriculturalists in the state were removed from the production lines. In retrospect, this episode in California's history is a chilling reminder of where prejudice and racism can lead.[3]

At the same time that Japanese Americans were being removed from the life of the state, massive population movements into the state were occurring. World War II demanded a maximum output from California. Defense plants in the south and shipyards in the north required enormous numbers of workers. With the exodus of Japanese Americans, large numbers of migrant Mexican laborers and blacks from the southern states moved into the vacated agricultural and industrial jobs. There was also a massive influx of military personnel. California held strategic value as a military area due to its open land, transportation facilities, and convenience as a debarcation point for the Pacific theater. Vast training bases drew huge numbers of people to the state, implanting in their minds the images of palm trees and balmy climate. Although most of these people were only temporary residents during the war years, many were haunted by their memories and returned after the war to become permanent citizens of the state.

The impact of the war on the state's economy was tremendous. The aircraft industry, already established in Southern California, experienced an unprecedented explosion and became the largest growth industry in the state. By 1941, orders for aircraft made this the key industrial base, with companies such as Northrop, Lockheed, Douglas, North American, and Hughes employing hundreds of thousands of workers. The population impacts on the state were as great as the economic effects. This period saw the beginning of a flow of immigrants from other states that barely slackened after the war was over. Indeed, the industrial profile of Southern California was well fixed during this period. In terms of both population and economics, the war years created massive changes in the state, with a built-in potential for even greater changes to come.

POSTWAR CALIFORNIA: *American Suburbia*

The boom in population that began during the war continued in the postwar years. In fact, the state population rose by over 1 million in the years immediately following the war, and the total growth exceeded 50 percent between 1940 and 1950. Much of the growth could be attributed to returning servicemen who had seen California during the war and had returned with their families. The attractions of the Golden State were also compelling enough to retain large numbers of workers who had migrated to the state during the war.

It is not surprising that the population growth figures exceeded those of most other areas of the country. The desirability of California as a place to live created new settlement patterns for the state, transforming much of the landscape from rural to predominantly urbanized settings. The necessity to house this growing population helped develop a somewhat unique California style, the "ranch house" in the uniform tract suburb. The checkerboard tracts in those areas came to typify the population sprawl of California (Figure 8-15).

In terms of the economic and industrial growth of the state, the postwar years also saw significant growth. The actual reconversion to a peacetime economy had begun in the state before the end of the war. The state government attempted to plan for this future by creating economic planning commissions and expanding tax bases and other revenues.

Several factors helped ease this transition and make the shift less traumatic. The population growth itself carried with it the germination of new economic activities. New housing stimulated the construction industry, and if assembly-line production of housing was monotonous, it was nonetheless profitable.

The immediate drop in defense activities was largely offset by a retooling for consumer products such as furniture, clothing, and automobiles. The heavy wartime industries had recognized earlier the need for diversified production capability and were already moving into needed peacetime activities. Former war plants now engaged in civilian-oriented steel production, food processing, and other useful manufacturing activities. Construction- and transportation-related industries alone accounted for a large share of this postwar surge.

Except for a brief lull, the defense industry continued to hold a strong position in the economic growth of California, at least into the 1980s. Although World War II had ended, the Cold War provided a continued impetus to military spending and development. California's defense industries soon began exploring uses of nuclear energy, missile systems, aerospace technology, and electronic weaponry. Federal defense contracts supported huge programs in research and development, and "think tanks" such as the Rand Corporation and Jet Propulsion Laboratory were funded by these grants. By the early 1960s, California's share of federal expenditures for military purposes was around 25 percent of the nation's total. By the late 1960s, at least one-third of the state's industrial production involved defense and space activities.

This military-related economic growth had both positive and negative effects. It placed heavy reliance on federal government largesse for continued favorable employment figures. It converted a large proportion of

[3] Belated recognition of the injustice of internment occurred in a series of U.S. Supreme Court decisions lasting into the 1980s. Then the Civil Liberties Act of 1988 mandated a redress grant of $20,000 and a formal letter of apology to each of the surviving internees (about 65,000 of a total 120,000 were still alive). The first of these payments and apologies took place in October 1990.

Figure 8-15 *Tract housing. Often strings along ridge lines like birds on a wire, such view lots usually are premium-priced.* (Roger M. Rhiner)

the California work force into white-collar workers. But when there was a sharp downturn in government spending, a concomitant slump occurred in the aerospace and defense-related industries. Thus, unemployment figures climbed during the 1960s before other industrial growth helped pick up the slack.

A similar and perhaps more lasting slump began in the late 1980s and continued into the 1990s. With the winding down of the Cold War, companies like McDonnell Douglas, Northrop, Lockheed, and General Dynamics were forced to scale back their activities. Employment in defense-related industries declined by up to two-thirds, contributing to a severe recession that dominated both the public and private sectors. Clearly, reliance upon defense-related industries alone could be dangerous; however, such reliance was far lower than many believed, with only about 5 percent of California workers employed in these industries by the early 1990s.

Significantly, other diversified industries have balanced out uneven employment in the aerospace business. For example, the petroleum industry remained a key employer in the state, as did transportation-related businesses. Agribusiness has also held a strong place in the economy of California. Likewise, the large population has acted as a constant source of demand for consumer goods and services. The growth of the computer industry, especially in the area of microprocessors, has given rise to the description of the San Jose–Santa Clara Valley area as "Silicon Valley."

The ability of former defense industries to shift to commercial production in the future is a partial index to the economic fortunes of the state. In the face of earthquakes and urban riots, the state's overall economy has remained relatively productive—at least in contrast to much of the rest of the world. Although the unemployment rate in the state rose in the 1990s, it is probably more accurate to say that an economic readjustment, rather than a decline, was occurring. Certain regions have experienced much higher economic dislocations than others; many of the types of available jobs have shifted more toward service and trade industries; and stratification of the work force has been accentuated by the continued steady influx of unskilled foreign immigrants. However, overall, California has continued to be successful with people, money, and attitudes. The climate, the lifestyle, and the promise of economic opportunity have all contributed to the appeal of the state. In addition to the goods, produce, and entertainment exported to the rest of the world, California has continued to export its image and style to millions of people. In spite of future uncertainties and past mistakes, it is still seen by many as the Paradise of the West.

PATTERNS FOR THE PRESENT AND FUTURE

One might conclude that, more than anything else, accidents of history combined with unique geography have shaped California's development. This conclusion provides a base for viewing the future evolution of the state. Many patterns begun in past years are likely to persist, while positive changes may result from the lessons of the past.

On the positive side, California seems destined to have an even greater influence in the life of the nation. Socially, lifestyles and behavior in the state, from skateboards to fashions, seem to fascinate and set the pattern for the nation. As the massive tourism, variety of recreation activities, and significant retirement-oriented industries indicate, the Golden State caters to myriad interests.

Politically, the state has assumed a key role in the nation's power structure. California's voting numbers, political leaders, and political trends are watched throughout the nation. One classic example is the "taxpayers' revolt" that began in California with passage of Proposition 13 and spread to the rest of the nation. Similarly, the governor of California, regardless of party, now is a perennial force to be reckoned with in presidential politics.

Economically, the state is a key indicator of the health of the country, and any economic slump in California is a problem of national import. Patterns of population growth and migration to California have changed. Now, rather than reflecting the dominance of the American Midwest, trends in migration and population growth indicate dramatically increasing Latin American and Asian immigration. With California poised on the Pacific Rim, this shift is entirely logical and predictable, adding new dynamics to the cultural milieu of the state. The potential for both economic prosperity and sociocultural confusion cannot be ignored.

On the negative side, the state has yet to face up to some of its most pressing problems. Steady population growth has also brought ever-increasing urbanization and congestion. Growth in certain areas such as Riverside, Sacramento, San Diego, and Contra Costa counties has frequently outstripped the ability of those localities to serve the needs of the population.

Although industrial and business enterprises dedicated to providing luxury goods and services have assumed a major position in the state's economy, the *basic* needs of many groups are still not being met. Particularly in the more urbanized areas of the south, racial groups are gradually emerging as a new majority. These groups, as well as other Californians, are demanding access to better schools, housing, hospitals, and jobs. With their growth in numbers, these minority groups will be able to exercise more authority—a position that is unique in California history. Certainly, as the Los Angeles riots of 1992 demonstrated, issues of justice and perceived equality (or lack thereof) will loom large. With census figures revealing the shift to a dynamic and diverse mix of peoples, ethnicities, and backgrounds, and with no single dominant majority culture, empowerment has become the watchword.

California also must come to grips with its long-standing tendency toward extremism of all sorts. Many people in the state have been striving to understand the extremes of devotion afforded to such groups as white supremacists or inner-city gangs or guerrilla environmentalists. Certainly, these groups have lent California a somewhat suspect air in the minds of many. One wonders whether their existence is a symptom or a cause of California's uncertain role in the future.

The certainty for California's move into the new century is that problems do exist and must be solved. The effects of growth on the landscape and habitats of the state must be addressed and controlled. The quality of life in the state must be stabilized in terms of environmental health, economic security, and social equality. Issues like smog, congestion, urban decay, ecological waste, water distribution, and costs of government must be dealt with directly and responsibly. Clearly, this provides a major challenge for the people of the state—a challenge that must be met successfully if California is to remain a "golden" rather than a tarnished state.

The diversity of the state's geography and culture guarantees that many creative solutions will be pursued. Some insight into the possible, and perhaps unique, approaches that will be taken may be found in various aspects of the cultural landscape of California, which we examine in the next chapter.

CHAPTER

9

Contemporary Folkways, Cultural Landscapes

California has acquired a reputation not only for diversity but also for unmitigated strangeness. One of the oft-quoted gibes is that if the whole country were tilted up on end, all the loose nuts would end up in California. This theme, echoed with some consistency throughout the state's history, continues to be a favorite of eastern writers. But it is a judgment that both exaggerates and oversimplifies the astonishing variety of geographical and cultural personalities found in the Golden State. Within the confines of California can be found much of the best and worst of all 49 other states combined. Large-scale immigration has contributed to the phenomenon, as has the wide diversity of climate and landform. Thus, it may safely be said that California does have something for everyone.

This diversity of culture takes many forms. Clearly, the various regions of the state differ, one from the other. The redwood country of the north is another world than the concrete-covered south. The quiet calm of the Mother Lode towns contrasts distinctly to the fast-paced life of the southern beach communities, as San Francisco does to Los Angeles. Diversity is more than regional, however.

The range of entertainment forms in the state is phenomenal. Californians can choose from many forms of theme parks, ersatz historical locations, zoos, wildlife parks, and night spots. The choice of restaurant styles and formats adds yet another dimension to the leisure-time activities of the state's residents. Apartments for active singles vie with retirement communities for space and attention, while artsy cemeteries offer marriage and burial in the same site. Ethnic populations vary from locale to locale, district to district, and county to county. In all, the state provides an anthropologist's heaven for the study of varied cultures, peoples, lifestyles, customs, and behaviors.

CULTURAL GEOGRAPHIC ODDITIES

Although the concept of regional differences has long fascinated scholars in many fields, geographers in particular have traditionally been committed to the idea that certain cultural regions or cultural landscapes can be identified and recognized as unique. The unique interplay between the land and the people who inhabit it was what French geographer Paul Vidal de la Blache described as the "spirit of place"—a regional "personality" that helps us recognize (without written clues) that we are not in Los Angeles when we drive down the streets of Bakersfield.

As might be expected, many different schemes have been developed to explain the concept of "region." In one very simplified and straightforward approach, we can arrange California's regions into two categories: (1) nodal and (2) uniform. A *nodal* region takes its character primarily from some central place or "node." The entire region thus becomes identified with the interactions that take place within the region *with* the focal point or node. Trade or commercial activities, media, employment focus, and similar phenomena tend to tie all parts of the region to the node. Thus, when someone asks you where you are from and you answer, "the Bay Area" or "L.A."

you are implicitly recognizing the appropriate nodal region (even if you actually live in Concord or Whittier). By contrast, a *uniform* region has no focal point, node, or central place that gives the region its character. Rather, the regional identity is taken from a widely shared set of values, ideals, lifestyles, and social and physical characteristics common to the entire area. Thus, for example, wine country may not have a nodal point to define it, but the commonality of style and interests throughout the area certainly helps to provide a "spirit of place" that is quickly recognizable.

Some observers are bothered by the lack of precision in delimiting the boundaries of these regions. However, as prominent American geographer Wilbur Zelinsky has noted, such regions are often perceptual in nature. People living within (and outside) the regions create a self-perception of their "sense of place" and act accordingly. That the boundaries may not be precise does not diminish the importance of such regional awareness for demographers, marketers, social planners, and others. Certainly, the long-standing debate over the possible division of the state reflects an awareness that regional differences do exist. Indeed, in 1993, voters in 27 central and northern counties approved the concept of dividing the state in two, and the California Assembly Rules Committee approved a statewide advisory vote for 1994 to ascertain the attitude of the people toward dividing the state into Northern California, Central California, and Southern California.

Putting aside the political, economic, and social implications of dividing the state, regional differences clearly do exist in the minds of most Californians; the fascinating challenge is to determine what complex interplay of geography and culture creates these different cultural landscapes. The correlation between geography and culture cannot be ignored or dismissed. Although geography by itself does not fix the cultural personality of a region or people, its impact on lifestyles, economic patterns, and social characteristics is nonetheless readily apparent. California provides an interesting study of varying cultural identities that can be traced in significant part to the shifts of geographic, climatic, and demographic conditions from one part of the state to another. Although the generalizations that follow necessarily pass over individual divergences, it may still be agreed that some common identity does exist. This collective personality or culture of a region does provide some fascinating insights into the motivations of people and the ways they interact with the environment in which they live.

Figure 9-1 *Redwoods stretching upward toward the sun in northwestern California.* (Richard Hyslop)

Logging Paul Bunyan–Style: Redwood Country, Northern Forests, and Plains

One of the first associations that comes to mind in connection with Northern California is giant redwoods. Certainly, the redwood forests of the north are dramatic, unique, and picturesque. The role of lumber in the state, however, does extend beyond that one imposing species. California's timber production ranks among the highest in the country, with fir, pine, and redwood constituting the major forest regions of the state. Compared with most of the rest of California, the northwestern portion of the state is a green and wooded world of its own (Figure 9-1).

Another rural element of the north is the preponderance of livestock, with ranching playing a key role in the lives of many citizens. The open ranges of the northeast, in particular, foster a way of life that is close to the land and reliant on its bounty. Clearly, the prime element and predominant theme in the northern portion of the state is nature (Figure 9-2).

Given the closeness of the people to nature and the land, it is not surprising that controversy has arisen over the issue of proper use of the land and resources. The bitterest controversy has centered on the expansion of national and state parks. Environmentalists argue the need to preserve and protect the priceless forestland, while logging and local economic interests advocate

Figure 9-2 *Mount Lassen region. Northeastern California is rural and often remote, with areas such as Lassen and Lake Alamor setting a tone of wilderness.* (Crane Miller)

jobs, economic growth, and the protection of the timber industry.

There is merit to the argument of each side. Few visitors to the extensive federal or state-owned parks, forests, campgrounds, or recreation areas would question the value, beauty, or utility of these areas. Yet with lumber providing employment in regions such as those around Eureka, and with the demand for forest products constantly rising, the timber industry's position must be equally respected. An uneasy balance will, undoubtedly, continue to exist for some time. Ultimately, the people of the north will be the critical decision makers, with their opinions and behavior determining the balance among competing uses.

Because there is a higher percentage of native-born residents in this region, a strong sense of local identity prevails. The population base is relatively low, and rugged individualism is still a cherished ideal. With an economy based primarily upon agriculture, tourism, recreation, dairy farming, ranching, lumber, and commercial fishing, the spirit of the people embraces nature and reflects individual responses to it. There are few cities of any size, Redding being the largest north of Sacramento. This demographic factor has contributed to the conservative, self-reliant attitude of the people, which, in turn, has occasionally taken the form of opposition to and hostility toward development in many parts of the north. There has been genuine reluctance to see expansion or improvement of the state highway system, as well as a related desire to protect and promote local control over undeveloped regions.

This is not to say that the north is bereft of any desire to improve, advance, or mature. The presence of Humboldt State University in Arcata has assured the region of a concentrated and respectable effort to promote growth in such areas as forestry, fisheries, and oceanography. Likewise, seashore development has not been totally lacking, as Sea Ranch in the Mendocino area demonstrates. Similarly, Chico with its state university campus and Redding with shopping, entertainment, medical facilities, and other amenities have both experienced growth. However, they are more the exceptions than the pattern for the region, which continues to be characterized by slow growth and evolution. But the decline in the lumber and fishing industries has forced the north to examine alternatives and consider some changes.

Nevertheless, from the foggy northwestern seacoast to the inland volcanic moonscapes, the predominant spirit is deliberate and calm. If expansion and change are to occur, the people are determined to control such development themselves. Accustomed as they are to a quiet, natural lifestyle, they are unlikely to see a need for precipitous change in the near future.

Argonauts and Ghost Towns: Gold Country

It is amazing that the most populous state in the Union also contains a backcountry of astonishing proportions. Ghost towns and Old West communities abound, carrying the colorful frontier past of the state into the present. For the most part, this region is concentrated in the highland block of California known as the Sierra Nevada, in the communities strung along routes U.S. 395 and California 49, and in parts of the Mojave Desert (Figure 9-3).

Part of the historical romance of California stems from its past role as home of boisterous gold camps and rugged mining sites. The names of present-day sleepy communities give eloquent testimony to the roistering expansion of the 1850s, and crumbling ruins attest to the transitory nature of these boomtowns. For the historian or romantic, however, the Sierra breezes sigh like ghosts of the past, reminding the visitor of an exciting period now vanished.

The greatest modern-day treasure of the region may in fact be just this ambience. Certainly, the Mother Lode area cannot be characterized as a bustling manufacturing or population center. Its charm lies in the very fact that it maintains a relatively slower pace of life that evokes the past. Ironically, it is this aspect of the region that has brought about a minisurge in population since the early 1980s, fueled by retirees and by urban professionals seeking escape from the pressures and frustra-

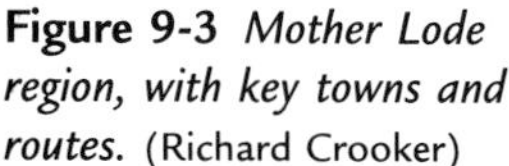

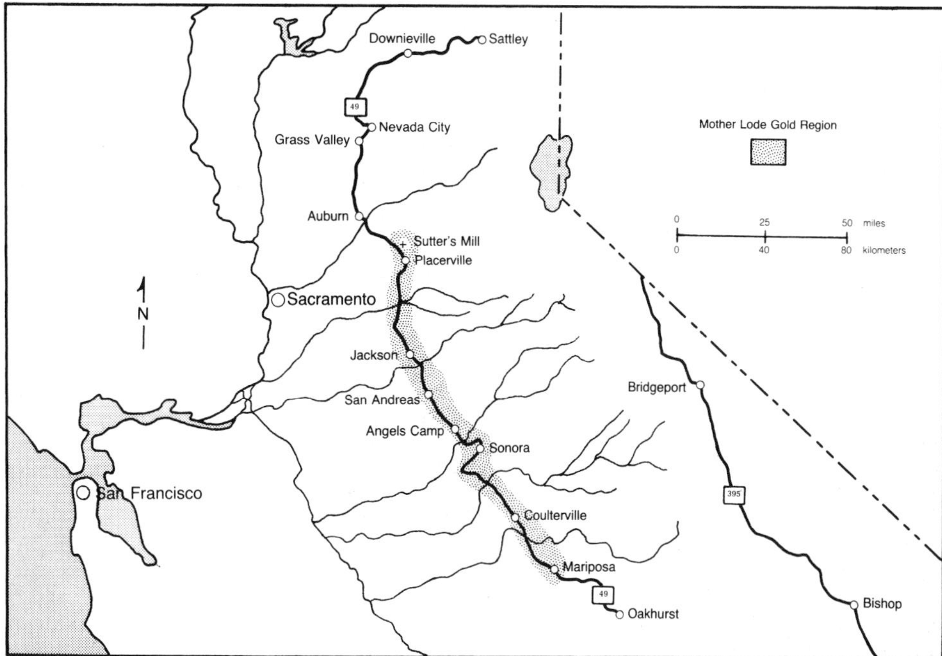

Figure 9-3 *Mother Lode region, with key towns and routes.* (Richard Crooker)

tions of city life. Because many of these new residents are employed by "footloose" industries that do not require any particular urban location, they have the luxury of seeking a place to live that provides them with a sense of country serenity and historical continuity. The subsequent growth in such small towns as Sutter Creek and Nevada City (in the foothills) and Hemet and Lancaster (in the desert) has been called "the growth of Penturbia" by some experts, referring to a fifth cycle of urban migration to smaller towns and communities.

Although there are some small pockets of high-tech industry in the foothill region, the primary income of the area comes from modest agricultural activities; from unromantic rock, gravel, and sand mining; from some ponderosa pine and white fir lumbering; and from the rapidly growing recreation- and tourist-oriented trade. The latter capitalizes on historical sites, winter sports, and wilderness recreation. The nature of this economic activity does not conflict with the yesteryear sense of simplicity and naturalness. This undoubtedly has contributed to the modest boom in foothill towns in recent years, with retirees and urban refugees finding the atmosphere they seek in the area.

The advantages of living in this region include distance from the turmoil of big cities, availability of open space and nature, and the presence of historically fascinating sites all around. The area encompasses forested foothill and barren desert, each with its own unique characteristics. The sense of peacefulness, however, pervades both foothill ranches and lonely desert and carries over to the little towns that still exist, such as Mariposa, Downieville, Garlock, Bodie, and Darwin. A visitor with a good imagination can populate the lonely stretches of U.S. 395 and California 49 with prospectors and cowboys (Figure 9-4).

Figure 9-4 *Gold Rush country. The romance of the Gold Rush survives in towns such as Columbia, Sutter's Creek, Sonora, and Nevada City.* (Richard Hyslop)

The sense of history is also evident in colorful places such as Fiddletown, Angels Camp, Mormon Bar, and Mokelumne Hill. Columbia and Calico (Figure 9-5) have been turned into restored western towns for tourist consumption exclusively, and their success is evidence of the popularity of commercialized history.

Many more of the Mother Lode cities now serve as gateways to the popular wilderness areas of Sequoia, Kings, and Yosemite national parks and other state and local wilderness settings. In the warm season, providing services for hiking, fishing, hunting, and camping constitutes the mainstay of these towns. In cold weather,

Figure 9-5 *Panoramic view of Calico, a mining ghost town restored by Walter Knott and then deeded to the county of San Bernardino.* (Richard Hyslop)

skiing becomes the prime attraction. Tourists and temporary visitors move on, but residents continue to enjoy their peaceful environment and colorful past. Architectural treasures of the past, magnificent wilderness, open landscapes, and low population density are attributes that hold the loyalty of the people. They take pride in their frontier heritage, and few would trade their lifestyle for one that required them to move to Los Angeles or San Francisco.

A State Without Wine Is Like a Day Without Sunshine: Wine Country

Winemaking has become a sophisticated, respected, and palatable business in California. From its early and simple beginnings with the vines brought by Franciscan padres, the California grape has matured into a well-traveled and widely consumed product. In fact, grape growing and wine making have become multimillion-dollar industries.

The term "wine country" popularly refers to the Napa–Sonoma area north of San Francisco. This is a somewhat misleading designation because respectable vineyards and wineries exist in many other locations throughout the state. In fact, the Central Valley and the southern counties now produce most of the state's wine. The Napa–Sonoma region, however, is still wine country to most Californians, and the region takes great pride in this identity even though it now produces only about 20 percent of California's total wine output.

The California wine industry is a remarkable success story. Currently holding a share of the total American wine market in excess of 80 percent and enjoying an extensive foreign export market, the business is solid and growing. Directly and indirectly, viticulture in the state owes much to the contributions of European winemakers. The immigration of experienced vintners brought first of all a new expertise to the state. During the 1850s, the German vineyards at Anaheim and the Italian Swiss Colony at Asti began operations. Contributions by Charles Krug, Etienne Thée, Charles Lefranc, and Agoston Haraszthy took the infant industry further toward fulfillment and demonstrated the significant impact of immigrants. The hard work of these new Californians resulted in expanded vineyards throughout the state, especially in the Napa–Sonoma region.

Indirectly, Europe helped by providing vine cuttings of quality and endurance. The infusion of this established stock enabled California wines to grow in popularity and quality. Continued expansion of the industry was assured, and new wineries sprang up in such diverse locations as Fresno, Madera, Modesto, Cucamonga, Rutherford, and Saratoga. Although Prohibition (1919–33) caused major financial damage to the California wine industry, it returned stronger than ever with the repeal of Prohibition.

Traditionally, wine country reflected mature respectability. Nowhere was this more evident than in the Napa–Sonoma region. In architecture, cultivated hillsides, and temperament, the area exuded a flavor of the Old World (Figure 9-6). Long-time residents prefer a region of cultivated serenity and traditional values. The plethora of family vineyards adds a superficial sense of a stable social order to the area. The feel of the traditional culture is found in the orderly sense by which the long rows of cultivated vines march across much of the available land.

The battle lines increasingly are being drawn between those who advocate growth and those who wish to maintain the Old World ambiance. Sonoma County has seen the greatest development pressures, with Bay Area commuters swelling the population base, wine-related tourist attractions expanding the demands on the infrastructure, and high-tech, white-collar, and ser-

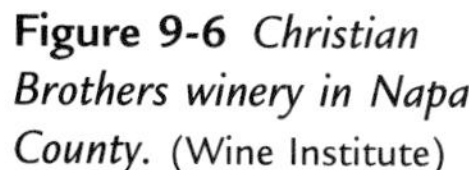

Figure 9-6 *Christian Brothers winery in Napa County.* (Wine Institute)

vice industries increasing their presence in the region. In response, controlled-growth advocates have fought a prolonged and bitter battle to roll back this powerful tide. Indeed, the city of Petaluma acquired some fame with the test case challenging its antigrowth ordinance (which was ultimately upheld by the courts) in the late 1970s. Although population growth and pressures for change have not been as intense in Napa, the tourist trade has created major problems for many residents, who resent what they see as exploitive commercialization of viticulture.

What is clear is that the romanticized, genteel, slow-paced wine country has been inexorably mutated. The once tranquil, scenic nature of the area has been significantly altered by the presence of visitors touring the many tasting rooms. Although the courtesy and charm characteristic of wine country is evident in the care with which the old-family wineries treat their guests, the underlying uncertainty over the future character of the region remains. The irony is that much of its supposed historical foundations have already eroded away. For example, if there is an old aristocracy in California, one is tempted to look for it here; however, many of these rustic, homey-looking wineries are now owned by international corporations.

Recently, as noted previously, the controlled nature of the agricultural landscape of the Napa–Sonoma area has been copied somewhat in the efforts of the people to prevent uncontrolled residential sprawl. Concern over spread of the Bay Area population has led to zoning efforts to retain the traditional mood, lifestyle, and flavor of the region. With its history of controlled environment, the region is likely to fight hard to retain its romantic and tranquil existence, or at least its facade.

Sophistication: The San Francisco Bay Area

San Francisco has acquired a rare and enviable reputation as an exotic, breathtaking, world-class city. To a certain degree, the entire Bay Area shares this romantic image; it is an environ of ethnic mixes and cultural plenty, refurbished Victorians, and cluster housing, self-conscious snobbery and experimental thinking, and scenic landscapes and dramatic skylines (Figure 9-7). It is at once historical and modern, hurried and calm, artsy and staid. It is unique, its self-image one of quintessential sophistication.

The San Francisco Bay Area is, of course, more than the simplified realm promoted by songwriters or chambers of commerce. Here is a greater metropolitan region that encompasses all or part of nine counties in an interconnected web of bedroom communities, residential tracts, shopping centers, diversified industrial-retail developments, and highways and railways. Though the Oakland–East Bay area and the San Jose–South Bay area form overlapping centers of residential and commercial activity, the main hub of this urban web is San Francisco, which provides the overall identification for the entire region.

The diversity of the Bay Area is phenomenal. It can be seen in the variety of architectural styles ranging from Asian influences, to refurbished Victorian

Figure 9-7 *San Francisco's changing face. High-density San Francisco seen from Twin Peaks, looking northeast along Market Street.* (Roger M. Rhiner)

splendor, to leapfrogging tracts, to palatial estates, to tidy, middle-class, ranch-style homes. It can be seen in the obvious extremes of economic affluence, from wealthy Hillsborough to troubled central Oakland, with every shade in between.

The ethnic makeup of the Bay Area is likewise notable. San Francisco's Chinatown is one of the most celebrated in the country. Oakland and Richmond contain high concentrations of African Americans. Inner San Francisco has enclaves of eastern European groups, Filipinos, blacks, Japanese and other Asians, and Mexicans and other Spanish-speaking populations.

The climate reflects the theme of diversity in an astonishing fashion. Owing to its watery surroundings, San Francisco itself enjoys climatic conditions entirely different than those experienced by most of the rest of the region. Meanwhile, across the Bay, temperatures and conditions represent a dramatic contrast—sun versus fog, heat versus cool, dry versus damp. Landscapes, too, vary immensely within a very limited geographic region. Urban settings alternate with agriculture; flat marshlands contrast with precipitous hillsides.

San Francisco itself has undergone some significant changes in recent years. Once a major center of commerce, trade, and finance, the city has gradually seen much of this activity shift to other Bay Area cities or to Los Angeles. Although it remains a highly popular tourist destination, the balance of ethnic, economic, and social forces has been replaced by a bipolar structure. San Francisco is now occupied primarily by wealthy, professional property owners (mostly white and Asian), with members of the middle and lower classes forced to commute in and out of the city to service this wealthy population. The effects of this demographic shift are significant, as the dynamics and diversity of new growth now stimulate other locales. This is not to say that San Francisco will not retain importance in the region. Its widespread reputation for charm and sophistication remains well deserved. Its international character endures in the culinary, architectural, and cultural choices available. Its political and social tolerance still fosters the open acceptance of alternate sexual preferences and lifestyles. And its climate, cable cars, gingerbread houses, theaters, restaurants, and fog continue to draw visitors to the city.

But where gentrification in San Francisco is creating a somewhat homogenized "ghetto" of wealthy white and Asian residents, outlying suburbs are recharging their ethnic and economic identities. At least partly because of this diversity, most experts see the East Bay and the South Bay gaining financial and trade dominance.

A brief tour of the region radiating outward from San Francisco, quickly reveals this diversity. To the east, across the Bay and connected by the Bay Bridge, is the Contra Costa–Alameda complex. This populous, urbanized region provides much of the industrial production of the Bay Area. Historically, the Oakland–Richmond complex was a key terminus for rail, ocean, and truck transport. However, Oakland has struggled in recent decades to survive the flight of residents and businesses to "ContraCostapolis" and its more attractive environs. Both economics and race play a role in this shift. Although metropolitan Oakland has a large minority population, there is a less obvious ethnic mix as one moves south and west from the Oakland core. Similarly, where Oakland appears alternately crowded and congested or boarded over and abandoned, shading out and away from the city are typically suburban cities like Walnut Creek, Hayward, Fremont, Concord, and Livermore. The Contra Costa County communities in particular have experienced growth in technical jobs, office complexes, and corporate expansions, as well as a resulting whiter, more middle-class population. The growth has not been without problems, with traffic jams, overpriced housing, unmanaged development, and shortages of local services among the more obvious.

Moving southward from San Francisco along the peninsula, the entire area stretching from South San Francisco to Palo Alto has been a steadily growing corridor. Neighboring peninsular cities such as Daly City, San Bruno, San Mateo, Redwood City, Menlo Park, and Palo Alto have struggled with the problem of expansion. They have had mixed success in controlling the development of the peninsula, which now reflects a blend of residential, retail, and diversified light industry.

To the extreme south of the peninsula begins the cluttered sprawl of San Jose and the Santa Clara Valley. Here is one of the most typically disarranged of California urban areas. A region of mixed development, decaying inner city, spreading tract housing, and bustling industry, the South Bay region is still experiencing problems coming to terms with itself. The successful industrial capacity of this locale was typified by the major aerospace establishment of Sunnyvale and by the explosive growth of the Silicon Valley computer industry. However, with the winding down of the Cold War, the reduction in military spending, and the effects of a prolonged recession, the fortunes of the region have been somewhat moderated. Still, within the greater Bay Area, the South Bay has enjoyed population growth, urban development, increases high-tech industrial, and overall economic prosperity. Meanwhile, the supporters of the rural, agricultural lifestyle battle to maintain a hold against the ever-expanding residential tract developments. Moving farther south and east from this point, the Bay Area identity begins to weaken until finally the agricultural Central Valley takes over.

To the north of San Francisco, across the famed Golden Gate Bridge, is the North Bay and the Marin County peninsula. The confused and unattractive urban sprawl is less evident in this area, which is a mix of mostly higher-income residential and open land. The wealthier socioeconomic structure here has helped to preserve much of the open marshland, recreation areas, and beautiful green terrain. A quieter, more peaceful section of the Bay Area, it gradually merges into the wine country, not far from the waters of the Bay itself.

Today, the greater San Francisco Bay region is undergoing some traumatic transitions. Long known for its commercial potential, the area continues to support rail, truck, and port facilities. The challenge of the growing southern portion of the state, however, has created serious competition for this economic base.

Like the rest of the state, the Bay Area continues to face serious questions about its future. Changes in the business climate, international politics, tax bases, demographic profiles, and economic structure present new realities. Also, events such as the Loma Prieta earthquake demonstrate that planning frequently falls short of necessity. With the growing demands of a burgeoning population, the Bay Area also faces many challenges to the old ways and assumptions. Faced with many of the same problems as the south, the region can no longer look with carelessly veiled contempt on Los Angeles. The growing pains of Southern California are increasingly becoming the reality in the Bay Area. The challenge will be to resolve these problems without mimicking and magnifying the mistakes of the south.

Nashville West: Bakersfield, the Central Valley, and the Farm Belt

Agriculture is an important economic activity in California; a substantial portion of the state's land is dedicated to food production. In few places is this more evident than in the Central Valley. Comprising most of the heart of the state, the Central Valley stretches for about 470 miles from Bakersfield north to Redding and lies between the Coast Ranges and the western Sierra Nevada. Fertile productivity is the hallmark of the region, and long, flat, open expanses of combed and cultivated land are the common landscape. It is a region populated by farmers, truckers, and field hands and characterized by wide expanses of cultivated fields, diverse crops, and agribusiness operations. It is the source of California's leading position in American agricultural production and income.

This is also a region of far more complexity and variation than imagined by the typical urban dweller of the megalopolis of greater Los Angeles or the *galactic metropolis* of the Bay Area.[1] The astonishing variety of sociocultural realities present within the Farm Belt in general, and the Central Valley in particular, lend a richness and depth to the face of California.

To begin to appreciate the magnitude of this richness, one might drive the length of California 99 or I-5 between Los Angeles and Sacramento. These routes pass through extensive agricultural land and are unbroken for miles and miles by any significant urban setting. The

[1] A galactic metropolis involves an influential core city surrounded by self-sufficient outlying communities loosely tied to the core city.

variety of crops visible along these routes is great enough to astonish the most sophisticated agriculturalist. A world apart from the populous bustle of the major urban centers, much of the Central Valley reflects yet another side of the state in both ecological and cultural terms. There are a few large cities located in the region. Most of them unmistakably reflect the priorities of the farm, although this is gradually beginning to change. The city of Bakersfield is a perfect case in point. Driving into Bakersfield, one encounters Montgomery Ward's, truck stops, farm implement dealers, gas stations, cafés, and honky-tonk bars. The uncomplicated strains of country-western music float through the air from bars and truck radios. Values in the songs are simple and direct, mirroring the attitudes of the residents of this region. Dubbed "Nashville West," Bakersfield has earned a reputation as the western home of country music.

Historically, the city also has served as a focal point for oil production and related industries, which has added to its rough-hewn aura. Interestingly, too, Bakersfield represented a cultural core of a substantial Basque population drawn to the region to raise sheep in the open spaces of Kern County. All of these activities provided the down-to-earth reputation prevalent throughout much of Bakersfield's history. This has been a city that was "close to the ground," without pretensions. It existed to serve the agricultural needs of the area with its shipping facilities and stores. Its merchants catered to the farm population, providing services, products, and entertainment for evenings and weekends off the farm. However, Bakersfield has begun to add a patina of urbanization and economic diversification to its cultural landscape. Although its country roots remain strong, the city no longer can be accurately characterized as agricultural only. The impact of suburbanization has become very evident, with some industries from Los Angeles moving here, some newer suburbs expanding, and a growing number of retirees choosing to locate in the region.

What is true of Bakersfield is also true of other Central Valley cities, such as Sacramento, Visalia, Fresno, and Stockton. Through the 1970s, although Sacramento was the state capital and contained immense government operations, it made no pretense of being a cosmopolitan center. In fact, it was an agriculturally oriented city located in the midst of a rich delta farming area. Sacramento was really a farm town; it just happened to be the state capital.

But this began to change in the late 1970s, and in the 1980s and 1990s, the region became one of the nation's fastest-growing metropolitan areas. This rapid urbanization has had both positive and negative consequences. Clearly, the Sacramento area has diversified greatly in terms of economic influences, cultural opportunities, and lifestyles—it is no longer dominated by an agricultural mentality. However, with the explosive growth, the trappings of urbanization have arrived as well, including uncontrolled sprawl development, traffic snarls, street gangs, and a bitterly divided local political structure. Indeed, the nastiest "street fights" seem to be between the advocates of suburban development and those who wish to return to the days of agricultural dominance. Thus, the greater Sacramento area fairly well typifies the rest of the Central Valley cities' problems.

Visalia came into being and continues to prosper in part due to its role as a farm service community. Likewise, Fresno, though it is a big city that offers many of the benefits and ills of a large urban area, also exists because it served the needs of the Central Valley farms that surround it. Stockton is equally a product of agriculture; it came into being due to the Gold Rush but stayed alive due to its role as a major agricultural service center.

These cities owe their existence to the agrarian wealth of the state, and most of their residents remain proud of the role of farming in California and the nation. If their values seem a little more basic and old-fashioned than those of the megalopolis, and if their music is the music of the country, and if their lifestyle is more sedate, that is what they cherish.

An interesting irony is that both old and new residents of the Central Valley tend to embrace the concepts associated with farm or country living. However, the rapid spread of suburbia into the Central Valley presents a challenge to the historically agricultural character of the region. Much of the development and growth is fueled by commuter residents who are willing to spend 4–6 hours a day traveling to and from work. The consequence of this phenomenon is that by the early 1990s, Central Valley farmland was being converted to suburban uses at a rate of approximately 20,000 acres per year. Furthermore, cities like Fresno, Visalia, and Stockton have experienced tremendous nonagricultural growth in recent decades.

Although not physically a part of the Central Valley, the Imperial Valley in the far south shares the same values, attitudes, and priorities. Like the Central Valley, the Imperial Valley is a highly productive agricultural region because of the long growing season, diverse soil conditions, extensive irrigation, and advanced technological applications. Many of the same problems face the area, including urban encroachment and agribusiness consolidation. Economically, socially, and culturally, it is a continuation of the Farm Belt.

Collectively, the farming portion of the state remains quite significant. It is responsible for mass production of cattle and dairy products, poultry, grapes, vegetables, hay, nuts, cotton, and fruit. The hard work and efficiency of the California farmer in producing these and other crops has made the state a world leader. Certain problems do exist, however.

Even with the increased demand for food, urban-residential land uses are cutting into productive agricultural land. The trend toward consolidation and incorporation in agribusiness has brought new complexity to the lives of the farm population. Unionization of farm workers and advancing agricultural technology are bringing about a shift in the number of persons attracted to this way of life. Furthermore, serious competition from abroad has cut into the profit margin of California agriculture. Add to this an increasing awareness of environmental issues pertaining to chemical fertilizers and insecticides, and the changing government policies relating to water pricing, and the future of agricultural California is challenging indeed. The California farming community, conservative, fundamental culture, may have to reexamine some of its goals. But even if some changes do occur, the Farm Belt will continue to exert significant influence on the economy, environment, and culture of the state.

Neon Glitter: Southern California and Los Angeles

On occasion, Southern California has had a rather shady public image. Characterized as the last stronghold for assorted faddists, nuts, kooks, and misfits, it cannot be accused of being dull! Congestion, freeways, smog, cars, trucks, recreational vehicles, and rows of identical tract houses typify much of the south. The urban tide of Los Angeles has spread in an almost continuous flow down through Orange County, up through part of Ventura County, and across through large portions of Riverside and San Bernardino counties.

The greater Los Angeles metropolitan area takes in a huge portion of the land of Southern California. The city alone covers more than 450 square miles, but the Los Angeles influence extends far beyond even these extensive bounds. In terms of its sphere of influence, the Los Angeles metropolitan region extends to the ocean in the west, to the San Fernando Valley and Ventura County in the north, to Orange County in the south, and to the "Inland Empire" of Pomona, Riverside, and San Bernardino counties in the east. The problems, concerns, and orientations of these areas are a part of the greater entity of Southern California. Los Angeles itself provides the unifying link and shared bond.

Over half of the state's total population is found in Southern California, with 30 percent in Los Angeles County alone. Although there has been migration outward from Los Angeles proper into the newer suburbs, (white flight), there is also continued growth within the city and county of Los Angeles (mostly new immigrants). In each location, however, the L.A. sphere of cultural influence is dominant.

As a culture of diversity, Los Angeles and Southern California represent an astounding range of urban-suburban environments. Within one metropolitan area can be found miles of beautiful beaches and blocks of decaying urban jungles. Cool mountain resorts look out over bleak desert landscapes. The plastic glitter of Hollywood contrasts sharply with the air-conditioned luxury of corporate Century City skyscrapers (Figure 9-8).

The mix of racial and ethnic groups is equally diverse. The south now has large and significant populations of African Americans, Hispanic Americans, Asian Americans, and others. Most of these are tossed together in an interesting cultural salad bowl called the Los Angeles Basin. Indeed, surveys by the Los Angeles Unified School District in the early 1990s revealed students speaking over 80 different languages. Ten years earlier, the *Los Angeles Times* had reported that 1980 marked the point of ethnic shift in Los Angeles, wherein the Anglo population now numbered less than 50 percent.

The southern portion of the state is clearly a land of plenty. There is an abundance of people, money, stores, freeways, motor vehicles, billboards, noise, shopping centers, gas stations, parking lots, restaurants, lights, schools, fast food chains, theaters, houses, condominiums, apartment buildings, ghettos, barrios, slums, street gangs, deteriorated buildings, office vacancies, paranoia, pollution, random violence, lawyers, and real estate salespeople.

The nature of this abundance can be traced to many of the factors that originally drew the population to the region. The defense and aerospace industries offered jobs; the movie industry promoted glamour; the Mediterranean climate promised health. The romance of the state was touted by the tourist and service industries, which relied on masses of people for their success. And once people arrived, they stayed.

In the 1980s and 1990s, many of these "older" immigrants decided to move out from the core—some to the surrounding suburbs, some to other states. However, to massive numbers of both legal and illegal immigrants from Latin America and Asia, the relative wealth of and opportunities in Los Angeles continued to act as a powerful magnet. This was also true of foreign businesses. In recent years, a boom in foreign ownership of major downtown L.A. properties has contributed to the international character of the area. Such prominent landmarks as the Bonaventure Hotel, Crocker Plaza, The Park, Pacific Financial Center, The Biltmore, One Wilshire Boulevard, California First Bank, and 800 Wilshire are now or have been owned by foreign corporations.

The Los Angeles area in particular, and the Southern California region in general, does have

Figure 9-8 *Famous intersection of Hollywood and Vine. The Capitol Records building, designed to look like a stack of records, is one of many fanciful buildings in Southern California.* (Greater Los Angeles Visitors and Convention Bureau)

problems. People's desire to pursue unique notions and lifestyles has not fostered a commitment to planned or controlled development, and monotonous and ugly suburbs attest to this fact. An exaggerated sense of individualism among area inhabitants has encouraged a political environment in which efforts to build mass-transit systems and establish clean-air standards for industry can be thwarted. Thus, air quality has become a crucial issue at the same time that the population clings to its imagined independence by commuting one-to-a-car. The inflated sense of self has made local government decision making a morass of confused, contrary, and involved arguments. The interaction of ethnic and racial groups is uncomfortable at best. The explosion of violence following the initial verdict in the Rodney King beating trial in April 1992 demonstrated the tenuous nature of race relations in Southern California.

All the clichés concerning the cultural wasteland of Southern California bear careful examination, however. In spite of the crush and confusion of a booming population, Los Angeles has achieved enviable progress as well. The abundance of quality private and public universities and colleges has made Southern California a recognized leader in the field of education. Here can be found some of the world's most impressive programs of collegiate extension courses. With a multitude of locations, subjects, and time slots, these programs have made higher education truly a public pastime. Southern Californians can study virtually any subject, take educational tours to most parts of the world, and participate in the learning experience for as long as they wish. Also, with the impetus of military and government contracts, the Los Angeles metropolitan area has become a center for scientific research and development, as evidenced by the Rand Corporation and the Jet Propulsion Laboratory in Pasadena.

Unfortunately, the end of the Cold War and the recession of the 1980s and 1990s caused a major reassessment of priorities within both corporate and educational establishments. Down-sizing and streamlining became the operational guidelines, and with tightened tax bases and reduced income, economic growth slowed. This is not to say that Southern California is moribund—quite the contrary. However, the region has been forced to evolve, with low-tech manufacturing and service-based industries becoming more evident.

There has been a serious commitment to the arts, with the Music Center, County Art Museum, and various dramatic companies the envy of much of the rest of the nation. Contrary to popular opinion, the Los Angeles area offers a wide variety of cultural activities. Numerous private museums supplement the public offerings, with the Norton Simon and the J. Paul Getty museums being only two obvious examples. On almost any sunny weekend, one can find outdoor art shows featuring the work of thousands of painters, sculptors, and other craftsmen and artists. Likewise, community support for the arts ranges from local theater groups, to shows sponsored by savings and loan institutions, to local library presentations. A slightly different form of art can be seen in the profuse displays of architectural experimentation found in both private and commercial edifices, from shopping centers to hideaway retreats.

Los Angeles has come to rival New York City as the news capital of the country, both in television and print media. In 1980, *Newsweek* identified the *Los Angeles Times* as one of the three best metropolitan newspapers in the country. In the electronic media, Los Angeles has become a top assignment.

The contributions of ethnic groups to the culture of the region are significant. Like San Francisco, the Los Angeles metropolitan area has an extensive mix of ethnic groups, and thus highly diverse lifestyles. Although the various ethnic cultures are mixed well into many areas of the Los Angeles Basin, there are some notable concentrations that have become largely synonymous with the various ethnic populations.

Watts is one of the sections clearly identified with the African American population. Receiving nationwide attention in the late 1960s for the riots that occurred there, Watts has subsequently attempted to achieve a more cohesive sense of African American identity, self-direction, and pride. Certainly, the annual Watts Festival and other events have focused attention on the contributions of African Americans to the culture of the state and the nation. And although the 1992 Rodney King riots flared strongly in Watts, community leaders exerted a concerted effort to try to prevent further damage to the residents and the reputation of the area.

East Los Angeles is dominated by the growing and important Mexican American community. The influence of this ethnic group is, of course, evident throughout California and the Southwest. East Los Angeles, however, has long been identified as a Spanish-speaking *barrio*, or community, and the area has provided a cultural haven for this population. And the growth of the Latino population in the area has accelerated with the influx of undocumented aliens who are drawn to this neighborhood by the familiarity of a Hispanic-based culture.

At the other end of the city is West Los Angeles, an area that reflects a strong Jewish influence. Delicatessens and synagogues are more evident here, and the names on stores and buildings reinforce this cultural identity.

Figure 9-9 *Buddhist temple in Hacienda Heights. Reportedly, this is the largest Buddhist temple in the United States.* (Richard Hyslop)

Many other sections of the greater Los Angeles region reflect diverse ethnic-cultural influences. A rapidly growing Southeast Asian population has swelled the size of the city's Chinatown and Little Tokyo and has also established itself in outlying regions such as Orange County. Indeed, these refugees were drawn in large numbers to such cities as Westminster, Garden Grove, and Santa Ana. By the 1990s, more Cambodians and Vietnamese had migrated there than to any other urban region in the United States and had established clearly identifiable ethnic enclaves for themselves (Figure 9-9).

Finally, Native Americans, immigrants from South Pacific islands, relocated English and Canadian citizens, and many others help provide the dynamic cultural diversity to be found in Los Angeles.

In all, Southern California has largely outgrown its provincial image and has moved to the forefront of the nation as a true center of culture, politics, economics, ethnic awareness, and power.

Beachboys, Boating, and the Body Beautiful: Southern Coastal Playgrounds

Perhaps the most common image of California, as seen in posters, movies, and other media, is the beach-and-seashore scene of the southern coastal playgrounds. A common scene is that of surfers (or lifeguards) dramatically swooping toward shore on foaming waves while bronzed and beautiful bodies lounge in the sand. The glamour of sunny California is also portrayed in pictures of trim sailboats lolling off the coast or bobbing lazily in tidy marinas.

This imagery is not created purely for export consumption; living within this particular culture has its effects on the residents as well. Frequently, the stereotype creates a sense of obligation to fulfill the expectations associated with the image. In spite of UV level warnings, what Southern Californian would not feel embarrassed to be seen pale and flabby in shorts or bathing suit? In fact, the region exudes a self-conscious pride in staying beautiful, bronzed, healthy, and trim. It is an

Figure 9-10 *Solvang, a picturesque, tourist-oriented town.* (Courtesy of Solvang Conference and Visitors Bureau)

attitude that permeates the culture and does not stop with youth, as casual observation at Venice Beach will quickly reveal.

The Southern Californian beach scene is a fascinating subculture with great outward display and little depth. There is a peculiar arrogance that encourages conspicuous displays of wealth and that rewards a tan acquired by wasting countless hours lying in the sun and by supporting tanning clinics that preserve the deep-baked tan. It is a wealthy, smug, sensual, gaudy, indulgent, and social subculture.

The area included in the southern coastal playground begins around Santa Barbara and extends southward along the coast through Los Angeles and Orange counties and into San Diego County. This long stretch of coast includes a remarkable variety of beach and coastal headlands, as well as an equally broad range of affluence and styles. Santa Barbara is a wealthy, exclusive, and self-contained community that retains a flavor of old Spain in its "Queen of the Missions" and its Moorish courthouse. Tidiness and order are the style here, and expensive ranchos and urban mansions are typical. The frantic pace of the big city is missing in both Santa Barbara itself and the surrounding countryside. The quaintness increases as one moves inland to the lush, coastal valley farmlands. The Scandinavian-flavored tourist town of Solvang (Figure 9-10) is just one additional interesting landmark of the Santa Barbara region, where stability, homely virtues, and good living are evident.

Other selected communities along the southern coastal shoreline share the exclusive, wealthy security of Santa Barbara. Corona del Mar, Lido Isle, Palos Verdes, and Emerald Bay take quiet pride in their expensive lifestyles. These are not transient, brash beach communities, but rather strongholds of the affluent.

In contrast to these orderly, quiet, wealthy beach towns are the popular beach areas where the fast life prevails. Malibu, Newport, Santa Monica, Venice, and Marina del Rey share a preoccupation with youth, parties, the body beautiful, and novelty. Sports cars, surfboards, sailboats, and sexual tension are evident in all quarters; transience and frantic motion are the norm.

Between these two extremes are various other styles of oceanfront atmosphere. Surfing, pier fishing, and oil wells share space with middle-class neighborhoods in Huntington Beach. Tidepools, tourists, art, and scuba diving typify such areas as Laguna Beach, Catalina, and La Jolla. All along the coast, high-rise condominiums and apartment complexes compete for space with exclusive restaurants and growing financial centers. The older beach areas of Seal Beach, Manhattan Beach, Dana Point, and San Clemente are feeling the impact of new money on old lifestyles, and shabby dwellings are being replaced by expensive residential and business edifices.

The continued growth and increased affluence have inevitably changed the face of the southern coastal area. The frantic pace at which residents party and dance, shop and dine, skate and boat, and try to make a buck has become the norm. Only the persistent pier fishermen and the rugged bluffs of Palos Verdes attest to the peaceful past.

Resort Mecca and Commercial Hub: The San Diego Region

Vacation, recreation, and retirement are the crucial draws of modern San Diego. They have made the city one of California's leading metropolitan areas while at the same time giving it a unique flavor. Building upon its climate, natural variety, and geographic location, San Diego made of them an art, a lifestyle, and a justification for existence. With one of the best ports on the coast, San Diego has a strong maritime orientation, with the extensive naval facilities exerting a strong social, economic, and political influence. Likewise, the port facilities have supported a significant shipping industry and a large commercial fishing fleet. The extensive naval and private marine activities, in turn, have helped foster related light manufacturing industries, as well as a respectable aerospace program.

Typically, this type of economic base is associated with a conservative outlook in the general population—and San Diego is no exception. A traditional no-nonsense attitude is evident in most of the cultural character of the area. Politically, the region has long been known as a conservative stronghold in the state. Economically, the *laissez faire* doctrine is still popular. Socially, the values of small-town America are highly cherished.

On the other hand, San Diego has a lighter side to its collective personality—a single-minded dedication to play! The same factors that encourage and support a maritime and naval presence have also created a strong recreational spirit. The seaside environment has nurtured numerous ocean-oriented pastimes and entertainments. Mission Bay has been developed into a posh and scenic locale. Here, Sea World and its tourist shows draw visitors by the thousands (Figure 9-11). In addition, the extensive marina, hotel, and restaurant facilities have made this area a major recreation center. Coronado Island caters to those who prefer Victorian-style buildings and fancy dining and lodging. Deep-sea sport fishing is a recognized tourist attraction. Recreational boating, water skiing, surfing, swimming, and skin diving round out the pursuits available in the area.

Figure 9-11 *Bottle-nosed dolphins somersaulting through the air at Sea World in San Diego.* (Sea World)

Location and climate are also part of the playground requisites of the San Diego region. Proximity to the Mexican border makes bullfights, *jai alai*, and foreign trade available to tourists. Mild weather enhances the appeal of scenic Del Mar racetrack, various golf courses, numerous state parks, famous Balboa Park with its Shakespearean Festival and museums, Torrey Pines State Park, world-famous San Diego Zoo, Wild Animal Park, Old Town, and many other attractions. The city and its surroundings cater to the tourist trade. The abundance of motels, hotels, restaurants, picnic grounds, shops, and amusement parks is eloquent testimony to its success.

Historically, San Diego has undergone some interesting changes. From its beginning as a sleepy mission and presidio, it has evolved into a more aggressive and sophisticated locale. In recent years, the expansion of the region has been dramatic, with much of this expansion occurring in outlying areas such as Escondido, Oceanside, and La Mesa. As a new rapid-growth area of residential settlement, the area has seen a shift in the average age pattern. Although retirement-age persons are still attracted to the region, the increase in young families is noticeable. This has created a rising demand for schools, playgrounds, and housing.

Indeed, one of the single greatest fears of many San Diegans is that the rapid growth will turn the region into a southern version of oft-reviled Los Angeles. In the 1980s, both San Diego County and San Diego City reached the number-two position in total population for counties and cities in the state (close behind Los Angeles County and Los Angeles City). The problems of growth have led to a strident antigrowth movement, which successfully pushed much of the new development outward along "transportation corridors," particularly in north county areas. The fact that much of the population growth comes from ethnic minorities (mostly Asian and Hispanic) has resulted in instances of racist rhetoric and behavior in the region, including highly publicized actions by white supremacist groups. It is abundantly clear, especially with the passage of the North American Free Trade Act (NAFTA), that this infusion of new ethnic groups will not stop. The critical issue for San Diego is how to blend social, economic, political, and cultural diversity into a more smoothly functioning whole—not unlike the state at large.

A final minor footnote can be found in some of the earlier pastoral nature of the region, which can still be found in the backcountry areas. Peaceful farm communities like Julian, Banner, Aguanga, and Santa Ysabel alternate with picturesque agricultural valleys and hillsides. However, even these areas are feeling constant pressure from growth and development. With the population projected to reach over 3 million people in the next decade, the area will continue to be dominated by constant expansion. Its role as residential and resort mecca will be augmented by its critical function as a natural hub for culture and commerce.

MAD DOGS AND CALIFORNIANS

To state that California has its unique and peculiar cultural phenomena is to express the obvious. Although this aspect of the state has been overplayed by eastern commentators, residents of the state do seem to take pride in forging new fashions and fads, social arrangements, and entertainments. Experimentation has never been a process feared in this state. When these experiments are successful, they are admired by outsiders. When they fail, they are cited as evidence of the basic looniness of Californians. What begins in the Golden State, however, frequently spreads to the rest of the nation and becomes part of American popular culture.

Creating Dreams: Disney, Knott, and Others

Technology has brought numerous benefits to American society. Greater leisure time is created with advancing technology. Likewise, technology usually brings a higher degree of affluence. With leisure and affluence, public literary and sophistication generally rise. The end in this chain of development is a demand for amusements to fill up the new leisure time. California has certainly met this challenge.

The forms that mass amusement and entertainment take in this state are wide-ranging. The traditional forms of television and movies remain popular, of course. In the major urban areas, cable and subscription television have become multimillion-dollar operations, providing first-run movies, sports and entertainment events, and soft-core pornography. But California also has become expert in providing less prosaic forms of amusement.

Theme parks are hugely popular attractions. Visitors have proved willing to spend millions of dollars to enjoy created experiences in nostalgia, adventure, the Old West, futurism, romanticism, or combinations thereof. Although the original Disneyland in Anaheim may be the best known of these parks, it is certainly not the only one. Knott's Berry Farm and Ghost Town (Buena Park), advertising itself as "America's oldest amusement park," offers a variety of adventures and experiences. Marriott's turn-of-the-century Great America in Santa Clara offers nostalgia and some amusement; Six Flags Magic Mountain in Valencia provides white-knuckle rides and experiences, as well as music and entertainment. Universal City Tours brings the movie world to the customer, or vice versa. Raging Waters in San Dimas and San Jose provide human-managed "natural" thrills. Other parks provide similar adventures in which technology and imagination combine to entertain customers (Figure 9-12).

A particular offshoot of theme parks is the wildlife-nature amusement park. In these seminatural settings, wildlife and sealife are displayed and often provide shows for the guests. San Diego has its Sea World and Wild Animal Park. The Bay Area has its Marine World Africa U.S.A. Such attractions typically attempt to create an illusion of natural setting, which distinguishes them from the more traditional and still popular zoos.

Still another type of popular amusement features historical themes. In many cities, major tourist attractions are "Old Towns" from the Spanish days (Sacramento, San Diego) or "roots" locations (Olvera Street, Cannery Row). Similarly, the California missions build upon this historical sense to draw tourists. Where true history may not be available, enterprising California businesspeople have created a subgenre of commercial *kitsch*—popularized, if not accurate or aesthetically appropriate, history. Here, stylized plastic history is merchandised in restaurant-shopping-entertainment complexes. San Francisco's Pier 39, Cannery, and Ghirardelli Square are joined by such consciously quaint tourist towns as Sausalito and Carmel-by-the-Sea and by Knott's Berry Farm's replica of Independence Hall and similar cultural expressions.

Finally, there are a variety of other forms of entertainment available in the state. These range from convention center presentations to concerts held in stadiums to programs put on by local theater groups and colleges. Elite culture can be found in opera houses, music centers, symphony orchestras, museums, and art shows. Various bars, country-western clubs, jazz halls, and rock concerts provide yet another dimension of the multilayered entertainment available in California.

Eating Your Way to Nirvana—By Railroad, Bistro, Pub, and Drive-in

Food is a special form of amusement in California. As with other forms of entertainment, this does not make the state different from the rest of the nation—except in degree. Based on the amount of time and space ded-

A

B

Figure 9-12 *Disneyland (photo A) and Knott's Berry Farm (photo B), two of California's major tourist attractions.* (A: © Disney Enterprises, Inc.; B: Knott's Berry Farm)

icated to eating, food apparently is one of the primal entertainment urges in American society, especially in its relation to all forms of recreation. Whether it be drinks at the concert, snacks at the game, or junk food at the mall, food seems to be a necessity for most recreation.

This becomes even more interesting when we consider that the act of eating itself has become major entertainment for huge numbers of people. By the late 1970s, as much as one out of every three food dollars was spent on dining out. According to *Nation's Restaurant News*, in the late 1990s over $290 billion was spent annually on dining out, and California accounted for a gigantic portion of this growth. Various factors have contributed to this growth, including increases in leisure time, growth in disposable income, higher percentages of single and childless adults, and the rise in the numbers of working women. California has some additional unique factors that have made dining out even more feasible and popular.

First, the weather usually does not pose any significant obstacle to an evening out. Mobility, an assumed and accepted part of California culture, facilitates the dining-out phenomenon. The casual and experiential lifestyle encourages culinary adventures; because Californians frequently seek or create novelty, the varieties of dining experiences are extensive.

Logically, in a state wed to the automobile, speed, and mobility, fast food restaurants head the list in popularity. The fast food establishments have been able to achieve such a prominent position because they fulfill certain needs: The menu is standardized and predictable; the food is relatively inexpensive; and the service is fast and, in many cases, does not even require leaving one's car. Because there are so many fast food spots, a wide variety of food choices are available.

The burger, fries, and soft drink establishments are king, with the Golden Arches leading the pack. California-style Mexican food is a major competitor, with chains selling what has been called "pasteurized Mexican food" for Anglo tastes. A variety of other fast food outlets offer fish, chicken, pizza, roast beef, hot dogs, a variety of Asian dishes, and other less familiar dishes. All share the vital characteristic of serving the lifestyle of a people always on the move who desire cheap, fast, and edible food (Figure 9-13).

Figure 9-13 *Typical California fast food outlet. The choice of fast food establishments and menus is almost unlimited—and while you eat, you might even be a part of a TV or movie filming.* (Greater Los Angeles Visitors and Convention Bureau)

Holding a more prominent position as entertainment venues are the vast array of theme restaurants. Here, the entertainment function or gimmick is often as vital as the food. Depending on whim, mood, and funds, diners can choose from railroad cars, sailors' havens, family boardinghouses, English pubs, ranchers' tables, mining grottos, gypsy dens, or plantation mansions. If you can imagine a theme, there is a restaurant in California catering to the concept. The key is the "experience" more than the food; if the food is tasty, so much the better.

California also reflects some unusual culinary styles that are made possible by its cultural and agricultural environment. Natural or health food establishments build on the idea of selling a California lifestyle. Thus, Golden State gastronomics support avocados, bean sprouts, alfalfa shoots, fruit, nuts, and grain as part of the healthy California image.

The plethora of ethnic groups in the state has made possible an ethnic restaurant business unsurpassed in the country. In Los Angeles alone, many hundreds of noteworthy ethnic restaurants compete favorably for the dollars of hungry diners.

The state's long coastline has made seafood a vital part of the diet. The choice of seafood restaurants up and down the coast is extensive, ranging from fancy, expensive bistros to stand-at-the-counter fish stalls.

Another category of serious eater is served by the dinner house approach, which usually features limited menu items and large quantities. Buffets also cater to this clientele by letting diners fill their plates with any number of salads or starches for a fixed price.

Plastic and neon are the identifying marks of another familiar category of restaurant. This is the coffee shop or café, which has the enduring advantage of always being open for business. For mobile Californians, this kind of 24-hour establishment provides an oasis in the night.

Another category is the cocktail and candlelight spot, which sells snob appeal and romance. These places can be recognized instantly by their French names or locations atop tall buildings. A toned-down version of this category is the fine-food-by-formula restaurant, where smiling, college-aged waiters and waitresses serve the specialty (turtle, lobster, or omelet) in a standardized "gourmet" decor.

Perhaps the most salient characteristic of the restaurant experience in California is the choice available. The state has made dining out a recreational activity equal to that found in any amusement, theme, or wild animal park. Speed, mobility, romance, ethnicity, variety, entertainment, convenience, and image may be purchased at a choice of restaurants usually within a half-hour's drive. All this variety, in addition to food, makes for quite a bargain. If, as the old saying goes, "We are what we eat," some interesting speculations could be made about California culture.

Isolating Age: Leisure World and Age Ghettos

Given the specialization of restaurants and amusement parks in the state, we should not be surprised to find specialized arrangements for living as well. One of the most noticeable of these is the segregation by age in

Figure 9-14 *The "safe haven" of Leisure World in Seal Beach: oasis for senior citizens or age ghetto?* (Ace Aerial Photography, Laguna Hills, CA)

communities around the state. In its compulsion to celebrate youth and beauty, California culture seems embarrassed by wrinkles, age, and grey hair. Thus has evolved a form of age ghetto and segregated living. It also seems that, once started on this process of segregating by age, people do not know where to stop.

Most identifiable throughout the state are the various restricted communities catering to the elderly. Formally structured closed societies of the aged, labeled euphemistically as "Leisure World," are found in several parts of the state. Here, locked safely behind security gates and fences, the elderly can pursue their interests free from outside distraction or bother (Figure 9-14).

Less structured, but equally real, are the informal age ghettos of mobile home parks. These parks are scattered in various locations throughout the state, including stark desert, cool seashore, and dusty urban jungle. Specializing in housing the elderly on fixed incomes, they provide a central gathering spot for the lower-income individuals. In these age ghettos, families can visit their grandparents on three or four special occasions a year but be spared the embarrassment and inconvenience of seeing them on a regular basis. Out of sight, the aged can safely be ignored, and California can perpetuate its myth of eternal youth.

Another interesting form of age classification is the over-40 adults-only townhouse and condominium development. Growing in popularity, these communities attempt to stratify and structure communal life in such a way as to exclude both the old and the young. The ideal residents are the mature couples who have achieved some measure of success, have moved past child-rearing age, and now seek to enjoy some of the pleasures affluence can bring. Together with other people of similar background and interests, these couples can effectively shut out the irritations of youth and childhood on one hand and senility and infirmity on the other. Of course, some of the segregated living patterns are not exclusively driven by age. Rather, the financial resources controlled by members of the over-40 group allow them to exercise selectivity in how and where they live.

The classic form of age segregation is the young adult or singles complex. Often recognizable by the fact that they are named after a tree (The Apple, The Aspen, The Pines), these dwellings are California's answer to the search for life in the fast lane. Characterized by relentless organized socializing, these communities provide a form of family security to otherwise independent youth. It is a snob appeal approach that promotes the healthy, free-and-easy good life of California—to those who live there, at least. Marriage usually spells the end of this residential period, because such a permanent relationship implies that one has become too sedate and staid for this community.

Finally, there is a growing form of age segregation that has not been sought by its residents: the community for families with small children. For those who can afford the rising costs of suburban living, new housing tracts frequently contain a high proportion of small children. For less affluent families, apartment or condominium complexes that will accept small children are becoming less common. In practice, those that do accept children eventually become age ghettos themselves, overcharging families for the privilege of living with small children. Although such segregated patterns have been challenged in the local courts, no definitive answer has been developed.

What the entire process of separating people by age implies about California is intriguing. The concept of the traditional nuclear family appears to be seriously undermined in the state. Although similar changes in living patterns have been observed throughout America, nowhere is it more apparent than in California. All the factors that might explain this shift are evident in the state, including mobility, technology, expanded recreation, and economic affluence. If California sets styles for the rest of the nation, here is one style in which it has moved far ahead of the rest of the country.

The Art of Arty California Burial

As might be expected, Californians have found a way to make even the process of death unique, unusual, and stylish. The traditional community cemetery is too tame for many in this state. Rather, California residents can experience the hereafter surrounded by rose gardens, art galleries, statuary gardens, and serene country meadows created especially for the benefit of the deceased. Price is no object!

The most familiar symbol of the California burial must be Forest Lawn. Planned in the early 1900s, this institution set the pattern that others would follow. Rather than conceding death, Forest Lawn and its descendants created a whole cycle of services ranging from marriage chapels to art galleries to final places of rest for the weary in "slumber rooms" and grassy knolls. In California, one is not necessarily buried in a cemetery; instead, one may be enshrined in a memorial garden or a heather-filled moor or a lawn of oaks or a devotion garden. How can anyone view such a prospect with terror, especially with the ever-present sun beaming down?

In a strange way, a California funeral can become a semigala event involving an automobile parade and a final chance to use the products for which one was so strongly encouraged to prepay. By selecting ahead of time, one can choose exactly how to spend eternity. This includes piped-in music for those who expect to be bored. Where else but in California would such an attitude exist?

There may be one sad side to the funeral business, however. It should be noted that memorial gardens are not limited to people. A thriving business in pet cemeteries marks the last word in devotion. Pet owners can buy permanent resting places for their dogs, cats, birds, horses, mice, hamsters, turtles, goats, and other exotic animal friends. With names like Good Shepherd and Pet Haven, these institutions cater to Californians' need to preserve the memories of faithful companions. The graves are frequently decorated with flowers, toys, and trinkets and are visited regularly by the lonely owners. In a culture dedicated to the pursuit of happiness, it is a fascinating commentary that many can find true friendship only with a pet.

California's Salad Bowl Culture

Concerning the ethnic makeup of the state, much has changed in the past 40 years. As recently as the 1960 census, the Anglo population constituted 92 percent of the total in California. By the 1980 census, that figure had dropped to 67 percent. Census figures for the 1990s revealed further shifts, with Anglos and blacks declining as a percentage of total population and Asians and Hispanics rapidly increasing their proportion. With the rise in numbers of different ethnic populations, Californians have become much more aware of their many contributions to the unique culture of the state.

In a state as diverse as California, the number of ethnic groups is large. In terms of total numbers alone, however, certain specific groups have become more influential and obvious in the affairs of the state. The largest single ethnic group is of Hispanic origins and includes Mexican Americans or Chicanos, Cubans, and other Latin Americans. Over the past three decades, this group has grown even more due to a significant influx of undocumented aliens who have snuck across the border between the United States and Mexico. The 1990 census showed that this group had grown by a rate of over 60 percent since 1970 (when the total California population grew at a rate of 18.5 percent and the Anglo population by less than 5 percent).

The Hispanic population tends to be clustered more in certain regions of the state. The highest concentrations of Hispanics can be found in the Los Angeles Basin, the San Diego region, and the San Francisco Bay Area. In part, this reflects the greater appeal of urban areas for Hispanics. The agricultural regions of Imperial County, San Benito County, and the San Joaquin Valley, however, have substantial Hispanic populations that are quite visible and influential in these areas. With its rapid growth rate, this distinctive and proud ethnic group has come to constitute 25 percent of the total California population.

The second largest ethnic minority group in California is of Asian derivation and includes residents of Chinese, Korean, Japanese, Vietnamese, and other Southeast Asian origins. The 1990 census revealed that this group was experiencing a significant growth rate. Although the San Francisco Bay Area, Los Angeles Basin, Orange County, and San Diego County account for significant concentrations, people of Asian background can be found throughout the state, primarily in urban settings. This consolidated category of Asian ethnic peoples now makes up 10 percent of the state's total population, a cultural force that will have increasing impact.

The 1990 census revealed that blacks, with 8 percent of the total, made up the third largest minority population in the state. This group saw an increase of 33 percent over the 1970–80 period but has declined in growth since then. As with most other ethnic groups, the black population tends to be found in greater numbers in the metropolitan regions such as Los Angeles or Alameda County. Although the traditional source of black immigration to California has been states in the Deep South, this pattern has broadened out somewhat in recent years. With its size and influence, the black population promises to be a force in the future of the state.

The last ethnic population of statistically important size in California is that of the Native Americans, a group that composes 1 percent of the state's population total. In California, these peoples are spread between discrete Indian Trust Lands (reservations) and urban locales. Holding approximately 450,000 acres of land in the state, the native California Indian population has begun to assert its identity and rights in various legal cases, public forums, and economic actions.

The categories noted here are, of course, overly broad clusters that reflect the needs of the Census Bureau. Subsumed within these categories are many other distinctive ethnic groups. These broad categories do, however, provide a convenient and simplified focus for discussion of the cultural contributions of non-Anglo Californians.

In spite of the fact that until recently the Anglo population exceeded 90 percent of the California total, the cultural identity of the state reflects many non-Anglo influences. One need only look around to see the impact of various ethnic groups on the life of California.

On the most obvious level, the physical landscape of California reflects many ethnic influences. Much of the architecture of the state is a tribute to these sources. The tile roofs, mission-style buildings, hacienda-theme houses and tracts, neo-adobe constructions, open-patio formats, and similar re-creations of the mood of early Mexican California abound. Likewise, Asian influences can be found in modified pagoda-style buildings and houses, in ornamental gardens, and in interior decor. In most major cities in the state, one can find echoes of many cultural styles in the buildings and dwellings.

Certainly, entertainment patterns and styles in California are shaped by various ethnic contributions. Music, restaurants, and dramatic productions can be found catering to every ethnic taste or desire. Thus, Mexican restaurants featuring mariachi music adjoin Jewish delicatessens, and Chinese theaters. *Ballet folklórico* vies for patrons with Japanese *kabuki* and African music festivals. Supermarkets and other stores stock a wide variety of ethnic foods and products, and certain neighborhoods in large cities specialize in uniquely exotic markets. Such specialized neighborhoods are particularly evident in the large metropolitan areas of the state.

Many of the businesses in the state reflect an ethnic character in their development and history. The Japanese and Hispanics traditionally have held key roles in the agriculture of California. The Italian influence is evident in the fishing, banking, and wine industries. Southeast Asians are increasingly assuming a larger role in the fishing industry. Native Americans control some important acreage in the state, including parts of Palm Springs, and play a significant role in California casino-style gambling. Entertainment and sports reflect the presence of a variety of ethnic cultures, with the growth in the popularity of soccer the most obvious example of this influence.

The growing impact of California's ethnic populations is also being felt in law, politics, literature, religion, life styles, language, and general culture in the state. If current demographic trends continue, California's culture will become even more multiethnic.

Finally, we should note the impact of general immigration into the state by foreign-born citizens. Historically, the countries that have contributed the greatest share of foreign born are, in descending order, Mexico, Canada, the United Kingdom, Ireland, Italy, Germany, and the former Soviet Union. Indeed, well over 25 percent of California's total current population is foreign born, and the pattern is holding firm. Such a continual influx has only served to enhance the dynamic and varied culture that has made California such a fascinating and ever-evolving society.

RECREATION OR ELSE

Although some of the more common forms of entertainment in the state were discussed previously, the more active forms have not yet been addressed. Sedate, family-style amusements are certainly vital, but Californians tend to shine most when they are actively pursuing physical excitement. Recreation is a huge business in the state, accounting for millions of hours of leisure and millions of dollars of profit.

Spectator Spectacles: There's More Than One Coliseum

Sports are big attractions statewide. In fact, California has been called the sports capital of the world. At the professional, college, high school, community, and individual level, athletics are widely popular throughout the state. Several things contribute to this advanced position in the world of sports. Climate permits almost year-round competition and activity. Affluence and mobility make most spectator and participant sports easily accessible to the majority of citizens. Some unique geographical characteristics expand the number of sports and recreation activities available. California can support any activity that desert, mountain, or ocean creates.

On the spectator level, there is much to choose from. The number of stadiums, arenas, speedways, parks, courses, and tracks in California is stupendous. Big cities and small towns alike offer events loyally patronized by fans. Professional sports draw wide followings in football, baseball (Figure 9-15), basketball, soccer, auto and boat racing, boxing, bowling, golf, horse racing, hockey, tennis, rodeo, and some lesser sports. Alumni, student, and popular support for amateur athletics makes possible extensive athletic programs at thousands of schools, colleges, and universities across the state. If spectator sports are one's particular love, there is no shortage of choices to that need. In person, or by television, radio, or cable transmission, interested Californians can satiate their appetites.

Figure 9-15 *Dodger Stadium.* (Los Angeles Dodgers)

Active Play: Everybody Is a Star

If spectator sports and activities are a passion, participant recreation is a compulsion. In a culture in which awareness of the body is a fetish and concern for health has become a religion, active play is a necessity. Happily, California can provide almost any recreational experience imaginable.

One broad category of active recreation involves the "cult of the body." This involves those activities that call for displays of the body and awareness of the strenuous exercises needed to keep in shape, including aerobics, weight training, and body sculpting. The widespread appeal of this is evident in the multimillion-dollar industry that has sprung up to satisfy these needs. Health clubs, spas, racquetball courts, and athletic clubs can be found almost any place where busy executives, workers, housewives, and students can slip away to the club for an hour's strenuous workout.

Another form of popular activity utilizes the wealth of natural settings available in the state. Numerous national and state parks offer opportunities to rough it in the open air. Hiking, swimming, camping, fishing, boating, and hunting can be pursued in forests, wilderness areas, deserts, and mountains all over the state. One can climb Mount Whitney or brave Death Valley, explore Yosemite National Park or wander through Lava Beds National Monument, ski at Mammoth or Tahoe or houseboat on the Sacramento River Delta. Between the

Figure 9-16 *John Muir Wilderness Area outside Big Pine in the western Sierra, a popular hiking and camping destination.* (Richard Hyslop)

federal and state public lands and waters, residents and tourists have a wide choice of natural settings to explore and enjoy (Figure 9-16).

A more common and accessible form of recreation familiar to California is the localized sports activity. Many Californians regularly schedule weekly golf games or tennis matches. Even more can be seen during early morning or late evening in their jogging togs. Neither rain, sun, smog, Thanksgiving, nor Christmas can stop these devotees from doing their daily miles. The relative importance of these sports can be assessed easily by comparing the price of golf, tennis, or jogging shoes with normal street shoes. That certainly reinforces the importance attached to recreational activities pursued by the majority of Californians.

Of course, other forms of active recreation are available. With as many tastes, styles, personalities, and preferences as exist within the state, no one needs to feel ignored. In the area of recreation, too, California is a culture of diversity. And, as we will see in Chapter 10, this remarkable diversity extends to agricultural California.

CHAPTER

10

The Farm: Agricultural California

The most remarkable attributes of the contemporary California cultural landscape are farms and cities. By any measure, be it value, yield, or variety of product, California farm output is unsurpassed among the 50 states. This agricultural dominance derives largely from the productivity of some 12.5 million acres of *prime land*—land having obvious locational and physical advantages such as proximity to urban markets and services, relatively level or gentle topography, good drainage, ample water supply, fertile soil, long growing season, optimal microclimatic condition, and minimal development costs.

At the same time, with 33 million people, California claims the nation's largest and most urban population: Nine out of every 10 Californians live in cities and towns of 2,500 or more population and consequently are classified as *urban places* by the U.S. Bureau of the Census. In the context of spatial expansion, urban California manifests itself in a scattered, hit-and-miss sprawl over the landscape—a housing tract here, a shopping center there, an industrial park somewhere else, and patches of undeveloped land and freeways everywhere. California cities seem to grow in every way imaginable, except compactly.

This *urban sprawl* might not be worth getting too excited about were it not for the fact that it is absorbing copious amounts of a precious and scarce resource: the state's 12.5 million acres of prime agricultural land. Estimates vary widely, but some place the rate of conversion of prime land from agricultural to urban use at close to 50,000 acres a year. For the reasons just cited and as Figure 10-1 shows, urban development and agriculture generally compete for the same prime land. But in a relatively free land market economy like California's, where urban land uses are considered "highest and best" (actually, most valuable in generating revenue), agriculture almost inevitably loses the competition.

Given the magnitude and geography of both agriculture and urbanization in California, it is little wonder, then, that the conflict between the two is escalating. In fact, the impact of rapid urbanization on land currently devoted to intensive irrigated agriculture may well be the single most significant problem facing the state's natural resource complex. The exceptionally favorable combination of landform, climate, and soil that is responsible for California's intensively cultivated, highly valued agricultural commodities is a limited and high-demand resource not only to the state and the nation but to the world as a whole. Once the paving of California reaches the point of rendering the state a net importer of food and fiber, then what? The answer is painfully clear: A world already suffering from hunger in too many places will have lost its single most productive agricultural region. Save for the nagging probability of droughts, there is no prime agricultural environment comparable to California's anywhere on earth. It seems unconscionable that this bountiful resource is slowly but surely being eaten away.

So that these complex issues may be better understood, this and the next chapter will examine land utilization on California's farms, in its cities, and in between in the rural-urban fringe.

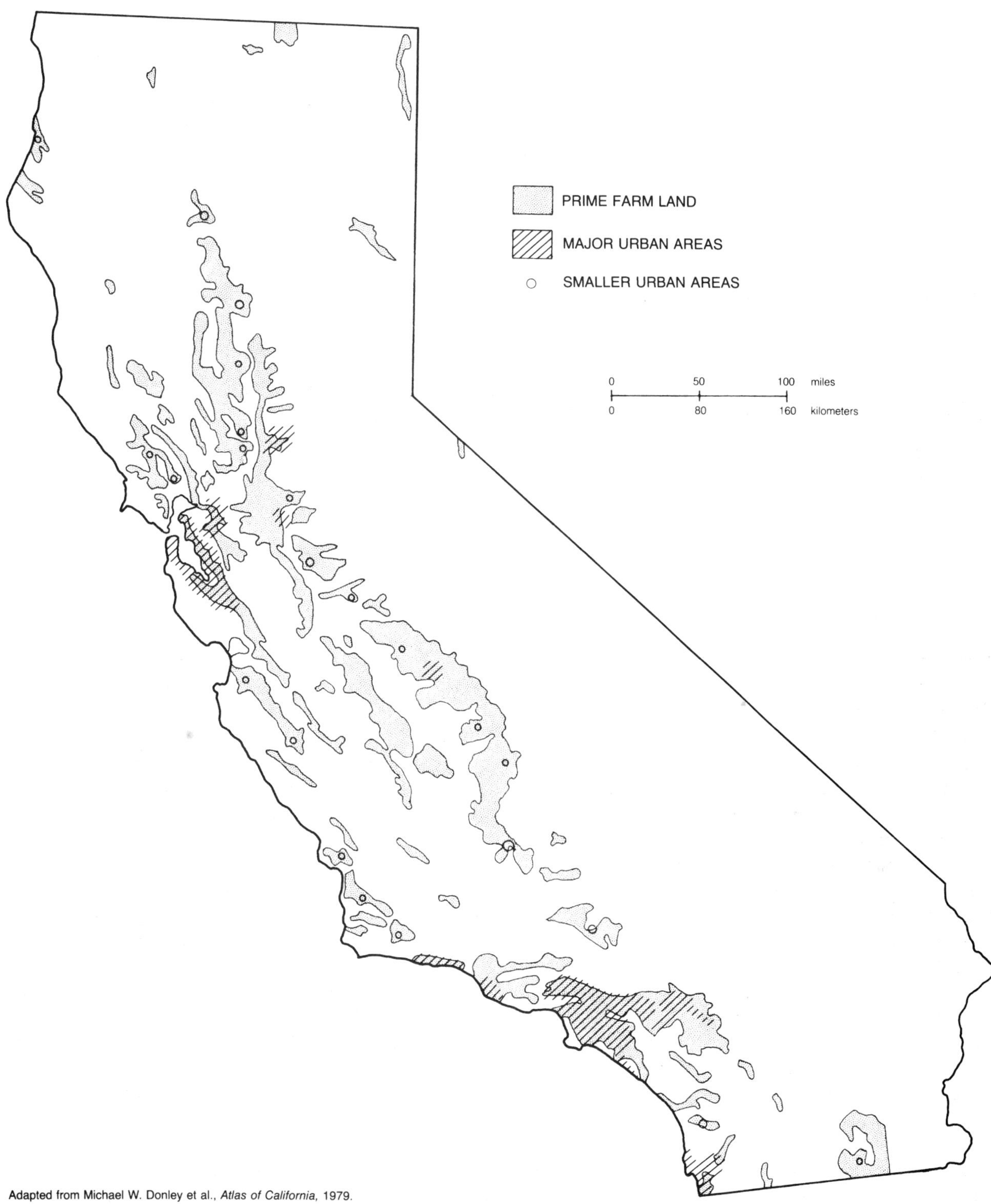

Figure 10-1 *Prime California farmland and urbanization.* (Richard Crooker)

AGRICULTURAL RECORDS, RESOURCES, AND COMMODITIES

The record-setting performance of California agriculture over the years can be viewed in a variety of ways. As Table 10-1 shows, California, with its farms producing nearly $25 billion worth of commodities in 1996, led the nation in agricultural output for the fiftieth consecutive year. Texas was second with $13.1 billion worth of farm output, followed by Iowa at $12.9 billion, Nebraska at

TABLE 10-1 Leading California farm commodities and counties by value, 1996.

COMMODITY	RANK	STATE TOTAL VALUE (BILLION $)	FIVE LEADING COUNTIES BY RANK AND PERCENTAGE OF STATE TOTAL (% OF VALUE) 1	2	3	4	5
Milk and cream*	1	3.54	Tulare 20.10%	San Bernardino 12.80%	Merced 12.50%	Stanislaus 10.80%	Riverside 9.10%
All grapes*	2	2.83	Fresno 19.60%	Tulare 15.00%	Kern 14.20%	San Joaquin 8.60%	Madera 7.80%
Nursery products*	3	1.66	San Diego 21.80%	Los Angeles 9.30%	Orange 7.80%	San Mateo 6.90%	Ventura 5.40%
Cotton lint	4	1.24	Fresno 37.30%	Kern 24.00%	Kings 17.00%	Tulare 7.90%	Merced 7.70%
All cattle and calves	5	1.15	Tulare 20.00%	Imperial 13.30%	Fresno 10.30%	Kern 6.30%	San Bernardino 5.70%
Almonds*	6	1.13	Kern 17.00%	Merced 15.70%	Stanislaus 14.90%	Fresno 11.50%	Madera 9.20%
Lettuce*	7	0.92	Monterey 55.50%	Imperial 11.60%	Fresno 9.40%	Santa Barbara 5.20%	San Luis Obispo 4.20%
Oranges	8	0.78	Tulare 50.70%	Kern 19.70%	Fresno 12.00%	Ventura 5.10%	San Diego 5.00%
Alfalfa hay*	9	0.75	Imperial 16.60%	Tulare 9.90%	Kern 9.30%	Merced 8.70%	Fresno 7.40%
Flowers and foliage*	10	0.69	San Diego 47.40%	Santa Barbara 9.30%	San Mateo 7.80%	Monterey 6.00%	Santa Cruz 5.10%
Tomatoes (processing)*	11	0.61	Fresno 33.70%	Yolo 19.40%	Colusa 9.50%	San Joaquin 7.80%	Solano 6.40%
Strawberries*	12	0.54	Monterey 33.20%	Ventura 26.10%	Santa Cruz 15.30%	Santa Barbara 11.80%	Orange 6.00%
Broccoli*	13	0.41	Monterey 55.50%	Santa Barbara 15.60%	San Luis Obispo 7.40%	Imperial 4.70%	Ventura 4.30%
Rice (excluding seed)	14	0.40	Colusa 23.40%	Butte 21.10%	Glenn 17.20%	Sutter 17.20%	Yuba 7.40%
Eggs (chicken)*	15	0.39	Riverside 30.00%	San Diego 16.70%	Stanislaus 15.60%	San Bernardino 10.80%	Merced 10.30%
Peaches (all)*	16	0.35	Fresno 33.80%	Tulare 18.00%	Stanislaus 9.80%	Sutter 7.60%	Merced 6.50%
Walnuts (English)*	17	0.32	San Joaquin 20.60%	Stanislaus 14.00%	Tulare 10.90%	Butte 9.50%	Sutter 7.80%
Lemons*	18	0.32	Ventura 62.70%	San Diego 8.30%	Riverside 7.80%	Tulare 6.20%	Kern 5.80%
Wheat (excluding seed)	19	0.29	Imperial 19.90%	Kern 10.40%	Tulare 8.50%	San Joaquin 7.00%	Solano 6.90%
Chickens	20	0.28	Merced 45.90%	Stanislaus 44.50%	San Bernardino 3.30%	Madera 1.90%	Sacramento 1.10%
Avocados*	21	0.26	San Diego 43.20%	Ventura 22.80%	Santa Barbara 12.50%	Riverside 11.80%	Orange 4.10%

Annual total: $24,788,855,000

* Commodity categories in which California is the nation's leading producer.

Source: California Department of Food and Agriculture, *California Agricultural Resource Directory, 1997* (Sacramento, December 1997).

$9.5 billion, and Illinois at $9 billion. Not only are Texas or Iowa usually a distant second or third, but no other state in the Union produces the variety and yield of crops California does.

The physical resource base for the state's national leadership in the production of 75 different commercial crop and livestock commodities is found in an abundance of prime valley flatlands, proximity to the ameliorating influences of the Pacific Ocean, lengthy growing seasons, extensive latitudinal range of climates, and other environmental amenities discussed in Chapters 3–7. Some of the state's agricultural environments are unique to the point of rendering California almost the sole producer in the nation of more than a dozen major crop commodities. Obviously, the conversion of prime land to nonagricultural uses in many of these rare environments is of the utmost concern.

In all, 250 different types of crops, including seeds, flowers, and ornamentals, are grown in California, and yields per acre for many of them are unmatched anywhere else in the world. California annually accounts for about 11 percent of the total value of U.S. agricultural production and by itself exported nearly $12 billion worth of food and fiber in 1996 to other states and nations. Agricultural exports each year make up about one-eighth of the state's total value of exports and thus significantly reduce a trade deficit caused by the import of largely nonagricultural products. Agriculture's role as the most valuable primary industry in the state looms all the more prominent when California's position as the seventh largest economy in terms of production and export in the world is brought to mind.

Production costs, and thereby net income, seems to vary more in farming than in most major California industries. Weather appears to influence the situation more than any other environmental, economic, or human influence factor. Although the Mediterranean fruit fly, or Medfly, crisis at times took precedence throughout the 1980s, and the poinsettia whitefly assumed prominence in the early 1990s, the latter destroying some $80 million worth of crops and idling 2,500 farm workers in the Imperial Valley late in 1991, weather remains the key. In the last year of the 1975–77 drought, for example, production costs soared as a result of expanded well drilling operations and increased investment in irrigation equipment. Consequently, for the first time in many years, net farm income in the state declined slightly. Increased energy, fertilizer, labor, machinery, and pesticide costs also contributed to the 1977 decline. It should be noted here that many fertilizers and pesticides, as well as fuels, are petroleum based, and thus energy costs as a whole will play a greater part than ever before in determining farm profits and losses. In 1978, however, the rains returned and so did farm profits, registering an estimated $3.06 billion out of $10.4 billion in gross sales. Although much protracted compared to the 1975–77 dry spell, the 1986–92 drought caused relatively little harm to irrigated agriculture through most of its first 5 years. It was primarily dry farming that suffered reductions in profit margins, mostly among cattle ranchers and dryland grain growers who had no way of compensating for lack of precipitation. However, as discussed in Chapter 4, the recent drought, which had already stressed many tree crops, was devastatingly amplified by the advent of a killer frost throughout the state in late December 1990. Although farm revenue grossed a record $18.9 billion in 1990, that year's late freeze caused a $209-million drop in orange revenues in 1991, mostly in Tulare, Kern, and Fresno counties (see Table 10-1). The citrus decline, coupled with losses from the whitefly infestation in the Imperial Valley and drought-caused cutbacks in irrigated cotton production in Fresno, Kern, and Kings counties (see Table 10-1), resulted in a rare annual decrease in overall gross farm income to $17.9 billion in 1991.

Perhaps of greater concern in a prolonged drought, though, is mounting pressure on agriculture to relinquish some of its lion's share of California's water resources. As noted in Chapter 5, agriculture annually accounts for 79 percent of total net usage in the state, and that proportion is expected to shrink to 75 percent by 2010. Municipal water users would like to see not only a sharper reduction for agriculture but also greater equanimity in water pricing for all users. The oft-cited gap between rice growers paying a few dollars per acre-foot of water and city dwellers shelling out upwards of hundreds of dollars per acre-foot aside, farmers are quick to warn of higher prices, reduced availability, and diminished quality for their products if their share of the state's water resources are cut. Moreover, they complain that municipal water districts reward their conservation efforts by raising the price of water to compensate for reduced revenue. This scenario became all too real as the drought dragged on into 1992 and Californians were relying on groundwater to supply 60 percent of their annual needs instead of the typical 40 percent.

To talk solely of gross cash receipts and net profits from the sale of farm products is to understate agriculture's broad impact on the California and national economy. As with any other industry, agriculture not only consumes goods and services from other sectors of the economy but also supplies them. For instance, when farmers purchase harvesting machinery and have it serviced regularly, their income is added to the income of the suppliers of these capital goods and services, who, in turn, increase their consumption outlays because of higher income. Thus, a chain reaction of spending and respending, known as the *multiplier effect*, is set off in the

TABLE 10-2 Number of farms, land in farms, and farm size in California, 1950–1994.

YEAR	NUMBER OF FARMS	LAND IN FARMS (MILLION ACRES)	AVERAGE FARM SIZE (ACRES)
1950	144,000	37.5	260
1960	108,000	39.0	359
1970	64,000	36.6	572
1975*	73,000	34.3	470
1980	81,000	33.8	417
1985	79,000	32.9	416
1990	85,000	30.8	362
1994	76,000	29.5	388

* The new definition (since 1975) of a farm in the U.S. *Census of Agriculture* is a place with annual sales of agricultural products of $1,000 or more.

Sources: California Department of Food and Agriculture, *California Agriculture* (Sacramento, 1992, 1995), and U.S. Bureau of the Census, *Census of Agriculture* (Washington, DC: U.S. Government Printing Office, annual).

form of increased demand for consumer goods and services.[1]

Agricultural economists have worked out quite different multipliers for crop agriculture and livestock agriculture; the two, however, average out to about 2.7:1. If this ratio is applied to the value of agricultural production in 1996 of $24.8 billion, it could then be said that the monetary impact of California agriculture that year was more than $67 billion. Such an estimate is rough at best and should be scrutinized in the light of economic variables such as propensity to consume, propensity to save, and taxation. But however it is expressed for an industry that year after year sets new income and expenditure records, taking into account agriculture's multiplier effect unquestionably presents a more holistic view of its economic impact.

Still another ancillary aspect of agriculture that impacts the state economy is *value added by manufacture* as is created by the multibillion-dollar-a-year food processing industry. Along with aerospace and electronics, food processing shares the manufacturing spotlight and for the most part derives its raw materials from California farms. Value added by manufacture of food and related products in 1995 in California amounted to $20.4 billion. During the recession of the early 1990s, employment in the food processing industry fell from 182,000 to 169,500 (1989–95), but not nearly as precipitously as in the aerospace industry, where employment dropped from 693,000 to 467,000 (1990–95).

Other indicators of California's agricultural success are found in a comparison of the number, size, and ownership characteristics of its farms with those of the nation as a whole. The nationwide trend to fewer and larger farms continues unabated, as it has now for decades, but the proportional changes are significantly different for California compared to the rest of the country. By 1990, the total number of farms in the United States had declined by more than 11 percent from the previous decade, from 2.44 to 2.16 million. As shown in Table 10-2, the number of farms in California dropped 41 percent in the four decades since midcentury, from 144,000 in 1950 to 85,000 in 1990, and by 1994 was down to 76,000. From 1950 through 1994, acreage in farms in the state shrank from 37.5 to 29.5 million while average farm size rose from 260 to 388 acres. Of California's total area of 100 million acres (158,693 square miles) in 1994, agriculture claimed 29.5 percent, and all other users 70.5 percent.

Farmland is far less vested in individual or family ownership in California than in the United States as a whole. Families are in charge of a little over half of the farm acreage in the state, which is well under the three-quarters they own and operate nationwide. Most of the rest of the state's farmland is held by limited partnerships and corporations. Corporate farming operations with relatively vast capital resources and huge economies of scale unquestionably have outcompeted the small family farm and thereby contributed to its demise, but this long-standing trend may be weakening. The U.S. Supreme Court's June 1980 ruling regarding the quarter-section (160 acres) limitation on the use of federal reclamation project water may be abetting the revival of smaller farm units. One possible indicator of a resurgence of family farming in California was the rise in the number of farms between 1975 and 1990, from 73,000 to 85,000 (see Table 10-2). In any case, as of 1990, whoever owned California farmland, its value by the acre of $1,753 was far above the national average of $693.

At the very heart of California agriculture's unprecedented productivity are its human resources, which represent a marshaling of expertise from as far away as Europe and Asia and from as near as Mexico and the Midwest. In 1995, nearly 300,000 Californians, or about 2 percent of the total work force, were employed in agricultural production and services: two-thirds as laborers, one-fourth as owner-operators and tenant farmers, and the remaining one-tenth in managerial and research and development roles.

Much of the field labor force migrates from place to place, depending on what seasonal harvest is at hand. Also, the ranks of the field work force swell considerably

[1] Technically, the multiplier effect is defined as the ratio of a change in output to a change in *aggregate demand*, or the total flow of cash expenditures in an economy during a year or other specified time period.

with both legally resident green card holders and undocumented alien workers during peak harvest seasons—there would undoubtedly be a serious labor shortfall if these people were not available for this back-breaking form of stoop labor. But passage of the Agricultural Labor Relations Act, increased mechanization, and other events of recent decades resulted in an increasingly industrial pattern manifesting itself in the farm labor market in the 1990s. Thus, even though seasonal demands for labor will continue with some crops, farm laborers in general will be more skilled, more steadily employed, and more organized. An unfortunate aspect of this emerging industrial pattern is the increasing displacement of farm workers by farm machinery and the attendant social costs, an issue we will address again as we next discuss the "what, where, and why" of California's major agricultural commodities.

LIVESTOCK PRODUCTS AND FEED CROPS

Since the dawn of commercial agriculture in California, livestock products have been the state's most valuable farm commodities. Two out of six of California's billion-dollar-a-year commodities, its first- and fifth-ranked commodities by value, and one-fifth of its agricultural revenues come from two livestock products: beef and milk. In 1996, beef cattle and calves grossed $1.15 billion in sales and milk and cream $3.5 billion (see Table 10.1). California has finally passed Wisconsin as the number one dairy state. California's other important livestock products are chicken eggs, averaging nearly $400 million in sales a year, and chickens (fryers and broilers) and turkeys, each grossing $250 million annually. California does lead the nation in egg production and is usually first or second in turkey raising. Because of a large local population of consumers, isolation from potential markets in the Midwest and East, and perishability, there is little export of any of California's various livestock products.

Livestock production of one sort or another exists in practically every county in the state. About 5 million head of beef and dairy cattle and 1 million sheep and lambs are likely to be found in California in any one year. This ubiquity owes partly to livestock's and feed crops' greater tolerance of marginal climatic, landform, and soil conditions than is the case with most of California's fruit, nut, and vegetable crops. For instance, Tulare County's hot, dry summers detract little from its being first in the state in feedlot beef cattle production. Likewise, Mono County's rugged Sierra rarely prevents range cattle from finding high-country summer pasture (Figure 10-2). If nothing else, such harsh environments usually do have plenty of nonprime grazing land. Add water, either by importing it via aqueducts or by drilling wells, and almost any habitat becomes suitable for raising livestock.

Although almost every county has sheep, Kern, Solano, Fresno, and Imperial claim the largest herds, with several hundred thousand head between them. Despite the impressive numbers of animals, wool and lamb

A

B

Figure 10-2 *Range cattle (photo A) and sheep (photo B) in Mono County.* (Crane Miller)

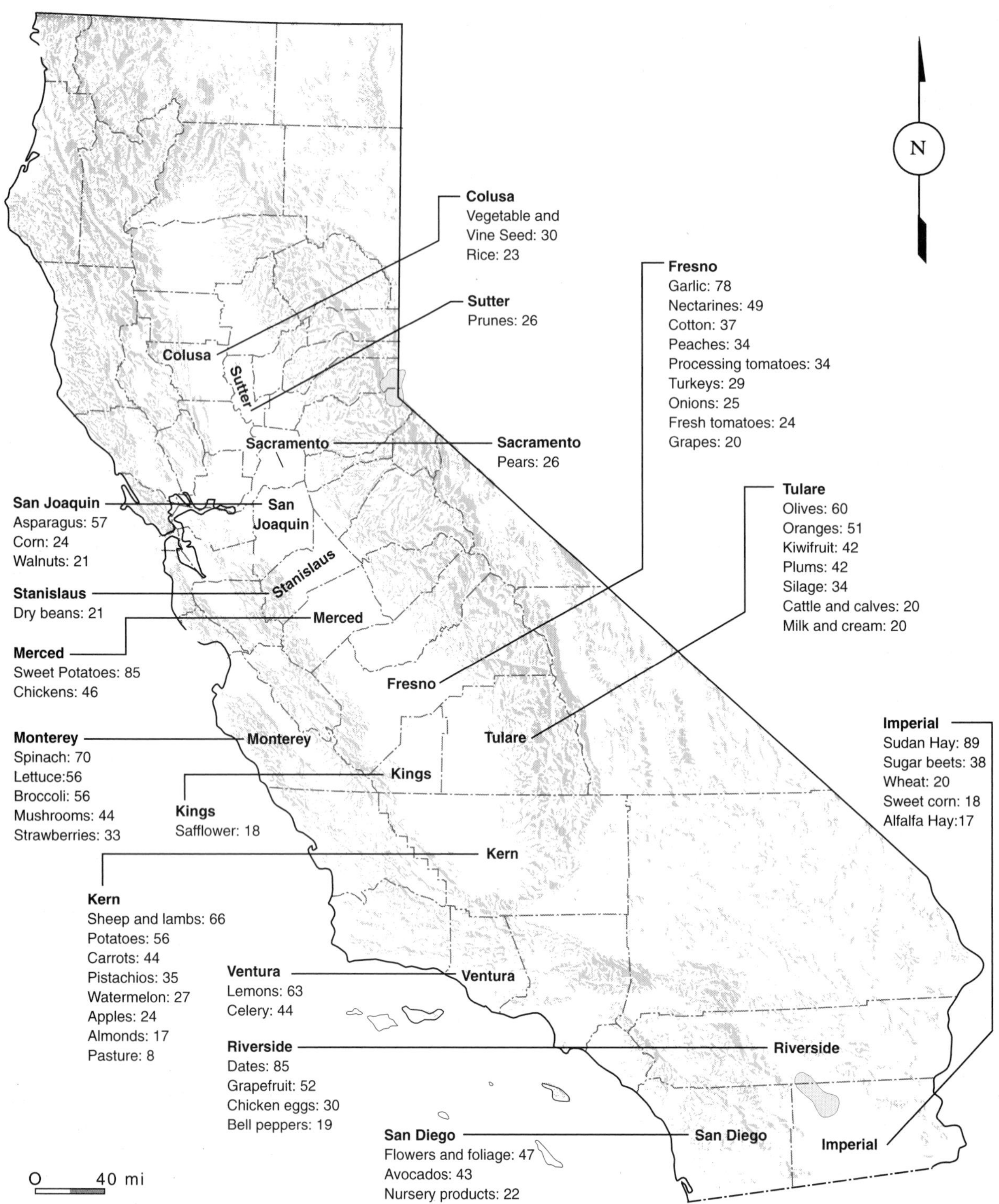

Figure 10-3 *Leading California counties in selected farm commodities, 1996. Percentage of total state value in 1996 is listed after each commodity.* (Source: Calif. Dept. Foods & Agric., *Calif. Agric'l. Resource Directory*, 1997.)

are not among California's more important farm commodities.

Actually, the fact that grains, hays, and other feeds are grown in all but a handful of counties in the state is the real secret to the widespread success of the livestock industry. Figure 10-3 illustrates the strong spatial ties between feed crop growing and livestock raising in the state. It's no accident, for example, that Imperial and Tulare counties are first and second in alfalfa production.

Another case in point is the concentration of dairy farming in Riverside and San Bernardino counties, where feeds are major local cash crops. Inclusion within the greater Los Angeles metropolitan area market has also contributed to San Bernardino's position as the leading milk-producing county by value in the United States for much of the 1980s and 1990s; however, in 1994, Tulare County assumed the lead position with Merced County coming in third behind San Bernardino County. A year earlier, and for the first time since 1915, California pulled ahead of Wisconsin as the nation's leading dairy state. With a higher risk of perishability in the transport of fresh milk and cream, the dairy industry tends to be more market oriented in location than the beef industry. Although some final feedlot fattening or *finishing* of cattle near slaughterhouses in the Bay Area and Los Angeles County does occur, it is not nearly as extensive as in midwestern cities.

The egg industry appears to favor metropolitan fringe production sites as well, with almost one-half ($393 million in 1996) of the entire state output of laying hens coming from Riverside and San Diego counties alone. Even so, chicken, turkey, sheep, and a large share of the state's dairy and beef production is concentrated in a dozen or so Central Valley counties, primarily because of abundances of locally grown feeds and irrigated pasturelands.

Alfalfa, other hays, barley, field corn, sorghum, and several other field crops form a raw materials base for the livestock industry that covers nearly half of all the cultivated land in California. In national commodity rankings by value, California is either first or second each year in hay production and second only to North Dakota in barley production. Although droughts diminish production, downpours wipe out whole crops, and inflation does its part, hay and barley annually bring in nearly $1 billion. At these rates, hay ranks as the seventh most valuable crop in the state. Nearly 90 percent of the state's annual barley crop is fed to livestock, with the remaining 10 percent being used in the brewing of beer and the making of other malt products.

Corn, most of which is used directly as grain feed or stored green as *silage* for later use as fodder, annually grosses just over $100 million. San Joaquin County is the leader in grain and seed corn production, accounting for one-fourth of sales and acreage (55,000 acres). Were California ever to give Iowa, Illinois, Indiana, and Minnesota a run for their money in hog production, corn would rank higher in the state. Sweet corn grosses about $25 million annually, with Riverside County accounting for about one-third of statewide production.

FOOD AND FIBER FIELD CROPS

Field crops, by definition, occupy vast acreages, yield best when irrigated, are generally sown and harvested by machine, and in California include all the aforementioned feed crops plus cotton, wheat, sugar beets, rice, Irish potatoes, dry field beans, and safflower. Cotton is California's third most valuable crop and one of only four billion-dollar-a-year crops. As such, cotton is the state's fourth-ranking agricultural commodity by value. Nationally, California is second only to Texas in cotton production.

Sugar beets and rice also loom large in the national picture, with California sometimes ranking first in year-to-year production of both commodities. In the growing of Irish potatoes, originally domesticated in the Peruvian Andes, California vies for second place with Maine and Washington while Idaho leads the nation in production. A century ago, California was the nation's leading wheat producer and flour exporter, but today it does well to place in the top 10 states each year. Higher yielding and more remunerative by the acre, irrigated field crops such as alfalfa, cotton, and rice have simply outcompeted wheat for prime acreage. The state's two major cereal grains, wheat and barley, continue to be *rain-fed* (dry farmed). Safflower, a thistlelike herb, yields both a low-cholesterol oil and red dyestuffs; the oil is currently very popular among dieters.

Cotton: King of the San Joaquin

A unique interaction of natural and human forces in the San Joaquin Valley has promoted the rise of cotton as king of California crops. Cotton growing is limited almost exclusively to the southern San Joaquin Valley, which accounts for about 90 percent of total state production. Fresno, Kern, and Kings counties (the leaders in that order both statewide and nationally) alone contain 800,000 acres of cotton.

Four facets of the physical geography of the southern San Joaquin Valley have contributed immeasurably to the success of cotton in the region. First, this southern part of the valley is essentially a basin of interior drainage in which the Buenavista and Tulare Lake basins have acted for millennia as sumps for the Kern, Kings, and other Sierran rivers and in the process collected a wealth of rich, loamy soils. Second, local surface runoff and groundwater resources (see Chapter 5) are adequate to meet irrigation demands, except in times of severe drought. Third, the dry, subtropical, summer climate of the region allows both a long growing season and a rainless harvest period. Finally, there was and still is relative freedom from the boll weevil and other pests.

Human ingenuity has added the final touch to modification of the natural landscape, resulting in an optimal environment for growing cotton. The initial reshaping of the land began a century ago with the building of levees by local farmers to reclaim the primeval lake beds for agriculture. Starting in the 1930s, the federal government stepped in, first with the Bureau of Reclamation's building of the Friant–Kern irrigation canal into the region as part of the Central Valley Project (see Chapter 5) and later with the Army Corps of Engineers' damming of the Kings (Pine Flat Reservoir) and Kern (Lake Isabella) rivers for purposes of flood control and water storage. The reservoirs and the Friant–Kern Canal supply thousands of small cotton farmers with irrigation water. But, again, the big growers had to lobby for an exemption from the 160-acre limitation on the use of federal reclamation project water or sell off most of their acreage in order to get to the limit. Despite such government help or hindrance, privately inspired flood control work continues, as witnessed during the 1969 floods when one large farming firm lined a Tulare Lake levee with several thousand wrecked, accordianized cars to keep it from being washed away.

Cotton cultivation and marketing date back to the early history of California. Originally domesticated hundreds, if not thousands, of years ago in Central and South America and perhaps in India as well, *Gossypium* was introduced to Alta California in the eighteenth century by Franciscan padres with plantings at several mission sites. But commercial cultivation and export of California cotton did not begin until the Civil War period, when the state government offered prizes for crops grown for the first time in the region. Once the Confederacy rejoined the Union, the demand for California cotton diminished. Worse yet, other crops proved more remunerative, and cotton all but disappeared from California until well into the twentieth century.

With reclamation of the Tulare lake bed already under way, the 1920s saw a resurgence of cotton growing in the southern San Joaquin Valley that persists to this day. Development of the Acala variety of upland cotton, which annually yields half a ton or more per acre in the lake bed *loams* (sand and clay soils high in organic content) and dry, warm climate of the San Joaquin Valley, was another factor in cotton's revival. Acala cotton bolls take well to machine harvesting, and since the 1950s, mechanization in the cotton fields has resulted in the elimination of an estimated 100,000 pickers' jobs.

Cotton *ginning*, or separation of the lint fiber from the seed, takes place in nearby towns such as Tulare and Corcoran. The lint is then baled and sent on its way to distant textile manufacturing centers in the southeastern United States and Japan. California's cotton is relatively clean, which has helped make the state number one in cotton gin throughput: 15,000 bales per year per gin compared to the number two state's (Arizona) annual throughput of 7,435 bales. The seeds are pressed into cottonseed oil and meal, which appeals to a multitude of buyers ranging from paint manufacturers to feedlot operators. Obviously, raw cotton leads all other field crops in value of export. (Rice is the only other significant export among California field commodities.)

Figure 10-4 Intercropping *(two or more different crops growing simultaneously) on the Oxnard Plain. Here, a lima bean crop is hosting soil nitrogen replenishment for a lemon grove.* (Crane Miller)

The state's cotton may find its way back to the garment manufacturing district in downtown Los Angeles or yard goods outlets throughout California. But its conversion to textiles within the state has been preempted by the lack of any long-standing textile manufacturing tradition, relatively high industrial labor costs, and dry climate. By contrast, along the Carolinas' Fall Line, where Atlantic-bound streams fall from the Appalachians and the Piedmont to the coastal plains, hydropower is always readily at hand, and high humidity facilitates the spinning and looming of fabrics such as cotton sheeting and denim. Nevertheless, California fed a virtual flood of raw material to the designer jean and urban cowboy fads of recent years.

Rotation Crops: Replenishing the Soil

Land producing cotton for too long becomes tired land. To restore moisture, nitrogen, and other soil constituents, fertilizers may be applied and/or alfalfa, barley, or Irish potatoes may be rotated with cotton. Alfalfa, a cloverlike member of the bean family, is especially effective in restoring nitrogen to soil through *nitrogen fixation*, which occurs in soil when one crop replenishes nitrogen depleted by another. Because neither plants nor animals can assimilate atmospheric nitrogen directly, microorganisms do the fixating (Figure 10-4). When grown on irrigated land, alfalfa yields additional benefits by allowing several crops in a year versus only one crop under dry farming. The advantages of such crop rotations notwithstanding, the demand for California cotton is so great that if federal cotton acreage limitations were lifted, cotton cultivation would in all likelihood spread rapidly at the expense of alfalfa and other field crops.

Rice: Automation in the Sacramento

Driving north out of the state capital on either California 99 or I-5 in the summertime, you can see flooded rice fields stretching out seemingly forever. What you are not likely to see are people tending those paddies, whether during the summer growing season, spring seeding season, or fall harvest. To be certain, sometimes a few workers are on the scene. But in the spring some of them will be aloft flying seeding planes, and in the fall, others will be nearer the ground driving *grain combines*—machine harvesters equipped with tracks instead of wheels so they can operate in the mud of recently drained rice paddies (Figure 10-5).

Figure 10-5 *Rice fields in Sutter County with Sutter Buttes in the background. Soil parent materials weathered from these volcanic plugs and tightly folded sedimentary formations surrounding the buttes have contributed to the area's agricultural capabilities. Traps in the sedimentary rocks also hold one of California's large working natural gas fields (see Chapter 3 and Figure 6-5).* (Crane Miller)

Contrast this rice-growing landscape with one in the Far East, where the paddies will be teeming with people sowing and later harvesting the crop. The unseen feature of this comparison is the difference in annual per-acre yields of rice: nearly 4 tons in California, but little more than 1 ton in Asia. In this country, Arkansas usually leads California in total rice production but is well behind in annual per-acre yield. Both states export some 60 percent of their production to Asia, California mostly through growers' cooperatives such as those that operate out of the Port of Sacramento.

Although cold-tolerant varieties of *Oryza sativa* had to be developed to allow for the possibility of an earlier- and cooler-than-normal fall, environmental conditions for rice culture are closest to ideal for California in the middle Sacramento River basin centering on Butte, Colusa, Glenn, and Sutter counties (see Table 10-1). Much of the four-county area is covered with dense, clayey soils and impermeable hardpan, which allows the land to hold water for months after flooding. Evaporation exacts its toll during the dry summer growing season, but the Sacramento and Feather rivers, with the help of state and federal water redistribution projects (see Chapter 5), supply sufficient water to maintain the continuous submergence of soil needed in rice production. On the other hand, the low relative humidity and heat of summer hastens the maturation of the crop. For these and other reasons, about 75 percent of the state's 446,150 acres of riceland is found in these four counties.

In recent years, however, California's rice industry has gone from boom to near bust. By 1981, rice had joined the list of billion-dollar-a-year agricultural commodities with 600,000 acres under cultivation. Thirteen years later, farm gate sales of rice had dwindled to a quarter of the 1981 figure, and cultivated acreage had been cut in half. Drying-up of export markets and water supplies were seen as the principal causes of the decline. Into the early 1980s, California was exporting 80 percent of its rice outpout, much of it to the Far East under the federally subsidized "Food for Peace" program. However, such food aid programs have limited lives, and new ones are fewer and leaner because of cutbacks in federal spending. Middle Eastern countries like Jordan and Turkey remain California's main overseas rice markets; but for the industry to revive, the Far Eastern market will have to be reopened while that of the Middle East is maintained.

Japan, which some time ago invoked a total ban on rice imports, is potentially California's biggest rice buyer. The Japanese prefer the sticky, medium-grain *japonica* variety to the drier, long-grain *indica* variety of rice. "Calrose" *japonica* is what California produces, and at a price far below that of growers anywhere else in the world. The Japanese appreciate this, and there exists the possibility that they will liberalize their agricultural import policy through the General Agreement on Tariffs and Trade (GATT). A small initial import quota was implemented in 1993, and Japan began importing California's and other states' rice for the first time since 1977. But recovery of the export market is proving to be anything but spectacular: Farm gate sales of rice in 1996 were a "flat" $395 million.

Water, too, has been a chronic problem for the rice industry. In 1998, even with the last drought a fading memory, rice growers were still being accused of getting water too cheaply and using it too lavishly, especially in times of severe shortage. At the least, they may pay considerably more for water than in the past, which could deter expansion of rice acreage. Still another deterrent is mounting concern for the ill effects of the herbicides growers use to control weeds and the air pollution caused by burning rice straw following the fall harvest.

VEGETABLES, CITRUS, AND AVOCADOS

Name a vegetable, any vegetable, and no doubt it is grown in California. The same can be said for most varieties of citrus (*Citrus*), avocados (*Persea*), berries, and deciduous fruit and nut crops. Specifically, California produces more than two dozen different vegetable commodities and leads the nation in the output of artichokes, asparagus, broccoli, Brussels sprouts, carrots, cauliflower, celery, chili peppers, garlic, lettuce, lima beans, onions, Oriental (Chinese) peas and other exotic vegetables, spinach, and tomatoes. In all, *truckcrops*, or vegetables alone, account for about one-fifth of California's annual farm revenue. Even more impressive is California's one-third share of the nation's total vegetable production.

In tropical evergreen tree crops, California is the nation's leading producer of avocados, figs, kiwifruit (Figure 10-6), lemons, olives, and pomegranates, but is a distant second to Florida in orange production: The state grows some 20 percent of the country's oranges while Florida boasts 70 percent of national output. Grapefruit, tangelos, and tangerines also number among California's commercial citrus commodities.

Although Figure 10-7 demonstrates the vegetable industry's geographic versatility, many types of produce prefer the mild temperatures and moist air of locations near the ocean, as the case study of Ventura County that follows demonstrates.

Figure 10-6 *Kiwifruit. Kiwifruit grows like a vine and must be supported by posts and heavy-duty wire, especially as the fruit begins to mature and increasingly weigh down the thin trunk and branches of the plant. California accounts for 100 percent of national production, with sales averaging more than $30 million annually in the late 1990s. Although Tulare County is usually first in production, Gridley in southern Butte County (the second-ranking county, with Yuba third, Kern fourth, and Fresno fifth) is considered the "Kiwifruit capital of North America."* (Crane Miller)

Ventura County: California's Fruit and Vegetable Industry in Microcosm

Ventura County offers a unique environment for agriculture, one that may seem common in California but that is altogether too rare on the face of the earth. Few regions anywhere encompass the combination of self-contained mountainous watershed, intermont basins, broad coastal plain, advantageously situated stream and groundwater systems, mild Mediterranean climate, and abundance of deep and fertile lowland alluvial soils that characterize Ventura County. As shown in Figures 10-8 through 10-12, this natural assemblage presents farmers with nearly limitless opportunities for intensive growing of a great diversity of high-value crops. In the 1990s, Ventura County consistently ranked in the top dozen California counties in agricultural production. Total farm gate sales for the county in 1996 were $851,847 (Table 10-3).

Long before the turn of the century, when Henry Oxnard established the nation's first permanent sugar beet industry in the county, Ventura was widely recognized for its inherent agricultural potential. Today, that potential has developed to the point that Ventura County mirrors the whole of California agriculture to a greater extent than any other single county. In all, Ventura County farmers produce 7 different livestock commodities, 6 kinds of fruit and nut crops, 12 major vegetable commodities, 14 minor vegetable types, 7 different field crop types, various cut flowers and nursery stock, and apiary products consisting of honey and beeswax. Although several counties far exceed Ventura County's $850 million in annual farm products sales (see Table 10-3), the county claims leadership in the state and the nation in growing celery, green lima beans, lemons, and Romaine lettuce, and second place in spinach and avocados (Table 10-4).

Lemons, which at $200 million in average annual sales are the county's most valuable commodity, find environmental conditions in Ventura County that exist in few other regions of like size in North America. A comparison of the physical geography of Ventura County (see Figures 10-8 through 10-11) with the distribution of lemon growing in the county (see Figure 10-12) reveals some of the critical relationships between the commodity and its environment.

Equally significant in understanding the where and why of the success of lemons in Ventura County is a

1. **Tulelake–Butte Valley**: Onions, potatoes
2. **Sacramento Valley**: Honeydews, persians, watermelons, other melons, tomatoes
3. **Delta**: Asparagus, sweet corn, lettuce, tomatoes
4. **Brentwood-Tracy**: Sweet corn, lettuce, tomatoes
5. **Santa Cruz–San Mateo Coast**: Artichokes, brussels sprouts, broccoli, cauliflower, peas
6. **Fremont–San Jose**: Broccoli, cauliflower, celery, sweet corn, garlic, lettuce, onions, peas, peppers, strawberries, tomatoes
7. **Patterson-Newman**: Broccoli, cantaloupes, cauliflower, honeydews, persians, other melons, sweet corn, lettuce, tomatoes, peppers
8. **Modesto-Turlock**: Carrots, honeydews, watermelons, other melons, strawberries, sweet potatoes, tomatoes
9. **Salinas-Watsonville**: Artichokes, snap beans, broccoli, cabbage, carrots, cauliflower, celery, garlic, lettuce, onions, peas, spinach, potatoes, strawberries, tomatoes
10. **Gilroy-Hollister**: Sweet corn, garlic, lettuce, onions, potatoes, peppers, peas, tomatoes
11. **West Side**: Cantaloupes, honeydews, lettuce, persians, other melons, onions, tomatoes
12. **Merced-Atwater**: Peppers, sweet potatoes, tomatoes, watermelons
13. **Kingsburg-Dinuba**: Sweet potatoes, watermelons
14. **Cutler-Orosi**: Tomatoes, other spring vegetables
15. **Kern-Tulare**: Sweet corn, cantaloupes, carrots, garlic, honeydews, lettuce, onions, peas, potatoes, sweet potatoes, watermelons
16. **Santa Maria–Oceano**: Artichokes, snap beans, broccoli, cabbage, carrots, cauliflower, celery, lettuce, peas, potatoes, strawberries
17. **Oxnard**: Broccoli, cabbage, carrots, cauliflower, celery, cucumbers, lettuce, spinach, strawberries, tomatoes
18. **Antelope Valley**: Cantaloupes, onions
19. **Los Angeles–Orange County**: Asparagus, snap beans, cabbage, carrots, cauliflower, celery, sweet corn, lettuce, peppers, strawberries, tomatoes
20. **Chino-Ontario**: Sweet corn, onions, sweet potatoes
21. **Perris-Hemet**: Cantaloupes, watermelons, other melons, carrots, onions, potatoes
22. **Oceanside–San Luis Rey**: Snap beans, cabbage, lettuce, peppers, strawberries, sweet potatoes, tomatoes
23. **Coachella Valley**: Asparagus, snap beans, carrots, sweet corn, cantaloupes, onions, peppers, tomatoes, watermelons
24. **Blythe**: Sweet corn, cantaloupes, honeydews, other melons, lettuce, onions
25. **Chula Vista**: Snap beans, cabbage, celery, cucumbers, lettuce, strawberries, peppers, tomatoes
26. **Imperial Valley**: Asparagus, broccoli, cabbage, cantaloupes, carrots, cucumbers, garlic, lettuce, onions, tomatoes, watermelons

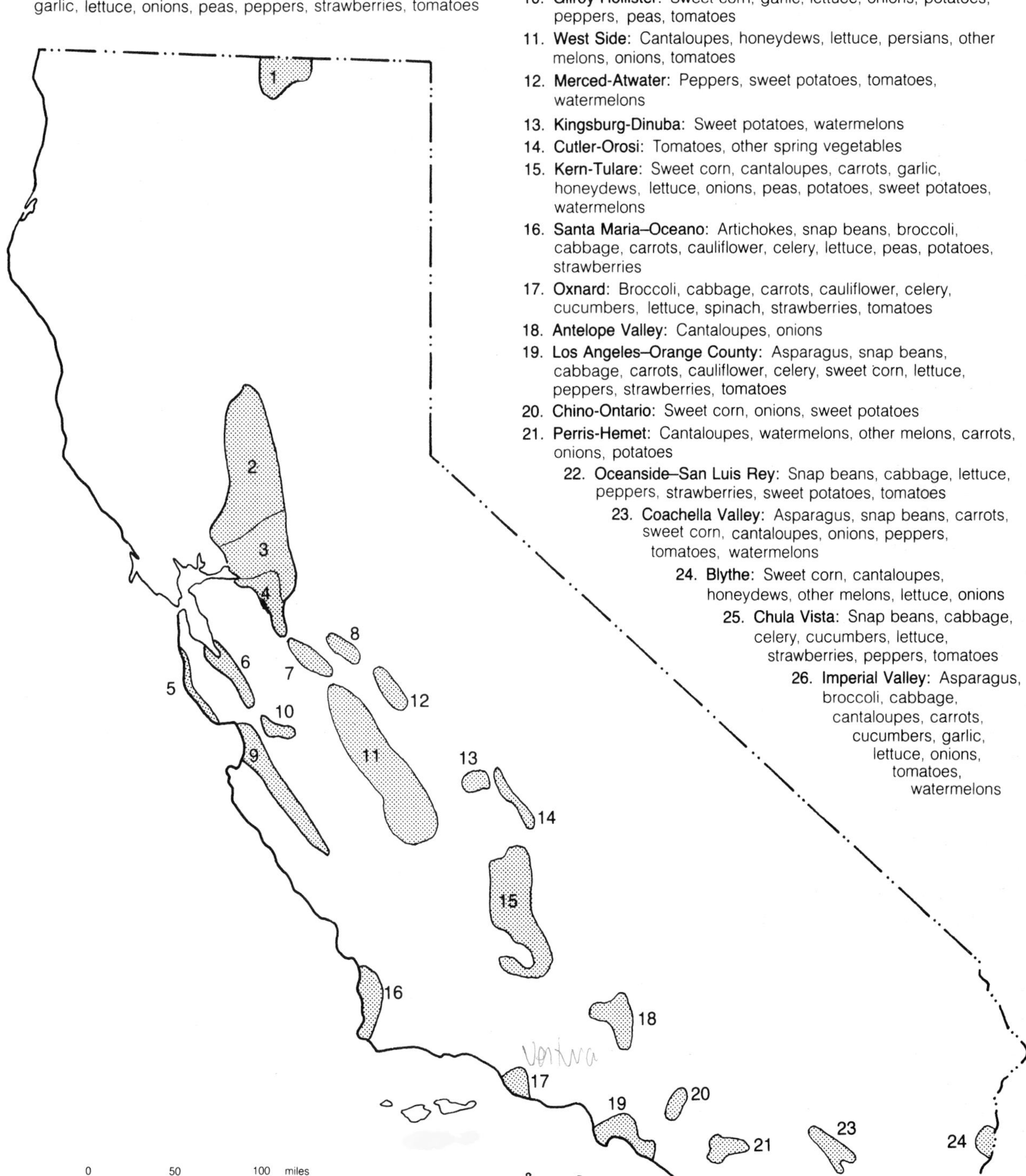

Source: California Crop and Livestock Reporting Service, 1980.

Figure 10-7 *Major vegetable, melon, and potato-growing regions in California. (See Figure 10-3 for county distributions.)* (Richard Crooker)

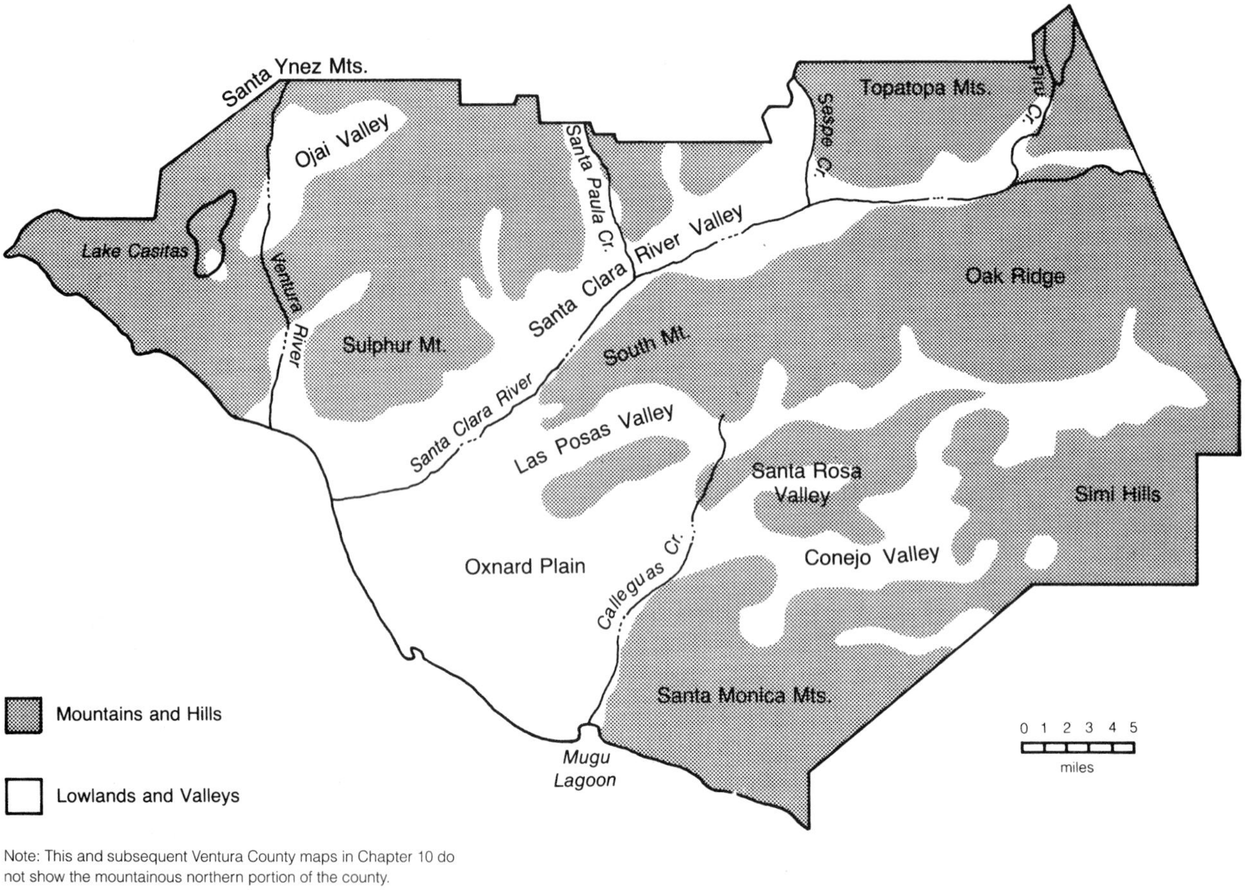

Note: This and subsequent Ventura County maps in Chapter 10 do not show the mountainous northern portion of the county.

Figure 10-8 *Ventura County, landforms and hydrography.* (Whitney Miller)

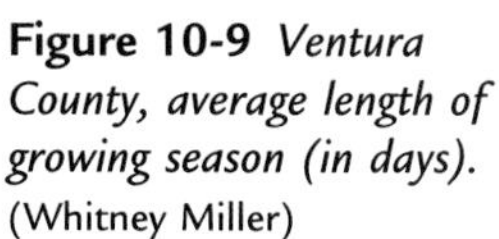

Figure 10-9 *Ventura County, average length of growing season (in days).* (Whitney Miller)

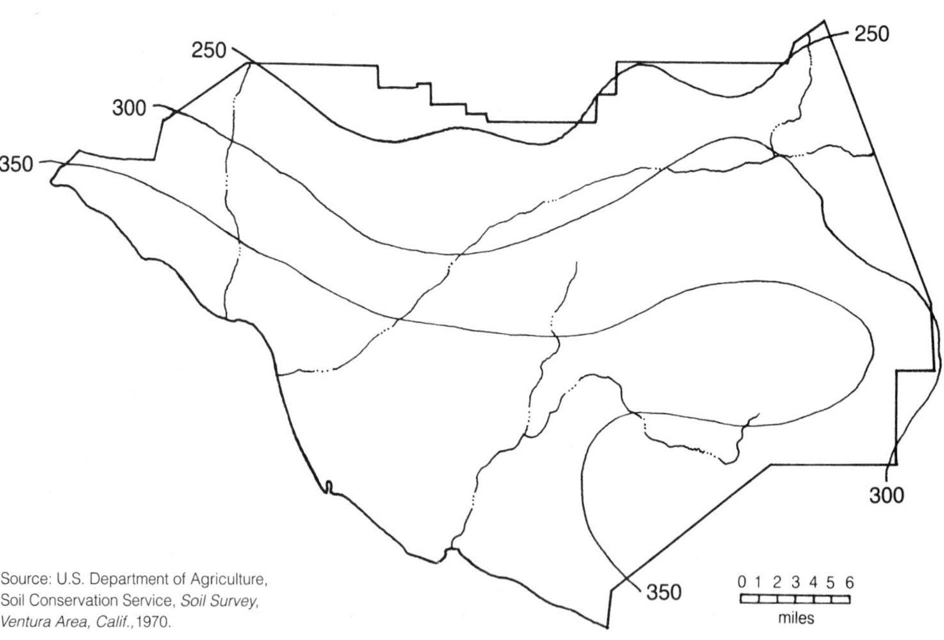

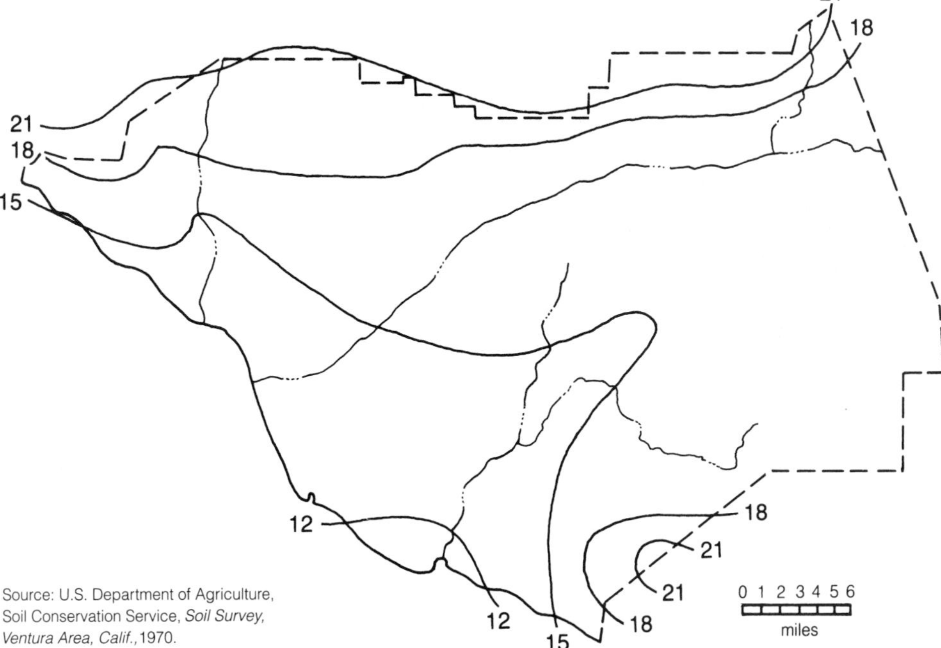

Figure 10-10 *Ventura County, average annual precipitation (in inches).* (Whitney Miller)

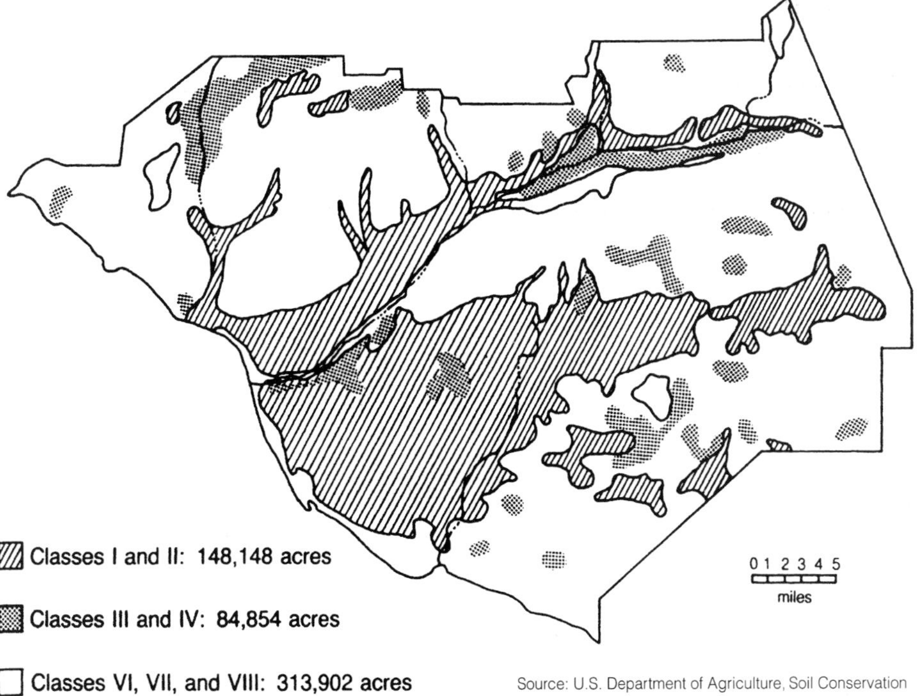

Figure 10-11 *Ventura County, soil capability classes.* (Whitney Miller)

knowledge of their unique climatic tolerances. Because lemons are the least frost-tolerant citrus fruit, most of the acreage is concentrated in the Oxnard Plain and lower Santa Clara River Valley (see Figure 10-8), where a nearly year-round growing season is the rule (see Figure 10-9).

Other aspects of the marine climate conducive to lemon production, especially of the Eureka variety, include (1) higher relative humidity, which promotes a thicker rind yet a juicier, more readily squeezed fruit; (2) greater exposure to cooling summer sea breezes, which delay maturity by preventing excessively high temperatures; (3) cooler summer temperatures, which often obviate the use of refrigeration in lemon warehouses; and (4) comparative freedom from wind damage, notably during the fall months when dessicating east winds or Santa Anas often wreak havoc on citrus and other tree crops deeper in the interior of Southern California. The dry winds not only blow down trees and "burn" their fruit, but also dry out soil, which

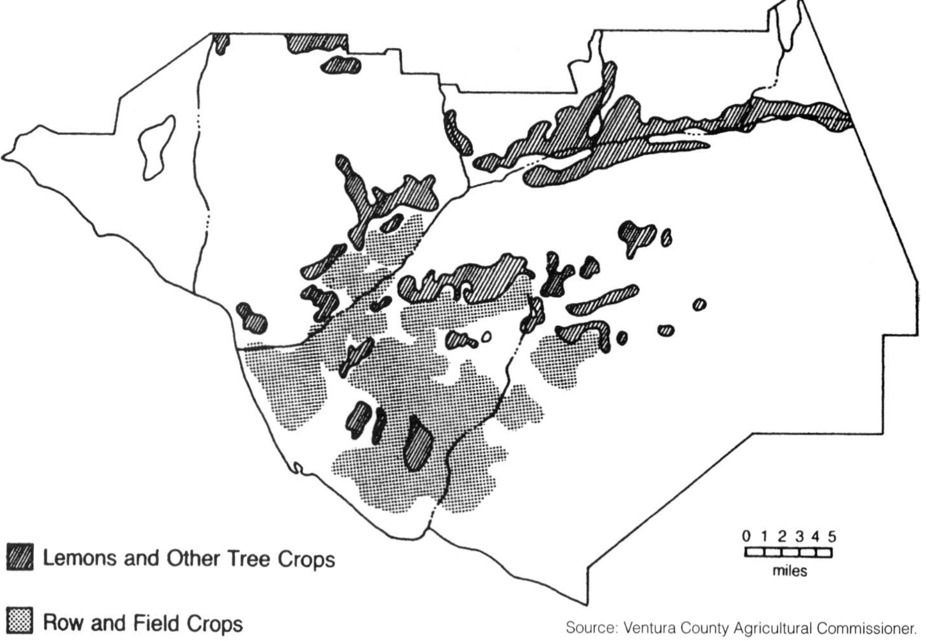

Figure 10-12 *Ventura County, crop distribution.* (Whitney Miller)

TABLE 10-3 Leading farm counties in California by total value of production, 1996.

RANK*	COUNTY	TOTAL VALUE (BILLION $)	LEADING COMMODITIES
1	Fresno	3.313	Grapes, cotton, milk, tomatoes
2	Tulare	2.802	Milk, oranges, grapes, beef
3	Kern	2.067	Grapes, cotton, almonds, carrots
4	Monterey	1.935	Lettuce, broccoli, strawberries
5	Merced	1.430	Milk, chickens, almonds, cotton
6	San Joaquin	1.352	Milk, tomatoes, wheat, grapes
7	Stanislaus	1.233	Milk, chickens, almonds, beef
8	Riverside	1.142	Milk, potatoes, eggs, lemons
9	San Diego	1.114	Avocados, flowers, eggs, nursery products
10	Imperial	0.957	Beef, alfalfa, lettuce, cantaloupe
11	Kings	0.884	Cotton, milk, beef, cottonseed
12	Ventura	0.852	Lemons, celery, avocados, strawberries
13	Madera	0.712	Grapes, almonds, milk, cotton
14	San Bernardino	0.681	Milk, beef, eggs, alfalfa
15	Santa Barbara	0.582	Strawberries, nursery products, broccoli

* State rank by total value out of 58 counties.

Source: California Department of Food and Agriculture, *California Agricultural Resource Directory, 1997* (Sacramento, December 1997).

TABLE 10-4 Rank, value, and acreage of selected Ventura County crops, 1996.

CROP	RANK*	TOTAL VALUE (MILLION $)	VALUE PER ACRE	TOTAL ACREAGE
Avocados	2	59.60	3,686	16,170
Broccoli	5	17.91	7,375	2,429
Nursery products	5	89.47	NA	NA
Celery	1	87.97	7,925	11,100
Spinach	2	5.77	NA	NA
Cucumbers	3	2.05	NA	NA
Lemons	1	201.36	9,034	22,285
Lettuce	7	25.60	2,741	9,342
Bell peppers	5	13.57	NA	NA
Strawberries	2	142.04	36,052	3,938

* State rank by total value out of 58 counties.

Sources: Values: California Department of Food and Agriculture, *California Agricultural Resource Directory, 1997* (Sacramento, December 1997); acreages: California Agricultural Statistical Service, California Agricultural 1989 Dot Maps, 1990.

necessitates more irrigation, and dehydrate leaves, which reduces photosynthesis. North-south-oriented cypress and eucalyptus windbreaks protect lemon groves from the Santa Anas that occasionally reach the coast and the brisk onshore winds that often follow winter frontal passages.

Higher humidity and soil moisture on the coastal plain do increase the chance of fungus disease occurring in lemons; fungicides and artificial soil drainage, however, have largely alleviated this problem. Some lemon groves are found in the interior valleys of Ventura County, but only on higher alluvial slopes where there is less likelihood than in valley bottoms of *cold air ponding*, or *drainage*, which occurs during subfreezing winter nights when there is little or no circulation in the local atmosphere. Even in these upslope *thermal belts*, wind machines and orchard heaters are on hand to break up unusually deep accumulations of dense, freezing air. Meanwhile, temperatures are warmer upslope, creating a classic case of *temperature inversion*, as explained in Chapter 4.

Lemon and other citrus trees generally bear their first fruit 4–6 years after initial planting as seedling trees. Once they have reached bearing age, most varieties of lemon tree will produce fruit year-round. Harvesting of the Eureka variety, however, usually occurs in spring and summer when the demand for lemonade (fresh and frozen) is greatest. Even lemon by-products, such as oils from the rinds that are used as fragrance bases in everything from furniture polish to shampoo, peak in demand during the warm season.[2] Nevertheless, the ability of the trees to bear throughout the year has strengthened the county's hold on the national market. With a near-national monopoly in production, the Ventura County lemon industry rarely feels the deleterious effects of competitive price bidding.

The high price of fresh lemons in the supermarket reflects the relatively high per-acre value of the crop back on the farm. In 1996, as seen in Table 10-4, a bearing acre of lemons produced $9,034 worth of the citrus. When compared to the harvest value per acre of avocados of $3,686 in the same year, it is easy to see why growers prefer lemons over other tree crops. Again, given what is tantamount to a national production monopoly and consistently high demand in the marketplace, Ventura County lemon growers can just about name their price. Only strawberries, which generated more than $36,000 per acre in 1996 (see Table 10-4), are more remunerative by the acre than lemons among all the county's fruit and nut crops.

While citrus production occurs both on the Oxnard Plain and in the interior valleys, vegetable or row crops are restricted essentially to the coastal plain, where they enjoy the best soils the Ventura County environment has to offer. For 10,000 years, the Santa Clara River has

[2] Frozen lemonade and all the by-products can be stored indefinitely; hence, summer peaks are of no real consequence. Because lemons cure for 2 weeks to 6 months, marketing time can be easily adjusted.

transported alluvium eroded from the mountainous northeast and deposited it to depths of several hundred feet over the Oxnard Plain. Together with other soil-forming agents, the recent alluvium provides the basis for the rich 5-foot-deep loamy sands and silty clay loams that presently make up almost 100 percent of the area under irrigation on the plain. When these soils are machine leveled, graded, and furrowed for irrigation, there is minimal loss of fertility because of their inherent organic content, great depth, lack of profile, effective moisture retention, and loamy texture. Few soils anywhere in the world possess this ready adaptability to machine cultivation. Yet, as will be addressed in the next chapter, this soil is being inexorably covered over by housing tracts, shopping centers, industrial parks, and the other artifacts of urban development.

Beside helping accumulate most of the county's prime soils in the Oxnard Plain (see Figure 10-8), the Santa Clara River acts as a vehicle for the supply of nearly all of the plain's groundwater. Most of the normal runoff from the mountains percolates into subsurface water-bearing sediments in the stream course and then flows by gravity down to the permeable alluvial beds of a broad *aquifer* (see Chapter 5) that underlies the entire Oxnard Plain. The pumping of confined groundwater from this and deeper aquifers has fostered the expansion of irrigation agriculture to all parts of the plain over the last half century.

Add to a dependable supply of well water a year-round growing season (see Figure 10-9) and plenty of dry-season fog, and optimal conditions for growing just about any kind of vegetable are found to exist. Advection fog, commonly caused by the upwelling of cold ocean water and the presence of the cold California Current (see Chapter 4), extends inland for 2–5 miles along the Oxnard Plain throughout many spring and summer days. In effect, the fog acts as a surrogate source of moisture for plants at a time when rainfall does not normally occur. This overcast condition is vital to the growing of summer vegetable crops requiring high atmospheric humidity and restricted sunlight. On the other hand, constant fog works to the detriment of most tree crops in that it promotes fungus and retards growth. Lemons, though, are more fog tolerant than are other citrus and avocados.

Of the some two dozen different vegetable crops grown on the Oxnard Plain, celery and tomatoes are by far the most valuable. Celery sales average $88 million per year, and tomatoes annually bring in about $14 million. Although the tomato (see the next section) is not a vegetable, it is cultivated and classified as such. It is a fruit, though, that requires copious amounts of irrigation water to prosper. Consequently, the price of both canning and fresh market tomatoes rose between 1989 and 1992, the last year of the drought. Thereafter through 1997, tomatoes returned to near 1989 price levels. El Niño–induced storms in 1998 damaged row crops generally and, in turn, caused tomato prices to rise.

Celery and green cabbage are Ventura County's principal winter vegetables and as such are marketed largely in the eastern United States and Canada at a time when supply and demand are at a maximum. An extended dry period from late spring through fall allows for celery transplants to be set out before the first heavy rains. Celery matures rapidly once the wet season arrives, with the harvest often starting before winter and sometimes extending into early summer. To thwart the spread of celery blight by leaf hoppers, celery land is taken out of production for at least 3 weeks in mid-summer. In the interim, quick-maturing rotation crops such as tomatoes, lettuce, and cabbage are grown. Obviously, rotation by double or sometimes even triple-cropping makes Oxnard Plain vegetable land all the more valuable.

Given the rotation and marked seasonality of row crops in general, it is not surprising to find that much of the vegetable land in the Oxnard Plain is operated by tenant farmers or on short-term leases. On the opposite side of the coin is tree cropland, which represents investment in growing a single crop commodity over many years. Consequently, orchard land is mostly owner-operated.

Tomato Technology: People Versus Machines

The tomato (*Lycopersicon*) is not a vegetable; it is a large, seedy berry that comes in two commercial varieties: (1) the hand-harvested fresh market or *pole tomato*, and (2) the machine-harvested, vine or *cannery* or *processing tomato* (Figure 10-13). These tomatoes are California's ninth most valuable crop, averaging more than $600 million annually over the last few years. Nationally, tomato processing is a multibillion-dollar-a-year industry, with California garnering 90 percent of the market. The state is also the nation's leading grower. Although Fresno is the leading county, production is centered in Yolo, San Joaquin, Colusa, and other counties peripheral to the Delta region of the Central Valley. Here, tomatoes find the irrigation water and well-drained soils they need.

The cannery tomato is one of several California fruit and vegetable commodities at the center of a storm of controversy over mechanization technology and its displacement of farm workers. On one side are growers and University of California (UC) agricultural researchers whose aim is to increase production while lowering costs. On the other are the farm workers and their collective voice, the United Farm Workers (UFW), who see mechanization permanently forcing more than 100,000 farm workers out of jobs and actually raising consumer prices.

Ancillary to mechanization displacing farm workers affiliated with the UFW or the Teamsters is the plight

Figure 10-13 *Cannery tomatoes being sorted by flotation.* (Crane Miller)

of illegal aliens who in 1987, one year after the Immigration Reform and Control Act (IRCA) of 1986 became law, were estimated to make up one-half of all aliens working on California farms. When 85 percent of agricultural employers hire one or more aliens, as they did in 1986 and 1987, it becomes obvious that California agriculture depends heavily on alien farm workers. As the IRCA becomes more widely observed, legal aliens will be hired at the expense of illegals. Thus, the ranks of those thrown out of work by machines will be swelled by unemployed aliens. As will be examined later in this section, the cost to society of a growing number of unemployed farm workers, regardless of the circumstances leading to their unemployment, could prove devastating.

There is no question that in the past 30 years, machines have all but eliminated human hands in the canning tomato harvest. From 1962 to 1972, during which time the Blackwelder tomato harvester was developed and put into operation, the number of harvest workers fell from 50,000 to 18,000. Significant expansion of cannery tomato acreage thereafter temporarily reversed the downtrend by adding nearly 10,000 to the work force by 1976. Since then, however, UC's labor-saving electronic-eye sorter has been introduced. There are fears that it, coupled with further innovations in machine harvesters, will diminish the number of tomato harvest workers to a scant few thousand by shortly after the turn of the century, or less than 10 percent of the 1962 labor force. If these displaced farm workers could simply switch to harvesting other crops, there would be no problem. But mechanization has invaded the harvesting of everything from almonds to sugar beets and consequently has fomented the growth of a class of unemployed workers perhaps more unfortunate than any to be found in the nation. Already, for example, there are second-generation hamlets of chronic unemployment in the San Joaquin Valley that grew out of the automation of the cotton harvest back in the 1950s and 1960s.

The social costs of a large class of able, working-age people on welfare are sobering to contemplate, whether they are as specific as increases in crime rates or as general as a deterioration of the quality of rural life. But the wrenching away by a machine of one's lifelong occupation and thereby one's dignity seems among the cruelest of setbacks. Machines do free us from toil, but they can also sever us from the only gainful occupation we know.

The plight of farm workers displaced by automation nothwithstanding, the living conditions of those out of work for whatever reason are often deplorable. Such conditions for unemployed or underemployed farm workers became commonplace in some parts of agricultural California during the 1990s, but nowhere more regrettably than in San Diego County. Although nobody can verify the numbers, possibly as many as 10,000 migrant farm workers, most of them Mexican nationals, reside in what has become a rural backwater of San Diego County. They live in crude wood-and-plastic shacks, many sleeping at night on pads or woven mats layed on dirt floors. Outhouses are the closest clump of trees or brush. Light is provided by candles, flashlights, or fires. Overpriced food is purchased from lunch trucks and, if not consumed immediately, is reheated later over the fires. And whenever it rains, these shantytowns become mud fields.

Local geography helps explain the severity of conditions for the rural homeless peculiar to San Diego County. The rugged, ravine- and ridge-laced topography of interior and northern San Diego County's Peninsular Ranges makes the illegal camps relatively inaccessible—except for their desperate residents. A scant few dozen miles to the south is the Mexican border, where spotlights on the border fence at night, increased border patrols during the day, and implementation of the IRCA have done little to stem the tide of illegal immigrants from Baja California seeking a better existence in California. Along with neighboring Imperial County, San Diego County also serves as the entry point for legal migratory farm workers heading farther north to pick lettuce in Monterey County, apples in Washington's Yakima Valley, or any number of seasonally harvested fruit and vegetable crops all the way up to the Canadian border. San Diego County's rugged hinterland and proximity to Mexico notwithstanding, its $1-billion-a-year farm income and ninth position among agricultural counties (see Table 10-3) render it a major employer of farm labor. But the county's agricultural and urban prosperity, especially the latter, have placed the cost of housing far beyond the reach of most local farm workers. Low-rent, legal migratory worker camps exist in much of the state, but there seems a dearth of them in San Diego County.

DECIDUOUS TREE CROPS

A certain southern state advertises on its license plates that it is the "peach state" when, in actuality, California is the leading producer of this deciduous fruit in the nation. In fact, California outproduces all other states in all major commercial categories of stone or pit fruits except for cherries. Thus, California's dominance in *Prunus* production embraces almonds, apricots, nectarines, plums, and prunes as well as peaches. Rounding out the state's first-place ranking in deciduous tree crops are English walnuts (*Juglans regia*) and pears (*Pyrus communis*).

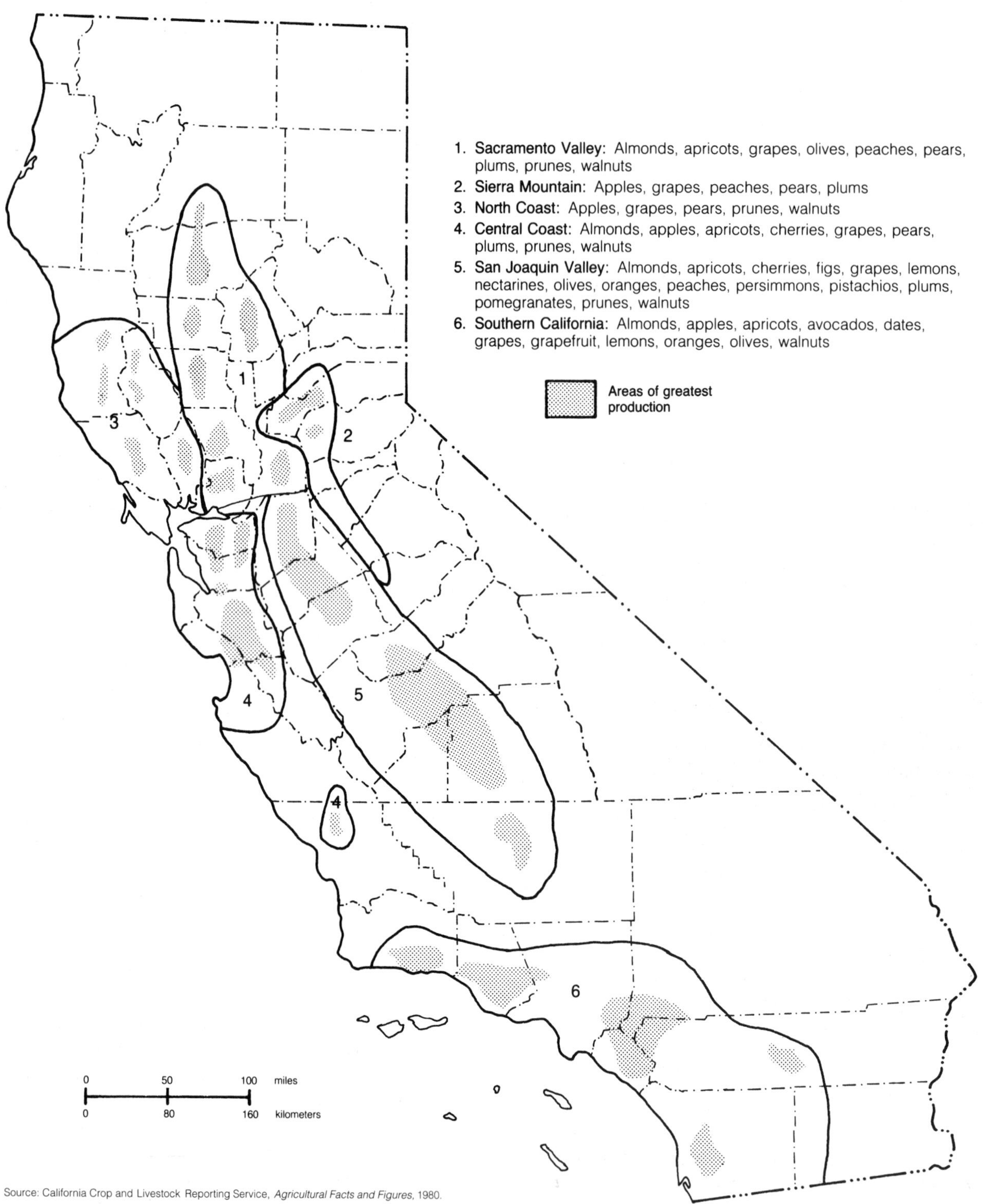

Figure 10-14 *California's major fruit- and nut-producing districts.* (Richard Crooker)

Annual farm gate sales of all these fruits and nuts hover around $3 billion dollars; almond sales alone exceed $1.1 billion annually (see Table 10-1). Export of these commodities depends on whether they are sold as fresh produce or canned, frozen, or dried, the latter three forms finding markets throughout the United States and Canada. This is especially true with bagged or bottled walnuts, for California grows nearly 100 percent of the North American crop.

A look at Figure 10-14 reveals the preference of deciduous fruit and nut crops for the cooler, more northerly climes of California. Most of these crops were introduced by the Franciscans in the late eighteenth century and prospered reasonably well as far south as Mission San Diego. A hundred years later, Ventura County, still in sunny Southern California, was on its way to becoming the leading apricot and walnut producer. As commercial production began to expand significantly in the mid-twentieth century, however, the focus of almost all new plantings of deciduous fruit and nut trees shifted to the San Joaquin and Sacramento valleys. Here in the northern interior, where the winters are longer and colder and the summers hotter and drier than in coastal Southern California, deciduous trees are more likely to lose all their leaves before the cold season has completely passed. With peaches and pears (Figure 10-15), for example, the lack of chilling that accompanies too mild a winter results in delayed spring foliation, few flower and fruit sets, and eventual death of the trees. Moreover, the dry summer heat of the interior valleys and high deserts encourages sugar or fructose production in the fruits and wards off bacterial blights, viruses, and fungi.

Figure 10-15 *Lake County Bartletts, reputedly the world's best-tasting pears. Bartlett variety production centers in the pear capital community of Kelseyville, located just west of Clear Lake and volcanic Mount Konocti. Lake, Mendocino, and Sacramento counties are the leaders in the output of all commercial varieties of pears in the state. Other important California varieties include Anjou, Bosc, Comice, and Winter Nelis.* (Crane Miller)

VITICULTURE AND WINEMAKING

California's national prominence in viticulture and winemaking is unrivaled. The state produces 100 percent of the country's raisins, about 95 percent of its table grapes, and some 75 percent of all the wine consumed in the nation, which all adds up to a $2.83-billion-a-year (see Table 10-1) industry utilizing 670,000 acres of vineyards (Figure 10-16). The majority of the grapes picked each year are crushed at California wineries, the latter in turn producing several billion dollars worth of wine, brandy, and brandy spirits annually. Thus, it is in the production of wines and brandies that viticulture ultimately generates the greatest revenues.

Winemaking

The production process of California winemakers is indeed impressive and warrants a brief review before we delve into the history of the industry. Here, we will focus on winemaking as it applies to the production of California's finest wines, the *table wines*—so named because they should be served with the day's main meal.

The process starts with the mechanical destemming and crushing (Figure 10-17) of the fall harvest of grapes, which involves breaking the grape skins but not the seeds. Next, the *must*, or juice and pulp, is conveyed to fermenting vats for the production of red wines, with dark-skinned grapes imparting natural color to the wine. If white wine is desired, the juice is separated from the skins before fermentation.

Although grape juice contains ferments or yeasts (*Saccharomyces* spp.), the wild yeasts are inhibited by the addition of sulfur dioxide (SO_2) to the must. The sulfured must is then pumped into a tank and inoculated with wine yeasts, which, when exposed to air, act to convert grape sugar to alcohol and carbon dioxide (CO_2).

While the temperature of the fermenting mass rapidly rises, the CO_2 escapes into the air and the ethyl alcohol content of the new wine builds to a maximum of about 14 percent. As far as the finished product is concerned, state regulations require that red wines contain 10.5–14 percent alcohol by volume, and white wines 10–14 percent. So-called soft wines can range down to 7.5 percent alcohol. At the prescribed alcohol level, the sugar supply is exhausted and fermentation stops.

Depending on the volume of fermenting must, the local air temperatures (the cooler the air, the slower the

Figure 10-16 *White Chardonnay grapes and Gravenstein apples growing in western Sonoma County. Both are deciduous crops and do well in this coolest (Region I, Figure 10-18) of grape-growing climates. Chardonnay and Pinot Chardonnay wines number among California's premium dry white table wines. One nearby winery produces an unusual white wine known as Gravenstein Blanc, but there is no apple juice in it.* (Crane Miller)

fermentation), and whether red or white wine is being made (white wine needs more time to ferment than red wine does), fermentation on average lasts several weeks. Fermentation can take place in vats made from one of a number of different kinds of material, including redwood, oak, stainless steel, or concrete. Given California's relatively warm and short vintage season (2–3 months) in which all the table wine for the year is made, large wineries usually require elaborate cooling systems to keep fermenting musts from exceeding 90°F for red wines and 70°F for white wines.

Following fermentation, the new wine is placed in storage cooperage to be clarified, racked (removal of sediment-freed wine to smaller containers), and aged prior to bottling. Thereafter, some premium table wines may be bottle-aged at the winery for up to 2 years. Wine, it is said, gains character with time, and a fine wine "should not be sold before its time." Even once a fine table wine is ready to be consumed, it should be uncorked and allowed to "breathe" somewhat before dinner: an hour before if it is red wine, and a half-hour before if it is white wine.

Figure 10-17 *Wilmes presses, which gently extract as much juice as possible from crushed grapes without breaking the seeds. The press on the right is ready for loading. A rubber bag in the center of the press is inflated to gently squeeze the grapes against the press's inside walls. A continuous screw conveyor brings the grape "must" from the crusher.* (Wine Institute)

Modern wine production as just described owes much to Louis Pasteur and his fellow chemists, who little more than a century ago helped transform the age-old art of winemaking into the science we know today as *enology*. But what was the state of the art and science of *winegrowing*[3] in California more than a century before Pasteur and his associates' discoveries?

The Mission Grape

Although native species of grapes had grown wild for millennia in California and perhaps even been pressed into wine by aboriginal peoples, it was the Franciscan missionaries who brought the first vines of the Old World wine grape into the region. Given the urgency of establishing production of a dependable supply of sacramental wine, Father Junípero Serra and the other padres no doubt wasted little time in planting the first cuttings and/or raisin seedlings of *Vitis vinifera* after their July 1, 1769, arrival at the Mission San Diego de Alcalá site. The abundance of wild grapes here and at subsequent mission sites as far north as Sonoma convinced the padres that Alta California had an optimal environment for viticulture.

While awaiting *vinifera*'s first fruits, they even attempted making wine from the native species of *Vitis*. The Monica variety of *vinifera* or Mission grape,[4] which came to California from Spain and Sardinia via Mexico, eventually became the viticultural mainstay of all 21 missions. One Mission grape vine, the Trinity Vine at Mission San Gabriel Archangel, continued to yield fruit for 170 years. The Old World grape had obviously found an ecological niche of grand proportions in California.

Commercial wine grape growing had somewhat less auspicious beginnings in California, although the Mission grape remained the principal raw material of the industry for many decades following the secularization of the missions in the mid-1830s. Joseph Chapman, who arrived at Monterey in 1818 and was only the third American to become a permanent resident of Alta California, planted the first commercial vineyard in the province. Chapman was a shipbuilder by trade, but a change in profession was in the offing when in 1834 he set out some 4,000 Mission vines near the pueblo at Los Angeles. Chapman was probably encouraged by Father Sanchez and other Franciscan viticulturists at the nearby San Gabriel mission, where 100,000 vines were annually yielding some 500 barrels of wine and 200 barrels of brandy. By the end of the decade, Chapman's and the padres' Mission wines had gained a favorable reputation among travelers from Mexico and the United States. However, remoteness from potential markets and a chronic shortage of bottles all but dashed hopes of significant exports of California wines.

The first professional *viticulturist* to come to California was Jean Louis Vignes (sometimes pronounced "Vines"), a native of France's Bordeaux wine district. His first sight of the vast vineyards of the Franciscans, plus those of Chapman and five other minor winegrowers, led him in 1833 to purchase 104 acres of potential vineyard land centered on what is now the site of the Los Angeles Union Station. His land was soon planted, mostly with Mission vines, but also with some experimental cuttings of French grape varieties. Vignes was the first person to bring foreign cuttings to California, for the Mission grape by that time was considered a California grape. He constructed a commercial winery near a large alder, the tree's presence inspiring him to name the rancho El Aliso, and his neighbors to call him Don Luís del Aliso.

Vignes's first vintage was in 1837; by 1843, he was producing 40,000 gallons of wine a year, the bulk of it from Mission grapes. His sole major competition was William Wolfskill from Kentucky, who in 1838 bought several vineyards in what is now the heart of downtown Los Angeles. By the time of statehood in 1850, Vignes and Wolfskill had practically cornered California wine production. Their wineries were producing nearly 60,000 gallons yearly, and the Mission grape was the raw material of their success.

The Green Hungarian

Count Agoston Haraszthy, a Hungarian, was no newcomer to the art and science of wine grape growing in California. The appellation "Green Hungarian" refers to a white table wine and the white grape that produces it, the latter of which may have been one of the 300 different varieties of *vinifera* that the count collected in Europe and brought to California in 1861. If Haraszthy was not Green Hungarian's originator in California, he was certainly its principal promoter. This was but one of many viticultural feats performed by Haraszthy, who is considered the father of modern California viticulture.

Like his predecessors, Haraszthy started with the Mission grape. He planted his first vineyard near San Diego shortly after his arrival in California in 1849. But

[3] The term *winegrowing* actually refers to a cottage-industry type of winemaking that still exists in many parts of the world, though largely absent from California since the advent of modern production techniques. Given that they are located in out-of-the-way rural areas and that their harvested grapes will ferment and spoil on their own, such winegrowers will often make their own bulk table wine and sell it, rather than their grapes, to wholesale brokers. In turn, the brokers sell the wine to local vintners, who blend it with other growers' wines. Obviously, quality control suffers in such an operation, and the wine is likely to be below par. See M. A. Amerine and V. L. Singleton, *Wine: An Introduction*, 2nd ed. (Berkeley: University of California Press, 1977).

[4] According to other accounts, Criolla (also developed in Mexico) was the first variety of *vinifera* introduced by the Franciscans and thus is said to be the forerunner of the Mission grape.

Haraszthy's stay in Southern California was as short-lived as his use of the Mission grape. In 1852, he headed north and bought 211 acres of ex–Mission Dolores land in San Francisco at the southern end of present-day Market Street. Here, he planted two new European grape varieties: (1) Zinfandel, which is of Italian origin, and (2) Muscat of Alexandria, direct from Málaga, Spain. By 1854, Haraszthy had more than tripled his grape acreage on the San Francisco peninsula, and his future appeared promising. But in the ensuing years, he was accused (and later acquitted) of embezzlement while in the employ of the U.S. Mint, and his vineyards suffered from both neglect and the frequency of peninsular fog. All this prompted him to sell out in the Bay Area and move even farther north to Sonoma.

The Sonoma Valley proved to be the environmental niche that Haraszthy and *vinifera* were looking for. Fungus-causing fog was not the problem it was nearer to San Francisco, yet the valley was close enough to the Pacific to allow a long growing season. Moreover, Franciscans and their Mission grapes, Native Americans working ex-mission vineyards, and General Vallejo's wines and brandies had already gained some measure of viticultural repute for Sonoma. But it was Haraszthy who would make the valley famous. By 1858, he had seen to the planting of 165 different varieties of *vinifera*, the building of a winery and storage cellars, and the founding of Buena Vista Vineyards. In 1863, Haraszthy and eight others incorporated the Buena Vista Vinicultural Society with holdings of 6,000 acres planted with nearly all the 300 varieties of *vinifera* he had earlier collected in Europe. The corporation's plan was to be producing 2.26 million gallons of wine annually by 1873.

Sadly, Haraszthy never saw his dream come true. In 1868, he traveled to Nicaragua to build a distillery and returned only briefly to San Francisco before his mysterious disappearance in 1869. Supposedly, he was devoured by alligators in a stream on his own Nicaraguan property. This Hungarian's "greenness" in a hostile tropical milieu undoubtedly led to his undoing, but not before he had forever modified California's winegrowing landscape.

Agoston Haraszthy's contemporaries and successors in California viticulture—Frohling, Gallo, Kohler, Krug, Masson, Mondavi, Petri, and Wetmore, to name but a very few—followed his lead in supplanting Mission grapes with all manner of European varieties. What Haraszthy had visualized as essential to the attainment of variety and quality in California wines has been accepted and implemented by the industry over the past hundred years.

Vinifera's Enemies

First there was phylloxera, and then there was Prohibition. The former was naturally and probably accidentally introduced to California in the 1850s, whereas the latter was strictly a cultural imposition, thrust upon the state with enactment of the 18th Amendment to the U.S. Constitution in 1919. Both dealt near-fatal blows to California winegrowing in their time, but both would eventually be vanquished as the industry rebounded to new levels of international prestige.

Phylloxeras is a vine disease transmitted to either the leaves or the roots of a grape vine by a species of vine louse or aphid known as *Phylloxera vastatrix*. The *Phylloxera* insect is a native of the eastern United States that was introduced to Europe in the mid-nineteenth century when native eastern American vines (*Vitis labrusca*) were sent there for grafting to European vines. The less resistant European *Vitis vinifera* quickly fell prey to phylloxera, and by 1865, most of Europe's vineyards had been ravaged. It is now known that phylloxera had indirectly made its way to California, via European cuttings, in the 1850s; but it was not until 1876 that the disease was considered a major threat to California wine growing. First and hardest hit was Sonoma County, where by 1879 more than 100,000 vines had to be uprooted. Still, by 1880, the majority of California winegrowers were apathetic about taking any meaningful action. Consequently, the *Phylloxera* insect, which by this time had evolved to its winged form, spread throughout the counties of Napa, El Dorado, and Placer, as well as Sonoma. Vineyardists affected were losing an estimated $1,000 a day. The disease struck the older vines the hardest, all but wiping out the Mission grape in Northern California. The newer European varieties of *vinifera* were hard hit, too, but eventually enough were saved to provide the base for a resurgent wine industry. Even so, all the European varieties would have succumbed to phylloxera had it not been for the efforts of three men.

Charles Wetmore, Professor Eugene Hilgard of the University of California College of Agriculture, and Professor George Husmann were, in the main, the developers of the only practical means of combating phylloxera. Led by Husmann, the three proved that if resistant native American vines were used as root stocks for *Vitis vinifera*, phylloxera would not attack such vines. At first *V. californica* and *V. labrusca* were grafted to *vinifera*; however, more recently hybridized species of *V. riparia, rupestria*, and *berlandieri* have proved most effective as rootstocks. Thus, by 1890, phylloxera had ceased to be a major problem, and a major relocation of grape varieties had been completed. European varieties became dominant in Northern California, but the Mission grape remained supreme in Southern California, which had remained relatively unaffected by phylloxera, mainly because of the sandy soils found in its vineyards.

No sooner had phylloxera been quashed than the rumblings of impending national prohibition of alcohol began to take their toll. Eight years before Prohibition became law, California wine production had peaked; thereafter, it went into a deepening slump. By 1919, the state's wine production had declined to less than half of the 1912 mark. However, raisin and table grape acreage increased until, by 1927, the state showed its largest acreage in history, 648,000 acres. The effect of this switch in cultivation was to create an imbalance of table and raisin grapes relative to wine grape varieties.

Prohibition's worst effect, then, was to cause a dearth of wine grape varieties by the time of eventual repeal. When the 21st Amendment was passed in 1933 and Prohibition was repealed, an oversupply of the wrong varieties of grapes, dilapidated wineries, unsound cooperage (storage tanks), inexperienced personnel, and lack of capital hindered the rebirth of the California wine industry. But reborn it was, for wine connoisseurs the world over remembered the excellence of California wines. The real problem through the rest of the 1930s was meeting the demand, especially in the midst of the worst economic depression in history. Eventually, such difficulties as too much sweet wine and not enough table wine production were overcome, and since the 1940s California has accounted for about 80 percent of all domestic production. *Viniculture* alone, or growing wine grapes, exclusively occupies 330,000 acres of the state's farmland. Today, New York is a distant second as producer of wine in the nation.

The needs for standardization within the industry and for advancement of sales of domestically produced wines were ameliorated in 1934 by three separate events: (1) organization of the Wine Institute as a nonprofit association of more than 80 percent of the nation's winegrowers to prevent the marketing of unsound and misbranded wines and to advertise wine as a food product; (2) establishment of the federal Alcohol Control Administration to originate and implement standards of competition and promote stability within the industry; and (3) assignment of the California Board of Public Health to the task of establishing minimum standards of quality for wine.

Although Prohibition is unlikely to return, phylloxera is once again back in the vineyards of California. This time it is a new strain, named Biotype B, which successfully attacks the rootstocks that were resistant to its predecessor. Biotype B probably first invaded Sonoma and Napa valley vineyards in the mid-1980s, but it was not perceived as a major problem until the early 1990s. Although milk cartons placed around individual vines may fend off some of the insects, they readily migrate through soil, under artificial barriers, and into rooting systems. As mentioned previously, the winged form diffuses more rapidly and over larger areas than the ground-bound form. By 1992, the disease had spread to the extent that growers were uprooting vines, fumigating the soil, and replanting with newly developed, resistant rootstocks, all at a cost of up to $25,000 per acre. These costs, coupled with diminished grape production, beset growers already plagued by the economic recession. For those who survive financially, the forecast is that wine production and revenues probably will not return to pre-outbreak levels until the twenty-first century. But phylloxera-beseiged growers may simply give up and sell their land to developers. Many Napa and Sonoma vineyards are protected from urban development under the California Land Conservation Act (CLCA) as agricultural preserves; however, as examined in Chapter 11, the taxes saved under the CLCA may never make up for the losses brought on by phylloxera.

Wines, Varieties, and Climates

Post-Prohibition standardization within the U.S. wine industry brought with it recognition of the classification of wines into three general categories: generic, proprietary, and varietal. *Generics*, or *semigenerics* as they are called under U.S. regulations, are identified by geographic origin, such as Burgundy from France or Rhine wine from Germany. Generic wines have obviously long since lost their original regional significance and are consequently labeled according to their actual place of production, such as California Mountain Burgundy. A *proprietary* wine, such as Thunderbird or Silver Satin, is narrowly exclusive by comparison, denoting on its label the name of a wine that no other winery can make. Of the three types, however, the *varietal* label is the one to be most trusted by the wary wine buyer. If the label reads "Cabernet Sauvignon" or "Pinot Noir" or "Sauvignon Blanc" or one of dozens of different varieties of *vinifera* grown in California, one is assured that the majority, and perhaps all, of the grapes used to produce that wine are of the variety on the label. By law, California varietals contain at least 75 percent of the grape variety listed (51 percent prior to 1983). However, more expensive varietal wines are produced exclusively or nearly so from the variety labeled. One can conclude that the environmental parameters for growing fine varietal wine grapes are more confining than those required of varieties used in the production of ordinary table wines and so-called jug wines.

Of the ecological constraints on wine grape growing, climate in general and temperature in particular are most significant. The white Palomino or Golden Chaselas variety, used mostly to make sherry, provides a case in point. Any grape, including the Palomino, requires a given number of sunny, warm days to grow to full size and then to ripen. During early growth, the acid content of the grape will increase until the fruit reaches half of its full size. Then, shortly before ripening begins, acidity will start to decrease. All during the ripening period, sugar content increases.

The Palomino follows this maturation pattern throughout except that it is an early-ripening variety

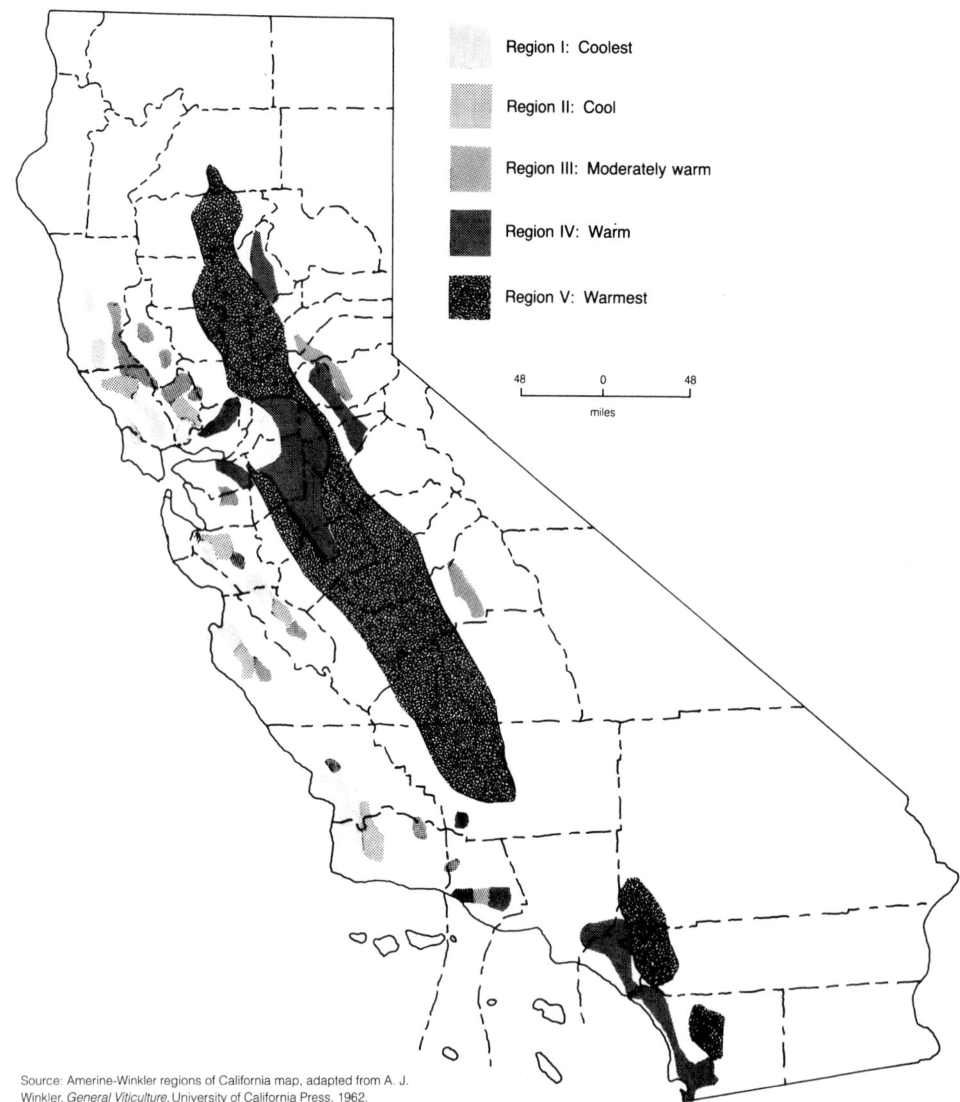

Figure 10-18 *California degree-day regions.* (Brett Wilson)

and therefore develops a comparatively low acid content. Its underacidity renders the Palomino poor material for making dry table wines. Given enough time, however, the Palomino will develop a relatively high sugar content and thus be suitable as the basic ingredient of fortified (with brandy) dessert wines and apéritifs, such as sweet and cocktail (dry) sherries. Consequently, most of California's acreage in Palomino grapes is situated in the Central Valley where the summers are long, hot, and dry. The remaining Palomino vineyards are found in inland coastal valleys where they must contend with summers that are just a bit too short, cool, and humid for optimal maturation of their fruit. Winter, on the other hand, is rarely a problem in either the coastal valleys or the Central Valley, for the deciduous vines have lost their leaves and are ready for some cold and wet weather.

Perhaps the best argument for temperature as the most critical climatic element in wine grape growing is shown in Figure 10-18, a map of M. A. Amerine's and A. J. Winkler's degree-day regions of California. Several decades ago, these two University of California enologists set about collecting different varieties from every principal grape-growing district in California. The grapes were converted to wine, which was analyzed chemically and subjected to periodic tastings. From thousands of analyses and tasting records, they were able to form a basis for classifying the capabilities of hundreds of grape varieties according to local temperature regions. In their studies, Amerine and Winkler considered other factors, such as rainfall, fog, humidity, and duration of sunshine, but found that these had only a minor effect on the balance of the composition of the fruit at maturity. The major factor was temperature, or *heat summation*, as they termed it.

Amerine and Winkler established five regions, based solely on the summation of heat as degree-days above 50°F for the period from April 1 October 31. Simply stated, this is the total amount of heat that is available to a vine during its growing season and is calculated as follows: If the mean temperature over a 5-day period were, for example, 70°F, the summation would be 100 degree-days, or $(70 - 50) \times 5 = 100$. The ranges of degree-days for each of the Amerine-Winkler regions shown in Figure 10-18 are as follows: Region I, the coolest, with less than 2,500; Region II, cool, with 2,501–3,000; Region III, moderately warm, with 3,001–3,500; Region IV, warm, with 3,501–4,000; and Region V, the warmest, with 4,001 or more.

A rough rule-of-thumb application of the heat summation system would probably find the drier, higher-tannin varietal wine grapes doing best in Regions I and II and the sweeter, less astringent types coming from Regions IV and V, with a mix of fruitier semidry and semisweet varietals growing in Region III. Table 10-5 lists some of California's better known varietal wines and the degree-day regions in which the growing of grapes used in their production is greatest, as well as their acreage.

TABLE 10-5 Common California varietals, degree-day regions, and acreage, 1989.

RED VARIETALS	REGION	ACREAGE
Barbera	III, IV	10,800
Cabernet Sauvignon	II, III	29,700
Grenache	IV, V	13,000
Merlot	I, II, III	3,000
Petite Sirah	II, III, IV	NA
Pinot Noir	I, II	9,100
Ruby Cabernet	III, IV, V	7,000
Zinfandel	II, III, IV	33,000
WHITE VARIETALS	**REGION**	**ACREAGE**
Chardonnay	I, II, III	48,000
Chenin Blanc	I, II, III	32,000
French Columbard	III, IV, V	62,000
Gewurztraminer	I, II, III	2,100
Pinot Blanc	I	NA
Sauvignon Blanc	I, II, III	13,800
Sylvaner or Franken Riesling	I, II	1,400
White or Johannesberg Riesling	I, II	6,000

Regarding these varietals, several points should be made: (1) Dates on bottles are significant in denoting the age of some varietals (a dated label indicates that at least 95 percent of the wine is from grapes harvested that year) but not in terms of *vintage* (best year or years), for practically every year is a vintage year in California; (2) either red or white generic wines are produced from a blend of different varieties of grapes growing in any of the wine districts in California; and (3) there are only two true colors in wine, red and white. So-called rosé wines are best made from red grape varieties, the skins imparting color to the wine while being allowed to remain in contact with the must during fermentation. But rosé wines can also be made by blending red and white grapes or red and white wines.[5] The more popular red varieties (and their growing regions) used to make rosé wines include Carnelian (IV, V), Grenache (IV, V), Napa Gamay (III), and Zinfandel. This does not necessarily preclude a rosé wine's having varietal qualities. It is true, though, that the varietal-like character of a rosé is diminished when the must is removed from the skins after only about 15 hours of fermentation in order to keep the wine from becoming too dark. Among the more esteemed rosé "varietal" wines are California Grenache and California Gamay. Yet, born in the 1900s and rising to become California's star wine of the 1990s is the "blush varietal" White Zinfandel.

In the realm of sparkling wines, California faces strong competition from New York. The Finger Lakes district of upstate New York is second only to the wine districts of California in producing the nation's finest champagnes. In both states, the better champagnes are produced in the same manner as those of Champagne, France—by means of secondary fermentation in the bottle. (The French do not recognize California sparkling wine as "Champagne," which may one day lead to California creating its own appelation, such as Italy did with "Spumante" and Spain with "Cava.") Such naturally fermented champagnes should develop in the bottle for at least a year before they are marketed; those with dates from several years back are usually at their peak of quality.

All this time and added expense can be saved by employing the *bulk* or *Charmat* process, wherein champagnes are made by fermentation in large tanks and then bottling under pressure. Cheaper and quicker still is artificial carbonation, accomplished by addition of CO_2 to the wine as in the manufacture of soft drinks. Bulk- or Charmat-process wines are champagnes but must be labeled accordingly, whereas labeling carbonated wines as champagne is expressly forbidden. In these poor man's sparkling wine types, the gas fails to become a component part of the wine, and consequently they go flat all too quickly after opening.

Naturally fermented champagnes enjoy the character that aging adds and are made from the choicest

[5] M. A. Amerine and V. L. Singleton, *Wine: An Introduction*, 2nd ed. (Berkeley: University of California Press, 1977), p. 124.

grape varieties. Depending on whether they are California champagnes, sparkling burgundies, or pink champagnes (sparkling rosés), the state's finest sparkling wines are usually made from Chardonnay, Pinot Blanc, or Pinot Noir grapes. When California champagne and sparkling burgundy are combined to produce cold duck, a concentrate of Concord grapes from the state of Washington may be added to produce a relatively sweet sparkling wine. Generally, though, champagne aficionados prefer their wines on the dry side, ranging from *brut* (very dry; less than 15 grams of sugar per liter) to *sec* (medium sweet; 17–35 grams of sugar per liter).

A New Wine Geography for California?

Many bottles of premium California wine are labeled with a county name, such as Napa or Monterey, indicating the geographic origin of at least 75 percent of the grapes used to produce the wine. By the early 1990s, however, a new system of appellation geography designating American Viticultural Areas (AVAs) was in the works. An AVA printed on the label tells the wary buyer where 85 percent or more of the grapes used to make the wine came from. Once the new system is fully developed, designated AVAs will far outnumber counties identified as producers of premium wines. In Napa County alone, for instance, several AVAs have already been registered, with more to be proposed. California AVAs, however, will be far less rigorously defined than appellation areas in France, where such determinants as control of grape variety plantings and production are employed. Thus, it appears doubtful that California appellations will ever gain the prestige of their French counterparts. Moreover, California wine buyers have come to rely on county appellations and will likely not bother to learn a far more complex geography. In any case, the quality of wineries and the grape varieties they use as raw materials are more reliable indicators of excellence than are appellations, at least in California. Is the wine industry simply spinning its geographical wheels on this one? "Reinventing the wheel" might be another way of putting it.

ODDITIES IN COMMODITIES

Marijuana cultivation, because it is illegal; turf production, because most of us perceive lawns as being born, growing up, and living out their lives in one place; Christmas tree farming, because nothing edible or buildable will come from the trees; and energy crops, because few people realize that food and feed crops are capable of yielding motor vehicle fuel—these are rarely thought of as everyday, garden-variety types of agriculture. Yet, not only are they bona fide agricultural products like beef and barley, but some of them are also beginning to nudge aside more familiar kinds of crops on the California agricultural landscape.

"Where's the Pot?"

While the subject of spirits is still fresh in mind, we might consider another of California's euphoria-inducing agricultural commodities, albeit an illegal one. Marijuana (*Cannabis sativa*, an annual herb of the mulberry family) has become one of California's most valuable crops. Because the cultivation and sale of marijuana are felony offenses in California, the annual value of the crop can only be guessed at. Such "guestimates" usually range from $800 million to more than $2 billion. If one accepts the higher-priced end of the range, then marijuana could be said to be in the same league as nursery products, cotton, and grapes in terms of the annual value of the crop. A much earlier and more conservative estimate by the U.S. Drug Enforcement Administration placed pot tenth among California farm crops with a value of $186 million in a single year late in the 1970s.[6] Ten years later, marijuana ranked fourth among all agricultural commodities with a cash crop estimated at $1.3 billion by the UCLA Drug Research Group.[7] Only milk, grapes, and nursery products were ranked higher in 1996 (see Table 10-1). Eradication efforts have increased, with authorities seizing some $657 million worth of pot in 1989, but probably twice that amount reaches underground markets.

For obvious reasons, no attempt is made here to map pot-growing regions, although such a geography of marijuana in California might become a bestseller.

Other than on condo balconies and in backyard gardens, however, where is most of the "commercial" crop grown in the state? Many people would point to the northwestern counties, where a highly potent variety of *sativa*, known as *sinsemilla* (meaning "seedless"), is widely cultivated. Actually, counties from one end of the state to the other that offer isolation, rugged terrain, clayey soils, and a reasonably lengthy growing season are likely candidates for marijuana cultivation. Although cities and suburbs constitute the major markets for the drug, individuals who grow it in urban areas do so more for private consumption than for profit. Be it in a rural or urban milieu, however, pot production is expanding, and California is rapidly moving toward self-sufficiency and a diminishing need for imports. Unfortunately, the same can't be said for some of the state's

[6] John Hurst and Phil Garlington, "Pot: The Price and Potency Are 'Way Up,'" *Los Angeles Times*, November 25, 1979, pp. 1, 20.
[7] California & Co./Daniel Akst, "Pot Still Remains Among Top Cash Crops," *Los Angeles Times*, 1991, pp. D1, D20.

Figure 10-19 *Turf grass growing in the Santa Clara Valley. The grass finds its major market just a few miles away in the San Jose metropolitan area.* (Crane Miller).

legal natural resources and manufactures, such as oil and natural gas or Chryslers and Fords.

While efforts at decriminalization of marijuana in California have made some inroads, dreams of eventual legalization have so far gone up in smoke and are likely to remain there in the foreseeable future. One wonders, though, to what degree the state's pot farmers would support legalization. At some point in any gradual legalization, income from marijuana production would become legally reportable, and the Franchise Tax Board, the Internal Revenue Service, and other public agencies would begin demanding their cut. Such revenues could conceivably lower the overall tax burden of nonsmokers, but more likely they would be spent on rehabilitation programs. Whatever its legal fate might be, the funny-smelling stuff has subtly become an important part of California's agricultural landscape.

Another Kind of Grass

This is the kind that is mowed, not smoked, and it too is increasing its share of the state's agricultural output. As seen in Figure 10-19, turf grass occupies relatively large tracts of land and must be irrigated. But turf grass is a highly remunerative crop, with several counties each producing more than a million dollars worth a year. Markets for rolled turf range from country clubs and nurseries to race tracks and city parks departments.

What is curious about nonedible nursery stock in general, which includes ornamental flowers and Christmas trees as well as turf grass, is that it is prospering from urbanization rather than being harmed by it. Obviously, urbanization has created new markets for horticultural products. The same could be said for agricultural products. What is really happening, however, as we will see in Chapter 11, is that agriculture is giving ground to urbanization while at the same time gaining increased sales from it. In a sense, Peter is being robbed to pay Paul.

Christmas Tree Farming

Commercial Christmas tree farming is more correctly deemed *silviculture* than agriculture, but whatever the terminology, it is increasingly taking place on prime land formerly used to grow food and feed crops. The economics of supply and demand and diminishing returns may check this trend; should it continue unabated, however, world food supplies would be that much more in jeopardy. Today, Christmas tree production on former agricultural land in California is becoming a multimillion-dollar-a-year industry.

From the farmer's viewpoint, though, the conversion from agriculture to silviculture has its merits. Take, for instance, the hypothetical case of a citrus grower turned Christmas tree farmer. No longer will our erstwhile citrus farmer need to drill wells or import water, for Monterey pines and most other Christmas tree varieties are rain-fed crops. Not only will water cost nothing, but the farmer will no longer have to maintain an expensive irrigation system. Labor savings will appear attractive as well, especially if the farmer charges the same price for every tree, regardless of size, and lets customers cut their own. The buyers attracted through local newspaper and radio ads will probably not have forestry degrees or experience, but all will no doubt gravitate to the largest trees on the farm. Our farmer can then sit back and watch customers go about the selective cutting of the forest. Drought or fire or disease might destroy this dream farm, but in the meantime the farmer is realizing higher profits and suffering fewer ulcers than when he or she was in citrus. And when winter frosts strike, it's not necessary to get up at night to light orchard heaters.

Energy Crops: A Sweet Solution?

Agrifuels, such as ethyl alcohol or *ethanol* distilled from plant sugars, may be California's best answer yet to high petroleum prices and diminishing supplies. Such carbohydrate-rich crops as barley, corn, potatoes, sugar beets, and wheat grow by the hundreds of thousands of acres in California, and their production could be doubled or tripled in rather short order. Historically, *gasohol*, a 90-octane mixture of 90 percent gasoline and 10 percent ethanol, was the only major agrifuel product available to the motoring public. But a revised Clean Air Act mandates that other such fuels be developed. Thus, we might soon be burning pure 200-proof ethanol in our engines. After all, racing drivers have been using alcohol fuel in their specially designed engines for years and getting up to 18 percent more power than from gasoline.

Another source of crop energy now being tested involves *biomass conversion*—in particular, the manufacture of producer gas from crop and other plant residues. If and when a major conversion from hydrocarbons to carbohydrates comes about, we would find that the latter also supplies a host of by-products ranging from lubricants to paint bases.

Before we become too enthusiastic about agrifuels, however, we should examine some of their shortcomings. For instance, environmentalists and air resource people point to more smog if gasohol replaces gasoline. Gasohol evaporates more readily than gasoline, with most of the additional evaporation emanating from gas tank spouts and carburetors. Fuel-injected engines unquestionably run more efficiently on gasohol, but carburetor-equipped cars are still around. Worse yet, emission control devices in nearly all cars are unable to capture the resulting increase in hydrocarbons from evaporating and burning gasohol. Consequently, the more gasohol used in existing cars, the greater the amount of hydrocarbon emissions reaching the atmosphere, although emissions of oxides of nitrogen will diminish. With agrifuel production in general, the other major difficulty will be coping with hidden energy costs. Think of how much energy might be spent on cultivating new cropland, running new equipment, and producing more pesticides and fertilizers. Lastly, much fuel is needed for converting crops into ethanol.

As we now leave the farm and venture into the city in Chapter 11, we would do well to recall that urban California constitutes the major market for the state's agricultural products and as such has contributed immeasurably to the prosperity of rural California.

CHAPTER

11

The City: Metropolitan California

The preceding chapter and this one might well be jointly dubbed "The Farm and the City: A Tale of Two Places," for the continued well-being and development of rural California is inexorably linked to that of urban California. More than 95 percent of California's land area is devoted to farms, forest, wilderness, and other nonurban uses, but most of the population and most decision making are found in the cities.

Regional agglomerations of cities have coalesced into metropolises, which may, in turn, one day form great, or not-so-great, megalopolises. It's important to gain a clearer understanding of how metropolitan California has grown, what its functions are, and where its impacts on prime agricultural land have occurred and are likely to take place. The connection between farm and city is, after all, a two-way street, and the sooner we appreciate the geographical bond involved, the better.

PATTERNS OF METROPOLITAN GROWTH

In California and elsewhere, the metropolitan community is largely a product of the twentieth century and, as such, represents a relatively new type of living unit. In comparing the metropolitan community to the city, Amos Hawley makes this distinction: "The city is the creature of the nineteenth century [in North America]; its successor in the twentieth century is the metropolitan community. This new urban unit is an extensive community composed of numerous territorially specialized parts, the functions of which are correlated and integrated through the agency of a central city."[1] In somewhat similar terms, Raymond Murphy defines the metropolitan area as consisting of "a recognized, substantial population nucleus and the adjacent areas of countryside and scattered urban development that have a community of interest with the nucleus."[2] In a sense, the metropolitan area is analogous to the central city and its suburbs, except that the former is not necessarily limited to a contiguous built-up area.

The idea of centrality applies in California metropolises to varying degrees, with San Francisco perhaps best epitomizing the concept and Los Angeles probably serving as the worst example. San Francisco's waterfront location, deep-water harbor, mild marine climate, function as a hub of finance, and other characteristics move geographers to compare its centrality with that of the urbanized northeastern seaboard of the United States: "Just as Megalopolis is America's hinge with Europe, San Francisco is its hinge with Asia."[3] Los Angeles, by contrast, is viewed as the least centralized of American metropolitan communities, in a very real sense "a city in search of a center." Unfocused as Los Angeles is,

---

[1] Marion Clawson, R. Burnell Held, and Charles H. Stoddard, *Land for the Future* (Baltimore: Johns Hopkins Press for Resources for the Future, 1960), p. 55.

[2] Raymond E. Murphy, *The American City: An Urban Geography* (New York: McGraw-Hill, 1966), p. 15.

[3] Stephen S. Birdsall and John W. Florin, *Regional Landscapes of the United States and Canada*, 4th ed. (New York: Wiley, 1992), p. 418.

though, "millions of people, Americans and foreigners alike, have behaved as though the region were indeed a legendary Terrestrial Paradise, the fabled Big Rock Candy Mountain."[4] As we will see in this chapter and the next one, their centrality or lack thereof notwithstanding, both metropolises are key components of the Pacific Rim.

Metropolitan Statistical Areas

The U.S. Bureau of the Census (actually the Office of Management and Budget) also has its definition of metropolitan, designating a county or group of contiguous counties that contain at least one city of 50,000 inhabitants or more a *Metropolitan Statistical Area* (MSA). Figure 11-1 maps 23 such MSAs in California as of the 1990 census; because of their delineation according to county boundaries, however, in many cases they cover huge tracts of land that are anything but metropolitan. For example, Riverside–San Bernardino is the nation's largest MSA in area but one of the smallest in population. Actually, most of it is sparsely settled desert. Obviously, the MSA serves more as a data-gathering unit with fixed boundaries than as anything else.

MSAs were formerly known as SMSAs, or Standard Metropolitan Statistical Areas. Since SMSAs were redesignated as MSAs in the early 1980s, the increase in the number of such areas (from 17 to 25) has occurred mainly in rural California, mostly in the agricultural heartland of the Central Valley and in the retirement mecca of the western Sierra Nevada. Glancing at Figure 11-1, the notable additions from north to south through the two regions are the Redding, Chico, Yuba City, Merced, and Visalia–Tulare–Porterville MSAs and expansion of the former Sacramento SMSA to include El Dorado as well as Placer, Sacramento, and Yolo counties. The land use competition implications of this metropolitan expansion, especially as concerns agriculture versus urbanization in the Central Valley, are far-reaching and will be discussed later in the chapter.

A comparison, one that could be regarded as either trivial or sobering, matches California's population of almost 33 million with that of Canada's nearly 27 million in 1990. Canada is the world's second largest country in area (3.85 million square miles)—though the Soviet Union no longer exists, Russia is largest with almost twice the area of Canada—and possesses water, energy, and land resources that turn Californians green with envy. Canadians have been lured to our great metropolitan areas by climatic, economic, and other expectations, so much so that the combined Los Angeles–Anaheim–Riverside MSA, in effect, constitutes the fourth largest Canadian city. That is, the several hundred thousand Canadians and former Canadians living in the greater Los Angeles metropolitan area number more than the inhabitants of all Canadian cities except for Montreal, Toronto, and Vancouver.

Historically, the development of the metropolitan community is a relatively recent phenomenon in North America and, as such, is to a large degree a function of modern technology. Until the turn of the century, urban growth was confined to compact *centripetal*, or inward-growing, communities. But with the twentieth century came dramatic advances in agricultural, industrial, and technological productivity that soon revolutionized the nature of cities. The growth of the automobile industry in an expanding industrial economy was a key factor in the areal explosion of cities and the emergence of suburbia. Notably in California, but throughout the metropolitan United States as well, the automobile quickly became the major mode of commuting to and from work. This is evidenced by the phenomenal rise in automobile registrations in the nation: from 8,000 in 1900 to 40,333,591 by 1950.

Accompanying technological progress was a rapid increase in personal income levels, with per-capita disposable income rising from $682 in the first year of the Depression in 1929 to $3,089 thirty years later. The standard of living of Americans improved so dramatically that residing in the suburbs was no longer the exclusive province of the wealthy but could now be enjoyed by people from a broad range of socioeconomic backgrounds. In short, the forces of technological progress and the concomitant rise in living standards allowed a relaxation of the centripetal forces that largely precluded metropolitan development in the nineteenth century.

Shrinking Central Cities

One of the unique demographic features of metropolitan growth is the propensity of the urban fringe or hinterland to experience a population growth rate that exceeds that of the central city. In fact, this tendency has so accelerated in recent decades as to suggest that the metropolitan area explosion is in essence a suburban explosion. Between 1950 and 1960, for example, 12 of the nation's 20 largest cities actually experienced a population decline within city limits, but all showed a significant increase in total metropolitan population because of suburban growth. In the same decade, 96 percent of the national population increase occurred in Standard Metropolitan Areas (SMAs were the predecessors of SMSAs and MSAs), and three-quarters of this increase took place outside the administrative boundaries of central cities.

Through the 1960s, 1970s, and 1980s, the central-city depopulation trend intensified. Among America's larger

---

[4] Howard J. Nelson, *Los Angeles Metropolis* (Dubuque, Iowa: Kendall Hunt, 1983), p. 2.

Figure 11-1 *California Metropolitan Statistical Areas. MSAs are named after the city or cities that qualified them rather than the counties that compose them—for example, the "Redding" MSA in Shasta County.*

central cities, Baltimore, Detroit, Philadelphia, and Washington DC all lost population between 1980 and 1990. Chicago is perhaps the biggest loser in the nation, at least in a relative sense, according to 1990 census data. Between 1970 and 1980, the Windy City lost 644,072 people, more residents than moved away during the two preceding decades. And by 1990, Chicago's population had decreased another 221,346, from 3,005,072 in 1980 to 2,783,726. With these losses, Los Angeles passed the Windy City to become the nation's second largest city,

Figure 11-2 *Aerial view of San Francisco. Except for the Presidio to the northwest (upper left) and rectangular Golden Gate Park, the University of California's San Francisco campus, Buena Vista Park, and Twin Peaks to the southwest (lower left), this vertical view displays wall-to-wall city at the northern end of the San Francisco Peninsula. With the main boulevard, Market Street, diagonally offsetting the directional orientation of the rectangular street pattern, the pattern itself tends to flatten or obscure the very hilly terrain with 40 hills and elevation changes ranging from sea level to 929 feet.* (NASA)

with a 1990 census count of 3,485,398. Even sprawling Los Angeles posted a modest gain of only 140,397 new residents during the 1970s, but it gained a half million in the 1980s. And San Francisco, once California's most populous city, lost population during the 1970s as it had during the 1960s. But by 1990, San Francisco had rebounded, climbing from 678,974 in 1980 to 723,959. Still, San Francisco slipped ever farther behind San Jose as the Bay Area's largest city, with the latter's population growing from 629,400 in 1980 to 782,248 by 1990. Statewide, San Francisco has long since fallen from second place, now ranking as the fourth largest city behind Los Angeles, San Diego (1,110,549 in 1990), and San Jose. It may no longer be California's "second city" in population, but many will always look on San Francisco as California's "first city" in all other regards. And few will quibble with San Francisco's status as the cultural and financial hub of the nation's fourth largest CMSA. Only the MSAs centering on New York City, Los Angeles, and Chicago, in that order, have larger populations.

San Francisco's Depopulation: 1960–1980

Originally founded in 1835 as Yerba Buena, which rather prophetically translates to "good grass," San Francisco reigned as the state's largest city from Gold Rush times until the 1920s. During those 70 years, the city was leveled on more than one occasion by fire and/or earthquake, yet each time it recovered to assume even greater urban prominence. San Francisco seemed destined to grow in spite of the worst disasters and hazards imaginable, and when it ran out of horizontal space (Figure 11-2), it reached for the sky. Today, Baghdad-by-the-Bay puts other California metropolises to shame as a skyscraper-filled, cosmopolitan urban center,[5] and one with hardly a trace of air pollution. Why, then, did San Francisco's population slip from 740,300 in 1960 to 715,674 in 1970 to 678,974 in 1980, while other California cities seemed to buck the national central-city depopulation trend? San Francisco is back on the growth road, but just barely, and for how long? In 1990, the city was still 16,000 short of its 1960 population.

Explanations of San Francisco's population losses abound, many of them symptomatic of depopulating central cities throughout the United States. Deteriorat-

[5] In 1974, San Francisco ranked sixth in the United States in skyline height, with New York first and Los Angeles nineteenth. See Larry Ford, "The Urban Skyline as a City Classification System," *Journal of Geography* 75 (1976), pp. 154–164.

Figure 11-3 *Hunter's Point to the southeast (right) and Market Street and downtown San Francisco to the northeast (left) as viewed from Twin Peaks. In Figure 11-2, the Twin Peaks vantage point is located in the southwestern (lower left) portion of the aerial photo.* (Crane Miller)

ing and overcrowded residences in inner-city neighborhoods constitute one of the most serious problems. In Chinatown, much of the housing is considered substandard. Out toward the Bay on Hunter's Point (Figure 11-3), where much of San Francisco's sizable black population resides, a mix of old and new low-cost housing exists.

Many Chinese have relocated in the more affluent northwestern part of the city close to Golden Gate Park, while black families have done likewise in the older Haight-Ashbury district. Appropriately dubbed "Hashbury" back in the 1960s, the district became a self-styled hippie haven, but by the 1980s that reputation had faded and Haight-Ashbury had reassumed an image of residential stability. More recently, Haight-Ashbury even became a fashion styling center of sorts, originating a neo–flower children look of the 1990s. Anywhere in San Francisco, affordable housing is at a premium, although the hilly terrain affords a plethora of spectacular views. Relatively high rents are undoubtedly a root cause of the exodus, as well as a deterrent to prospective residents. An apparently increasing number of blacks who work in San Francisco now commute to work from the East Bay, whereas the Chinese population by and large seems to prefer continuing both employment and residence entirely within the city.

Inner-city crime could be said to be a major deterrent to settling in San Francisco, were it not that every American city seems similarly plagued. In the evening, in fact, the streets of San Francisco are probably safer places to be than those of downtown Los Angeles or Sacramento. The city is rich in restaurants, theaters, museums, and clubs, well diffused throughout the city and thus reducing the presence of dark, lonely, and out-of-the-way streets and back alleys. The old adage that "when the lights go up, crime goes down" seems to apply here.

San Francisco's position nearly on top of the San Andreas fault has been cited as a contributor to depopulation. But again, residents seem "unmoved" by this environmental hazard. This finding was borne out in a hazard perception study in which not a single respondent in the sample group identified earthquakes as a disadvantage of living in San Francisco.

Finally, there is the simple lack of space, which is probably a more substantive reason for San Francisco's population loss than any other. Bounded on three sides by water and on the fourth by its border with San Mateo County, San Francisco City and County long ago saw almost every square inch of its 60 or so square miles developed to urban land uses, including parks. The city then had nowhere to go except up. But vertical development has its limitations, too, especially when building finances become scarcer and the danger of the San Andreas fault acting up becomes greater. Los Angeles, Sacramento, San Diego, San Jose, and other big California cities each have four or five times or more area for development than does San Francisco. Moreover, none of these cities sits right on the San Andreas, although San Jose may be a bit too close for comfort.

The October 1989 Loma Prieta earthquake is a case in point. That 7.1 temblor and attendant liquefaction (see Chapter 2) wreaked havoc in the city's Marina District, displacing thousands of residents from the district and elsewhere in the Bay Area for varying periods of time. Yet, in ensuing years, there's been no apparent exodus of San Franciscans from the city, although many interviewed right after the quake threatened "to get out as soon as possible." Nevertheless, San Francisco's high-density population and high environmental hazard risk are evidently proving increasingly oppressive to residents, even though some won't admit it until after they've moved away.

Where Did They Go?— A Bay Area Rebound

Where have former and even would-be San Franciscans gone? Some have moved across the Bay to established cities like Oakland, Berkeley, and Richmond, and then down the East Bay to populate newer bedroom communities such as Hayward and San Leandro. Others have migrated directly down the peninsula, populating a string of old and new cities from Burlingame and San Mateo southward to Sunnyvale and San Jose. During the 1960s and 1970s, Sunnyvale was one of the nation's fastest-growing cities, and San Jose became the state's second largest city in population. There seemed an inexhaustible supply of former prune orchard land available for urbanization in the South Bay–Santa Clara Valley region. And northward from San Francisco and mountainous Marin County lay thousands of acres of valley bottomland just waiting for the bulldozer and urban

TABLE 11-1 California's "Green Cities Index" and cities' populations and ethnicity 1990.

OVERALL RANK	CITY	POPULATION/ RANK	POPULATION DENSITY	AIR QUALITY (OZONE COUNT)	ENVIRONMENTAL STRESS	ETHNICITY* INDEX
8	San Francisco	723,959/51	63	2	7	2
13	Oakland	372,242/26	50	40	1	NA
27	San Diego	1,110,549/59	32	59	54	42
40	Sacramento	369,365/24	36	52	54	NA
50	Fresno	354,202/18	33	56	54	18
53	San Jose	782,248/54	40	40	43	20
54	Anaheim	266,406/6	44	61	51	NA
61	Los Angeles	3,485,398/63	51	63	43	3
62	Long Beach	429,433/33	54	63	43	3
64	Santa Ana	293,742/13	58	62	54	NA

* Ethnicity index (1 = ethnically most diverse, 50 = least diverse) rank by county where proportions of non-Hispanic whites, non-Hispanic blacks, Hispanics, and other non-Hispanics are nearest to being equal; based on the 1990 census.

Sources: World Resources Institute, *The 1992 Information Please Environmental Almanac*, "Green Cities" (Boston: Houghton Mifflin, 1992), pp. 169–186; U.S. Department of Commerce, Bureau of the Census, *1990 Census of Population and Housing*, Final Report, January 1991; and *The World Almanac and Book of Facts 1992*, "The 50 Most Racially Diverse Counties in the U.S." (New York: Pharos Books, 1991), p. 133.

development. In the 1970s and 1980s, the San Francisco–Oakland metropolitan area spread to Napa and Sonoma, with vineyards and orchards being replaced by housing tracts and shopping centers. Physically binding this supercity together is a complex network of trans-Bay bridges, commuter rail lines, and freeways. Politically trying to keep the whole thing under control is California's very first Council of Governments (COGs), the Association of Bay Area Governments (ABAG). The more than 4 million residents, 85 incorporated cities, and seven counties within ABAG's original jurisdiction give us some sense of the enormity of metropolitan San Francisco's earlier expansion. By 1990, the Bay Area MSAs contained 6.25 million residents and 10 counties, their inland penetration carrying them all the way to the university town of Davis in northeastern Solano County.

Actually, only a small proportion of the greater Bay Area's gain was San Francisco's loss. In 1988, for instance, 421,000 people moved to California from out of state or country, more than in any single year since World War II. Those and other immigrants dispersed throughout the state, but most settled in one of the two largest metropolitan areas. Moreover, families moved around within the state, many reasoning that Northern California was a better bet economically and environmentally than the region south of the Tehachapis. If one is seeking residence somewhere in metropolitan California, the Bay Area certainly seems to offer a better overall quality of life than does the Los Angeles area. The Bay Area can lay claim to more water, cleaner air, less environmental stress, fewer people, better mass and rapid transit systems, and less street crime than in metropolitan Los Angeles. The Loma Prieta earthquake and the Oakland Hills fire might discourage a few people from moving to metropolitan San Francisco, but southland cities are no safer from the devastation of earthquakes and fires than those in Northern California.

The thesis that San Francisco and other Bay Area cities offer a better all-around urban environment than Los Angeles and other Southern California cities gained credence in the 1991 Green Cities Index (GCI) published by the World Resources Institute.[6] Developed by geographer Susan L. Cutter at Rutgers University, the GCI ranks 64 of America's largest cities (1 being best and 64 being worst) based on 24 different measures of environmental quality. Table 11-1 shows the 10 California cities appearing in the 64-city GCI ranked according to a compilation of values for 14 different environmental measures, as well as several other measures. Lower rankings indicate better environmental performance, but "these measures are not weighted as to their relative importance or overall contribution to making a city more livable, from an environmental perspective. Rather, they

[6] World Resources Institute, *The 1992 Information Please Environmental Almanac* (Boston: Houghton Mifflin, 1992), pp. 169–186.

serve to illustrate how each city fared when compared to others."[7] As the table shows, San Francisco ranked highest among California cities, at 8, while Santa Ana ranked lowest, at 64. For the sake of comparison with other U.S. cities, Honolulu was ranked at the top, Portland (Oregon) at 12 (just ahead of Oakland), Tulsa at 26 (just ahead of San Diego), Denver at 39 (just ahead of Sacramento), New York at 57, Chicago at 60 (just ahead of Los Angeles and Long Beach), and St. Louis (Missouri) at 63 (just ahead of Santa Ana). Before making light of Santa Ana's poor overall showing, remember that only 64 of America's largest cities were ranked. If the list were longer, Santa Ana would look better.

Table 11-1 also ranks the "California 10" by GCI in population, population density, air quality by ozone count, level of environmental stress, and ethnicity. For example, San Francisco, with its population on the rebound and already high population density thereby increasing, ranked next to last in the nation in the GCI for density. As noted earlier in the chapter, San Francisco's Chinatown has a chronically acute problem in this regard. According to an April 1991 television news report, 80 percent of Chinatown's elderly were living in tiny, substandard, rat-infested rooms with little possibility of help from the city because of a worsening recession.[8] In the context of the GCI, however, San Francisco's low population density ranking was somewhat counterbalanced by its relatively high standings in air quality (second in ozone occurrence) and lack of stress on selected components of the environment (seventh in overall air, water, and other component qualities). Although not listed in Table 11-1, one GCI category in which 9 of the 10 California cities can take great pride is "Energy Use." With the exception of Fresno, they are ranked first through ninth among all 64 U.S. cities in being the "smallest users of energy for heating and cooling," starting with San Diego and ending with Sacramento. Finally, San Francisco's second-place standing, out of 50 counties nationwide, in ethnic diversity will be addressed later in this chapter.

Suburbanization

The suburbanization movement represents an escape from congestion like San Francisco's, as well as a move to create more space for burgeoning city populations. The population explosion alone would have caused such a movement, but the concomitant migration to lower-density areas makes the spreading of cities proceed at a more rapid rate and causes it to cover more territory. As a result, the main characteristic of suburbanization is the thin, often discontinuous distribution of population over broad expanses of land. This low-density development, which often occurs in a mixture of developed and vacant land, is the essence of urban sprawl.

Los Angeles: Suburbs in Search of a City

Founded in 1781 as a Spanish pueblo, the growth of *El Pueblo de Nuestra Señora La Reina de Los Angeles*, both as a city and a metropolitan area,[9] contrasts markedly with that of San Francisco. The differences in physical setting (see Chapter 3) alone would no doubt have caused Los Angeles to develop quite differently from San Francisco. Originally situated more than two dozen miles upstream from the Pacific along the intermittent Los Angeles River, the City of Angels could expand for miles in almost any direction over relatively level land. But in its first 100 years, Los Angeles hardly took advantage of any of its breathing room, although it did become the most populous pueblo during both the Spanish and Mexican periods, reaching 1,500 residents by 1836. A dozen years later, though, San Francisco, with its proximity to Gold Rush country and a huge natural deep-water harbor, temporarily stole the urban limelight from Los Angeles.

In contrast to its first century, Los Angeles started its second 100 years with the "Boom of the Eighties,"[10] a boom that would ebb and flow right to the present as the metropolis heads into its third century. The advantages of Los Angeles's physical geography, principally an abundance of developable terrain (see Chapter 3) and the mildest of Mediterranean climates (see Chapter 4), had become legend to midwesterners and easterners by the 1880s. So westward they came, most via recently completed transcontinental railroad lines, but some by

---

[7] World Resources Institute, *The 1992 Information Please Environmental Almanac*, (Boston: Houghton Mifflin, 1992), p. 177.

[8] CBS Television, *This Morning*, April 10, 1991.

[9] Although the Los Angeles–Long Beach MSA and Los Angeles County are one and the same, with an area of 4,000 square miles and a 1990 population of 8.9 million, the Los Angeles metropolitan area could be said to at least include all the counties that border Los Angeles County except Kern County, which is physically separated by a formidable mountain range. Such a metropolitan area, then, would also include Orange County (2.4 million), Riverside County (1.2 million), San Bernardino County (1.4 million), and Ventura County (669,016). Riverside County is California's fastest-growing county, nearly doubling in residents from 1980 to 1990. Ventura County, on the other hand, is one of the slowest gainers, with an increase of only about 140,000 for the 1980s. Early in the 1970s, Los Angeles County actually lost population, dropping from 7.0 million to 6.9 million, but it recovered by about 500,000 between 1975 and 1980. The 1971 San Fernando earthquake may have driven some people away from L.A. County, but not so the 1994 Northridge quake. The COG for Los Angeles and adjoining counties is the Southern California Association of Governments, or SCAG.

[10] See Glenn S. Dumke, *The Boom of the Eighties in Southern California* (San Marino, CA: Huntington Library, 1944).

Figure 11-4 *Aerial view of Santa Monica Bay and the Los Angeles Basin. The dark area left of center and fronting on the bay is the chaparral-covered Santa Monica Mountain range. The eastern extension of the range, the Hollywood Hills, separates Beverly Hills, Hollywood, and downtown Los Angeles on the south from the urbanized San Fernando Valley to the north. The three intermittent rivers flowing out of the southeast (lower right) portion of the basin are, from left to right, the Los Angeles, San Gabriel, and Santa Ana. Concrete-lined flood control channels for the most part, the river courses are sometimes mistaken for freeways in aerial and satellite images.* (NASA)

old wagon trails, others by ship around Cape Horn, and a few by a trek over the isthmus of Panama. Spurred on by news of the hastened subdivision of vast Mexican-era ranchos, everyone from real estate speculators and retirees to fledgling citrus farmers and petroleum wildcatters flocked to the southland.

By about the turn of the century, the newcomers had firmly set the pattern for America's first sprawling metropolis. Far-flung irrigation colonies from Anaheim to Pasadena had begun their transition to fashionable suburbs. Edward Doheny had struck oil west of Figueroa near downtown Los Angeles. William Mulholland had launched a search for desperately needed water that would end successfully in faraway Owens Valley (see Chapter 5). The southland's first and last interurban rail transit system, the Pacific Electric Railway Company, had incorporated. Fashionable Bunker Hill had been connected to the rest of downtown Los Angeles by a funicular railway, later known as Angel's Flight; the funicular operated from 1901 until 1969. And the city of Los Angeles proper had passed the 100,000 mark in population.

Los Angeles attained the 1 million mark by the 1920s, when it again assumed the status of the state's most populated city. The city reached the 3 million plateau during the early 1980s, as it continued to maintain its number one position. By 1990, Los Angeles's residents numbered 3.5 million.

Today, the city of Los Angeles spans 465 square miles and has indeed radiated outward from its civic center in every direction imaginable (Figure 11-4). To the west, many a boulevard has helped bring the city all the way to the shores of Santa Monica Bay. Truly the most incredible of these arteries is Wilshire Boulevard, a 25-mile corridor of hotels, office buildings, high-rise condominiums, banks, and exclusive department stores stretching from downtown Los Angeles (Figure 11-5) through affluent Beverly Hills to the palisades of Santa Monica.

Southward, the city again reaches for the sea, culminating its southerly development in the finest of man-made harbors at San Pedro.

Occupying much of the older residentially developed territory between the city and its port is Watts-Willowbrook, home of California's largest black community, as well as a growing Hispanic population. The Coliseum (Figure 11-6) Sports Arena, Museum of Science and Industry, Exposition Park, and the University

Figure 11-5 *Downtown Los Angeles skyline. The Department of Water and Power reflecting pool is in the foreground. To the left of center is the city's tallest building, the 1,017-foot-high First Interstate World Center. Immediately to the east (extreme left side of the photo) is the ever-expanding Little Tokyo district of downtown Los Angeles.* (Crane Miller)

Figure 11-6 *Los Angeles Memorial Coliseum, site of the 1932 and 1984 Summer Olympic Games and home to the USC Trojans football team. The Coliseum was also home to the professional Los Angeles Raiders football team, but in 1995 the Raiders moved back to Oakland. The UCLA Bruins and the Los Angeles Rams football teams also played in the stadium but have long since changed their venues, the former to Pasadena's Rose Bowl and the latter to Anaheim in 1981 and then to St. Louis, Missouri, in 1995.* (Crane Miller)

of Southern California also occupy considerable territory as a contiguous unit in south Los Angeles.

East of downtown Los Angeles, longer-established Hispanic neighborhoods, or barrios, dominate the cultural landscape, although the occasional eastern European and Asian enclaves are also to be found. Many suburbs east of the Los Angeles River originated along transcontinental rail lines in the days when passenger trains were in vogue. Pasadena, Arcadia, Downey, Norwalk, and other bedroom communities to the northeast and southeast first developed quite independently of any early transportation linkages with Los Angeles or the rest of the nation.

Finally, far to the north of downtown Los Angeles lies the city's last and largest residential frontier, the San Fernando Valley. Readily accessible by freeways over the Hollywood Hills and through Glendale Narrows, the San Fernando Valley was quickly transformed from a verdant farmscape of citrus groves and row crops to an indifferent expanse of tract houses, commercial strips, schoolyards, and parking lots. In the quarter century following the end of World War II, the valley's population exploded, from less than 100,000 to more than a million.

Although the ethnicity of the valley population has diversified in recent years, whites are still much in the majority in the outlying, more affluent suburbs. This situation, and distances sometimes exceeding 50 miles between San Fernando Valley public schools and those in inner-city minority neighborhoods, hindered efforts to implement court-ordered busing of students throughout the Los Angeles Unified School District in the 1980s. Institution of bilingual education, especially for the more than half of the district's student body who are Hispanic, represents still another effort at improving integrated education throughout this impossibly large city.[11] Whether these and other attempts at diffusing Los Angeles's highly polarized ethnogeography ever work, the day is fast approaching when once again a majority of the citizenry will be Hispanic.

Getting Los Angeles Back on Track

Navigating Los Angeles's vast urban geography is generally accomplished by car or bus—and far more by the former than the latter. Unlike the Bay Area with its commuter train and BART systems, metropolitan Los Angeles was nearly devoid until recently of rail transit

[11] Larger in area than the city of Los Angeles, the district serviced 710 square miles, 637 schools, and 540,000 students in the early 1980s. It is the nation's second largest school district in population. In 1981, mandatory busing was demandated when the state Court of Appeals and the California Supreme Court ruled that antibusing Proposition 1 (1980) was constitutional and applicable in the Los Angeles case. By the opening of the 1981–82 academic year, district enrollments were approximately 45 percent Hispanic, 25 percent black, 23 percent white, 7 percent Asian-American, and less than 1 percent Native American. Ten years later, in 1991, total K–12 enrollment had grown to 637,017. Hispanic representation had soared past 64 percent; black and white proportions had fallen to 15 percent and 13 percent, respectively; the Asian-American increment remained at 7 percent; and Native American students remained at less than 1 percent.

systems linking its far-flung suburbs or closer-in ethnic neighborhoods. Angelenos are married to freeways and surface streets. Of course, San Franciscans have them, too, but they can ride the rails if they so desire. Greater Los Angeles, though, has not always lacked in mass rail transit and, as a matter of fact, is seeing to its revival.

Although streetcars made their final runs in Los Angeles in 1963 and the last of the Red Cars was retired from the Los Angeles–Long Beach line in 1961, old-timers can recall the days when Southern California boasted an electric railway system containing 2,000 miles of track and serving nearly a million area residents. Born in 1887, when electrified streetcars began replacing horsecars in Los Angeles, and greatly expanded once Henry Huntington bought into it, electric rail service reached its heyday by the 1920s. Not only was the Los Angeles Railway carrying hundreds of thousands of local riders daily in its streetcars, but the Pacific Electric was moving another 100,000 to and from cities dozens of miles from downtown Los Angeles on its interurban lines.

The demise of these systems began with the Great Depression of the 1930s, when operating costs remained constant while ridership and revenues plummeted. Another nail in the coffin of mass rail transit came with the dedication of Los Angeles's first freeway—the Arroyo Seco or Pasadena Freeway—in the late 1930s. With freeways touted as the panacea to Southern California's urban transportation problems, sentiment to replace streetcars and interurban Red Cars with automobiles and buses intensified. The Los Angeles Railway and Pacific Electric won a few years' reprieve from World War II and attendant gas rationing, but cars and freeways were the wave of the future, and an aging transit system was not about to stand in the way. The beginning of the end came even before the war was over, when, on January 10, 1945, Huntington's estate sold the Los Angeles Railway "to the Los Angeles Transit Lines, whose parent company, National City Lines, was owned by General Motors, Standard Oil of California, Firestone and Rubber Company, among others."[12] Although Los Angeles Transit Lines initially ordered new streetcars and otherwise improved the system, it wasn't long before streetcars were being replaced by a few trackless trolleys and a lot of diesel buses. The final blow came in 1958 when the Los Angeles Transit Authority took over what was left of the streetcar system and phased it out completely in 1963. A similar fate befell the Pacific Electric system when its proposal in 1947 to build a Red Car right-of-way down the middle of the yet-to-be-built Hollywood Freeway failed to gain sufficient funding. At the time, there were no funds legally alloted for the median rail lines. The addition to the freeway would have cost $20 million and was said to be capable of handling more passengers than eight lanes of freeway, but the decision makers of the day and an increasingly auto-oriented public turned a deaf ear. Moreover, the Pacific Electric system as a whole was decaying, losing money and abandoning line after line through the 1950s.

In retrospect, the conventional wisdom of the day—be it from the automakers, the petroleum companies, the transit authority, the motoring public, or whomever—dictated that all of Los Angeles's eggs be put in the freeway basket and that the existing mass rail transit system be junked. Obviously, hindsight will not resurrect the old system; suffice it to note that the new system is being built at a phenomenally greater cost. Four decades ago, costs for upgrading the Pacific Electric system were spoken of in terms of millions of dollars. Today, as we will see, development of the new Metro rail system is costing billions of dollars. It seems fair to say that lack of foresight in the past, even if it had been solely to preserve some of the old Pacific Electric rights-of-way, is costly in the present as well.

Few would argue with the statement that "rail transit can move people faster and cheaper than any other travel alternative."[13] With approval by Los Angeles County voters in 1980 of a half-cent sales tax increase to help pay for it, the Southern California Rapid Transit District (SCRTD) is developing the 150-mile Metro system mapped in Figure 11-7. Completed and put into operation in the early 1990s as the first of several light-rail lines radiating out from downtown Los Angeles, the new system's Blue Line has lived up to its advanced billing. Not only have opinion polls demonstrated rider satisfaction with and loyalty to the Blue Line, but the Los Angeles–Long Beach line has removed up to 4,000 cars from the roads each weekday. The Red Line (see Figure 11-7), which when completed will link Los Angeles with North Hollywood in the San Fernando Valley, will be capable of serving 300,000 passengers a day, or 10 times the Blue Line's daily ridership. The Metro system will eventually expand to 300 miles of railways and busways, with $49 billion budgeted for its development. But the territory it covers will still be a far cry from what the discarded system covered. Furthermore, in January 1998, mounting cost overruns threatened to halt construction of the Red Line in its tracks. Besides skyrocketing costs, the Metro system plan has been beset by controversy. Regarding driverless trains proposed for the Green Line from Norwalk to El Segundo, for example, debate raged over such matters as copying proven systems, like Vancouver's highly successful "Sky Train," and building the cars in America rather than overseas. In January 1992, Los Angeles even went

[12] Howard J. Nelson, *The Los Angeles Metropolis* (Dubuque, Iowa: Kendall/Hunt, 1983), p. 272.

[13] Southern California Rapid Transit District, "The Metro Red Line" brochure, no date.

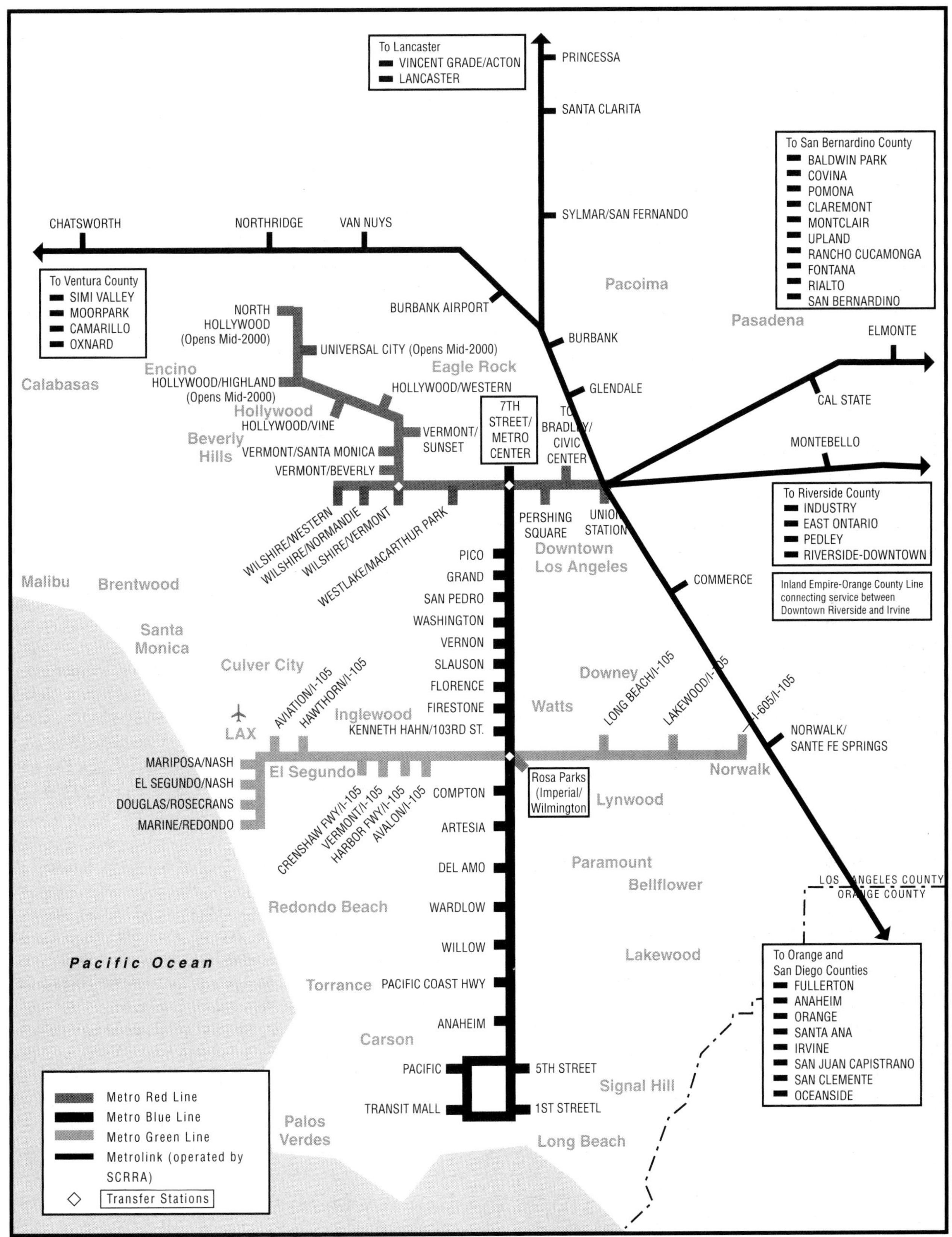

Figure 11-7 *Metro light-rail system. Like the old Pacific Electric system, the new Metro system pivots on downtown Los Angeles. With the Blue Line from Los Angeles to Long Beach completed in 1990 and the Red (to North Hollywood) and Green (to El Segundo) lines scheduled for completion early in the next century, some 150 miles of track and dozens of stainless steel trains will be serving several hundred thousand county commuters by the turn of the century.* (Source: Courtesy of Metropolitan Transit Authority, Los Angeles.)

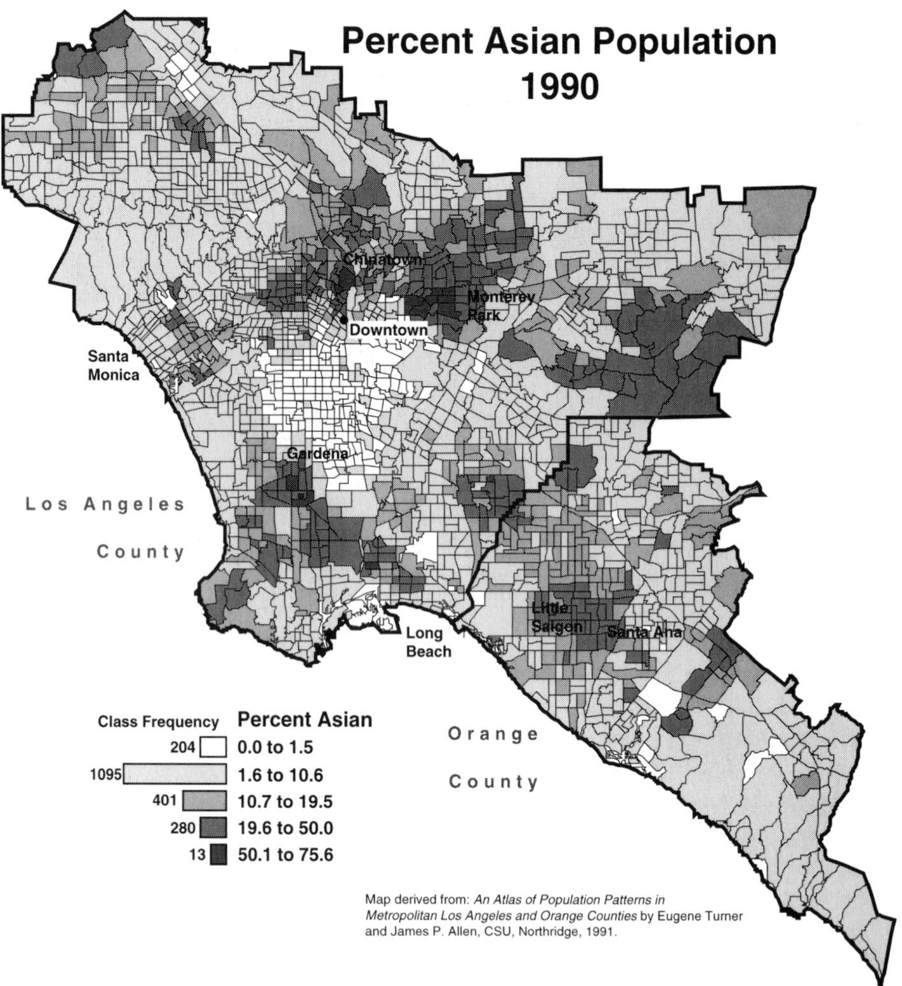

Figure 11-8 *Asian population in metropolitan Los Angeles County, 1990.* (Source: Computer-drawn map by Professor Eugene Turner, Department of Geography, CSU Northridge, from U.S. Census Bureau, *1990 Census of Population and Housing*, Public Law 94-171 data.)

so far as to revoke its contract with Sumitomo of Japan for building the driverless cars, and accusations of "Japan-bashing" quickly followed. In 1999, it came to light that a stretch of the eastern portion Green Line and the Century Freeway, in which the rail line occupies the center divider space, is being worked on to compensate for possible subsidence caused by a high water table beneath the freeway and rail line. As much as anything else, the El Nino rains of 1998 may be a culprit, since they contributed to the rise of the water table level.

Los Angeles and San Francisco: A Tale of Two Ethnicities

Viewed from almost any perspective, Los Angeles and San Francisco live up to their reputations as two of the most ethnically diverse counties in the nation. Only Queens County, New York, ranked ahead of them among "The 50 Most Racially Diverse Counties in the U.S." (see Table 11-1). Researchers view Los Angeles and San Francisco as meccas for the study of urban growth and ethnic diversity, noting that "unprecedented waves of immigration from all parts of the world are fast turning Los Angeles into the nation's greatest urban laboratory"[14] and that "the dramatic population changes that are remaking San Francisco touched every neighborhood and district during the 1980s, from Portola to the Sunset, from Nob Hill to the Ingleside. Fueled by 43 percent growth among Asians and 20 percent among Latinos, the changes recorded by the 1990 census also showed a chipping away of old ethnic enclaves. What emerged is a picture of a city evolving into one whose well-known ethnic diversity is more geographically spread out than ever before."[15]

Criss-cross Los Angeles County on any day, in any direction by surface street, and not freeway, and it will be difficult to come away with anything but a sense of the county as a paragon of ethnic diversity. One such route might be the north-south, 20-mile stretch of Western Avenue. Starting in the north, just a few miles northwest of downtown Los Angeles, a drive down the entire length of Western takes one through several growing Asian neighborhoods (Figure 11-8) and the largest and

[14] Tracy Wilkinson, "L.A.'s Turn as Urban Laboratory," *Los Angeles Times*, December 16, 1991, p. A1.
[15] Dexter Waugh, "City's Districts Gain in Diversity," *San Francisco Examiner*, May 8, 1991, p. A1.

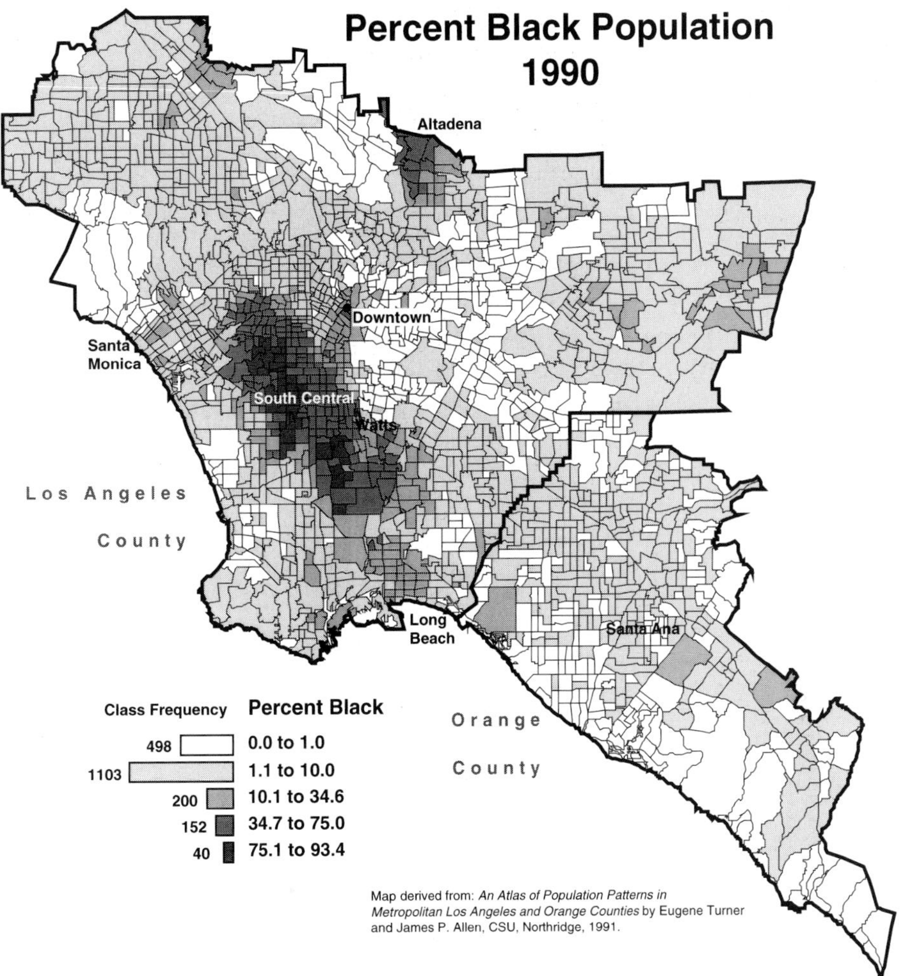

Figure 11-9 *Black population in metropolitan Los Angeles County, 1990.* (Source: Computer-drawn map by Professor Eugene Turner, Department of Geography, CSU Northridge, from U.S. Census Bureau, *1990 Census of Population and Housing*, Public Law 94-171 data.)

A

B

Figure 11-10 *(A) Intersection of Western Ave. and 6th St. after the 1992 Los Angeles riots. In western Koreatown, every business in the neighborhood shopping complex in one corner of the intersection was gutted by fires set by outsiders during the riots. (B): Older, three-story apartment building on Olympic Blvd. just west of downtown Los Angeles, torched during the riots. Note the downtown Los Angeles skyline in the background.* (A & B: Crane Miller)

longest-standing black community (Figure 11-9) in the county. Most apparent along the southbound transect is the sudden transition from Asian to black community at the southwestern edge of Koreatown, or just past Olympic Boulevard. South of here, one cannot help but notice the sudden dearth of signs in Korean, Tagalog (Filipino) Thai, and other Asian languages that decorate storefronts, restaurants, and billboards northward along so much of Western Ave. (Figure 11-10). Least apparent along such a cultural cross-section are tensions between the Asian American and African American communities that have led to conflict—for example, over the owner-

TABLE 11-2 Population and ethnic group distribution, Los Angeles and San Francisco counties, 1980 and 1990.

	LOS ANGELES COUNTY				SAN FRANCISCO COUNTY			
ETHNIC GROUP	1980 POPULATION	PERCENTAGE	1990 POPULATION	PERCENTAGE	1980 POPULATION	PERCENTAGE	1990 POPULATION	PERCENTAGE
American Indian*	—	—	29,159	<1	—	—	2,635	<1
Asian†	452,232	6	907,810	10	145,461	21	205,686	28
Black	926,360	12	934,776	11	84,857	13	76,343	11
Hispanic	2,066,103	28	3,351,242	38	83,373	12	100,717	14
White‡	3,953,603	53	3,618,850	41	355,161	52	337,118	47
Other§	79,205	1	21,327	<1	10,122	>1	1,460	<1
Total	7,477,503	100	8,863,164	100	678,974	100	723,959	100

* In 1980, American Indian, Eskimo (Inuit in Canada), Aleut (Aleutian Islander), Pacific Islander, and Asian were all under "Asian."

† In 1990, "Asian" category included only Asian and Pacific Islander.

‡ Expanded designation for White is Non-Hispanic White.

§ Some small Asian and Pacific Islander groups were reclassified as Asian in 1990. Therefore the "Other" category showed a paper loss and the "Asian" category tended to be inflated, but ever so slightly.

Source: U.S. Census, 1980, Summary Tape File 1, and U.S. Census, 1990, Public Law 94-171 Redistricting File.

ship of retail establishments by Korean Americans within the black neighborhoods to the south.

Table 11-2 affords some insight into this multicultural tapestry by noting the phenomenal growth of the county's Asian population and the leveling-off of the black population between 1980 and 1990. With both populations approaching the 1 million mark, African Americans remain concentrated in the South Central L.A. area while large communities of Asian Americans are found in such widely separated places as Koreatown and Gardena, which border South Central Los Angeles on the north and south, respectively, and Monterey Park and Long Beach, which are farther away from the African American community (see Figures 11-8 and 11-9).

Nowhere in the county has the Asian population grown as dramatically or as quickly as in Koreatown. A generation ago, this mid-Wilshire district of aging, two-story, single-family houses and low-rise apartment buildings was predominantly non-Asian, inhabited largely by white retirees. Today, though many of the original dwellings remain, albeit the worse for wear, the rest of the human landscape has become overwhelmingly Asian, and the appelation "Koreatown" aptly describes the district. Population densities have been high for decades, but they have recently reached some 46,000 persons per square mile. This concentration of resident population, coupled with bus ridership at a record 200,000 boardings per day along the Wilshire corridor, prompted the Southern California Rapid Transit District to run one of its new Metro Red Lines through Koreatown (see Figure 11-7). Koreatown is bursting at the seams and thereby inevitably spilling over into neighboring districts, especially southward into the African American community, where residential and commercial property is more affordable than eastward toward the downtown central business district, northward into the Hollywood Hills, or westward in the direction of Hancock Park, The Miracle Mile, and eventually Beverly Hills. Los Angeles's black community is long established and may view "the new kids in the neighborhood" with some degree of trepidation and hostility. But Asian populations are booming and expanding their geography everywhere in the county, be it in Long Beach, where both Southeastern Asian and Hispanic populations are growing and competing for territory, or in Hacienda Heights–Walnut, which has a newer Chinese population.

According to Figures 11-11 and 11-12 and Table 11-2, both Hispanic and white populations are more widespread and numerous than those of Asians and blacks. A drive "born in East L.A." and taken along the entire length of Whittier Boulevard (Figure 11-13) into La Habra in Orange County could be duplicated on any number of other routes through Los Angeles County, and all would reveal the ubiquitous geography of Hispanic and white settlement. Each group numbers about 3.5 million and is found to varying degrees in every corner of the county; however, between 1980 and 1990, the Hispanic population grew by 1.3 million while the number of whites declined by more than 300,000 (see Table 11-2). "White flight" has been to Orange, Riverside, and San Bernardino counties and even out of state, especially to the Pacific Northwest. The Hispanic influx has come mainly from Mexico, but Cubans, Guatemalans, Puerto Ricans, Salvadorans, and other Latin Americans have immigrated in sizable numbers as well. For instance, there are about 400,000 Salvadorans in greater Los Angeles, which is the largest such population outside of the country of El Salvador itself. Language and religion distinguish the cultural geography of Hispanics from that of non-Hispanic whites. Although most Hispanic-Americans are bilingual, their principal language is Spanish, and their religion Roman Catholic. Some 300,000 Hispanics in Southern California speak Portuguese, many of them tracing their roots back to Latin America's largest country, Brazil, and before that, Portugal. Although descended from a myriad of Indo-European and Finno-Ugric or Altaic linguistic stocks, the vast majority of non-Hispanic whites speak only English. Their religious affiliations, on the other hand, vary greatly, from Roman Catholicism to Protestantism to Judaism.

Looking only at the all-encompassing categories of Asian, black, Hispanic, and white merely skims the surface of Los Angeles County's ethnicity—as if seeing the forest but not the trees. Yet, to produce such an intricate ethnogeography would entail examining the where and why of everyone from Fairfax Avenue's Jewish community and Los Angeles Unified School District students who speak some 80 different languages to the Islamic faithful and the nation's and county's least remembered people, urban Native Americans. But we must move on to Bagdad by the Bay and its eminent ethnicity.

San Francisco County is as ethnically diverse as Los Angeles County, but on a much smaller scale. Compared to Los Angeles County's 4,079 square miles and 8,863,164 residents, San Francisco (city and county) covers only 91 square miles and is home to 723,959 people (see Table 11-2). The closeness of San Francisco's districts to one another, as shown in Figure 11-14 and the ethnic diversity of the districts and relative lack of tensions between them (compared to Los Angeles) speak favorably for the county both as a Third World center and melting pot.

The centerpiece of San Francisco's ethnogeography is Chinatown, which barely occupies 16 city blocks but ranks with Toronto and Vancouver among the most populous Chinatowns in North America. As is the case

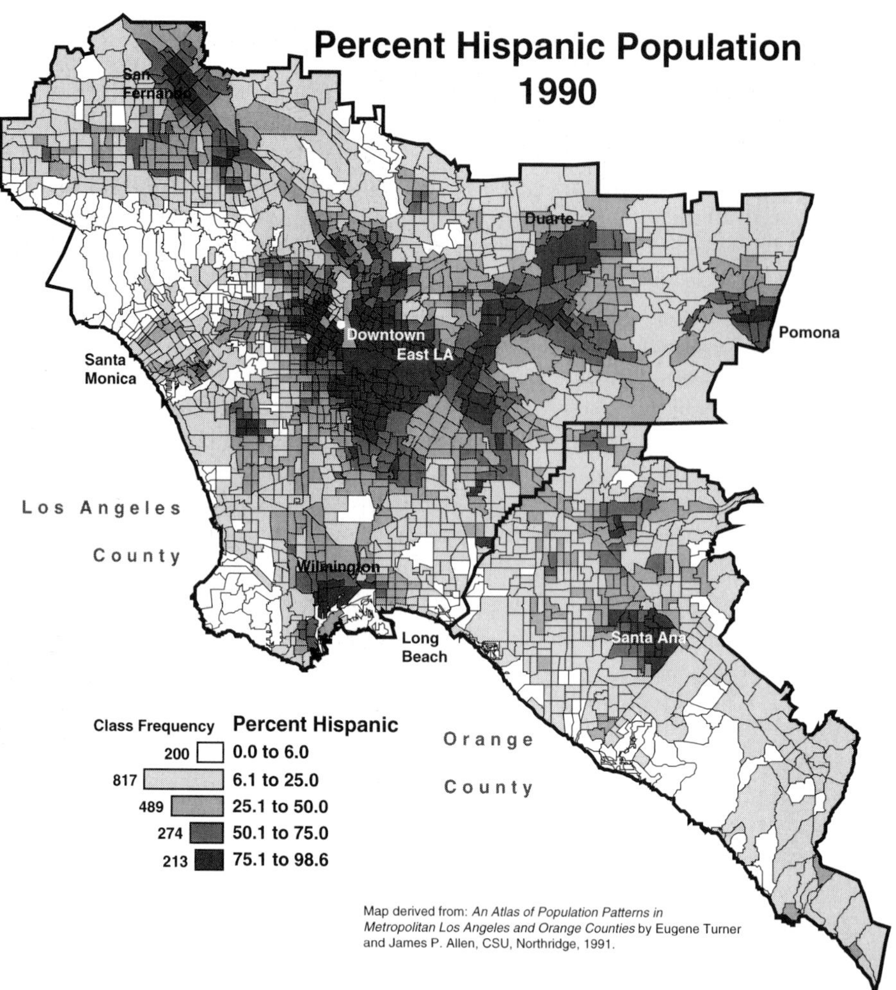

Figure 11-11 *Hispanic population in metropolitan Los Angeles County, 1990.* (Source: Computer-drawn map by Professor Eugene Turner, Department of Geography, CSU Northridge, from U.S. Census Bureau, *1990 Census of Population and Housing*, Public Law 94-171 data.)

in the two Canadian Chinatowns and as mentioned earlier in the chapter, San Francisco's Chinatown is woefully overcrowded, its population further swollen by recent emigrés from Hong Kong. In addition to the Hong Kongese, other Chinese, principally from Taiwan and from mainland China, are among the newcomers to Chinatown and other districts in San Francisco. An equally significant influx of Hong Kongese and Taiwanese has occurred in metropolitan Los Angeles, but it has been more diffuse and less focused on downtown Los Angeles and its nearby Chinatown than in the more centralized Asian districts of San Francisco.

A distinctively Chinese cultural landscape pervades the confines of Chinatown but is also evident, albeit more subtly, in San Francisco's other Asian American neighborhoods. Seeing and hearing Mandarin Chinese, as well as English, is the rule in the ethnically Chinese districts, although Cantonese may be spoken by Hong Kongese. Buddhist temples are in evidence, but Christian missions in Chinatown stand out, perhaps because they seem out of place. More likely to catch the eye than religious structures, though, are exterior building walls painted red by their owners for good luck. Inside these buildings, one might find living quarters facing any direction but west, the supposed direction of hell.

With the proportion of Asians in San Francisco County rapidly approaching 30 percent (see Table 11-2), their representation in districts outside of Chinatown has increased dramatically. In the Richmond, a predominantly white district until a decade or so ago, Asian representation has pulled even or slightly ahead of the white percentage of population, with each group accounting for about 47 percent of the total district population of 45,000. In the nearby Sunset District, the Asian and Hispanic populations are on the rise while the white population is decreasing. Many of the newcomers are young and middle-age Chinese, while some of those departing the Richmond District are older individuals of Irish and Italian descent. Somewhat larger populations and nearly equal proportions of Asians, Hispanics, and whites live in the Bernal Heights, Crocker Amazon, and Excelsior districts.

As seen in Table 11-2, both "black flight" and "white flight" dominate emigration patterns in San

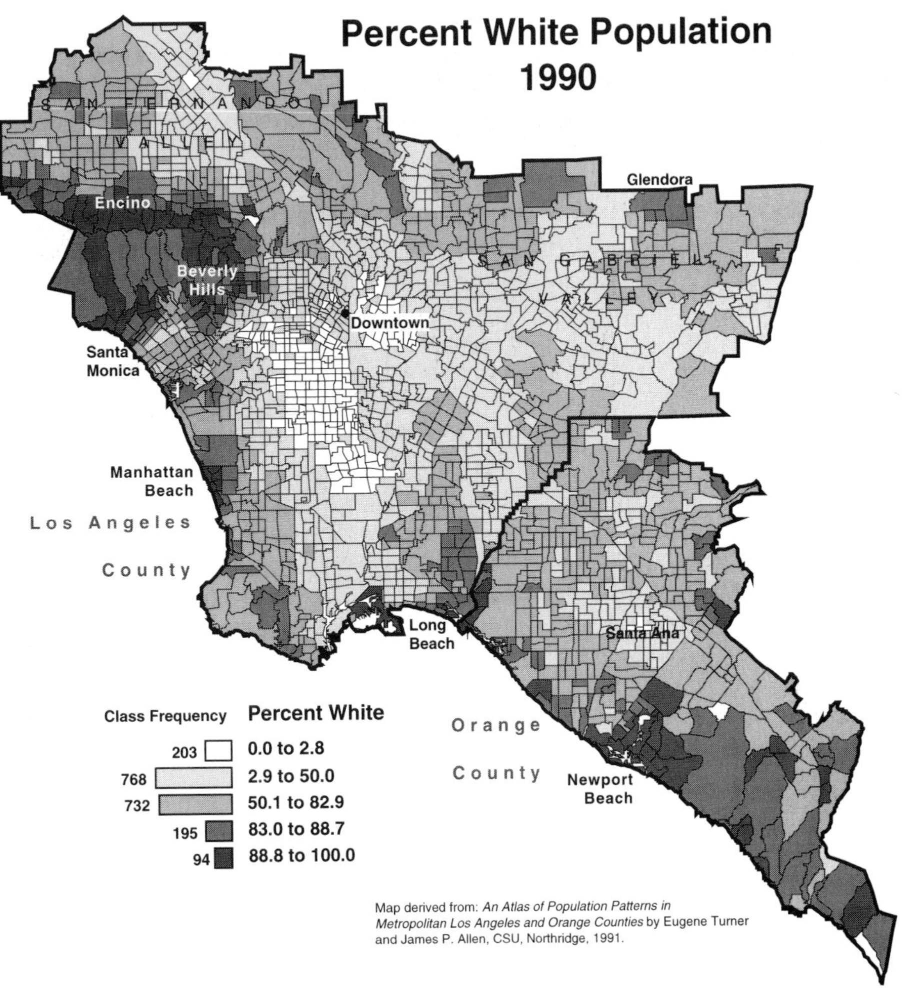

Figure 11-12 *Non-Hispanic white population in metropolitan Los Angeles County, 1990.* (Source: Computer-drawn map by Professor Eugene Turner, Department of Geography, CSU Northridge, from U.S. Census Bureau, *1990 Census of Population and Housing,* Public Law 94-171 data.)

A

B

Figure 11-13 *(A) Whittier Boulevard in East Los Angeles. A mix of Spanish and English signs and the occasional street vendor can be seen along this stretch, with the downtown L.A. skyline barely visible a few miles to the west. (B) Whittier Boulevard in east Whittier. Predominantly white Friendly Hills and La Habra Heights lie just to the north, and about the only Spanish sign seen around here is on the "El Camino Real" bell in the foreground, which marks the route of the Spanish-era "Royal Road" or "King's Highway" along this section of Whittier Boulevard.* (Crane Miller)

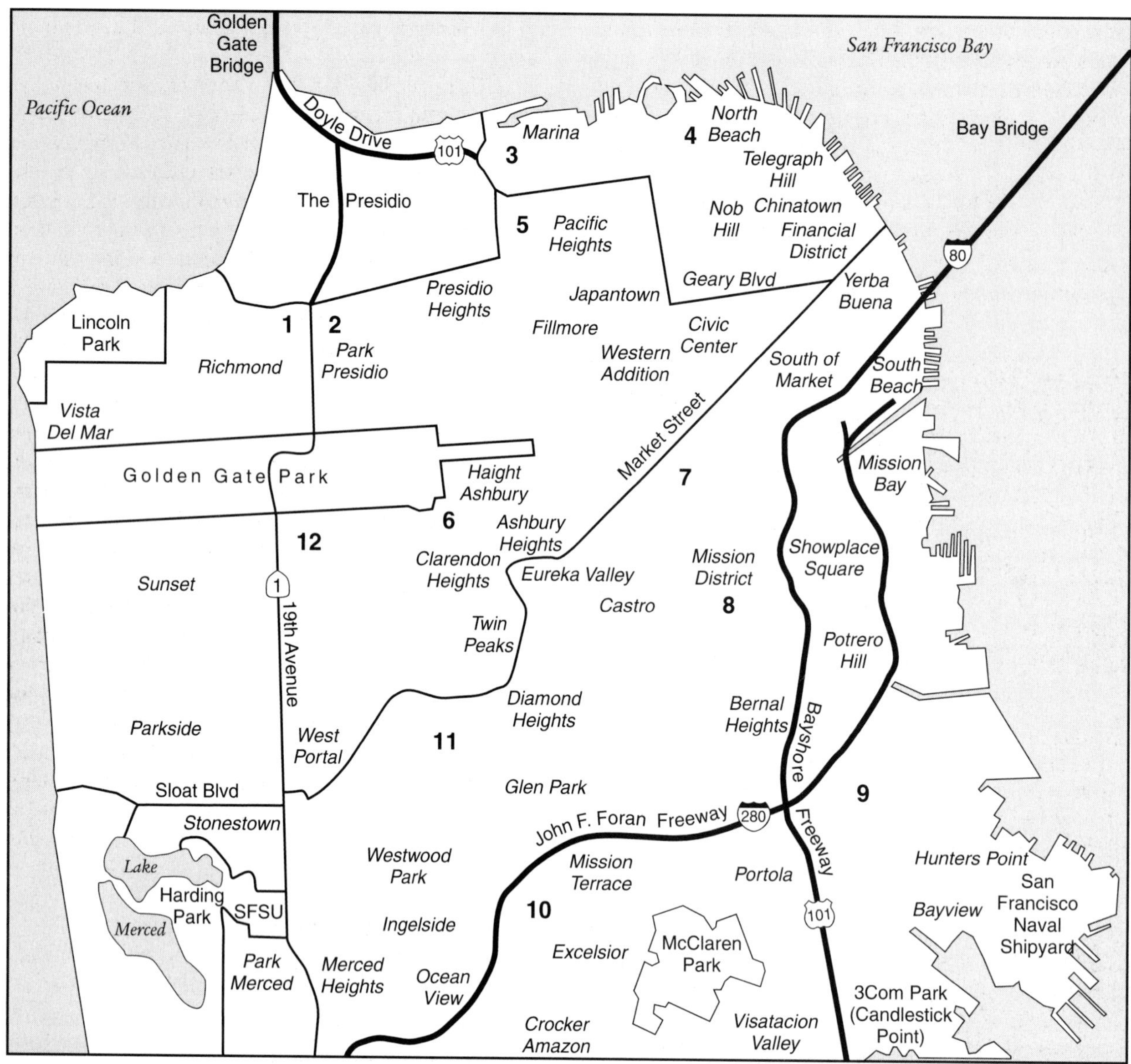

Figure 11-14 *San Francisco's districts (named) and communities (numbered). As of the 1990 census, Chinatown was over 80 percent Asian; the southeastern portion of the Mission District (principally, Potrero Hill and Bernal Heights) was over 80 percent Hispanic; the Bayview District, Hunter's Point, and a small southeastern corner of the Western Addition was over 60 percent black (all three districts were over 80 percent black in the 1980 census); and South of Market, South Beach, the Marina, and portions of the northwestern Mission and Buena Vista districts were over 80 percent white.* (Source: San Francisco Department of City Planning, Office of Analysis and Information Systems [October 1991].)

Francisco. The county's African American population peaked at nearly 100,000 in the early 1970s but had declined by more than 20 percent going into the 1990s. About two-thirds of San Francisco's African Americans, many of them second- and third-generation residents whose families originally came from the Deep South, live in Bayview–Hunter's Point, the Western Addition, Hayes Valley, Visitacion Valley, and OMI (Ocean View–Merced heights–Ingleside) (Figure 11-15). A large proportion of blacks traditionally have commuted to work in the East Bay and are increasingly relocating closer to their place of employment. The East Bay's Alameda County, which includes the cities of Berkeley, Oakland, and San Leandro, has absorbed many of these former San Franciscans and now claims the second largest black population among California's 58 counties: 222,873 blacks, or 17.4 percent of the total county population of 1,279,182 in 1990. Only Los Angeles County has a larger black population (see Table 11-12). And as in Los Angeles, where Asian "towns" fringe the long-established South Central black community and Hispanic barrios have made deep inroads into it, formerly predominant black districts in San Francisco have undergone ethnic changes. The transition is not an easy one for many of the long-time residents who must adjust to everything from a spreading trilingualism (Chinese and Spanish increasingly being spoken, as well as English) to a new wave of food vendors dispensing Chinese, Central American, and Mexican dishes in the neighborhoods.

Figure 11-15 *Residential street in the West Portal District of San Francisco, which lies just to the north of OMI. OMI includes the Ocean View, Merced Heights, and Ingleside districts in the southwestern part of the city. High-rent, single- and multiple-family housing dominates the West Portal and neighboring districts.* (Crane Miller)

If there is any district in San Francisco that epitomizes the ethnic diversity of the city, it is the Western Addition (Figure 11-16). The district earned its name as nineteenth-century San Francisco expanded into the hilly terrain west of Market Street and the original central business district. Much of the pre-twentieth-century Victorian architecture remains there, having survived the San Francisco earthquake and fires of April 18–21, 1906, and the Loma Prieta earthquake of October 17, 1989. Residing in many of the century-old multifamily structures, some of which have been renovated and others of which have been replaced by new public and private housing, is a virtual melting pot of blacks, eastern Europeans, Latinos, Filipinos, Japanese, and other ethnic groups. Located a little more than a mile west of downtown, the five-acre Japan Center stands out as a focus of the Western Addition's cosmopolitan milieu with its Japanese baths, galleries, gardens, pagodas, and restaurants. Enclaves of African Americans and whites of western European descent persist in the district, but whites presently predominate in greater San Francisco only in the pricey Marina, Pacific Heights, and Sea Cliff districts to the north and northwest of the Western Addition.

In reflecting on Los Angeles's and San Francisco's ethnic makeup, it seems likely that the numbers and proportions of Asians and Hispanics will continue to rise rapidly while those of blacks and whites will level

Figure 11-16 *The Western Addition, a vestige of nineteenth-century San Francisco. The Western Addition can be seen in the distance to the west (left) of Market Street, which is downtown's main thoroughfare and appears just to the right of center in the photo. The view is northeastward from Twin Peaks, which soar over 900 feet above sea level. San Francisco Bay and the hills overlooking the East Bay cities of Oakland, San Leandro, and Hayward can be seen in the background.* (Crane Miller)

TABLE 11-3 Population and ethnic group distribution by county, 1990.

	TOTAL POPULATION	WHITE	PERCENTAGE	BLACK	PERCENTAGE
California	29,760,021	17,029,126	57.2	2,092,446	7.0
Alameda	1,279,182	680,017	53.2	222,873	17.4
Alpine	1,113	772	69.4	5	0.4
Amador	30,039	25,129	83.7	1,670	5.6
Butte	182,120	158,242	86.9	2,238	1.2
Calaveras	31,998	29,288	91.5	180	0.6
Colusa	16,275	10,105	62.1	81	0.5
Contra Costa	803,732	560,146	69.7	72,799	9.1
Del Norte	23,460	18,302	78.0	853	3.6
El Dorado	125,995	113,053	89.7	579	0.5
Fresno	667,490	338,595	50.7	31,311	4.7
Glenn	24,798	18,461	74.4	131	0.5
Humboldt	119,118	104,671	87.9	934	0.8
Imperial	109,303	31,742	29.0	2,272	2.1
Inyo	18,281	14,819	81.1	71	0.4
Kern	543,477	340,892	62.7	28,851	5.3
Kings	101,469	54,426	53.6	7,747	7.6
Lake	50,631	44,603	88.1	924	1.8
Lassen	27,598	21,920	79.4	1,699	6.2
Los Angeles	8,863,164	3,618,850	40.8	934,776	10.5
Madera	88,090	52,974	60.1	2,294	2.6
Marin	230,096	194,665	84.6	7,529	3.3
Mariposa	14,302	12,771	89.3	120	0.8
Mendocino	80,345	67,775	84.4	482	0.6
Merced	178,403	96,701	54.2	7,889	4.4
Modoc	9,678	8,479	87.6	78	0.8
Mono	9,956	8,329	83.7	41	0.4
Monterey	355,660	186,166	52.3	21,506	6.0
Napa	110,765	89,453	80.8	1,167	1.1
Nevada	78,510	73,697	93.9	172	0.2
Orange	2,410,556	1,554,501	64.5	39,159	1.6
Placer	172,796	152,601	88.3	987	0.6
Plumas	19,739	17,996	91.2	151	0.8
Riverside	1,170,413	754,140	64.4	59,966	5.1
Sacramento	1,041,219	721,932	69.3	93,970	9.0
San Benito	36,697	18,793	51.2	167	0.5

ASIAN AND PACIFIC ISLANDER	PERCENTAGE	AMERICAN INDIAN, ESKIMO, ALEUT, AND OTHER	PERCENTAGE	HISPANIC	PERCENTAGE
2,710,353	9.1	240,158	0.8	7,687,938	25.8
184,813	14.4	9,674	0.8	181,805	14.2
5	0.4	257	23.1	74	6.6
200	0.7	520	1.7	2,520	8.4
4,961	2.7	3,073	1.7	13,606	7.5
187	0.6	629	2.0	1,714	5.4
321	2.0	344	2.1	5,424	33.3
73,810	9.2	5,695	0.7	91,282	11.4
433	1.8	1,458	6.2	2,414	10.3
2,318	1.8	1,268	1.0	8,777	7.0
54,110	8.1	6,840	1.0	236,634	35.5
773	3.1	475	1.9	4,958	20.0
2,255	1.9	6,269	5.3	4,989	4.2
1,632	1.5	1,722	1.6	71,935	65.8
172	0.9	1,683	9.2	1,536	8.4
14,879	2.7	6,860	1.3	151,995	28.0
3,408	3.4	1,337	1.3	34,551	34.1
431	0.9	1,040	2.1	3,633	7.2
293	1.1	803	2.9	2,883	10.4
907,810	10.2	50,486	0.6	3,351,242	37.8
1,084	1.2	1,338	1.5	30,400	34.5
9,064	3.9	908	0.4	17,930	7.8
113	0.8	601	4.2	697	4.9
866	1.1	2,974	3.7	8,248	10.3
14,109	7.9	1,597	0.9	58,107	32.6
35	0.4	385	4.0	701	7.2
114	1.1	346	3.5	1,126	11.3
25,365	7.1	3,053	0.9	119,570	33.6
3,391	3.1	813	0.7	15,941	14.4
615	0.8	757	1.0	3,269	4.2
240,756	10.0	11,312	0.5	564,828	23.4
3,635	2.1	1,702	1.0	13,871	8.0
112	0.6	573	2.9	907	4.6
38,349	3.3	10,444	0.9	307,514	26.3
92,131	8.8	11,642	1.1	121,544	11.7
653	1.8	284	0.8	16,800	45.8

(continued)

TABLE 11-3 (*continued*)

	TOTAL POPULATION	WHITE	PERCENTAGE	BLACK	PERCENTAGE
San Bernardino	1,418,380	862,113	60.8	109,162	7.7
San Diego	2,498,016	1,633,281	65.4	149,898	6.0
San Francisco	723,959	337,118	46.6	76,343	10.5
San Joaquin	480,628	282,766	58.8	24,791	5.2
San Luis Obispo	217,162	176,246	81.2	4,325	2.0
San Mateo	649,623	392,131	60.4	34,000	5.2
Santa Barbara	369,608	244,309	66.1	9,379	2.5
Santa Clara	1,497,577	869,874	58.1	52,583	3.5
Santa Cruz	229,734	171,203	74.5	2,330	1.0
Shasta	147,036	134,001	91.1	1,045	0.7
Sierra	3,318	3,060	92.2	6	0.2
Siskiyou	43,531	38,246	87.9	682	1.6
Solano	340,421	207,476	60.9	43,858	12.9
Sonoma	388,222	327,429	84.3	5,268	1.4
Stanislaus	370,522	261,323	70.5	6,109	1.6
Sutter	64,415	46,140	71.6	987	1.5
Tehama	49,625	43,049	86.7	246	0.5
Trinity	13,063	11,881	91.0	53	0.4
Tulare	311,921	170,283	54.6	4,305	1.4
Tuolumne	48,456	41,887	86.4	1,529	3.2
Ventura	669,016	440,555	65.9	14,559	2.2
Yolo	141,092	96,825	68.6	2,975	2.1
Yuba	58,228	42,924	73.7	2,341	4.0

Source: California Department of Finance, Demographic Research Unit (Sacramento, 1991).

off and/or continue to decline slowly. As mentioned previously, much of the black and white population is leaving Los Angeles and San Francisco for neighboring counties where housing is cheaper, jobs are more plentiful, and the commute to work is less time consuming. For instance, as enumerated in Table 11-3, in 1990 there were 109,162 blacks (7.7 percent) and 862,113 whites (60.8 percent) in Los Angeles's neighbor to the northeast, San Bernardino County. San Bernardino was California's second fastest-growing county from 1980 to 1990 (see, Table 1-2), with a larger percentage of the 523,364 new arrivals (see Table 1-3) coming from Los Angeles County than anywhere else. In the East Bay, in addition to Alameda County, Solano and Contra Costa were the ninth and thirty-seventh fastest-growing counties during the 1980s (see Table 1-3), with 1990 black population of 43,858 (12.9 percent) in the Solano and 560,146 (69.7 percent) in Contra Costa.

Table 11-3 also reveals diminishing ethnic diversity with increasing distance from the two great metropolitan centers. Calaveras, Nevada, Plumas, Shasta, and Trinity counties, which are among the state's most remote, were all more than 90 percent white in 1990. Calaveras and Nevada counties, which are mountain meccas (western Sierra) for white retirees, were the fourth and fifth fastest-growing counties in 1990 (see Table 1-2). It is noteworthy that some of these rural counties are all the less ethnically diversified for their paucity of native North American residents. On the other hand, urban counties show the largest numbers and propor-

ASIAN AND PACIFIC ISLANDER	PERCENTAGE	AMERICAN INDIAN, ESKIMO, ALEUT, AND OTHER	PERCENTAGE	HISPANIC	PERCENTAGE
55,387	3.9	13,136	0.9	378,582	26.7
185,144	7.4	18,912	0.8	510,781	20.4
205,686	28.4	4,095	0.6	100,717	13.9
55,774	11.6	4,624	1.0	112,673	23.4
5,774	2.7	1,894	0.9	28,923	13.3
105,559	16.2	3,306	0.5	114,627	17.6
15,050	4.1	2,671	0.7	98,199	26.6
251,496	16.8	9,060	0.6	314,564	21.0
7,690	3.3	1,714	0.7	46,797	20.4
2,610	1.8	3,728	2.5	5,652	3.8
8	0.2	60	1.8	184	5.5
351	0.8	1,703	3.9	2,549	5.9
40,494	11.9	3,076	0.9	45,517	13.4
10,234	2.6	4,068	1.0	41,223	10.6
18,146	4.9	4,047	1.1	80,897	21.8
5,748	8.9	948	1.5	10,592	16.4
325	0.7	881	1.8	5,124	10.3
99	0.8	599	4.6	431	3.3
12,468	4.0	3,972	1.3	120,893	38.8
362	0.7	952	2.0	3,726	7.7
32,665	4.9	4,285	0.6	176,952	26.4
11,455	8.1	1,655	1.2	28,182	20.0
4,625	7.9	1,610	2.8	6,728	11.6

tions of native North Americans as residents. More than a fifth of the total of 240,158 people in this category in the entire state lived in Los Angeles County in 1990. Most of these 50,486 individuals originally were from other parts of the United States and Canada. They live mostly in southeastern Los Angeles County and at the lowest rung of the socioeconomic ladder. They are the native peoples of the continent, yet in North American urban settings they are the forgotten people. One explanation for their plight is their inability to claim strength in numbers. As Table 11-3 shows, less than 1 percent of California's 1990 population was listed as "American Indian, Eskimo, Aleut, and Other," while 9.1 percent (2,710,353) were listed as "Asian and Pacific Islander," 7.0 percent (2,092,446) as "Black," 25.8 percent (7,687,938) as "Hispanic," and 57.2 percent (17,029,126) as "White."

Looking ahead to 2020 in Table 11-4, we see a virtual tie projected between "White" and "Hispanic," with each group representing 41 percent of the total population, "Blacks" 6 percent, and all "Other" (mostly Asian American) 12 percent. Many factors play into the future geodemography of California, not the least of which is the aging of the baby boomer *cohort* (persons or groups who share something in common, usually their year of birth) mentioned in Chapter 1. Yet, despite the baby boomers born in the 1950s and 1960s attaining grandparenthood by 2020, California's total population is projected to reach 49 million, with more than 97 percent of it living in cities, large and small.

TABLE 11-4 Population and ethnic distribution by county, 2020 projection.

COUNTY	WHITE	PERCENTAGE	BLACK	PERCENTAGE	HISPANIC	PERCENTAGE	OTHER	PERCENTAGE
California	20,061,693	41	3,118,197	6	20,076,972	41	5,719,656	12
Alameda	617,882	37	308,117	19	357,102	21	381,045	23
Alpine	1,085	54	4	<1	228	11	696	35
Amador	60,368	86	1,807	2	4,888	7	3,446	5
Butte	234,906	76	4,077	1	49,233	16	22,705	7
Calaveras	86,792	85	945	<1	8,262	8	6,244	6
Colusa	16,575	50	148	<1	14,763	45	1,378	4
Contra Costa	687,204	57	122,287	10	214,856	18	188,441	15
Del Norte	31,769	53	3,505	6	15,858	27	8,400	14
El Dorado	220,097	83	1,257	<1	34,151	13	8,374	3
Fresno	462,092	29	72,984	5	796,378	50	258,211	16
Glenn	22,249	47	302	<1	18,555	39	6,127	13
Humboldt	125,970	76	3,629	3	20,341	12	15,000	9
Imperial	32,669	15	5,842	3	176,737	81	4,030	1
Inyo	21,013	73	278	<1	3,743	13	3,579	13
Kern	576,638	44	80,298	6	601,242	46	51,872	4
Kings	78,652	40	17,270	7	102,547	49	9,037	4
Lake	86,606	80	2,451	2	14,724	14	4,767	4
Lassen	32,222	73	2,923	7	6,717	15	2,098	5
Los Angeles	2,900,887	22	1,008,622	7	7,510,091	58	1,496,952	13
Madera	89,261	42	5,492	3	115,795	54	3,549	1
Marin	187,162	78	9,074	4	30,729	13	13,045	5
Mariposa	24,910	84	258	<1	2,212	7	2,197	8
Mendocino	91,368	67	1,014	<1	33,395	25	10,264	8
Merced	128,083	32	12,619	3	191,626	48	69,619	17
Modoc	12,084	83	255	2	1,336	9	824	6
Mono	12,460	67	167	<1	5,309	28	762	4
Monterey	213,743	37	30,645	5	290,528	51	39,166	7
Napa	96,031	65	1,959	1	43,406	29	6,440	5
Nevada	151,721	90	532	<1	12,351	7	3,768	2

URBAN CALIFORNIA AT WORK AND PLAY

By some measures, agriculture dominates the economic landscape of California, but by other parameters, and certainly in urban California where most of the people live and work, it is outpaced by retail and wholesale trade, manufacturing, and tourism. Despite rural California's geographic attractions, such as a great number and variety of resource-based factory sites (near sources of raw materials and energy) and a seemingly boundless diversity of scenery, trade, manufacturing and tourism are largely urban-oriented industries in the state. The reasons for this range from the existence in the city of a large, affluent buying public and skilled labor force to the availability of more and better convention, entertainment, lodging, eating, and transportation facilities. In a very real sense, Californians' hard work and hard play have earned them national leadership in most aspects of trade, manufacturing, and tourism.

TABLE 11-4 (*continued*)

COUNTY	WHITE	PERCENTAGE	BLACK	PERCENTAGE	HISPANIC	PERCENTAGE	OTHER	PERCENTAGE
Orange	1,684,190	51	54,957	2	1,206,308	36	360,928	11
Placer	315,027	85	2,826	<1	37,241	10	13,956	4
Plumas	23,237	87	122	<1	2,206	8	1,235	5
Riverside	1,210,755	38	155,170	5	1,580,373	50	200,638	7
Sacramento	1,008,783	55	217,564	12	328,887	18	284,295	15
San Benito	40,052	48	214	<1	41,181	49	1,765	2
San Bernardino	1,351,745	40	281,291	8	1,505,556	45	217,852	7
San Diego	1,946,197	49	263,210	7	1,333,925	34	437,141	10
San Francisco	278,411	36	92,047	12	152,368	20	254,565	32
San Joaquin	399,903	42	46,681	5	310,210	32	199,662	21
San Luis Obispo	255,820	73	8,081	2	74,849	21	12,650	4
San Mateo	312,001	38	37,363	5	256,963	31	219,300	26
Santa Barbara	244,986	46	13,220	2	250,522	47	27,781	5
Santa Clara	861,360	44	72,165	4	584,161	30	440,917	22
Santa Cruz	184,799	57	3,550	1	121,904	38	12,076	4
Shasta	230,539	86	2,574	1	15,017	6	19,096	7
Sierra	3,397	88	4	<1	271	6	179	5
Siskiyou	50,465	82	1,155	2	6,429	10	3,360	6
Solano	305,765	49	86,892	14	126,085	20	106,605	17
Sonoma	416,828	72	11,175	2	125,440	22	27,460	4
Stanislaus	447,708	53	18,441	2	318,458	38	55,584	7
Sutter	83,144	49	5,297	3	54,895	33	25,264	15
Tehama	61,041	74	658	1	18,682	22	2,314	3
Trinity	16,041	84	45	<1	1,186	6	1,669	9
Tulare	217,373	34	8,513	1	378,767	59	39,704	6
Tuolumne	76,000	78	1,645	2	16,452	17	2,998	3
Ventura	506,661	49	23,042	2	441,127	42	69,626	7
Yolo	160,510	56	7,634	3	86,556	30	31,183	11
Yuba	66,201	54	3,900	3	23,850	20	27,808	23

Source: California Department of Finance, Demographic Research Unit, Report 93-P-3, *Projected Total of Population of California Counties* (Sacramento, May 1993).

Retail and Wholesale Trade

California's prominence carries over into almost all other aspects of the national urban economy as well. For instance, the 1992 Census of Retail Trade showed California leading the nation with $224.6 billion in retail sales generated by 162,111 retail outlets, with New York a distant second. Much of the retail trade industry's outstanding performance is owing to the popularity of suburban shopping centers; shopping centers, however, have prospered at the expense of older downtown and ribbon (along boulevards) retail trade areas.

In the closely related category of wholesale transactions, the 1992 Census of Wholesale Trade showed the Golden State as having $432.9 billion in sales. Although New York once vied with California for leadership in wholesaling, the latter now claims national prominence. In 1992, "motor vehicles and automotive parts and

Figure 11-17 *Hollywood Bowl. The bowl has a 17,599-seat capacity, and architect Frank Lloyd Wright was retained to design the orchestra shell. The Bowl, the nearby Greek Theatre, and the Universal Amphitheatre form a triumvirate of Hollywood Hills outdoor concert areas that features internationally known entertainers, bands, and symphony orchestras every summer.* (Greater Los Angeles Visitors and Convention Bureau)

supplies" led the wholesale "durable goods" category with sales of $69.1 billion while "groceries and related products" dominated "nondurable goods" at $66 billion. To say that automobiles and agriculture pervade wholesaling, as they do so much in the rest of California's economic geography, seems borne out by the numbers. Second in 1992 in wholesale nondurables were "petroleum and petroleum products," which posted sales of $27.8 billion and further confirmed California's standing among the top three or four petroleum-producing states. California wholesalers, as in the case of the state's retailers, increasingly prefer to locate their distribution centers in suburban rather than central-city sites. Given the largest and one of the most prosperous populations in the United States, a wealth of natural resources, and other geographic amenities, is it any wonder that California is at or near the top as a revenue and/or employment producer in the above and other urban-dominated economic activities, including construction, education, entertainment (Figure 11-17), museums (Figure 11-18), financial institutions, government, medical services, sports (Figure 11-19 and 11-20), and transportation?

Manufacturing

Every 5 years, when the national census of manufacturing establishments is conducted, California registers impressive, and sometimes unprecedented, gains. The 1977 Census of Manufactures for California was no exception, showing value added by manufacture increasing from $31.2 billion in 1972 to $54.9 billion in 1977, or by 76 percent in the 5-year period.

Eighteen years later and despite massive cutbacks in the defense industry in California, the 1995 Annual Survey of Manufactures for California recorded more than a tripling of value added by manufacture, to $178.4 billion. The Golden State continues its dominance in value added by manufacture, with New York, Ohio, Texas, Illinois, Michigan, Pennsylvania, North Carolina, and other major industrial states lagging far behind. As tabulated in Table 11-5, the number of production workers in the state in 1995 stood at 1,146,000, which was up slightly from the 1,115,400 reported in the 1992 Census of Manufactures, as shown in Table 11-6.

Five counties, two in Northern California (Alameda and Santa Clara) and three in Southern California (Los Angeles, Orange, and San Diego) control the lion's share of manufacturing in the state. In 1992, as tabulated in Table 11-6, the five counties accounted for 74 percent of all manufacturing employment, 72 percent of total value added by manufacture, and 70 percent of all new capital expenditures in manufacturing in the state. With one exception, *diversified* manufacturing is the rule in these counties; that is, manufacturing in four of the counties was fairly well spread over most, if not all, of the 19 different Standard Industrial Classification (SIC) categories listed in Table 11-5. The one exception, of late, has been Santa Clara County with its Silicon Valley and the trend toward *specialized* manufacturing of computer hardware and software.

All five counties boast a mix of both *capital-intensive* and *labor-intensive* industries, the former being more automated and employing fewer people and the latter being less automated and employing more people in the

A

B

Figure 11-18 *(A) The Getty Center, which opened in 1997 in the Santa Monica Mountains just to the north of West Los Angeles. The 110-acre center houses one branch of the J. Paul Getty Museum and the Getty Research, Conservation, Information, Education, and Leadership Institutes, as well as the Getty Grant Program. The J. Paul Getty Museum, along with its Getty Villa location (a few miles to the northwest off of Pacific Coast Highway), which will reopen in 2001, is home to one of the continent's finest collections of European art. (B) View southeastward from "The Getty." The San Diego Freeway can be seen in the foreground, and Brentwood, the UCLA campus, Westwood, the west Wilshire Boulevard high-rise strip, and Century City loom in the background.* (Crane Miller)

Figure 11-19 *Bay Meadows race course. Since satellite wagering at racetracks where racing is not actually being held became legal, attendance has fallen at the races themselves. Another source of competition is the dozens of Indian gaming casinos all over the state. Still, this and other spectator sports in California, whether amateur or professional, generate impressive crowds and revenues.* (Bay Meadows Race Course, photo by Doug Murchison)

Figure 11-20 *Toyota Grand Prix of Long Beach, North America's number one street race and the largest special sports event on the West Coast. Seen here, the April 1996 race drew some 300,000 spectators, surpassing the record-setting 1995 attendance of 284,000. Although preceded by an inaugural Formula 5000 event in September 1975, the first Formula One World Championship saw the green flag on Ocean Boulevard in Long Beach in April 1976.* (Whitney Miller)

manufacturing process. For instance, in comparing Alameda and San Diego counties in Table 11-6, it could be said that Alameda County was relatively capital-intensive with 47,600 workers producing some $9.3 billion in value added by manufacture in 1992, whereas San Diego County was comparatively labor-intensive with 69,900 workers producing some $6.8 billion in value added by manufacture in 1992. In making these comparisons, it should be recalled from Chapter 10 that San Diego County in the late 1990s entered into the billion-dollar-a-year category in agricultural commodity production, whereas the annual value of farm production in Alameda County is far below that figure. Tempered by cutbacks in defense spending during the 1990s, manufacturing in the four counties lost momentum, with Santa Clara County the lone exception. By 1998, though, the four counties were back on track, sharing in the statewide economic recovery. An interesting side effect of Santa Clara County's persistent prosperity in manufacturing through fluctuating economic cycles is the maintenance of a strong residential real estate market. While housing prices slipped throughout most of the state into the late 1990s, they remained high in and around Santa Clara County. Going into the twenty-first century, San Jose, Santa Clara County's largest city, ranked fifth in the nation among the most productive manufacturing cities. Ahead of it were Kokomo, Indiana; Detroit, Michigan; Wilmington, Delaware; and Houston, Texas.

Tourism

California is a fun place. It would have to be to attract $56 billion in tourist dollars in 1994 and thereby perpetuate its leadership in travel and tourism over second-place Florida, third-place New York, and fourth-place Texas. Because tourist dollars are here defined as revenue generated from the expenditures, employment, and tax revenues by travel 100 miles or more away from home, $56 billion is undoubtedly an understatement of the total impact of the tourist industry on the California economy in 1994. But regardless of any additional impact a multiplier effect might have had, $33.8 billion was spent by 276 million travelers in and to California in 1994, several million of whom were foreigners, mostly from Canada, Europe, Japan, and Mexico. California tourism is said to generate more than half a million jobs and a billion dollars in state tax revenues each year.[16]

Nearly 70 percent of visiting travelers to California arrive and depart by air. Four busy airports, including Los Angeles International (LAX) in Los Angeles County, San Francisco International (SFO) in San Mateo County, San Diego International (SAN or Lindbergh Field) in San Diego , and John Wayne Airport in Orange County, handle most of this passenger traffic.

According to the 1987 Census of Service Industries, California's 1,800 hotels, 2,987 motels, and 15,795 amusement and recreation service businesses generate the bulk of the tourist revenue ($6 billion from hotels and motels and $19 billion from amusement and recreation services). California is ranked far ahead of second-place Florida in these three accommodation and service categories. In the realm of restaurants and the like, or so-called eating and drinking places, taxable transactions amounted to nearly $26.8 billion in 1996, which was a national record.

In the realm of urban amusement attractions, Disneyland (in north Orange County), Knott's Berry Farm (in northwest Orange County), San Diego Zoo and Wild Animal Park (two locations in southern San Diego County), and Universal Studio Tours (in the southern San Fernando Valley in Los Angeles County) were among the 10 leading amusement parks in the nation in attendance. The four parks annually draw more than 21 million visitors, which are more visitors than frequent the nation's gambling capital, Las Vegas, Nevada, in a year.

Although tourists in California often chance "get rich quick" side trips to Las Vegas, Laughlin (along the lower Colorado River in Nevada), Tahoe–Reno, and reservation casinos (see Chapter 3), urban California has its own legalized gambling attractions in the form of city-licensed poker palaces, such as in Gardena in western Los Angeles County, and parimutuel wagering at horse-racing tracks, such as Santa Anita and Hollywood Park in Los Angeles County, Los Alamitos in Orange

[16] Except where sources are actually quoted, most of the data in this section are derived from the California Office of Economic Research, *Number and Characteristics of Travelers to California.*

TABLE 11-5 California manufacturers' employment, payroll, and value added by industry group, 1995.

INDUSTRY GROUP	ALL EMPLOYEES NUMBER (000)	ALL EMPLOYEES PAYROLL (MILLION $)	PRODUCTION WORKERS NUMBER (000)	PRODUCTION WORKERS HOURS (MILLION)	PRODUCTION WORKERS WAGES (MILLION $)	VALUE ADDED BY MANUFACTURE (MILLION $)
California	1,927.7	68,288.1	1,146.0	2,305.3	27,877.0	178,358.4
Food and kindred products	169.5	4,753.9	122.8	249.7	2,909.3	20,383.3
Textile mill products	14.1	338.6	11.6	26.3	217.2	743.7
Apparel and other textile products	166.5	2,655.1	138.3	250.3	1,756.6	6,813.9
Lumber and wood products	51.0	1,194.3	42.1	86.9	871.7	2,721.2
Furniture and fixtures	51.9	1,142.6	41.8	86.5	748.5	2,414.4
Paper and allied products	38.7	1,258.9	30.8	63.2	823.5	3,550.3
Printing and publishing	155.9	4,857.9	75.9	156.3	2,018.8	13,242.0
Chemicals and allied products	59.0	2,364.6	29.4	60.0	821.7	10,856.5
Petroleum and coal products	14.6	754.8	9.5	20.5	455.3	4,607.3
Rubber and miscellaneous plastics products	88.7	2,308.1	66.9	135.0	1,349.5	5,956.6
Leather and leather products	6.8	125.1	5.7	11.6	85.5	263.8
Stone, clay, and glass products	43.3	1,306.8	34.2	68.5	929.9	3,469.5
Primary metal industries	27.8	880.5	21.7	42.9	606.8	2,300.6
Fabricated metal products	134.0	3,999.0	96.5	202.5	2,399.6	8,942.4
Industrial machinery and equip.	181.7	7,648.9	98.7	202.9	2,847.3	24,922.6
Electronic and other electric equipment	257.7	10,398.7	142.0	298.0	3,605.6	33,010.7
Transportation equipment	160.8	7,237.1	80.8	151.0	2,731.5	13,914.3
Instruments and related products	150.9	6,837.3	67.0	133.4	2,149.3	17,439.7
Miscellaneous manufacturing industries	44.0	1,075.4	30.4	59.7	549.5	2,805.7
Auxiliaries*	110.9	7,150.3	—	—	—	—

* Auxiliary units (central administrative offices, warehouses, research and development laboratories, etc.) of multiestablishment companies. These units serve the manufacturing establishments of a company rather than the general public.

Source: U.S. Department of Commerce, Bureau of the Census, Annual Survey of Manufactures (1995).

County, Del Mar in San Diego County, Bay Meadows in San Mateo County (see Figure 11-19), Golden Gate Fields in Alameda County, and numerous county fair tracks.

San Diego as the Tourist Mecca of the Twenty-First Century

When Portuguese pilot Juan Rodriguez Cabrillo first sailed into San Diego Bay on September 28, 1542, he was escaping a big storm. His two ships, sailing under the Spanish flag, laid anchor in the shadows of the cliffs of Point Loma, and he named the protected embayment San Miguel. Having discovered the finest natural deep-water harbor of his voyages along Baja and Alta California, he duly noted in his journal that "they discovered a port, enclosed and very good."

San Diego's geography lures more tourists and their dollars than anywhere else in California except for the greater Los Angeles and San Francisco areas. Already California's second largest city in population, by the early twenty-first century, San Diego may well be the state's premier tourist attraction. It is less than 100 miles

TABLE 11-6 California manufacturers' employment, payroll, value, and capital expenditures for selected counties, 1992.*

COUNTY	NUMBER OF ESTABLISHMENTS	ALL EMPLOYEES NUMBER (000)	ALL EMPLOYEES PAYROLL (MILLION $)	PRODUCTION WORKERS NUMBER (000)	PRODUCTION WORKERS HOURS (MILLION)	PRODUCTION WORKERS WAGES (MILLION $)	VALUE ADDED BY MANUFACTURE (MILLION $)	NEW CAPITAL EXPENDITURE (MILLION $)
California	50,490	1,947.4	65,255.3	1,115.4	2,249.2	26,867.7	154,678.2	9,729.1
Alameda	2,591	83.8	2,961.9	47.6	95.5	1,310.1	9,275.6	725.9
Butte	255	5.2	114.0	4.1	7.8	78.3	330.6	23.8
Contra Costa	783	27.7	1,098.1	13.2	26.8	439.2	2,878.4	559.7
Fresno	708	25.9	626.5	18.9	36.4	376.2	1,824.6	145.4
Humboldt	275	6.4	178.8	4.9	10.2	127.4	559.3	23.2
Kern	391	10.9	278.7	7.4	15.5	152.3	1,087.5	64.1
Los Angeles	18,439	725.4	22,617.7	443.8	879.2	10,141.7	48,775.9	2,561.5
Marin	405	5.6	161.1	3.2	6.6	71.6	381.4	12.8
Merced	127	8.5	181.8	7.0	14.0	139.4	534.2	33.1
Monterey	285	8.3	260.7	4.8	9.8	119.0	805.7	30.9
Napa	240	6.6	197.9	3.6	6.9	90.3	647.1	70.9
Orange	5,798	239.7	7,890.1	136.6	277.5	3,356.0	16,557.1	810.4
Placer	249	7.6	238.0	4.2	9.8	99.8	938.9	32.1
Riverside	1,256	37.9	1,030.8	25.1	49.9	552.8	2,642.3	117.0
Sacramento	952	28.3	872.9	15.1	29.7	355.5	2,031.2	109.3
San Bernardino	1,885	56.4	1,561.7	38.4	79.0	858.6	3,772.1	244.4
San Diego	3,351	126.0	3,969.0	69.9	137.8	1,630.3	6,821.5	575.6
San Francisco	1,447	35.2	1,193.6	19.9	38.3	382.9	3,250.0	63.5
San Joaquin	568	25.8	705.9	18.9	39.1	459.5	2,264.2	174.1
San Luis Obispo	287	5.6	129.3	3.3	6.2	65.6	369.4	69.7
San Mateo	1,034	34.6	1,493.6	16.4	32.6	485.6	3,743.6	221.1
Santa Barbara	543	19.2	657.1	9.6	18.4	230.9	1,363.9	74.0
Santa Clara	3,455	260.9	12,320.7	94.0	193.3	2,846.7	29,548.0	2,157.2
Santa Cruz	392	11.7	358.9	7.6	13.9	174.8	1,392.4	66.2
Shasta	251	5.0	152.1	3.7	7.5	98.8	287.6	16.4
Solano	268	7.9	259.3	5.1	10.3	144.1	1,153.3	64.8
Sonoma	785	19.9	611.4	12.2	24.2	312.9	1,418.4	82.5
Stanislaus	409	25.0	664.4	18.3	35.4	422.4	2,571.8	120.1
Tulare	290	12.8	335.9	9.7	19.9	208.2	979.4	63.1
Ventura	981	33.0	1,064.0	19.2	57.8	448.2	3,414.6	236.1
Yolo	191	6.8	194.0	4.7	9.6	109.8	620.2	33.5

* Selected counties have 5,000 or more employees.

Source: U.S. Department of Commerce, Bureau of the Census, 1992 Census of Manufactures.

Figure 11-21 *View eastward of San Diego Bay, the downtown San Diego skyline, and Embarcadero. Besides serving as a major cruise ship port and U.S. Navy and Coast Guard base, San Diego also harbors a tuna clipper fleet. But hard times have befallen the U.S. tuna fishing industry, in large part because of U.S. compliance with regulations banning the taking of dolphins in fishing for tuna and competition with the Mexican industry, which has been slower to observe the constraints.* (Crane Miller)

from the state's largest and the nation's second largest city in population, Los Angeles. It is nearly as large as Los Angeles in area, but with far more developable land. It is farther removed than Los Angeles from the San Andreas and other worrisome earthquake fault zones. It's geographically on or close to more of the Pacific than Los Angeles could ever dream of. Its winters, summers, and transitional seasons are milder than anywhere else in the state, by any climatic measure one chooses. Its air is far cleaner, more of the time, than that of smog-bound Los Angeles or most other cities in the continent. Its protected, deepwater harbor (Figure 11-21; see also Figure 3-3) is the finest natural one in the state, save for San Francisco's. Its beaches, such as at Coronado, Pacific Beach, and La Jolla, are sandier, whiter, and broader than any found in or near any other California city. Its lodging facilities are the equal of those of coastal resorts anywhere. It is steeped in prehistory and history, claiming artifacts dating back thousands of years and to the first European settlement in California in 1769. It is America's largest city whose corporate limits lie along an international boundary. And its amusement and recreation establishments are without compare. For all these reasons, San Diego should soon become California's principal tourist attraction.

THE IMPACT OF URBAN SPRAWL ON AGRICULTURE

Judging by the plethora of articles, books, editorials, and other commentaries on the subject, there is steadily mounting concern over the urban encroachment on California farmland and its ultimate impacts on the agricultural resources of the state and the nation. Views on the situation range from alarm to complacency, and consequently, a number of controversial questions have been raised over the issue: Does the metropolitan area explosion and the conversion of farmland pose any real threat to the food supply, or at least to some elements of that supply? Or has the growth of population and cities so broadened the market for agricultural products that more land has come into production than been taken out, so that, if anything, greater food surpluses have resulted?

In another sector, has the forced retreat of farms to areas farther and farther removed from urban markets noticeably affected the cost of food by putting more distance between producer and consumer? And, too, hasn't the urban market for agricultural products spread? Or have capabilities in processing and transporting of food so improved that the actual location of agricultural production in relation to its markets is of little importance?

In this regard, are there not vast areas of potentially good cropland in California literally waiting to be put into production? Yet doesn't the distinct possibility exist that the production of some specialty crops, which require optimum conditions of soil and climate found only in or near certain metropolitan areas, may soon cease altogether in the nation? Or, if the market continues to demand such products, will it in essence refuse to let them disappear?

Lastly, what do we make of the environmental implications of urban sprawl and the disappearance of agricultural activity? Should we allow the land and the air in rich farming regions, such as once existed in the San Fernando, San Gabriel, and Santa Clara valleys, to be paved over and become befouled with smog? To preclude this possibility, should agriculture be preserved and thus subsidized in metropolitan areas as an open-space or greenbelt amenity? Or has agriculture already received more than its share of subsidies such as those gained through price supports, special reclamation and irrigation projects, and property tax breaks?

The answers to all these questions are likely to be subjective and to depend to a large degree on one's point of view. Chances are, developers and urban planners would voice opinions markedly different from those of ecologists and farmers. But one thing is certain: Prime farmland is objectively definable, and there is really not that much of it left on earth.

Land Economics and Land Taxes

Many economists view the conversion of farmland to urban uses as an expression of the free land market functioning very efficiently. They maintain that supply and demand is affecting the orderly transfer of fringe land simply because the land usually has more value for urban than rural uses. And if agricultural land does eventually become scarce, prices offered for it by those interested in agricultural development could conceivably exceed those offered by urban interests. According to Robert O. Harvey and W. A. V. Clark, "If the price of citrus fruits or some other agricultural specialty became sufficiently high to yield a return on the land higher than that earned under an urban use, then a transfer from urban to agricultural uses would take place in contrast to that which typically occurs."[17] James Gillies and Frank Mittelbach look with disfavor on suggestions for preserving agricultural land that would constrain operation of the land market in metropolitan areas, noting that "these efforts should not be such that they encourage the continued use of land for agricultural production if there is some higher and better alternative use of the land as measured by the capitalized value of its earning capacity."[18]

A point often not stressed by either agricultural conservationists or urbanists is the fact that urbanization, depending on where and how it occurs, usually does *not* pay its social costs. This is especially true "in the case of residential development (the major suburban land use), where new expenditures (particularly for schools and other public facilities) might in fact more than offset the benefits of the increased tax base. Even more so if development occurs at random. Because of leap-frogging development, utilities, sewers, and roads have to be extended in even more uneconomic fashion."[19] On the other hand, agriculture requires comparatively little in the way of services and usually more than pays its own way with regard to social costs. Unfortunately, this contribution of agriculture to the welfare of the metropolitan community is generally ignored in the operation of a free land market. Indeed, in ignoring the social benefits of agriculture, the land market may in fact *not* be doing its job as efficiently as is generally assumed.

Rising land values accompanied by rising taxes on the land come sharply into focus when an expanding urban society extends into a retreating farmland base, a farmland base that has been losing ground in metropolitan California at an annual rate of almost 50,000 acres since the end of World War II. The farmland base in question exists at the rural-urban fringes of metropolitan areas where land supply and demand are greatest. The supply function owes to the availability of more and cheaper land in the rural fringe than in areas closer to metropolitan centers, a circumstance that encourages outlying development. The demand function is attributed largely to a desire on the part of a majority of the home-buying public to reside in a low-density, single-family suburban environment.

Since 1982, when the state Farmland Mapping and Monitoring Program began keeping closer tabs on farmland conversion, the rate has increased from 50,000 to 74,000 acres per year.[20] All the more portentous for agriculture is the expansion of urbanization from the fringes of the Los Angeles, San Francisco–Oakland–San Jose, and San Diego metropolitan areas into the Central Valley, the state's, nation's, and world's single most prolific food basket. "So far, the Central Valley's vast agricultural economy remains intact, but the trend toward urbanization is clear. If growth in the lower two-thirds of the Valley continues at the same rate as in the decade before 1986, an American Farmland Trust survey showed that farmland acreage equal to virtually all the cropland in Stanislaus County will be lost within 20 years."[21] By today's prices, if all of Stanislaus County's agricultural production were suddenly to cease, the state would lose about a billion dollars worth of farm income annually and one of its top 10 food-producing counties (see Table 10.3).

Stanislaus County, whose population grew from 194,506 in 1980 to 370,522 in 1990 and which ranked thirteenth in growth for the decade, further serves as a Central Valley case in point in that its rare, 6-foot-thick prime topsoil is rapidly being covered over by thousands of single-family tract houses. The county seat and largest city, Modesto (pop. 122,200 in 1991), has attracted most of the new residents. But the Sierran foothills region to the east promises to attract increasing numbers of retirees and affluent younger families, and the area northwest of Modesto is luring residents willing to make the 2-hour commute to work in the East Bay region. The land most likely to accommodate the new

[17] Robert O. Harvey and W. A. V. Clark, "The Nature and Economics of Urban Sprawl," *Land Economics* 41, 1 (February 1965), p. 8.

[18] James Gillies and Frank Mittelbach, "Urban Pressures on California Land: A Comment," *Land Economics* 34, 1 (February 1958), p. 82.

[19] Association of Bay Area Governments, *Bay Area Regional Planning Program–Agricultural Resources Study*, Berkeley, August 1969, pp. 2.15–2.16. Any growing city needs a broad *tax base*, that is, one that includes generous proportions of revenue-producing commercial, industrial, and perhaps even mineral extraction land uses, as well as deficit-producing single-family residential land use. With such a broad land, property, or real estate tax base, a city could then theoretically build an adequate *infrastructure* or internal framework of public service systems (law enforcement, schools, smog abatement, transportation, waste disposal, and the like). Exclusively residential communities, as are commonly found in suburbia, have very narrow tax bases, and thus their property owners are subject to relatively high tax rates. Such cities also depend on state and federal aid to a greater extent than do cities with a broad tax base.

[20] Maria L. LaGanga, "Cash Crop or Cash Out?" *Los Angeles Times*, August 5, 1990, pp. D1, D9–10.

[21] Harold O. Carter et. al., "Keeping the Valley Green: A Public Policy Challenge," *California Agriculture* 45, 3 (May–June 1991), pp. 10–14.

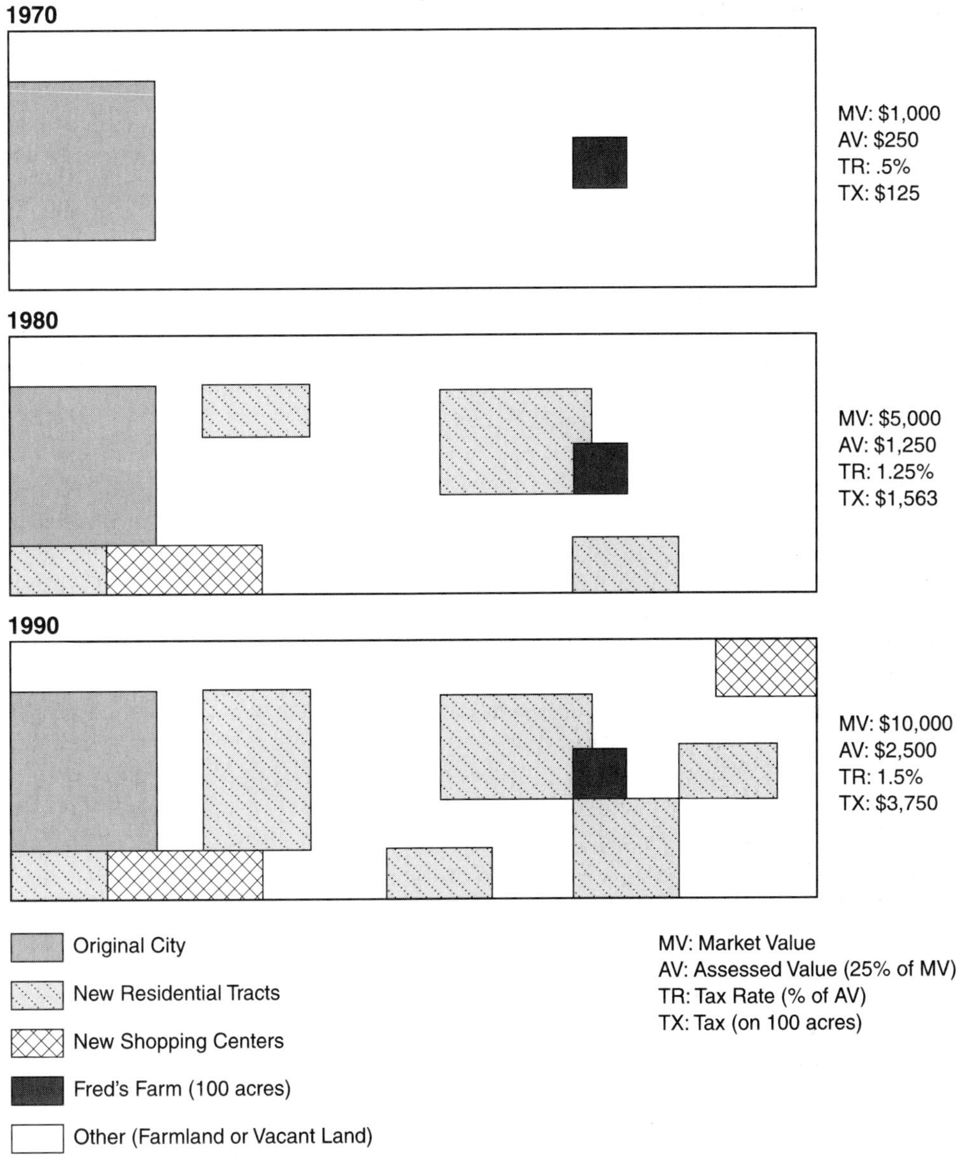

Figure 11-22 *Rising property taxes on farmland: The hypothetical case of Farmer Fred from 1970 to 1990.*

arrivals is the 27 percent of the county's nearly 1 million acres that is classified as prime soil, for its location in level but well-drained bottomlands and inherent versatility make it as attractive to developers as to farmers. Some farmers have proposed placing more than 270,000 acres of prime land in an agriculture-only zone that would serve to redirect urban development to agriculturally marginal areas. Were agriculture forced onto such marginal land, more fertilizers and chemicals would be needed to maintain production, crop yields would no doubt diminish, costs would rise, and organic farming would become even more elusive. Unfortunately, many farmers are opposed to an agricultural zone, arguing for the right to sell their land at any time and to anybody, be it another farmer or a developer.

Both supply and demand are also affected by the recent skyrocketing index of average farm real estate value versus the relatively declining index of farm income. This is yet another sign of impending prime land scarcity at the rural-urban fringe. A dynamic rural-urban fringe land market operates to the benefit of farmers who want to get out of agriculture, for they can now reap a higher price for their land than at any time in the past. But for farmers committed to staying in agriculture, it represents a hardship in the form of increasing tax assessments, which can rise along with other expenses to the point at which it is uneconomical to continue farming.

Take, for instance, the hypothetical plight of Farmer Fred illustrated in Figure 11-22. Between 1970 and 1990, the market value of his 100 acres of cropland appreciated 10-fold, from $1,000 to $10,000 per acre. His property taxes, however, increased 30 times, from $125 to $3,750 for all his land. Most of the $3,750 Farmer Fred paid in 1990 went for urban services from which he derives little or no benefit. If anything, Fred probably

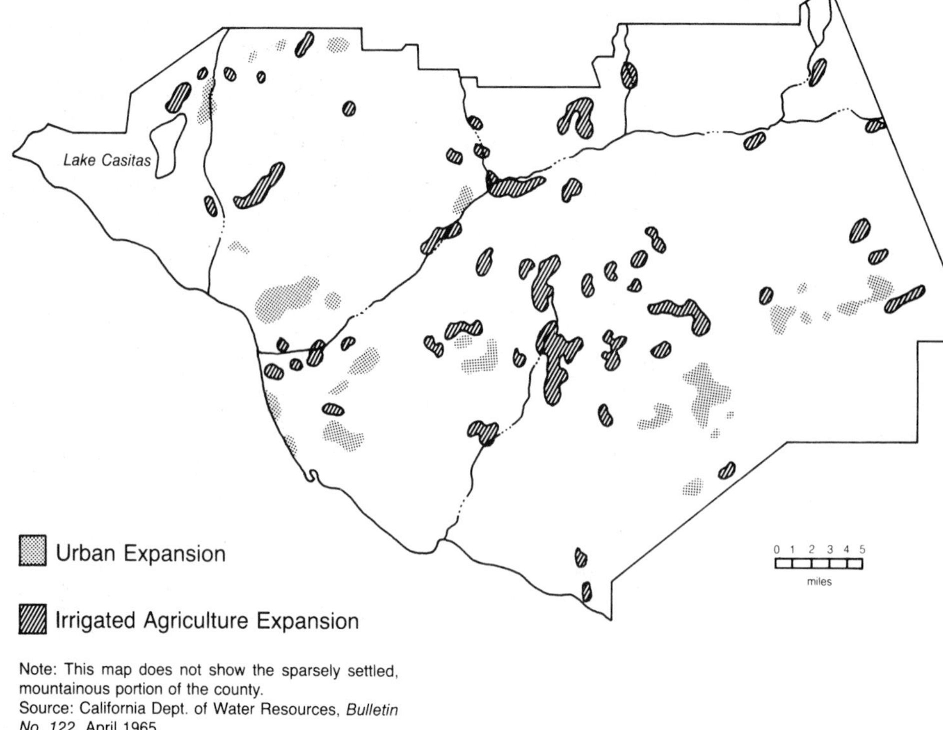

Figure 11-23 *Ventura county land-use changes, 1950–1961. Even though we are now at the turn of the century, suburban sprawl has plagued California agriculture since the end of World War II.*

realizes more disbenefits than benefits from his new suburban neighbors. Tract house owners complain about his noisy tractoring and noxious spraying on weekends while he tries to keep their kids from trampling his vegetables and picking his citrus after school. Worse yet, the city will no doubt annex the neighboring tracts and his land as well, so that the city tax rate and those of more special districts will be added to his tax bill. Fred is also paying more than he ever did before for labor, machinery, fertilizers, pesticides, irrigation systems, water, fencing, ad nauseam.

In the face of increased taxes and operating costs and thus declining profits, Fred and thousands of other urban fringe farmers must either sell their land, which is often the only profitable commodity left to sell, or revert to growing only the highest-value crops. Even this latter alternative is economically feasible only to a certain point. Is it any wonder, then, that thousands of farmers, especially smaller operators, are getting out of the food supply business? Urban sprawl is hardly the sole reason the number of farms in California diminished from 144,000 in 1950 to 85,000 in 1990 (see Table 10-2), but it may take the heaviest toll.

Urban Shadows

Another pattern to be noticed in Figure 11-22, as well as in Figures 11-23 and 11-24, is urban sprawl's haphazard invasion of rural land and the resultant chopping up of agriculture. This not only disrupts the continuity of farm production but also creates an *urban shadow* where, in anticipation of capital gains, land speculators have taken large tracts of farmland out of production long in advance of their planned development. Perhaps the farmer abandoned cultivation of the land well before title passed to the buyer, or maybe the developer saw mortgage money suddenly dwindle in supply. Moreover, even with the average selling price of new tract houses now dipping well below $200,000 in several metropolitan fringe counties in California and mortgage interest rates at their lowest in 20 years, the number of qualified buyers has shrunk because the 1990s' recession threw so many out of work.

Obviously, a seriously diminished market will influence developers' decisions on whether to build on their land. Higher-density residential development may offer some hope, though even condominiums and townhouses are increasingly out of reach of the buying public. But whatever originally caused the land's vacancy, it may simply sit there for years doing nothing more useful than growing weeds to control erosion and providing an investment tax shelter. Some developers rent newly acquired land back to farmers or farm it themselves until the day of actual development is at hand, although such practices are usually limited to companies with significant financial resources and situations in which highly remunerative crops, such as strawberries, can be grown.

Figure 11-24 *Aerial view of Ventura County. Sprawl in Ventura County takes its toll of productive farmland even before urban development begins. Note, for example, the decline of the citrus groves adjacent to the new housing tract in the upper left, while more remote crop areas appear better irrigated and generally better maintained. The incompatibility of agricultural and suburban land uses bordering each other is not always apparent at first glance, but it's there.* (Crane Miller)

A more insidious shadow cast by urbanization is its *irreversibility*. Prior to suburbanization in metropolitan California, when land use change usually meant that one agricultural use replaced another, the results were not necessarily irreversible. That is, an irrigated parcel of land could readily revert to being dry-farmed or even to supporting its original natural plant cover. Not so with urban development of land, for residential subdivisions, shopping centers, industrial parks, freeways, and the like are relatively permanent. It would be nothing short of miraculous if any agricultural lands were reclaimed once urbanization had occurred. Instead, abandoned urban land all too often becomes dead land with a bombed-out look to it.

Where Will New Suburbanites Live?

There were years during the 1970s when it seemed California had attained zero population growth, but by the 1980s immigration was up and a minor population boom was in progress into the 1990s. Where will more Californians live, especially those eager for detached residences out in the countryside? In other words, how and where should urbanization be directed in the future?

Admittedly, the prime-land alluvial valleys will remain the most profitable land for the spread of suburbia. But what would the benefits and costs be if urbanization were diverted away from prime irrigable land to surrounding benchlands, foothills, and areas of lower-grade soils? There is ample land of this sort throughout metropolitan California, as exemplified in suburban developments like Mission Viejo and Westlake Village. It may be that higher grading and construction costs on hilly, marginal agricultural lands such as these may be partially or even fully offset by savings in social overhead for flood control, drainage, and sewage disposal. Other benefits possibly accruing to hillside suburban residents are availability of view sites and structures above the smog level. Best of all, there might still be some orchards and other useful greenery left to see in the valleys below, instead of wall-to-wall rooftops and pavement with an occasional park thrown in for aesthetic embellishment.

CONSERVING AGRICULTURE IN SUBURBIA

Short of proposing the retention of farms in suburban areas as subsidized open-space or greenbelt amenities, a practice that would interfere with operation of a free, competitive real estate market, several methods of institutionally preserving agriculture have been tried. The expedients submitted thus far in various metropolitan regions in California and the nation include incorporation of exclusively agricultural cities, establishment of conservation easements, enactment of rural or agricultural zoning, and preferential assessment of farmland. All these proposals are interrelated to varying degrees and in some cases have been employed as tools in compiling general plans containing an agricultural element. Such an agricultural element, for instance, has been proposed for addition to the Stanislaus County General Plan to protect the 270,000+ acres of prime land mentioned earlier in the chapter.

Agricultural Cities

Incorporation of cities to function primarily as agricultural entities has been attempted in California, but with little success. Probably the best-known case is that of Dairy Valley, incorporated by dairy operators, mostly of Dutch ancestry, in 1956 on several parcels of land in southeastern Los Angeles County. The idea was to preserve an agrarian way of life and keep suburbia out. By barely a decade after incorporation, however, suburbs had surrounded Dairy Valley and driven property values up to $15,000–$30,000 an acre for residential development and to $35,000 an acre for the 135-acre Cerritos regional shopping center site. At these prices, it is hardly surprising that dairy farmers began selling more land and less milk, with some moving their operations 40 miles east to Chino. By the end of the 1960s, Dairy Valley was renamed Cerritos, and the city began acting more suburban than agricultural. Faced with operational obsolescence, incompatibility with neighboring urban land uses, and high bids for their land, the four dairies that remained as of 1980 were gone well before the end of the decade. The last agricultural holdout to bite the dust in Cerritos was the Artesia Milling Company, which was originally known as Holland Dairy Feed and more recently made feed pellets for horses and rabbits. The company's property was condemned in 1988 to make way for a mall expansion, with the local redevelopment agency rumored to have spent about $1 million an acre for the land.[22] Except for some remaining "rurban" ranchettes built by dairymen decades ago, Dairy Valley days are long gone.

Conservation Easements

Assuming it is in the public interest to preserve agricultural land, cities or counties in California are empowered to obtain easements through voluntary sales or by *eminent domain* (the right of government to take private property from a landowner for the "public good" by paying fair market value). Many urban planners, however, believe that acquiring the *fee-simple*, or total, rights to real property is less complicated and accomplishes the same ends as easement acquisition. Probably the major methodological drawback to introducing a program of agricultural easement procurement in California would be that of simultaneous acquisition of large, contiguous blocks of prime developable land and the negative influence such acquisition would have on the value of development rights sacrificed. In any event, acquisition of easements or fee-simple rights solely for purposes of preserving agriculture seems an unacceptable tactic in a free land market economy.

Agricultural Zoning

Zoning ordinances constitute the oldest and most widely used method of control over land use that can be exercised by a local government. They can also serve as the most effective tool in the hands of planners for promoting orderly development of a community. But critics of rural zoning argue that it is a stop-gap measure at best. They contend that if the market value of land soars high enough, a *variance*, or new urban zoning ordinance, can be obtained and the land in question will convert to highest and best use.

No more classic case of implementation of agricultural zoning to thwart urban sprawl exists than in Santa Clara County in Northern California. Near the city of San Jose in the 1950s, subdivisions became scattered over 200 square miles of fruit land while actually occupying only 12 square miles. This left 188 square miles of land beyond financial reach because of its overinflated speculative value, thus serving as a classic example of urban sprawl.

In 1954, before too much of the prune and pear land became vacant, a group of growers banded together and acquired from the Board of Supervisors the first exclusively agricultural zone in Santa Clara County. A general pattern of rural zoning soon followed to protect farmland from subdivision, but this did not preclude strip annexing by San Jose and other municipalities. As a result, the farmers persuaded the state legislature to pass an interim law to forbid poaching by cities.

But zoning in this case proved to be only a stopgap. In 1958, William H. Whyte observed that "while the farmer continues to pay relatively low taxes, the surrounding land keeps soaring in value, and so, potentially does his own. What will happen when the land goes to $15,000 to $20,000 an acre? Farmers got the zoning approved; they can get it disapproved. The local Farm Bureau is quick to admit it, and now feels additional means must be sought to preserve farmland."[23]

Preferential Assessment and the CLCA

Laws aimed at preventing assessment of farmland at subdivision values have been on the books of several states since the 1950s, and in some cases these and re-

[22] Bettina Boxall, "Islands of the Past: Suburbia Engulfs Last of Cerritos' Dairy Days," *Los Angeles Times*, July 31, 1988, Southeast Section, pp. 1 and 5.

[23] William H. Whyte, Jr., "Urban Sprawl," *Fortune* 52, 1 (January 1958), p. 106.

Figure 11-25 *Fallbrook prime land under siege from suburbia. This region of northern San Diego County is first in the state and nation in the growing of Hass avocados, the leading commercial variety. The first Hass tree ever developed still stands in La Habra Heights, some 60 miles north of Fallbrook.* (Crane Miller)

lated laws have been declared unconstitutional. In 1966, California voters approved an amendment (Proposition 3) to the state constitution allowing for preferential assessment of land used solely for the production of food or fiber and other so-called open-space uses.

William H. Whyte looked with disfavor on this means of preserving agricultural land as well, noting that "when the price is right the farmer will sell out, low taxes or no. And, why not? He is not going to foreswear a large capital gain so suburbanites will have pretty scenery. Unless there is some compelling incentive, he is going to relocate."[24] Whyte went on to cite the example of a Maryland county that lost over $2.3 million in tax assessments because of preferential assessment. With this money, the county could simply have bought 1,500 acres of farmland and thus done a more permanent job of preserving the land for agriculture.

Despite its shortcomings, the state government in essence supported preferential assessment as the best tool for controlling conversion of farmland to nonagricultural uses with passage of the California Land Conservation Act (CLCA) of 1965, or Williamson Act (AB 2117, sponsored by Assemblyman John Williamson). The basic goals of the CLCA are (1) to preserve a maximum amount of farmland in order to maintain the state's agricultural economy and ensure adequate food supplies for the nation, (2) to discourage premature and unnecessary conversion of farmland to urban uses, and (3) to maintain farmland in developing areas as valuable open space. Prime land under the Williamson Act is defined as (1) Class I or II in the Soil Conservation Service land use capability classification (I = most capable), (2) 80–100 in the Storie Index rating (100 = best), (3) land used for livestock with an annual carrying capacity of at least one animal unit per acre, (4) land used for trees and vines that earns income during the commercial bearing period of at least $200 per acre, and (5) land used for unprocessed agricultural products that earned at least $200 per acre in 3 of the previous 5 years.

The key feature of the CLCA has landowners sign annually renewable contracts with their respective counties to keep their land in agriculture for a minimum of 10 years. In return, the county agrees to assess the land on the basis of agricultural use rather than on a typically higher valuation based on what it would bring for subdivision use. In other words, *use value* assessments instead of *market value* assessments are made on the land if the owner restricts its use to agricultural production for 10 or more years. Rarely, if ever, in metropolitan California will the agricultural use value of land come anywhere close to its subdivision market value.

A parcel of land receiving a property tax break under the CLCA is referred to as an *agricultural preserve*, and depending on the county party to the contract, such a preserve will be of a specified minimum size. A farmer can get out of an agricultural preserve simply by filing notice of nonrenewal at the end of the first year of a contract, but he or she will have to keep the preserve in agricultural production for the remaining 9 years of the contract. A farm owner who wants to cancel a contract altogether must pay a penalty equal to 50 percent of the reassessed value of the preserve unless waived by the state director of agriculture as being in the public interest.

The CLCA has been in force now for over three decades, yet there remain serious doubts about its appeal in prime-land metropolitan areas. The act was slow to catch on in its early years, and even by fiscal year 1975–76, only about one-third of California's 12.5 million acres of prime land (Figure 11-25) was in agricultural preserves. By the 1990s, prime land in preserves had levelled off at barely over 5 million acres.

Making the CLCA's performance even more questionable was a case study of land under contract in 11

[24] William H. Whyte, Jr., *The Last Landscape* (Garden City, NY: Doubleday, 1968), pp. 116–117.

central California counties that found that farmland near incorporated areas was much less likely to be under contract than was more remote land. Another study determined that initial preserve signups were concentrated in nonprime areas located some distance from cities. Referring to these studies, University of California agricultural economists Hoy F. Carman and Cris Heaton noted that "much of the land under contract was in little or no danger of being converted to nonagricultural use, whereas much land not under contract is viewed by its owners as having development potential."[25]

Because property tax revenues are either lost or shifted to other taxpayers, perhaps the most glaring defect of the CLCA is the negative fiscal impact it has on local governments and school districts. Compounding the problem in 1978 was overwhelming approval by the voters of the property tax–slashing initiative, Proposition 13. Not only have local government's revenue-producing capabilities been further reduced, but in looking to state government for fiscal bailouts, local governments may be witnessing their own political clout being weakened beyond recourse. The concept of *home rule*, whereby local communities plan their own destinies, has indeed suffered a severe blow in the implementation of Proposition 13.

Proposition 13 is a mixed blessing insofar as the CLCA in particular and agriculture in general are concerned. Farmers for the most part favored it, for they realized a rollback of their land taxes from about 3 percent to 1 percent of 1975 market values. It is noteworthy that agricultural landowners are in the dubious position of being the highest per-capita property taxpayers in the state. Obviously, where tax savings are greater under Proposition 13 than under the CLCA, there will be a sharp drop in CLCA participation. Furthermore, the farmers enjoying bigger tax savings under Proposition 13 will no longer have to worry about tying their land up in 10-year contracts; they can now sell their land to a developer whenever they please. Whether all this will encourage landowners to stay in agriculture or get out is difficult to predict. But one thing seems fairly certain: Proposition 13's attenuation of home rule will open the door for state government's playing a tougher role in planning conservation of agricultural resources. Whether state government ever assumes such a role, however, is purely speculative at this time.

Property taxes lost to county and city governments because of the Williamson Act have exceeded $100 million annually in recent years; in 1988–89 alone, losses amounted to $120.4 million.[26] Williamson Act assessed valuations in 1988–89 were about half of what Proposition 13 valuations would have been. The state compensates partially for tax revenues lost by local governments under the CLCA, but such *subvention* usually is far less than what was lost. In 1988–89, for example, the state subvented only $14.5 million to counties statewide. Obviously, county tax collectors are not overjoyed with the CLCA. On the other hand, participating farmers realized a $120.4-billion benefit in 1988–89.

Perhaps even more significant than the impact of Proposition 13 on the CLCA was the February 1981 California Supreme Court ruling that agricultural preserve contracts may be canceled only in extraordinary and unforeseen circumstances. Conservationists agree that the decision strengthens the CLCA by making a contract much more difficult to cancel. Entering the twenty-first century, more than 15 million acres of California farmland are now protected under the act.

In this and the previous chapter, we considered the rural and urban landscapes of California. In so doing, we also gained an appreciation for the state's economic geography and thereby set the stage for an examination in Chapter 12 of California's role in the broader economy of the Pacific Rim nations. As we enter the twenty-first century, or what is dubbed the "Pacific Century," a brief look at California's recent performance in the Pacific Rim trade scene seems appropriate. The recent Asian economic downturn notwithstanding, California's exports exceeded $100 billion for three consecutive years, 1996–98, and the state strengthened its positions as the world's seventh largest economy and the leading exporting state in the nation. While exports to Asian nations in the Pacific Rim declined slightly in 1998, they increased by 10 percent to Canada ($12.7 billion) and Mexico ($13.7 billion). Canada's, Mexico's, and the United States' signing of the North American Free Trade Agreement (NAFTA) has provided the principal impetus for growing trade among the continent's three largest economies. On the threshold of the Pacific Century, this portends well for all of North America's economies, as well as for those of Asia, Australia, and South America.

---

[25] Hoy F. Carman and Cris Heaton, "Use-Value Assessment and Land Conservation," *California Agriculture* (March 1977), pp. 12–14.

[26] Harold O. Carter et al., "Land in the Balance: The Williamson Act Cost, Benefits, and Options, Executive Summary," Davis, University of California–Agricultural Issues Center (December 1989).

CHAPTER

12

California on the Threshold

Reportedly, when the Chinese wish to see someone experience difficulties and problems, they extend the following "benediction": "May you live in interesting times!" As the new century dawns, Californians certainly live in interesting times, in both positive and negative senses. In many ways a microcosm of the nation, California represents at the same time the best and the worst of a challenging future. In an even broader sense, we can recognize in the state the embodiment of the rapid evolution of the larger global environment—social and physical. The fascinating reality is that many of the "problems" perceived by critics of California are merely manifestations of the inexorable *changes* a rapidly shrinking world is experiencing. Indeed, when the United Nations held a global summit on world development in the mid-1990s, the topics of discussion could well have been taken from a town hall meeting in almost any community in California: chronic unemployment, pervasive poverty, degradation of the physical environment, deterioration of the social order, and the widening gap between rich and poor. Yet, while we recognize these issues, we must also concede the progress that has been made: advances in medicine, which have extended life expectancy; improvements in education, which have resulted in a higher literacy rate; and technological progress, which has brought overall increases in quality of life.

Perhaps what is most important is to identify the changes, understand the context, and then try to create a coherent plan for the future. Certainly, one of the major reality shifts for California (and the world) has been the end of the Cold War, which has brought with it significant changes in economics, politics, and social focus. Rather than military preoccupation, economic issues have now become paramount. The security and prosperity of California, the United States, and the world are highly dependent on the operations of the world economy, and foreign policy is being conducted with different presumptions than before. In an increasingly "borderless" world, trade is a key issue at the national and local levels, and the line between domestic and foreign policy is somewhat blurred. Trade, immigration, narcotics, education, jobs, culture, health, environment, transportation, communications, entertainment, agriculture—all are at once both international and local in scope. In the area of immigration, California has become the single most popular destination in the United States and thus has a crucial stake in the discussion of national policy on the issue. In the realm of narcotics, Los Angeles is a (if not *the*) major entry port in the country. In the employment sector, the *maquiladora* program has moved many jobs out of the state and into Mexico. The impact of immigration on education has been notable for the accompanying demand for multicultural and multilingual teaching. The health and medical establishment of California is now faced with significant problems such as tuberculosis, once deemed an all-but-eliminated disease. And continual population growth has meant increasing traffic congestion, strains to urban infrastructures, and environmental difficulties.

THE PROSPECTS

Many years ago, Curt Gentry wrote a book entitled *The Last Days of the Late, Great State of California*[1] in which he envisioned a catastrophic end to the state brought about by a devastating earthquake. More importantly, he used this natural disaster as a starting point to reflect on the problems the state had created for itself. Gentry questioned whether California would have destroyed itself even without a natural disaster. His conclusions were troubling. In essence, he portrayed a state in crisis, populated by dysfunctional residents who were constantly sowing the seeds of their own physical, cultural, and spiritual destruction. His imaginary cataclysmic quake merely brought quicker closure to the process.

Several decades later, novelist Cynthia Kadohata, in her *In the Heart of the Valley of Love*,[2] continued the tradition of envisioning California in the future. Rather than attributing the decline of the Golden State to a natural disaster, Kadohata built upon the theme of social-political-environmental decay. Taking Los Angeles as her symbolic representative, she described a situation of moral decay and corruption in government and society, of environmental pollution so bad that cancer and disfiguring skin diseases were the norm, and of class riots and daily violence so common as to evoke no surprise from citizen or official alike. Though lacking a catastrophic earthquake, Kadohata's fictional destruction of the state is no less troubling in its implications.

Of course, California has not yet been destroyed by human, social, or natural disaster. The thought of self-destruction, however, remains an unsettling idea. Little progress has been made toward creating an efficient mass transit system in California. In a state where mobility and effective transportation are critical needs, the dominant mode is still the economically inefficient and energy-wasteful automobile. Although some control has been imposed in the area of pollution, standards of air and water quality, noise regulation, and chemical waste disposal are still far from satisfactory. Indeed, revelations concerning hazardous chemical dumps in residential areas have exposed a whole new category of physical maladies suffered by unfortunate and unwary Californians.

Violence has become an accepted fact of life in large urban areas, and the increase in the numbers of guns being purchased reflects the level of confidence many residents have in current standards of police protection. Extremism of various types continues to threaten the social fabric of the state. The persistence of groups like urban gangs, religious cults, terrorist groups, racial hatred groups, neo-Nazis, and the Ku Klux Klan raise doubts about the overall stability of California culture and society. The predominant philosophy, however, remains apathy. The majority of Californians persist in ignoring the pressing problems of the present and future. In their pursuit of pleasure, citizens of the Golden State tend to gloss over the cracks in the social, political, and environmental structures.

Certainly, many of the difficulties are a function of size. California has a yearly budget, a resource base, and a population greater than those of many countries. The complexity of maintaining an orderly society in a population that numbers 10 percent of the total population of the United States is often overwhelming. The challenge of coordinating interests in a geographical and political entity that contains approximately 158,693 square miles of area is enormous. This sort of diversity has frequently led to discussions about dividing the state. To date, these discussions have never moved past the proposal stage, although the legislature has debated and acted at counterpurposes more than once, and it is still a remote possibility in the future.

In short, the prospects for California are quite mixed. Problems exist, and solutions are possible, but actions are often confused. The key to the future lies in the interaction between people and the natural environment. Whether they are in conflict or in harmony will decide the destiny of the state.

THE PEOPLE

The people of California face many changes. The increase in population is both a growing problem and an exciting challenge, the solution to which can dramatically affect both the prosperity of the state and the lifestyle of the people. The changing ethnic character of California also will impact upon cultural styles, social stability, and political power (Figure 12-1). As with other states, the development of California will be influenced by the overall strains on the social fabric of the nation as a whole. The people, however, are the first resource of the state, and it is up to them to find the solutions.

Growth and Population

The first reality about people involves the number of them who call the state their home. Around 30 million persons resided in the state as of the beginning of the 1990s, and this figure rose to close to 33 million by the end of the decade. Census Bureau statistics show no indication of a decline in the constant growth rate, and the upward trend suggests an area of concern.

Although California's population density is not yet at a critical juncture, distribution of the people is a problem. Concentration of population in the south is a continuing phenomenon that severely stretches existing resources. The south already has a water shortage po-

[1] New York: Putnam, 1968.

[2] New York: Viking Press, 1992.

Figure 12-1 *Nob Hill in San Francisco. California is a state of vast ethnic and social diversity, as the mix of buildings on this San Francisco street shows.* (Roger M. Rhiner)

tential. The south already has too many cars on the roads. The south already strains to provide employment for its population. The south already suffers a housing problem. Unfortunately, there appears to be little relief in sight. The north is in no better position to absorb excess population and at times expresses open reluctance to do so. This is a major issue, and so far, too little thought has been directed toward its solution.

Perhaps the reasons behind the gridlock in planning for population increases have something to do with changing economic patterns, shifting political attitudes, and tenuous ethnic balances. Whatever the reasons, the need to deal with continuing growth is a given. Depending upon what authority is cited, over the next several decades, the state's population is expected to surge by 10–20 million persons (around 40 percent from foreign immigration).

Of course, the significance of these population statistics is found in the reality of lifestyles of Californians. The means by which we respond to demographic change will help determine the quality of life for the future. Such issues as job availability, housing, water and energy supplies, waste management, education and social services, and human civility pose daunting challenges for Californians who seek to maintain the state's golden image. Certain issues, in particular, provide focus for consideration, including immigration, migration, and ethnicity, as they relate to population growth.

Immigration, Migration, and Ethnicity

At once one of the most difficult and yet most exciting elements of the modern California scene is the dramatic interplay of ethnic groups. Reflecting its position on the

Pacific Rim, California is experiencing an astonishing readjustment in its ethnic mix, from an Anglo-dominated majority to a more international model. Driven by such factors as economic opportunity, geographic juxtaposition, NAFTA-induced freer trade policies, and political instability, new immigrants have been steadily moving (legally and illegally) into the state. Especially evident in urban settings, is the multiethnic pattern of Buddhist temples and Korean churches, Little Saigons and Little Tokyos, Chinatowns and Hispanic barrios (representing most nations of Central America), and enclaves of Russians, Armenians, Iranians, Jews, Cambodians, Laotians, and many others.

This increase in new immigrants has generated negative reactions in many quarters. Some groups, feeling socially and economically threatened, have initiated anti-immigration activities in an effort to halt the tide. The passage of controversial Proposition 187 in the mid-1990s (ostensibly denying education and certain social services to undocumented aliens) demonstrated the intensity of the debate, as did Governor Pete Wilson's unsuccessful but symbolic lawsuits against the federal government for reimbursement for and protection from the costs of such immigrants to the state. Of course, the actual impact of immigrants on California is by no means clear. Some vocal authorities allege that the immigrants are a drain on the welfare system and refuse to assimilate into the mainstream; other experts point to directly contradictory evidence suggesting that these new immigrants have a dynamic and positive effect on the socioeconomic fabric in California. What many fail to recognize is that this is hardly a new debate, but rather reflects a national (and Californian) tradition of cycles of immigration and hostile reactions dating back two centuries. Californians have long struggled to resolve their ambivalent racial-ethnic attitudes toward one another.

Although the state has a long history of mixed races, it has also had a rather poor record of promoting social equality and harmony. The continuing existence of tension between ethnic groups cannot be denied, and the 1992 Los Angeles race riots were neither unprecedented nor totally unexpected. As the state moves into a new century dominated by regional and global changes, the citizens of California represent a cross-section of reactions. Some white flight from the state has taken place, although this is occurring more among lower socioeconomic classes than among the better educated (and thus more employable). Some experts have also noted an interesting minor trend of "black flight" to the suburbs, as the newer immigrants preempt upwardly mobile black Californians. Thus, it must be emphasized that racial tension is not a whites-only phenomenon. Tension among black, Asian, Hispanic, and Middle Eastern populations are just as real. With California becoming the first Third World state in the United States, the measure of progress may be found in how modern Californians deal with this issue.

A brief summary of the crucial demographic data underscores the changes and challenges facing the state in the twenty-first century. First, we are still growing. The U.S. census projects that approximately 60 percent of the nation's total growth well into the new century will take place in a handful of states in the South and West—with California showing the single largest gain. This would bring California's share of the nation's total population to 15 percent within the next 30 years.

Second, as with the nation as a whole, California's population continues to grow older. With current life expectancy in the United States estimated as high as 82 years (compared to 65 only a half-century ago), the profile of California's population is maturing. Indeed, the U.S. census also projects that the 65-and-older population of the state will double over the next two decades. Contributing to the overall pattern, California's fertility rate will also continue to decline.

Third, ethnic and migration patterns will significantly modify the state's makeup. The realities of national immigration statistics are particularly notable in California, with the state leading the nation in percentage of foreign born immigrants. By far the largest groups are Asians and Hispanics, who by the late 1990s accounted for almost 85 percent of California's overall population growth (from both net migration and natural increase). It might be added that during the 1990s, there was a modest out-migration of whites and Native Americans. All of this reflects national adjustments, with predominant international "migration magnets" remaining the ports of entry (e.g., Los Angeles, San Francisco, San Diego) and internal "migration magnets" shifting to mostly noncoastal destinations (e.g., Phoenix, Las Vegas, Atlanta).

Finally, urbanization of the state will not abate. Already reflecting a 95 percent urban population, the increased immigration, natural increase, and internal readjustments promise to maintain large urban clusters, as well as boost secondary and tertiary population centers. Contrary to views of pessimistic observers of the 1990s, the results will continue to be growth rather than stagnation and change rather than decline.

The Social Fabric

An additional, broad, people-oriented difficulty involves strains on the general social fabric of the state. Various stresses of modern life are taking their toll on Californians. Sociologists have expressed concern over the fallout of "future shock" in America. As society moves faster and faster, people become more confused, alienated, and maladjusted. Unfortunately, California exhibits some of the more dramatic manifestations of this dysfunctional social order. Suicide and divorce rates both exceed the national averages. The state ranks in the top 5 in crime rate, and its expenditures on prisons and law enforcement lead the nation. California also ranks in the top 5 in homelessness, in drug abuse, in unemployment, and in welfare recipients, and it is in the top 10 for persons falling below the poverty line. It ranks number 1 in highest costs per patient day in hospitals. Its expenditures for the arts places it among the bottom 5 states, as does its percentage of eligible voters who bother to register. Stories in the media detail the escalating death rates from gang-related violence in virtually every metropolitan region. Add to these sobering statistics the disturbing images of the 1992 Los Angeles riots (following the verdict in the Rodney King beating trial) and the 1995 circus performance of the California legal system in the O. J. Simpson case, and the Golden State appears to have a rather brassy tarnish. Further damage to the perception of quality of life for California can be found in the too-common evidence of infrastructure decline, most dramatically chronicled by the bankruptcy of Orange County in the mid-1990s.

These statistics are disturbing and raise important questions about the overall quality of life in the state. Certainly, California can ill afford to ignore the social, economic, and political imbalances that feed these problems. The actions taken to create a healthier and more stable society will help determine how many people wish to continue to live in the state.

THE PACIFIC RIM—THE PACIFIC CENTURY

Many experts are now suggesting that the global balance of power is dramatically shifting from a European focus to the Pacific Rim. This is quite credible if we merely contemplate the approximately 50 countries with nearly half of the world's population that border the Pacific. Certainly for the United States, the shift has already occurred, with trans-Pacific trade now exceeding trade with Europe (a change that began in the mid-1980s). In an increasingly "borderless" world, trade has assumed a central role in international relations. The significance of this fact for California is immense.

Geographically, California sits squarely in the center of the new international economic coordinates. The immigration, economic, and political patterns emerging in the state are seen by many to represent the future of the country. As a prime gateway to the Pacific Rim, California has already attracted much foreign capital. Real estate investment in Los Angeles is a prime example, with roughly 30 percent of choice downtown sites now owned by foreign corporations based in Pacific Rim countries. San Francisco real estate also reflects a remarkable pattern of Hong Kong Chinese ownership. In terms of product movement, thousands of cars move both ways between Californian and Asians ports daily; California farm products are major profit items in this market; and trans-Pacific communications (business and personal) have created an extensive and continual linkage between Californians and virtually every site in the Pacific Rim. Likewise, north-south ties encouraged by the North American Free Trade Act (NAFTA) are particularly evident in California. Economic partnerships between San Diego and Tijuana and between Calexico and Mexicali have helped create new jobs and markets on both sides of the border. Chain migration ties families throughout California to Mexico, El Salvador, Guatemala, and many other nations to the south. The value of trade flowing through California ports is astonishing; California airports also account for billions of dollars of Pacific trade. As noted previously, California in general (and Los Angeles in particular) is *the* gateway for swelling immigration from Asia and Latin America, and projections for the near future envision even more Pacific Rim immigrants, most of whom will be drawn to California.

As Asian consumer demand continues to rise, due both to sheer numbers and to increasingly sophisticated tastes, the potential for California businesses and products also rises. China, Korea, Taiwan, Singapore, Malaysia, and others represent growing opportunities for a rapidly changing California economy. Similarly, the increase in U.S. exports to Latin America has been dramatic, with some experts predicting that this market will soon surpass exports to western Europe. Thus, if the historical bases of California's economy (aviation, agriculture, defense, oil, and apparel) have stabilized or declined, the emergent state economy will be increasingly buoyed by Asian and Hispanic entrepreneurial initiatives, driven in part by the fortuitous geographic location of the Golden State.

THE NATURAL ENVIRONMENT

The natural environment is as critical to the state as its people. A fine balance must be struck to maintain maximum utility from the air, land, and water without creating irreparable damage. Nature has proved to be relatively tolerant of abuses perpetrated by humans, but the mistreatment must cease. Recent developments in environmental control have had mixed results. Several acts are of particular interest.

Federal Regulations

The *National Environmental Policy Act of 1969* is the basis for most modern environmental protection laws. It states that U.S. national policy is to foster "harmony between man and his environment." The act mandated the establishment of a Council on Environmental Quality, directed the president to present an annual state-of-the-environment message, and led to the use of environmental impact reports on federal plans that would affect the quality of the environment.

Of course, the major shortcoming of the act is that it does not apply to individual, corporate, or state actions. Over the course of its existence, even enforcement of environmentally responsible actions by the federal government has been fraught with inconsistencies and compromises. At best, the act can be said to establish a guiding principle that *should* be followed. Other federal enactments pertaining to environmental protection tend each to be addressed to one type of environmental problem and may involve regulation of states, private industries, and individuals. Thus, various acts attempt to regulate water quality, air quality, drinking water, endangered species, solid waste disposal, toxic substances, mining, coastal zones, insecticides, scenic rivers, and various other issues. Frequently, the federal government "assigns" authority to the individual states to enforce certain of the acts. However, in addition to the federal government, state governments have also engaged in efforts to regulate the environment. California's record of protecting the natural environment has followed a somewhat predictable pattern.

State Regulations

The *California Environmental Quality Act of 1970* was patterned closely after the federal act. It committed the state to maintaining a quality environment for the people of California, "now and in the future." Following this act, the state legislature passed a number of laws that addressed specific problems. Over half of the state codes soon contained sections on environmental problems and controls. Laws were passed that dealt with general pollution, water quality, air quality particulate pollutants, open space, parks and recreation areas, wildlife and wilderness areas, and nuclear power. These laws have attempted to set reasonable standards to protect the environment from excessive polluting or encroachment on natural areas.

California's efforts to regulate and clean up the air are well known throughout the nation. Recognizing that air pollution transcends boundaries, the state established regional air basins and placed responsibility at the local and regional levels for nonvehicular sources of pollution. It maintained primary statewide control over vehicular pollution, with some shared authority with regional agencies. Examples of the regional agencies include the South Coast Air Quality Management District, the Bay Area Air Quality Management District, the Sacramento Metropolitan Air Quality Management District, the Mojave Desert Air Quality Management District, and the San Joaquin Valley Air Quality Management District.

Figure 12-2 *California's efforts to seek new, clearer sources of energy can be seen in this windmill farm outside of Palm Springs.* (Richard Hyslop)

Regulations at both state and regional levels attempt to deal with agricultural and other burning, toxic air contaminants, emission-emitting equipment, fuel vapor and fuel-burning emissions, and other sources of air pollution. With the strictest requirements in the nation, California has been accused by automakers and others of setting draconian standards for air quality. However, as with many environmental rules, the issue seems less a question of feasibility than one of reduced profit margins. Apparently, Californians will have to continue to ponder their priorities: decent air quality versus corporate profits (Figure 12-2).

Water quality has also posed serious problems for the state. Both the *quantity* and the *safety* of water for human and animal use are crucial issues. Various enactments under state authority prohibit dumping or otherwise allowing noxious objects in waters, establish standards for drinking water, control groundwater contamination, regulate coastal marine environments, provide for toxic cleanup, and require regional plans to achieve compliance. Reflecting the historical economic and political influence of agribusiness, both the allocation of waters for environmental purposes (fish and wildlife) and the enforcement of quality standards often have been delayed and sidetracked.

Yet another area of growing concern for Californians involves environmental hazards. In recent decades, it has become more evident that plans for dealing with environmental hazards have been woefully inadequate. These hazards may be broken down into two major categories: human-created and natural.

In the realm of human-created hazards, one of the most evident is toxic waste sites. By the mid-1990s, California ranked second to only New Jersey in the number of toxic cleanup sites targeted for action by the Environmental Protection Agency. Spread all around the state, over 260 sites are listed as high priorities requiring billions of dollars to correct problems created by mining companies, industries, oil producers, commercial landfill/dump operators, military bases, and agribusiness. Soil and groundwater contamination, wildlife and aquatic life destruction, and human illnesses have been directly attributed to various of these sites. Yet progress on correcting these catastrophies has been modest, at best.

In the area of natural hazards, California has demonstrated a surprising lack of coherence in planning. In the early 1990s, the Oakland Hills firestorm demonstrated the danger of casual treatment of land use in an explosive urban/wildland interface. Over 3,000 dwellings and 25 lives were lost, with damage estimates approaching $2 billion. Two years later, firestorms swept through chaparral-covered residential areas stretching from Malibu to the Mexican border, destroying million-dollar homes in Laguna Beach, Malibu, Thousand Oaks, Altadena, and other communities. Yet, in spite of the huge property losses, the massive expenditures for firefighting, and the overall damage, homes were being rebuilt on the exact same sites within a year or two. Likewise, despite massive flooding along the Russian River in the north and flooding and mudslides in Malibu in the south, homes were quickly rebuilt in the same neighborhoods. Even after the 1989 Loma Prieta and the 1994 Los Angeles/Northridge earthquakes, life (and construction) began anew with little regard for the existence of the earthquake hazard zones. The logical question at this point is not *whether* California's public policy should address this problem, but *when*!

The necessity for planning to mitigate hazard losses is a fairly modern concern, not just for California but for the nation as a whole. Several factors have brought about this change. Population growth has led to increased demands for land that would have been ignored in the past. Aesthetics also have evolved over time: Where prime "bottomland" may have been the ideal of the past, now rocky cliffsides, waterside locales, "rural" brushlands, and mountain slopes exert strong appeal for their scenic qualities. Similarly, economics may make less costly land in fault zones, desert regions, slump zones, or outwash plains acceptable residential choices. Thus, many of the hazard losses noted previously occurred precisely because people located their homes and businesses in inherently hazardous areas. Certainly, the desire to occupy one's own house and property is well established in the American psyche. The importance of this desire is reflected in the "new housing starts" statistics collected and reported by the federal government. Typically, a decline in these numbers is viewed as a negative sign in the national (or state) economy. Yet, while the desire to see these numbers increase is understandable, the reality of a declining base of safe, usable land is ever-present. The scope of the problem is exacerbated by the preference of Americans to live in single-family, detached residences. Clearly, this places even greater pressures on static land resources, resulting in the increased use of potentially risky sites.

Set against the desire to build in questionable areas is the societal mandate to protect citizens from harm. When the government acts on behalf of the "public good," does this mean it should protect people from their own possible folly? That question has placed California's government (and other states' as well) in the middle of a classic political debate: How much regulation should government exert over its own citizens? There are two basic choices.

First, government may choose (by benign neglect) to allow persons to build on the San Andreas Fault, in fire-prone Malibu, along the banks of the Russian River, in the Santa Ana River floodplain, or anywhere else. Thereafter, if losses occur, the government may (1) do nothing, (2) provide public assistance, or (3) encourage private/insurance assistance. However, if the government knows of the risk, is it then liable to the injured homeowner (as some court rulings have held)? And if the property owners know the risk, should other citizens be forced to spend their tax dollars to "bail out the stupid"?

Alternatively, the government could attempt to prevent the losses in the first place. This would involve prohibitions against certain uses of land in hazardous areas. Of course, this immediately raises constitutional questions of private property rights, deprivation of property without compensation, and similar issues. Of course, the logic of preventing people from engaging in activities that might be harmful to them seems reasonable. However, as may be seen from the heated arguments over motorcycle helmet or seatbelt laws, not everyone readily accepts such "parental" wisdom. Challenges to land use regulations are spirited and instantaneous, so California government agencies have hesitated to establish even reasonable restrictions on building in hazardous zones. However, the irony is that even where these same agencies do *not* restrict land use for fear of legal challenge, they may face subsequent lawsuits for failing to *prevent* homeowners from building in the hazardous zone. It is a seemingly circular problem for which neither the courts nor legislatures at state or federal level have been willing to provide a responsible policy. Until such responsibility is exercised, losses such as in the Oakland Hills, Malibu, and Laguna Hills will recur with monotonous regularity. Can such preplanning ever be accomplished? The answer to this is a qualified yes. There has been a very modest effort to establish reasonable land use rules in the state along its vast coastline.

The *California Coastal Zone Conservation Act of 1972* and the *Coastal Act of 1976* established control over development of the state's coastline. These measures set up a Coastal Commission responsible for all development within the coastal zone. These acts have been relatively effective in preserving natural coastal environments from further erosion and have placed severe limitations on high-density development and modifications of the natural landscape.

Various specific regional planning programs have also been added to state law. The Tahoe Regional Planning Compact saw five California counties and the state of Nevada join together to create a regional development plan that would transcend state jurisdictions. The Tahoe Regional Planning Agency has met with mixed success in its efforts to provide a comprehensive approach to the area. Similarly, the San Francisco Bay Plan was set up to protect the delicate balance of the Bay and to coordinate all development bordering its waters. As a result of its efforts, overall water quality has steadily improved, and the Bay is gradually becoming an attractive recreation and wildlife environment again.

The trend for the future in the area of environmental legislation is encouraging. Further refinements of both federal and state laws have occurred on a regular basis. California has recently dealt with tobacco smoke, noise, engine emissions, hazardous wastes, and beverage containers. Federal legislation has likewise updated and refined the procedures established to protect the environment. In the area of legislation, the people occasionally have shown a willingness to pay the price for improvement.

THE WATER PROBLEM

The persistent problem of water allocation in the state has not been resolved. Few other issues can raise emotional and political hackles as effectively as this one, and Californians recognize that in their state, water is power. California has been able to achieve its enviable position because of the riches made possible by extensive irrigation. Agricultural and residential properties appreciate in value precisely because water can be brought to otherwise barren regions.

In spite of a comprehensive water management system, however, a crisis is present. As various and regular droughts illustrate, even one dry year has far-reaching consequences throughout the state. In vast portions of California, failure of water supplies not only impinges upon normal agricultural and household uses but also exacerbates a perennial fire danger. The brushfire season is one of growing concern as more people build in former brush areas. Many Californians have learned, to their dismay, that disaster can strike at home.

The water future is truly a question mark due to several factors. Water from the Colorado River declined drastically in the mid-1980s as a result of the Central Arizona Project, but new sources have yet to be found. Long-term solutions have not been determined or implemented as Californians continue to argue about sources, rationing, and other issues. While drought threatens one year and floods the next, residents seem unable to agree on viable alternatives. Unless the state can find a satisfactory approach soon, the water crisis will haunt Californians well into the new century.

There are several water-related issues in particular that warrant attention, each of which has an important role in the future of the state. First, increased population and growth means that existing supplies must be better managed. It seems obvious that land use planning and water supplies are closely linked. Indeed, some legislative proposals have specifically suggested that one should not occur unless the other is part of the process. As early as the mid-1990s, the Department of Water Resources projected that even in nondrought years, supplies would be inadequate to meet the growing demand. This fact suggests a built-in limit to future growth unless some viable answers are found. Another growth issue relates to urban versus agricultural uses. As residential uses expand, the amount of water available for agriculture must necessarily decline. Mindful of this trend, many farm operations are beginning to install drip irrigation systems, as well as converting from crops such as alfalfa, corn, and cereals to less water-intensive crops such as vine and fruit trees. This has also inserted the water supply issue into the debate about farmland preservation, agricultural profits, and unchecked urban sprawl. The overall unlikelihood of new interbasin transfer projects has significant implications for ongoing water management. Undoubtedly, better use and cleanup of major existing aquifers will be necessary, as will better methods of capturing runoff for recharging groundwater sources.

The second issue is the pollution of water resources, particularly by salts from irrigation, fertilizers, and pesticides, which compromises the health of humans, wildlife, and the land itself. The problem occurs when semiarid land is brought into production through irrigation. As the water, fertilizers, and pesticides are applied to the soil, plants absorb the waters but leave much of the chemical salts in the soil. Where there is inadequate additional water to leach the salts from the ground, a steady buildup results. Depending on the underlying soil layers, the salts eventually percolate into and pollute the groundwater or build up to the point that the surface plants cannot survive the salt content of the soil. Furthermore, if runoff occurs, the salts can find their way into surface waters, pose a hazard to birds and aquatic life, and compromise the quality of this source. Traditional solutions to this problem have included evaporation ponds, diversion drains, and removal of land from production. Unfortunately, each solution has its shortcomings, and to date, conflicting goals have prevented effective resolution. What is certain is that unless serious attention is paid to this problem, a decline in both environmental quality and in farm production will follow.

The third issue involves the distribution of existing supplies of water in the state. Specifically, the Sacramento–San Joaquin Delta is the pivotal supply point for most of California's drinking water. This fact has engendered much debate over the years concerning pumping and diverting water from the Delta to Southern California and to San Francisco. The oft-proposed Peripheral Canal project to divert water around the Delta has led to some of the most bitter political and environmental battles in the state. Clearly, a balance must be struck between the needs of the Delta for sufficient water to maintain its delicate ecosystem and the needs of the state for water to supply its growing population. Unfortunately, planning for the future has been irregular and inconsistent, and no responsible management plan for the Delta has emerged. Unless and until that happens, the situation promises only to get worse.

The final issue is the need for water recycling and conservation. With projected shortfalls, as noted previously, improving efficiency becomes crucial. Periodic droughts have brought with them increased awareness of the need for such efforts, with various water agencies around the state using recycled water for irrigating crops or for recharging local aquifers. Likewise, public education programs have attempted to convert domestic users into nonwasteful consumers, with low-flow toilets and showers and drought-resistant landscaping. Assuming the projections of continual population growth are accurate, conservation and recycling efforts will become a fact of California life.

THE PROBLEM OF MOVEMENT

The enduring difficulty of handling movement of persons and goods around this huge state will only intensify in the future. Californians have made some efforts toward resolving this problem, but a coordinated and committed approach is still needed. Population pressures, air pollution concerns, energy scarcity, and other problems mandate development of viable alternatives and expansion of experimental techniques that will provide greater transportation efficiency in the state.

Most discussion categorize California's transportation question into the three areas of land, sea, and air. Because of the varied topography of the state, different problems have arisen and different usages have developed for these forms.

In terms of transportation by water, California has become a leading international trade focus, particularly for the nations of the Pacific Rim. An extensive shoreline with many excellent ports has encouraged ocean transportation, and the ports of Los Angeles, San Francisco, Oakland, Long Beach, San Diego, Richmond, Sacramento, Stockton, and others serve the varied needs of this trade. The rise of container ships has also acted as a spur to this area of transportation in California. Although travel by water does not account for a major portion of people movement, it does constitute a significant mode of industrial mobility.

The revolution in air travel has certainly affected California in a major way. Access to the state is immediate and easy for travelers from throughout the country and world. Major airlines serve most large cities in the state, with San Francisco and Los Angeles connected by air with almost any place on the globe. As far as intrastate transportation is concerned, the extreme size of California is no longer as formidable an obstacle for the casual traveler or business executive who needs or desires to move from one end of the state to the other in a short time span. Regional airlines now can provide connections almost anywhere in the state. To a lesser degree, the airlines now also provide rapid cargo movement for the state. Although volume may be lower than land or ocean transport, air cargo does serve as an immediate and efficient means of prompt delivery.

Surface or land transportation is by far the dominant form in California. Here, the choice is movement by rail or by road, with the road the primary choice in recent years. As noted previously (see Chapter 8), the railroad played a key role in the development of California. In addition, the growth of San Francisco and Los Angeles was facilitated by the San Francisco Key Route Electric Railway and Henry Huntington's Pacific Electric line in Southern California. Railroads, however, have largely surrendered their role as passenger carriers and now concentrate on the transport of goods. The freight-hauling function of railroads has been augmented by piggybacking, container cargo, and other techniques for maximizing the efficiency of rail shipping. Passenger carriage is not viewed as a profitable venture, and most railroads prefer to leave this aspect of rail transport to Amtrak or short-distance commuter lines. The once popular passenger train is now largely a romantic tale from the past.

California's road system is another story: This is *the* heart of transportation in the Golden State. As energy inefficient and costly as it may be, the private automobile is perhaps the predominant icon of the state. From the early beginnings of El Camino Real connecting the 21 missions, the California road system has grown into a phenomenal network of arteries, freeways, expressways, and roads. Financed by a state gasoline tax, the highway system has become a thing of wonder, reflecting the fact that California is the heaviest user of automobiles of any state in the country.

To provide some organization and structure, the California Master Highway Plan was adopted in 1959. This plan envisioned an eventual pattern whereby approximately 60 percent of the state's total road travel would move on the freeways and expressways. This system of rapid thoroughfares was targeted for completion in the early 1980s as a coordinated means of dealing with the transportation needs of the state. Unfortunately, the system was already overused and inadequate long before the 1980s arrived. California's heavy urban commuter traffic placed a severe strain on the system. The additional fact of California's position as the leading trucking state placed an impossible burden on the highway system and demonstrated that alternatives were sorely needed.

Some experiments have been undertaken in an attempt to solve some of the commuter pressure. One of the most notable efforts is the Bay Area Rapid Transit (BART) system (Figure 12-3). Recognizing the peculiar and unique topographical problems of the area, with its peninsulas, growing population, and limited space, planners sought a system whereby commuters could be moved more rapidly in and out of the urban work centers of the region. Traffic snarls on the various bridges in the Bay Area emphasized the need to get individual motorists out of their cars and into a mass transit system. What eventually emerged was an automated electric diorail train/subway system that connected downtown Oakland and San Francisco with outlying areas such as Concord, Berkeley, Hayward, Richmond, Fremont, and Daly City. The system was built to provide rapid service and dealt with some geographic problems by constructing a subway tunnel under the floor of the Bay and through the Berkeley hills. Although BART has been plagued with difficulties and was almost inadequate from its date of completion, it nonetheless demonstrated that such a system could operate and would be patronized by the public.

Another effort to solve the problem of moving people about the Bay Area is found in the CalTrain commuter lines, which extend light-rail connections from San Francisco southward to San Jose and beyond. A similar internal line (the Muni Metro) helps move commuters around within San Francisco itself. Given the continuing traffic congestion in the region, along with the uniquely configured Bay and shoreline traffic patterns, mass transit seems particularly useful in this region. However, like its neighbors to the south, the Bay Area's transportation mode is still dominated by the automobile.

The case of Southern California is even more problematic in terms of traffic congestion and gridlock on the one hand and viable solutions on the other. Various studies have confirmed that commuter traffic in California is the worst in the nation, with the greatest problems occurring in the Los Angeles–Orange County region. Indeed, in recent years, the Federal Highway Administration identified the 10 busiest highway interchanges in the nation and found that the top 9 were in California. Six of these were in Los Angeles County, and two in Orange County; the other was in Alameda County. The

Figure 12-3 *BART train gliding past the Oakland skyline. The BART system is one of California's few attempts to provide mass rapid transit to its citizens.* (BART)

statistics behind these facts are staggering. In Southern California, peak commuter hour speeds drop to around 20 miles per hour; most commuters drive alone (around 70 percent by most estimates); there are over 31 million motor vehicle trips per weekday; and the number of cars on the road just keeps increasing.

Compared to the Bay Area, how has the greater Los Angeles area responded? Los Angeles has had much less success in the realm of mass transit. The Southern California Rapid Transit District historically has relied on a bus system to serve public transportation needs. It has been exceptionally difficult, however, to entice Southern California drivers out of private automobiles. Indeed, private automobiles still account for the overwhelming bulk of commuter traffic, as rush hour freeway traffic shows. Some experiments have proven partially successful, such as Commuter-Computer, Park-n-Ride, bus and car pool lanes, special lanes on metered ramps, and similar devices. In certain heavily traveled corridors, CalTrans has even added toll lanes to speed affluent commuters along.

Political pressure has resulted in commuter train runs (Amtrak) between Los Angeles and San Diego, Los Angeles and Orange County, and Los Angeles and Riverside/San Bernardino. Perhaps more noticeably, the 1990s saw the initiation of specifically commuter-oriented Metrolink trains to connect outlying areas with the downtown core. Even more dramatically, the county transportation agency undertook a comprehensive effort that resulted in the planning and partial construction of multiple light-rail routes connecting various areas in the Los Angeles basin. Planners envisioned 400 miles of light-rail, subway, and commuter lines and similar connections by 2010, labeled the Blue Line, the Green Line, the Red Line, and so on. To date, those portions of the system already completed are considered qualified successes, particularly by frustrated commuters who are now able to escape the crowded freeways. Nonetheless, it remains a major challenge to pry drivers out of their cars. Even various incentives to carpool, use public transit, or find alternative travel arrangements have not been overwhelmingly successful.

What does the future hold in the area of transportation for Californians? Indications are that some Californians are beginning to recognize the need for alternatives. San Diego's "Tijuana Trolley" is an interesting recent effort. These bright red electric trolley cars cover a 16-mile route between downtown San Diego and the Mexican border, carrying in excess of 10,000 people daily. This light-rail system was conceived and funded at the local and state level and has filled a real need in the area. Depending on its continuing success, extension of the system is contemplated for the future.

Sacramento remains heavily reliant on automobile transport, but efforts to expand a light-rail system have also been pursued. As the Sacramento metropolitan area continues to grow, such a system may well be crucial if gridlock is to be avoided.

Meanwhile, the state highway system apparently has reached critical mass. Few additional projects are planned, and maintenance of existing freeways and expressways has proved a substantial chore. Some modifications are in process, including a proposed system of rail and bus corridors along center medians of existing freeways. Other modifications have included video camera monitoring of heavily traveled routes, computerized signboards for traffic information, metered ramps, and other electronically controlled devices.

Basically, however, these are attempts to keep a transit system of private automobiles in operation and so represent short-term answers only. It remains for long-range planners to develop more comprehensive solutions. With the continuing air pollution problem, the increasingly limited energy sources, and the population pressures in the state, alternatives to the internal combustion engine and private auto are needed.

Some discussion has revolved around electric- or solar-powered vehicles. Other planners have envisioned totally automated freeways. More interest has been generated in modified residential-work patterns. "Old" abandoned systems of mass transit are being reexamined with an eye to learning and profiting from lessons of the past, such as the Los Angeles Pacific Electric Red Cars (trolley) or Angel's Flight cable system of downtown L.A.'s Bunker Hill area. The one certainty is that creativity, innovation, and dedication are needed in this vital area of California life.

THE FUTURE

What can California expect from the future? So far, the interplay between people and environment has been quite uneven. On the positive side, Californians are generally healthier, younger, better educated, more affluent, and more advanced scientifically and technologically than people anywhere else in the nation. On the negative side, however, they often seem confused about how to use and control the technology that has made their lives better. Furthermore they often are myopic when it comes to recognizing the impact of their actions on the environment.

One obvious area of interaction between people and environment is population size. With demographers projecting a population continuing to grow into the foreseeable future, Californians still cannot fully deal with the impact that their current population is having on the environment. Given a doubling of the current problems, how will California respond? How will land use policies be set? How will the already critical water problem be resolved? How will strained energy sources serve future needs? How will the economy of the state absorb additional workers? Most importantly, how will quality of life in the Golden State be affected? These are open-ended questions, whose solutions will depend on the responses of thoughtful and concerned Californians.

Certain selected issues bear attention as a final commentary on California's future. Although not meant to be a comprehensive list, the following points are worth brief attention. First, can the economy of California sustain its rank of eighth largest GNP in the world? Although the economic downturn of the 1980s and early 1990s suggested a dislocation of prosperity, subsequent events have suggested *change* rather than *decline*. Certainly, some companies have relocated in recent years to other states. Certainly, job decline has been substantial in the aerospace, defense, computer, and manufacturing industries. However, the influence of GATT (General Agreement on Tariffs and Trade) and NAFTA (North American Free Trade Act) have placed California in the forefront of economic growth. In fact, job creation by smaller, more adaptable, knowledge-intensive companies has been substantial. Areas such as telecommunications, medical enterprises, and foreign trade are growing segments of the new California economy. Much of this growth has been buoyed by a state government that oversees and officially supports offices of tourism, business development, technology, foreign investment, filmmaking, and world trade. Reflecting the direction of the future, the state has overseas trade offices in Germany, England, Mexico, Japan, and Hong Kong. All of this activity continues to be fueled by easy access to universities, ports, and airports, by high-tech firms, and by ambitious new residents.

Second, can California continue to meet the challenge of educating its citizenry? As a gateway to the nation, California plays a particularly crucial role in providing appropriate education to established and new residents alike. To answer the multitude of problems facing the K–12 (precollegiate) systems, the state has turned its attention to better funding, reduced class sizes, and alternate methods of training teachers. The very diversity of California's population creates unusual challenges in the classroom, and satisfactory solutions have been difficult to find. Higher education in the state has found itself facing a crisis brought about by increasing enrollments, declining state revenues, and rapidly changing technologies. Planning efforts at both the University of California and the California State University systems have attempted to address the problems by rethinking delivery systems, increasing private sector support, and challenging the faculty to find better ways of educating students. Given the critical importance of education to the state's prosperity, these will continue to be central issues well into the future.

Finally, can California manage to understand and establish priorities for its own well-being? For example, is it more important to build additional prisons or to address the issues of education, health, welfare, and the environment? Can the state avoid being buried in its own garbage by adopting a serious waste management plan? Can the ponderous, inefficient, and poorly respected court and legal system ever reform itself? These are serious questions requiring all Californians to rise to the challenge of keeping the state a dynamic and positive force in the global environment.

BIBLIOGRAPHY

American Automobile Association. *California/Nevada Tourbooks*. Annual editions. Heathrow, Florida: American Automobile Association, 1978–1999.

Automobile Club of Southern California. *Winery Tours*. Los Angeles: Travel Publications Department, 1991.

America's Most Endangered Rivers of 1999. Washington, D.C., American Rivers, 1999.

Amerine, M. A., and Singleton, V L. *Wine: An Introduction*. 2d ed. Berkeley and Los Angeles: University of California Press, 1977.

Association of Bay Area Governments. *Bay Area Regional Planning Program—Agricultural Resources Study*. San Francisco, Aug. 1969.

Atherton, Gertrude. *California: An Intimate History*. New York: Harper and Bros., 1914.

Bailey, Harry P. *The Climate of Southern California*. Berkeley and Los Angeles: University of California Press, 1966.

Bakker, Elna. *An Island Called California: An Ecological Introduction to Its Natural Communities*. Berkeley and Los Angeles: University of California Press, 1971.

Bancroft, Hubert Howe. *History of California*. San Francisco: The History Co., 1886–1890.

Baur, John E. *The Health-Seekers of Southern California, 1870–1900*. San Marino, California: Huntington Library, 1959.

Bean, Walton. *California: An Interpretive History*. 3d ed. New York: McGraw-Hill, 1978.

Beck, Warren A., and Haase, Ynez. *Historical Atlas of California*. Norman: University of Oklahoma Press, 1974.

Blackburn, Thomas C., and Anderson, Kat, eds. *Before the Wilderness: Environmental Management by Native Californians*. Menlo Park, Calif.: Ballena Press, 1993.

Boesch, Donald F. *Oil Spills and the Marine Environment*. Cambridge, Mass.: Ballinger, 1974.

Brodine, Virginia, ed. *Air Pollution*. New York: Harcourt Brace Jovanovich, 1971.

California (magazine; formerly New West). Various issues.

California Coastal Zone Conservation Commission. *The California Coastal Plan*. Sacramento, 1975.

California Department of Agriculture. *California Agriculture—A Report of California Principal Crop and Livestock Commodities*. Annual editions, 1969–1981. Sacramento: California Crop and Livestock Reporting Service.

California Department of Education. *The Central Valley Project*. Sacramento, 1942.

California Department of Employment. *Report to the Governor on Labor Market Conditions*.

California Department of Food and Agriculture. *California Agriculture*. Sacramento, 1992, 1995.

——. *California Agricultural Resource Directory, 1997*. Sacramento, December 1997.

California Department of Finance. *California Statistical Abstract, 1989*. Sacramento, 1989.

——. *California Statistical Abstract 1997*. Sacramento, November, 1997.

——. *Projected Total Populations of California Counties*. Report 93-P3. Sacramento, May, 1993.

California Department of Parks and Recreation. Various publications, including *A Guide to the California State Park System, California Historical Landmarks, and Golden Days of San Simeon*. Sacramento.

California Department of Water Resources. *California's Ground Water-Bulletin No. 118*. Sacramento, Sept. 1975.

——. *California Water: Looking to the Future*. Bulletin No. 160–81. November, 1987

——. *Bulletin 160-87*. November, 1987.

California Department of Water Resources; California Department of Fish and Game; U.S. Bureau of Land Management; U.S. Forest Service; Mono County; Los Angeles Department of Water and Power. *Report of Interagency Task Force on Mono Lake*. Sacramento: California Resources Agency, Dec. 1979.

California Division of Mines and Geology. *Earthquake Planning Scenario for a Magnitude 8.3 Earthquake on the San Andreas Fault in Southern California*. Special Publication 60. Sacramento, 1982.

——. Fault-Rupture Hazard Zones in California. Special Publication 42. Revised. Sacramento, March 1980.

——. Geologic Atlas of California. San Francisco, 1958–1966.

——. Geology of California. Compilation of various monthly issues from volumes 21–23 of Mineral Information Service. Sacramento, March 1968–Oct. 1970.

——. Mammoth Lakes, *California Earthquakes of May 1980*. Special Report 150. Sacramento, 1980.

——. "Mammoth Lakes/Long Valley Microearthquake Project." *California Geology* 36 (Sept. 1982): 187–191.

——. *San Fernando, California, Earthquake of 9 February 1971*. Bulletin 196. Sacramento, 1975.

——. *Studies of the San Andreas Fault Zone in Northern California*. Special Report 140. Sacramento, 1980.

——. *Urban Geology: Master Plan for California*. Sacramento, 1973.

California Energy: The Economic Factors. Invited Papers on California's Future Energy Sources. San Francisco: Federal Reserve Bank, 1976.

California Energy Commission, QFER Form 1, October 18, 1990.

California Governor's Office of Planning and Research. *The California Water Atlas*. Ed. by William L. Kahrl. Los Altos, Calif.: William Kaufmann, 1979.

California Office of Historic Preservation. *Forging a Future with a Past*. Sacramento, December, 1997.

California Office of the Lieutenant Governor. *Council on Intergroup Relations Intern Research Project: Third World Population in California*. Sacramento, June 1977.

California Water Resources Board. *Summary of Draft of Mono Basin EIR*. May, 1993.

"California's Lost Tribes," *The Sacramento Bee*. June 29–July 2, 1997.

Carlson, Oliver. *A Mirror for Californians*. Indianapolis, Ind.: Bobbs-Merrill, 1941.

Casebier, Dennis G. *The Mojave Road*. Norco, Calif.: King Press, 1975.

Caughey, John Walton. *California*. 2d ed. Englewood Cliffs, N.J.: Prentice-Hall, 1953.

Caughey, John, and Caughey, La Ree. *Los Angeles: Biography of a City*. Berkeley and Los Angeles: University of California Press, 1977.

Clarke, Thurston. *California Fault: Searching for the Spirit of a State Along The San Andreas*. New York: Ballantine Books, 1996.

Cleland, Robert Glass. *The Cattle on a Thousand Hills: Southern California, 1850–1880*. San Marino, Calif.: Huntington Library, 1951.

——. *From Wilderness to Empire: A History of California, 1542–1900*. New York: Alfred A. Knopf, 1944.

Continents Adrift. Readings from *Scientific American*. San Francisco: W. H. Freeman, 1971.

Crain, Jim. *Historic Country Inns of California*. San Francisco: Chronicle Books, 1977.

Dana, Richard Henry. *Two Years Before the Mast*. New York, 1840. Available in various editions.

Darley, Ellis F.; Nichols, Carl W.; and Middleton, John T. "Identification of Air Pollution Damage to Agricultural Crops." *The Bulletin, California Department of Agriculture* 55, no. 1 (1966): 11–19.

Delano Historical Society. "The Lakeside Story." *The Plow* vol. 12, no. 1 (Sept. 1976).

De Voto, Bernard. *Year of Decision: 1846*. Boston: Houghton Mifflin, 1950.

Donley, Michael W. Stuart Allan, Patricia Caro, and Clyde P. Patton. *Atlas of California*. Portland, Ore.: Academic Book Center, 1979.

Dumke, Glenn S. *The Boom of the Eighties in Southern California*. San Marino, Calif.: Huntington Library, 1944.

Durrenberger, Robert W., ed. *California: Its People, Its Problems, Its Prospects*. Palo Alto, Calif.: National Press Books, 1971.

Durrenberger, Robert W., and Johnson, Robert B. *California: Patterns on the Land*. 5th ed., rev. by California Council for Geographic Education. Palo Alto, Calif.: Mayfield Publishing Co., 1976.

Early Man in America. Readings from *Scientific American*. San Francisco: W. H. Freeman, 1973.

Englehardt, Zephyrin, O. F. M. *The Franciscans in California*. Harbor Springs, Mich.: Holy Childhood Indian School, 1897.

Fay, James S., ed. *California Almanac, 7th ed*. Santa Barbara: Pacific Data Resources, 1995.

Federal Writers Project. *California: A Guide to the Golden State.* St. Clair Shores, Mich.: Somerset Publishing, 1939.

Fiero, Bill. *Geology of the Great Basin.* Reno: University of Nevada Press, 1986.

Finklestein, Charles, and Baxter, Laurence D. *Planning and Politics: A Staff Perception of the Tahoe Regional Planning Agency.* Davis: University of California Institute of Government Affairs, Nov. 1974.

Fradkin, Philip L. *California: The Golden Coast.* New York: Viking Press, 1974.

Gentry, Curt. *The Last Days of the Late Great State of California.* New York: G. P. Putnam's Sons, 1968.

Gilliam, Howard. *Weather of the San Francisco Bay Region.* Berkeley and Los Angeles: University of California Press, 1962.

Gillies, James, and Mittelbach, Frank. "Urban Pressures on California Land: A Comment." *Land Economics* vol. 34, no. 1 (Feb. 1958): 80–83.

Gold Is the Cornerstone. 1948. Reprinted under the title *California Gold Rush.* Berkeley and Los Angeles: University of California Press, 1976.

Gribbin, John R., and Plagemann, Stephen H. *The Jupiter Effect.* New York: Vintage Books, 1976.

Grossman, Harold J. *Grossman's Guide to Wines, Beers, and Spirits.* 6th ed., rev. New York: Charles Scribner's Sons, 1977.

Gudde, Erwin G. *California Place Names.* 3d ed. Berkeley and Los Angeles: University of California Press, 1969.

Guiness, R. "The Changing Location of Power Plants in California." *Geography* 65, no. 288, pt. 3 (July 1980).

Hardy, Thomas K. *Pictorial Atlas of North American Wines.* San Francisco: Grape Vision Pty., Ltd., 1988.

Hart, James D. *A Companion to California.* New York: Oxford Press, 1978.

Hartman, David N. *California and Man.* 4th ed. Santa Ana, Calif.: Pierce Publishers, 1977.

Haslam, Gerald W. *The Other California: The Great Central Valley in Life and Letters.* Santa Barbara: Joshua Odell Editions, Capra Press, 1990.

Harvey, Robert O., and Clark, W A. V. "The Nature and Economies of Urban Sprawl." *Land Economics* vol. 41, no. 1 (Feb. 1965): 1–9.

Haslam, Gerald W. *The Other California.* Santa Barbara: Joshua Odell Editions, Capra Press, 1990.

Heizer, R. E, and Whipple, M. A., eds. *The California Indians: A Source Book.* 2d ed., rev. and enl. Berkeley and Los Angeles: University of California Press, 1971.

Helen, Alfred, ed. *The California Tomorrow Plan.* Los Altos, Calif.: William Kaufmann, 1972.

Hill, Mary. *Geology of the Sierra Nevada.* Berkeley and Los Angeles: University of California Press, 1975.

Holdman, Marks, ed. *The World Almanac and Book of Facts, 1992.* New York: Pharos Books, Scripps Howard Company, 1991.

Holtgrieve, Donald. *The California Wine Atlas.* Hayward, Calif.: Ecumene Associates, Environmental Research, 1978.

Howard, J. Nelson. *Los Angeles Metropolis.* Dubuque, Iowa: Kendall Hunt, 1983.

Howard, Robert B., and Shiroma, Debra. "Windmill Sites in Mountainous Areas of the United States." In *The California Geographer*, Vol. 18, 85–93. Rohnert Park: California Council for Geographic Education, 1978.

Huber, Walter Roy. *California Real Estate Principles.* 2d ed. Glendale: California Real Estate Publications, 1977.

Hunt, Charles B. *Death Valley: Geology, Ecology, Archaeology.* Berkeley and Los Angeles: University of California Press, 1975.

Hutchinson, W. H. *Oil, Land, and Politics: The California Career of Thomas Robert Bard.* Vols. 1 and 2. Norman: University of Oklahoma Press, 1965.

Hyslop, Richard S. *Hazards and Land Use Planning: The Unresolved Dilemma.* Ann Arbor, Mich.: University Microfilms, 1990.

Jackson, David D.; Lee, Wook B.; and Liv, Chi-Ching. "Aseismic Uplift in Southern California: An Alternative Interpretation." *Science* 210 (Oct. 31, 1980): 534–536.

Jackson, Helen Hunt. *Ramona.* Boston, 1884. Reprint. Boston: Little, Brown, 1939.

Jackson, W Turrentine. *Early Planning Efforts at Lake Tahoe: The Role of Joseph R MacDonald.* Davis: University of California Institute of Government Affairs, Jan. 1974.

Jaeger, Edmond C. *The California Desert.* Stanford, Calif.: Stanford University Press, 1965.

——. *North American Deserts.* 1957. Reprint. Stanford, Calif.: Stanford University Press, 1967.

Johnson, Hugh. *Hugh Johnson's Modern Encyclopedia of Wine*, 3d ed. New York: Simon and Schuster, Inc., 1991.

Johnson, Hugh. *Hugh Johnson's Pocket Encyclopedia of Wine*, 1992. New York: Simon and Schuster, Inc., 1991.

Johnson, Hugh. *The World Atlas of Wine.* London: Mitchell Beazley, 1971.

Kadohata, Cynthia. *In the Heart of the Valley of Love.* New York: Viking Press, 1992.

Kerr, William S. "Impact of Urbanization on Agriculture in Orange County, California." In *Yearbook of Association of Pacific Coast Geographers*, Vol. 34, ed. by John F. Gaines, 161–170. Corvallis: Oregon State University Press, 1972.

Kroeber, Theodora. *Ishi in Two Worlds.* Berkeley and Los Angeles: University of California Press, 1961.

Lake Tahoe Area Council and Engineering-Science, Inc. *Comprehensive Study on Protection of Water Resources of Lake Tahoe Basin Through Controlled Waste Disposal.* Lake Tahoe, Calif., June 1963.

Lantis, David W.; Steiner, Rodney; and Karinen, Arthur E. *California: Land of Contrast.* 3d ed., rev. Dubuque, Iowa: Kendall/Hunt, 1981.

Lee, W. Storrs. *California: A Literary Chronicle.* New York: Funk and Wagnalls, 1968.

Lewis, Oscar. *The Big Four.* New York: Alfred A. Knopf, 1938.

Los Angeles Department of Water and Power. *Draft Environmental Impact Report on Increased Pumping of the Owens Valley Groundwater Basin.* Los Angeles, Aug. 1978.

——. *Eastern Sierra Cloud Seeding Program: Draft Environmental Impact Report.* Los Angeles, Oct. 1981.

——. *Draft EIR: Water from the Owens Valley to Supply the Second Los Angeles Aqueduct, 1970 to 1990 and 1990 Onward Pursuant to a Long Term Groundwater Management Plan.* Vol.1.1, SCH 89080705, September, 1990.

——. *Final Environmental Impact Report on Increased Pumping of the Owens Valley Groundwater Basin.* Vols. 1 and 2. Los Angeles, June 1979.

——. *Los Angeles' Mono Basin Water Supply.* Briefing Document. Los Angeles, May 1982.

——. *Water Rights and Operations in the Mono Basin.* Los Angeles, 1973.

Los Angeles Times. Various daily issues, 1962–1999.

McWilliams, Carey. *Southern California: An Island on the Land.* Santa Barbara, Calif.: Peregrine Smith, 1973.

Metropolitan Water District. *Aqueduct* (magazine). Various quarterly editions, 1977–1981. Los Angeles.

Miller, Crane S. *Agriculture in an Urbanizing Environment: A Study in the Impact of Metropolitan Growth on Specialized Agriculture in Ventura County, California.* Ann Arbor, Mich.: University Microfilms, 1971.

——. "The Changing Agricultural Landscape of Simi Valley from 1795 to 1960." *Quarterly of the Ventura County Historical Society* vol. 13, no. 4 (Aug. 1968).

——. "The Southern Great Basin." In Association of American Geographers Field Trip Guide, 1981, 24–34. Washington, D.C.: Association of American Geographers, 1981.

——. "Spectral and Spatial Signature Recognition in Urbanizing Areas of Southern California from U-2 Color Infra-Red Imagery." In *Proceedings of the International Symposium on Image Processing*, ed. by Franz W. Leberl, 141–147. Graz, Austria: Graz Technical University, Oct. 1977.

——. "California Demographics," California Studies Institute, California State University, Sacramento, 1996.

Monkhouse, F J., and Small, John. *A Dictionary of the Natural Environment.* New York: Halsted Press, 1978.

Montgomery, Richard H., and Budnick, Jim. *The Solar Decision Book.* New York: Wiley, 1978.

Muir, John. *The Mountains of California.* Berkeley: Ten Speed Press, 1977.

Munz, Philip A., and Keck, David. *A California Flora.* Reprint. Berkeley and Los Angeles: University of California Press, 1973.

Murphy, Raymond E. *The American City: An Urban Geography.* New York: McGraw-Hill, 1966.

Nadeau, Remi, A. *The Water Seekers.* Garden City, N.Y.: Doubleday, 1950.

National Academy of Sciences and National Academy of Engineering. *Urban Growth and Land Development: The Land Conversion Process.* Washington, D.C., 1972.

Nelson, Howard J. "The Spread of an Artificial Landscape over Southern California." *Annals of the Association of American Geographers* vol. 49, no. 3, pt. 2 (Sept. 1958): 80–99.

Norris, Frank. *The Otopus*. Available in various editions.

Norris, Robert M., and Webb, Robert W. *Geology of California*. New York: Wiley, 1976.

Oakeshott, Gordon B. *California's Changing Landscapes: A Guide to the Geology of the State*. New York: McGraw-Hill, 1971.

Oliver, John E. *Perspectives on Applied Physical Geography*. North Scituate, Mass.: Duxbury Press, 1977.

Plane, David, and Peter A. Rogerson. *The Geographical Analysis of Population*. New York: John Wiley & Sons, Inc., 1994.

Ponte, Lowell. *The Cooling*. Englewood Cliffs, N.J.: Prentice-Hall, 1976.

Pryde, Philip R., ed. *San Diego: An Introduction to the Region*. Dubuque, Iowa: Kendall/Hunt, 1976.

Reisner, Marc. *Cadillac Desert*. New York: Penguin Books, 1987.

Rieff, David. *Los Angeles: Capital of the Third World*. New York: Simon and Schuster, 1991.

Robinson, W. W. *Land in California*. Berkeley and Los Angeles: University of California Press, 1948.

Rolle, Andrew F. *California: A History*. New York: Thomas Y. Crowell, 1969.

Ross, Michael J. *California: Its Government and Politics*. North Scituate, Mass.: Duxbury Press, 1979.

Salvador, Martinha, and Salvador, Jose. "Portuguese Pioneers of Southern California." Senior project for California State Polytechnic University, Pomona, 1978.

San Francisco Department of City Planning, "San Francisco's Districts" (Map). San Francisco Atlas. Office of Analysis and Information Systems, October, 1991.

Schaffer, Jeffrey P; Schifrin, Ben; Winnett, Thomas; and Jenkins, J. C. *The Pacific Coast Trail*. Vol. 3, California. Berkeley: Wilderness Press, 1977.

Schoenman, Theodore, ed. *The Father of California Wine: Agoston Haraszthy*. Santa Barbara, Calif.: Capra Press, 1979.

Security Pacific National Bank. Research Department. *Monthly Summary of Business Conditions for Southern California and Northern Coastal California*. Various issues, 1972–1982. Los Angeles.

——. *Northern California: Economic Issues of the Eighties*. March 1982.

——. *San Francisco Bay Area Report: A Study of Growth and Economic Stature of the Nine Bay Area Counties*. April 1971.

——. *Southern California: Economic Issues in the Eighties*. Sept. 1981.

——. *Southern California: Economic Trends in the 1970s*. May 1977.

Sedway/Cooke (firm). *Land and the Environment*. Los Altos, Calif.: William Kaufmann, 1975.

Sharp, Robert P. *A Field Guide to Southern California*. 3rd ed. Dubuque, Iowa: Kendall/Hunt Publishing Company, 1994.

Sierra Club. Angeles Chapter. *Southern Sierran*. Various monthly issues, 1974–1981. Los Angeles.

Sinclair, Upton. *Oil*. 1972. Reprint. Cambridge, Mass.: Robert Bentley, 1981.

Smith, Genny Schumacher, ed. *Mammoth Lakes Sierra: A Handbook for Roadside and Trail*. 4th ed. Palo Alto, Calif.: Genny Smith Books, 1976.

——. ed. *Deepest Valley*. rev. ed. Los Altos, Calif.: William Kaufmann, 1978.

Southern California Association of Governments. SCAG-76: *Growth Forecast Policy*. Los Angeles, Jan. 1976.

Southern California Edison Company. *Education Service Division*. E2 Report: Energy Education (newsletter). Various editions, 1978–1982.

Spencer, J. E., ed. *Day Tours In and Around Los Angeles*. Los Angeles Geographical Society Publication, no. 3. Palo Alto, Calif.: Pacific Books, 1979.

Starkey, Otis R.; Robinson, J. Lewis; and Miller, Crane S. *The Anglo-American Realm*. 2d ed. New York: McGraw-Hill, 1975.

Steinbeck, Elaine, and Wallsten, Robert, eds. *Steinbeck: A Life in Letters*. New York: Viking Press, 1975.

Steinbeck, John. *The Grapes of Wrath*. New York: Viking Press, 1967.

Steinhart, Carol E., and Steinhart, John S. *Blowout: A Case Study of the Santa Barbara Oil Spill*. North Scituate, Mass.: Duxbury Press, 1972.

Steiner, Rodney. *Los Angeles: The Centrifugal City*. Dubuque, Iowa: Kendall/Hunt, 1981.

Stobaugh, Robert, et al. *Energy Futures*. New York: Random House, 1979.

Storer, Tracy I., and Usinger, Robert L. *Sierra Nevada Natural History*. Berkeley and Los Angeles: University of California Press, 1963.

Strahler, Arthur N., and Strahler, Alan H. *Elements of Physical Geography*. 2d ed. New York: Wiley, 1979.

Taylor O. C., principal investigator. *Oxidant Air Pollutant Effects on a Western Coniferous Forest Ecosystem. Task B Report: Historical Background and Proposed Systems Study of the San Bernardino Mountain Area*. Riverside: Statewide Air Pollution Research Center, University of California, Jan. 1973.

Thomas, William L., Jr., ed. "Man, Time, and Space in Southern California." *Annals of the Association of American Geographers* vol. 49, no. 3 (Sept. 1959), supplement.

Thrower, Norman J. W. "California Population: Distribution in 1960." *Annals of the Association of American Geographers* 56, no. 2 (1966). See also map supplements for 1960 and 1970.

Time Inc. Magazines Special Issue: "California: The Endangered Dream." *Time*, November 18, 1991.

Twain, Mark. *Roughing It*. Available in various editions.

University of California, Agricultural Experiment Station and Cooperative Extension. *California Agriculture*. Various issues, 1969–1982. Berkeley.

U.S. Census, 1980, Summary Tape File 1.

U.S. Census, 1990, Public Law 94-171 Redistricting File.

U.S. Department of Agriculture. Soil Conservation Service. Soil Survey, Ventura Area California. Washington, D.C., April 1970.

U.S. Department of Commerce, Bureau of the Census. "California: Final Population Counts," PC(VI)-6, and "California: General Population Characteristics," PC(V)-6. In *1970 Census of Population*. Washington, D.C., Feb. 1971.

——. *1980 Census of Population and Housing: California*. Preliminary Reports, Feb. 1981.

——. *1990 Census of Population and Housing*, January, 1991.

——. *1978 Census of Agriculture*.

——. *1988 Census of Agriculture*.

——. *1995 Annual Survey of Manufactures*.

——. *1992 Census of Manufactures*.

——. *1977 Census of Manufactures for California*.

——. *1992 Census of Retail Trade*.

——. *1977 Census of Service Industries*.

——. *1987 Census of Service Industries*.

——. *1992 Census of Service Industries*.

——. *1992 Census of Wholesale Trade*.

U.S. Department of Commerce and Environmental Protection Agency. Council on Environmental Quality. *The Economic Impact of Pollution Control: A Summary of Recent Studies*. Washington, D.C., March 1972.

U.S. Department of the Interior. Bureau of Land Management. *The California Desert Conservation Area: Plan Alternatives and Environment Impact Statement*. Washington, D.C., Feb. 1980.

U.S. Geological Survey. *The Loma Prieta Earthquake of October 17, 1989* (pamphlet). November, 1989, rev. January, 1990.

——. Open File Report 88–398, 1988.

Vance, James E., Jr. "California and the Search for the Ideal." *Annals of the Association of American Geographers* 62 (1972).

Ventura County Agricultural Commissioner. *Agricultural Crop Report, Ventura County*. Annual issues, 1968–1981.

Ventura County Planning Department. *The Economics of Conserving Agriculture in Ventura County*. Ventura, Calif., Dec. 1970.

Watkins, T. H. *On the Shore of the Sundown Sea*. San Francisco: Sierra Club, 1972.

West, Nathanael. *The Day of the Locust*. 1939. Reprint. Cutchogue, N.Y.: Buccaneer Books, 1981.

Western Water Education Foundation. *Western Water*. Various 1982 editions.

Winchester, Simon. *Pacific Rising: The Emergence of a New World Culture*. New York: Touchstone, 1991.

Winkler, A. J.; Cook, J. A.; Kliewer, W A.; and Lider, L. A. *General Viticulture*. Berkeley and Los Angeles: University of California Press, 1974.

World Almanac and Book of Facts 1992, "The 50 Most Racially Diverse Counties in the U.S." New York: Pharos Books, 1991.

World Resources Institute. *The 1992 Information Please Environmental Almanac*. Boston: Houghton Mifflin Co., 1992.

Yeadon, Anne, and Yeadon, David. *Wine Tasting in California*. Los Angeles: Camaro, 1977.

Zedler, Joy B. *The Ecology of Southern California Coastal Salt Marshes: A Community Profile*. Washington, D.C.: U.S. Fish and Wildlife Service, Biological Services Program, 1982.

INDEX